Analysis and Design of Tall Reinforced Concrete Buildings

The global construction landscape is witnessing a continued surge in high-rise development with increasing sophistication and height. These buildings are complex and expensive structures, built on varied ground conditions from poor soil to rock. While existing books on tall building design typically focus on fundamental structural analysis and design principles, there remains a significant gap in practical guidance.

Analysis and Design of Tall Reinforced Concrete Buildings addresses this critical gap by drawing on the authors' extensive practical experience and cutting-edge research to create seamless connections between foundational design principles and contemporary engineering practice.

Designed to serve the entire spectrum of construction professionals, this comprehensive guide provides essential design guidance throughout every phase of construction projects. Whether you're a graduate student building foundational knowledge or a junior practitioner seeking practical insights for real-world applications, this book offers the depth and breadth necessary to navigate the complexities of modern tall building design and construction.

Chi Wai Law is a structural design engineer based in Hong Kong with 40 years of experience working on high-rise buildings, deep basements, various types of foundations, and is a co-author of *Analysis, Design and Construction of Foundations.*

Yung Ming Cheng has retired from The Hong Kong Polytechnic University, Hong Kong, and is now director of two engineering companies and part-time professor at the University of Hong Kong, with a special interest in geotechnical engineering and computational method, and is the author of *Slope Stability Analysis and Stabilization* and is a co-author of *Analysis, Design and Construction of Foundations.*

Analysis and Design of Tall Reinforced Concrete Buildings

Chi Wai Law

and Yung Ming Cheng

CRC Press
Taylor & Francis Group
Boca Raton London New York

CRC Press is an imprint of the
Taylor & Francis Group, an **informa** business

Designed cover image: Chi Wai Law & Yung Ming Cheng

First edition published 2026
by CRC Press
4 Park Square, Milton Park, Abingdon, Oxon, OX14 4RN

and by CRC Press
2385 NW Executive Center Drive, Suite 320, Boca Raton FL 33431

CRC Press is an imprint of Informa UK Limited

British Library Cataloguing-in-Publication Data
A catalogue record for this book is available from the British Library

ISBN: 9781041065968 (hbk)
ISBN: 9781041066309 (pbk)
ISBN: 9781003636380 (ebook)

DOI: 10.1201/9781003636380

Typeset in Times
by Deanta Global Publishing Services, Chennai, India

Contents

Preface

In Hong Kong and many other densely populated regions, the proliferation of taller and more numerous buildings has become a defining feature of modern urban landscapes. To minimize construction costs and timelines, designers and engineers must navigate a complex array of considerations throughout the planning, modelling, analysis, and execution phases. While numerous textbooks adequately cover the foundational principles of structural analysis and design, they often fall short in addressing the nuanced details of structural modelling, foundation analysis, and practical implementation. This gap is a recurring concern voiced by engineers – particularly younger professionals – who seek deeper insights into these critical areas.

The authors bring decades of hands-on experience in building and foundation projects across Hong Kong, complemented by their contributions to local design codes and handbooks. Recognizing the limitations of existing resources, they have refined their teaching materials to bridge these gaps, offering comprehensive coverage of both theoretical underpinnings and real-world design and analysis practices. A persistent issue they have observed is the opaque explanation of commonly used design methods, equations, and figures in standard references. As a result, many engineers default to computer software without fully understanding the underlying assumptions or exercising informed judgment – a practice that undermines robust engineering outcomes.

This overreliance on tools without contextual knowledge alarms the Hong Kong Institute of Engineers and seasoned professionals alike, who note that while junior engineers excel at software operation, they often lack the foundational expertise to interpret results critically. The authors share this view, deeming it an unhealthy trend in the field. It is against this backdrop that they developed the technical seminar "Do Not Fully Believe in Computer Programs," which draws on real-world case studies from engineers grappling with pitfalls in computational analysis, post-processing, and related design challenges. This book extends that mission, equipping readers with the clarity and discernment needed to harness technology effectively while prioritizing engineering integrity.

This book covers the foundational theory of structural analysis and finite element methods, along with the design of reinforced concrete buildings. It incorporates relevant provisions, advanced topics, and the latest theoretical developments. The authors hope this text will enable practicing engineers to master analysis and design techniques with ease, while fostering a solid grasp of both theoretical principles and practical applications. Recognizing that many engineers rely on finite element-based software in their daily work, the book also examines the limitations of the finite element method and offers essential precautions for using such tools effectively.

C.W. Law
Y.M. Cheng

About the Authors

Chi Wai Law is a structural engineer who has ample structural design experience in high-rise buildings, deep basements, and various types of foundations throughout his over 40 years of working experience. In addition to practical design work, he has also carried out research work on structural element and foundation design and published a number of academic papers, design manuals, and handbooks. He also has teaching experience in various higher diploma and undergraduate courses of civil engineering.

Chi Wai Law is an experienced structural design engineer in high-rise buildings, deep basements, and various types of foundations who also possesses teaching experience and author of a number of academic papers, design manuals, and handbooks.

Yung Ming Cheng is the director of Smart Site Limited and part-time professor in the University of Hong Kong.

Yung Ming Cheng has extensive engineering experience in soft, rock and immersed tunnels (MTR and EHC), steel, reinforced and prestressed concrete bridges design and construction, ground treatment, deep excavation in poor soils with different types of retaining wall types, slopes and piles engineering before joining the Hong Kong Polytechnic University, Hong Kong. His research interests include different numerical methods: finite element, distinct element and meshless methods, plasticity theory, slope stability, pile foundation, field tests, and IOT monitoring in construction. Cheng has published about 220 papers and 11 books based on the research works and has served in various engineering committees and government departments as advisor and as editor in several international journals. Cheng is rated world's top 2% scientist by Stanford University from 2020 to 2025, and has received several best paper awards from engineering journals.

1 Introduction

1.1 AIM AND SCOPE

The growing demand for space in commercial and residential sectors, particularly in Asia, has driven the design and construction of increasingly tall buildings. These structures represent significant investments and involve complex considerations throughout their design and construction phases. They must assume various forms to meet diverse functional requirements and strict constraints, including cost, and are often built on foundations situated in challenging ground conditions.

However, a critical gap exists in the available literature. While current texts on building and foundation design cover basic principles of structural analysis, they often lack the depth required for modern engineering practice. Specifically, they frequently omit detailed guidance on the sophisticated modelling of structures and foundations. This superficial treatment creates a significant disconnect between academic theory and the practical expertise required by professional engineers, a shortcoming consistently noted by recent graduates.

Building structures for residential, commercial, and industrial purposes are the most common and numerous human-made structures on earth. Compared to other structures, buildings are characterized by their complex forms, featuring numerous interconnected elements in their superstructure and foundation. These elements function together as an integrated system to ensure the overall structural adequacy, stability, and safety of the building. An effective design is distinguished by its efficient performance at both the individual member level and the holistic structural level.

Consequently, effective building design requires an in-depth understanding of individual structural members, their interactions, and their integrated performance. This understanding must be grounded in fundamental theories and their application through analytical methods. Today, proficiency in finite element methods and computer modelling techniques is essential, given their extensive use. Furthermore, designers must be fully aware of the limitations inherent in these computational tools.

The authors possess extensive practical and research experience in building and foundation engineering, including contributions to Hong Kong design codes and handbooks. Recognizing the limitations of existing texts, they have structured this book to bridge the gap between theoretical principles and actual design practice. A key motivation is the observation that many common design practices, methods, and equations are often applied by engineers without a clear understanding of their underlying principles. This concern is echoed by an examiner for the Hong Kong Institution of Engineers, who noted that while many engineers can operate complex software, they frequently lack the fundamental knowledge behind the analysis. This

DOI: 10.1201/9781003636380-1

book aims to address this critical shortcoming and foster a more profound and competent understanding of the discipline.

This book presents a systematic approach to the conceptual design of buildings, from low-rise to super-tall, by exploring fundamental theories and their practical applications.

1.2 REFERENCE TO CODES OF PRACTICE

Building structural design is governed by codes of practice, which provide designers with guidelines to ensure standards of safety, serviceability, durability, and ductility are met. This book references and, where appropriate, compares reputable national and international codes – including the Eurocode, American, Chinese, and Hong Kong standards – to broaden the designer's perspective.

From the early 2020s, the Eurocodes have been evolving into their second generation, with revised versions being published. The European Committee for Standardization has stated that the current first-generation codes may be used until 28 March 2028, allowing a transition period where both generations are applicable. Accordingly, this book references both generations as appropriate to provide comprehensive and up-to-date guidance.

1.3 HIGHLIGHTS OF CONTENTS

The following elements of the book are highlighted:

(i) Chapter 2 presents a historical review of the evolution of structural analysis methods for buildings. The review begins with the fundamental assumptions of structural analysis as background knowledge to the designer for his understanding of the basic principles and methods in structural analysis. The chapter then traces the progression from primitive, hand-calculation methods which rely on broad assumptions and approximations in the pre-computer era, to the first-order linear analytical methods that are now standard with computational tools. Subsequently, advanced non-linear methods are explored, including buckling, P-delta, large deformation, stage, and plastic analyses. The underlying theories of these methods are outlined, with emphasis on their superior simulation of actual structural behaviour. Applications and illustrative examples are provided to enhance comprehension, and the limitations of each method are highlighted. For further clarity, practical worked examples and theoretical derivations are included in Appendices A, B, and C. Collectively, this chapter aims to provide a clear understanding of the basic theories of structural analysis through their historical development.

(ii) Given the finite element method's extensive use and the knowledge gap many structural designers face, Chapter 3 outlines its fundamental principles and the formulation of various basic elements. It discusses the critical importance of selecting appropriate shape functions and the degree of meshing for accurate structural simulation, illustrated with practical examples. The chapter also emphasizes the method's limitations.

(iii) In the discussion of the various building systems, from simple to complex ones, Chapter 4 contains typical worked examples for readers to grasp the design concepts in a quantitative manner.

(iv) The origins of common types of loads applying to a building structure are discussed in Chapter 5, together with the determination of the magnitudes of the loads with respect to the underlying theories including dynamic loads on floors creating vibrations and seismic loads determined in accordance with the popular response spectrum method.

(v) Chapter 6 focuses on computer modelling, including basic techniques comprising the right choice of the type of finite element, required degrees of meshing. Optimization and Building Information Method have also been discussed.

(vi) With the extensive use of computer methods, dynamic analysis of building structures can now be accessed conveniently. However, general structural engineers may not be familiar with the method which is required for assessment of structural vibrations and seismic effects. In Chapter 7, the fundamental theories of structural dynamics are outlined by which the modal and numerical integration methods are developed. Applications are also demonstrated through worked examples by simple structures including plane frames. Both the vibration analysis of a building under seismic effects and floor vibrations under movements of occupants have been included.

(vii) Discussions of design for ductility, robustness, durability and fire resistance are included in Chapter 8, which are often not found in books for general building design. The design concepts of these topics are now dispensable in building structural design.

(viii) Chapter 9 contains section design for reinforced concrete members. The basic stress-strain relationships of various design codes have been outlined, upon which the design formulae of them are derived for axial load, bending, shear, torsion, and combinations of them. In addition to the simplified codified methods for section design for axial load coupled with biaxial bending in which the biaxial moments are decoupled, the accurate method in combining the biaxial moments and axial loads in the determination of stress-strain profile for accurate design has also been formulated in this Chapter.

(ix) The finite element method can produce complete sets of stresses including axial and shear stresses in in-plane structures and bending and twisting moments in out-of-plane structures. More accurate and economical design can then be furnished through complete design of these forces/moments. The approach is particularly useful in capturing rapid changes of stress for detailed examination of local effects. Clark's method for in-plane design and the Wood-Armer Equations for out-of-plane bending are described in detail in Chapter 10, under illustration by worked examples. Design to the 3-dimensional stress field comprising 6 components of axial and shear stresses forms the final part of the chapter based on Law et al. as an extension of Clark's approach which can be used in structures aiming at sophisticated design.

(x) Chapter 11 highlights rules for reinforced concrete detailing with respect to the Eurocode, American codes, etc.

(xi) Precast concrete construction has been gaining popularity in the construction of reinforced concrete buildings in recent decades. There are, however, inherent problems in the concern of robustness and rigidity in comparison with in situ concrete construction. Chapter 12 outlines the design of precast concrete components in concrete building structures with respect to construction. Emphasize has been placed on the design and construction of the connections.

(xii) An illustrative example in the analysis and design of a hypothetical reinforced concrete building is contained in Chapter 13 for demonstrating the application of the theories and design methods discussed in the foregoing chapters. The building is designed to resist both wind and seismic loads. The analysis and design are based on the Eurocode Codes.

(xiii) Fundamental theories and the underlying principles of foundation analysis and design are contained in Chapter 14. These theories and principles cover the classical ones by simple calculations adopting rigid cap/rigid footing to the advanced methods involving extensive use of computer software. The shortcomings and limitations are unveiled and commented. Discussions of the excavation and lateral support system for foundation construction have been included. Applicable background soil mechanics theories are contained for a better understanding of the working principles of the various systems.

2 Structural Analysis of Tall Buildings in General

From the Pre-Computer Era to Present

2.1 INTRODUCTION

The field of structural analysis has a long and rich history, with foundational theories like elasticity, energy principles, and stiffness and flexibility methods being well-established since the early twentieth century. However, applying these methods accurately required tedious mathematical manipulations, even for simple structures, in the era before computers. While many analytical solutions were developed, they often resulted in long and complex equations.

For instance, analyzing a simple one-bay, one-storey 2D portal frame using stiffness or flexibility methods involved at least six degrees of freedom, requiring the solution of six simultaneous linear equations which was a laborious task without modern computing power. For complex structures like tall buildings with thousands of members, analysis was impossible without significant simplifications, idealizations, and broad assumptions.

To overcome these challenges, engineers historically relied on design tables, graphs, and expertly formulated approximate methods. These techniques were widely practiced, and many older structures standing today were successfully designed using them. While most of these traditional methods are seldom used or taught now, their underlying concepts remain valuable. They are particularly useful for designers who need to perform quick checks on computer-generated results, helping to identify potential errors from incorrect boundary conditions, connections, or mesh design.

This chapter will explore these important alternative methods. Furthermore, it will discuss the revolutionary impact of computer methods, which have enabled the stiffness method to be used for accurate analysis of large, complex structures. While most common analysis is still first-order and linear, advancements in hardware and software now facilitate more sophisticated approaches, including buckling analysis, P-delta effects, large deformation analysis, and plastic analysis, all of which will be covered subsequently.

To facilitate readers' understanding of the evolution of analysis methods for buildings, the following Section 2.2 outlines and discusses the common assumptions made in structural analysis of buildings. It is important to note that these assumptions inevitably introduce errors, which are typically compensated for by factors of safety incorporated into design codes (e.g., partial load and material factors).

DOI: 10.1201/9781003636380-2

2.2 COMMON ASSUMPTIONS IN STRUCTURAL ANALYSIS FOR BUILDINGS

This section outlines and discusses the common assumptions adopted in structural analysis for buildings.

2.2.1 Linear Structural Behaviour

In reality, the structural behaviour of concrete is non-linear, which is characterized by a non-linear stress-strain relationship and cracking (detailed in Chapter 9). Despite this, the common practice of first-order linear analysis simplifies the process by assuming linear-elastic material behaviour, by which a direct proportionality between load and deformation is assumed. In addition, no deformation limit is imposed in the analysis. This simplification facilitates analysis while yielding results of acceptable accuracy so long as the structure is working with stresses well below the yielding limit.

2.2.2 Small Displacement

In conventional analysis, the displacements of a concrete structure are assumed to be sufficiently small that the geometry remains unchanged. This premise of small-deflection theory enables the use of a constant stiffness matrix across all load cases. When coupled with the assumption of linear-elastic material behaviour, it validates the principle of superposition, allowing for linear combinations of loads and significantly facilitating the analytical process.

2.2.3 Rigid Joint, Pinned Joint, and Semi-rigid Joint

By this assumption, joints where structural members meet are taken to be "rigid." The "rigid joint" assumption presumes that connections between structural members are perfectly rigid. Although the entire joint can translate or rotate as a single unit, the angles between the connected members remain unchanged. Consequently, all members meeting at the joint undergo identical displacements and rotations, as illustrated in Figure 2.1.

This assumption significantly reduces the number of degrees of freedom in a structural model. Otherwise, each member meeting at a joint would have its own rotation, rather than all sharing a single rotation at a rigid joint. Furthermore, without this assumption, a finite rotational stiffness would need to be assigned to each member end at the connection, which is a difficult task.

The assumption's validity is based on the fact that the joint region (e.g., a beam-column junction) is typically larger and stiffer than the adjoining members, which generally provides sufficient accuracy for analysis. Although some computer programs can model semi-rigid joints, this capability is not commonly used in the analysis of concrete buildings.

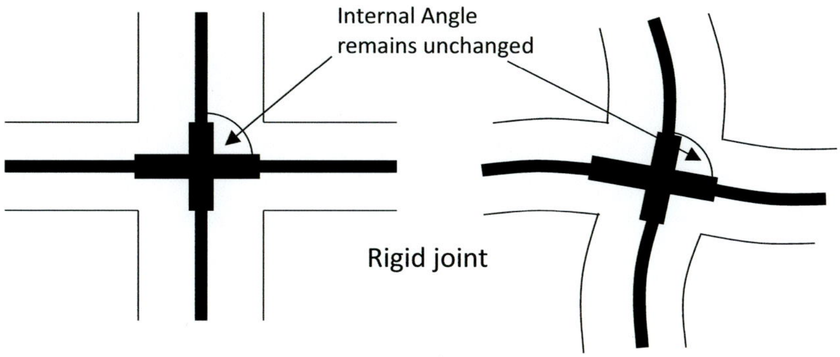

FIGURE 2.1 Rigid joint performance.

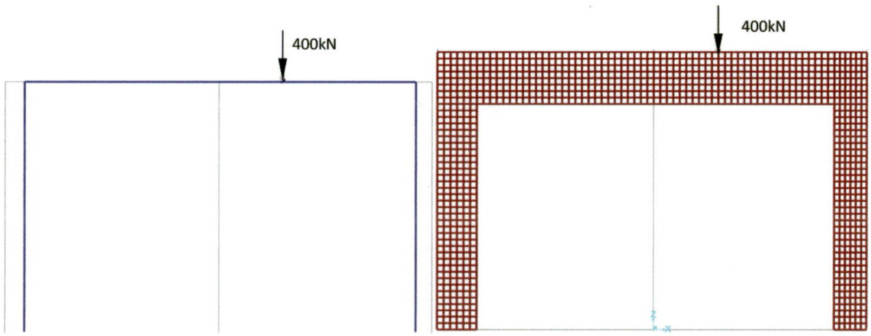

FIGURE 2.2 Testing rigid joint assumption by Worked Example 2.1.

To evaluate this assumption, the authors have tested a simple portal frame, modelled in two ways as shown in Figure 2.2 and detailed in Worked Example 2.1.

Worked Example 2.1
A portal frame is modelled by line elements on the left of Figure 2.2 and as an assembly of membrane elements by the finite element method on the right of the same figure.

The analyzed portal frame features a 400 × 800 beam on a 400 × 600 left column and a 400 × 400 right column, all with fixed supports. It was modelled in two ways: with rigid-jointed line elements (left) and with 400 mm thick membrane finite elements (right).

From the finite element model, centreline deflections at the joints were used to calculate rotations, determined as the initial tangent per Figure 2.3.

The rotations of the centrelines of the beam and the column at the joint differ by some 6% to 8% so that the joints are not perfectly rigid. Comparing with the line model with rigid joints, the line model has smaller rotations.

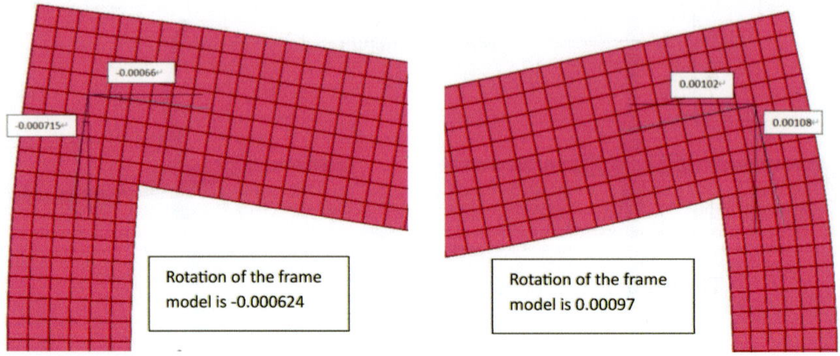

FIGURE 2.3 Joint rotations of the two models.

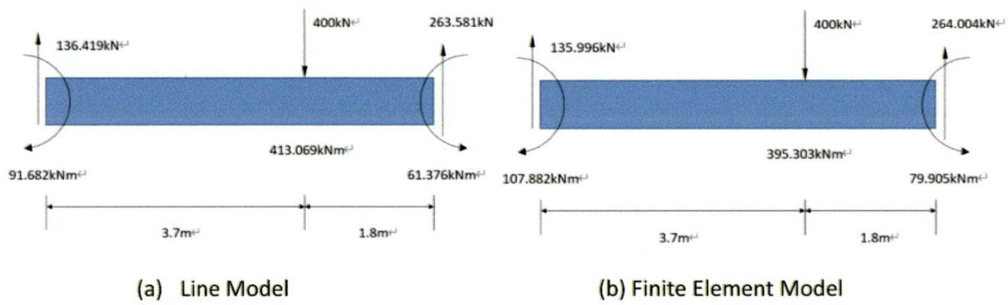

FIGURE 2.4 Comparison of internal forces between the line model and finite element model.

The internal forces of the beam between the column faces in the two models are examined and listed in Figure 2.4, showing some differences.

Nevertheless, to improve the accuracy of the line model, there is a practice of adding a "stiff arm" of large stiffness within the joint which extends from the centre of the joint to the column or beam face as shown in Figure 2.5. The use of the stiff arm is justified by the very large sections of the portion and therefore is taken as very rigid.

Full lengths stiff arms extending from the centre of the joint to column and beam faces as in Figure 2.5 have been tried, together with another model having only half-length arms. The internal forces at the column faces and mid-spans are listed as in Figure 2.6.

The results of the four models are summarized in Table 2.1.

The finite element model which is based on the actual structural configuration is considered to have achieved the highest accuracy and is therefore used as a basis for comparison. It turns out that the half-length stiff arm model agrees very well with the finite element model. The phenomenon is of course not conclusive by this single model. But the model with no stiff arms assumption underestimates stiffness of the members while the one with full length stiff arms over-estimates the member stiffness. It is not surprising that the half-length (or other with fractional lengths of stiff

FIGURE 2.5 The use of stiff arm in line model.

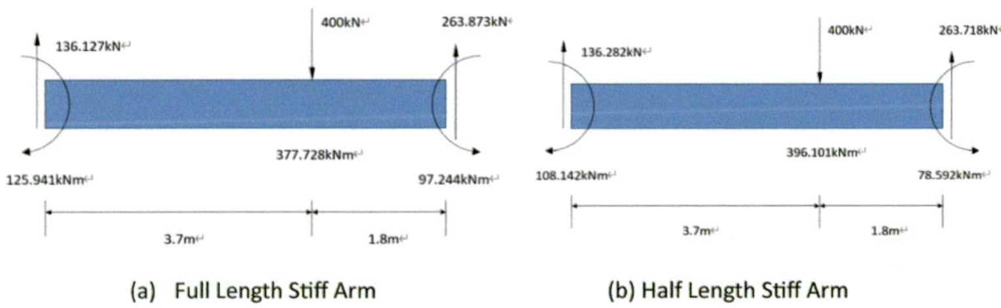

FIGURE 2.6 Line models with full and half lengths stiff arm.

TABLE 2.1
Summary of Internal Forces of the Beam in the Four Models

Model	Left Support Moment (kNm)	Maximum Span Moment (kNm)	Right Support Moment (kNm)
(A) Finite Element Model	−107.882	395.303	−79.905
(B) Frame Model – No Stiff Arm	−91.682 (−14%)	413.069 (4.5%)	−61.376 (−23.2%)
(C) Frame Model – Full-Length Rigid Arm	−125.941 (16.7%)	377.728 (−4.5%)	−97.244 (21.7%)
(D) Frame Model – Half-Length Rigid Arm	−108.142 (0.24%)	396.101 (0.2%)	−78.592 (−1.64%)

arms) can achieve better accuracy which may be a choice for general use. It will, however, be left to the designers to choose the right model on individual cases.

It is also a common practice that, under certain circumstances such as a beam spanning in the out-of-plane direction of a wall being supported on the wall, a member is assumed to have a "pinned" connection at a joint. The "pinned connection" allows free rotation. As such, one more degree of freedom is required for the analysis as shown in Figure 2.7(a).

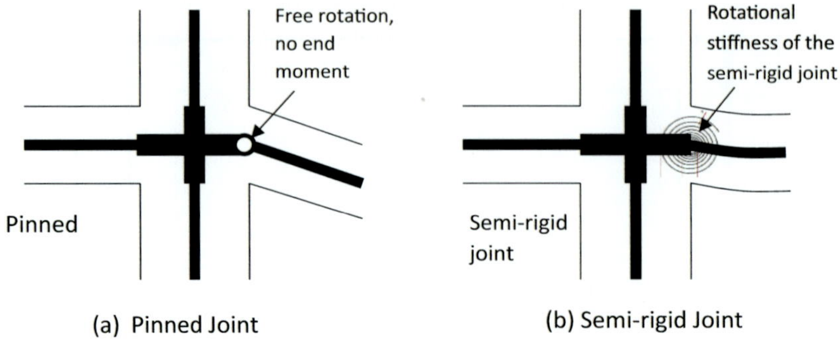

FIGURE 2.7 Pinned joint and semi-rigid joint.

Nevertheless, instead of being pinned or a rigid joint connection, a "semi-rigid joint" with stiffness in terms of the moment required to produce unit rotation can now be used for analysis in many computer software. But the stiffness has to be assessed by the user as shown in Figure 2.7(b). So far, it is not easy to assign stiffness, which is possibly a case or loading pattern dependent parameter. Nevertheless, an estimation can be made by isolating a free body, applying a unit moment M_u to the location and obtaining the rotation θ. The stiffness can be estimated as $k_\theta = M_u / \theta$.

The Eurocode BSEN1993-1-8 has set up some rules for determining whether a connection joint is rigid, semi-rigid or pinned for a structural steel structure. However, the same is not found for reinforced concrete structures.

2.2.4 Rigid Diaphragm

In the elastic analysis of structures, the axial and in-plane stiffness of members is typically much greater than their bending stiffness. Consequently, traditional analytical methods like moment distribution often neglect these axial deformations.

This high in-plane stiffness allows continuous floors to be idealized as "rigid diaphragms," which are assumed to have infinite in-plane rigidity. This simplification means all vertical elements on a given floor translate and rotate as a single, rigid body with no relative movement between them.

The rigid diaphragm concept is particularly useful for distributing lateral forces. For example, the lateral shear force between two connected cantilever walls will distribute proportionally to their respective stiffness values (proportional to $3EI / L^3$), since their lateral displacements at the diaphragm level must be equal.

The concept of rigid diaphragm has also been commonly adopted in the early computer era where computer storage and calculation speed were low. By constraining all nodes on a floor to move as a single unit, this idealization drastically reduces the number of degrees of freedom in the analytical mode, thereby minimizing computational demands.

Although modern computing power has largely eliminated these constraints, the rigid diaphragm simplification remains prevalent in commercial design software due to its convenience for applying lateral loads. Designers can simply apply a single point load representing the total floor-level shear from wind or seismic forces. The rigid diaphragm then automatically distributes this load to the supporting vertical elements (columns and walls). Otherwise, the loads have to be spread onto the various locations of the floor structures as uniformly distributed or linearly varying loads.

Other than rigid diaphragm, semi-rigid and flexible diaphragm can also be defined. Semi-rigid diaphragm refers to the actual in-plane stiffness of the floor structure, while flexible diaphragm is one without in-plane stiffness. To determine whether a floor structure can be qualified as a "rigid diaphragm," some rules have been set by the following codes:

(i) ASCE 7–22 (2022), 12.3.1.2 (section 12 titled "Seismic Design Requirement for Building Structures") – Diaphragms of concrete slabs or concrete-filled metal decks with span-to-depth ratios of three or less for seismic load in structures that have no horizontal irregularities are permitted to be idealized as rigid. For "horizontal structural irregularities," elaboration is given in table 12.3.1 under "diaphragm discontinuity irregularity" as a diaphragm with an abrupt discontinuity or variation in stiffness, including one having a cutout or open area greater than 50% of the gross enclosed diaphragm area, or a change in effective diaphragm stiffness of more than 50% from one storey to the next. The Code further permits diaphragm to be flexible where the computed maximum in-plane deflection of the diaphragm under lateral load is more than two times the average storey drift of the adjoining vertical elements of the seismic force-resisting system as shown in its figure 12.3-1 extracted as Figure 2.8.

Nevertheless, 27.4.5 of ASCE-16 (titled Wind Loads for Buildings) permits diaphragm constructed of untopped metal decks, concrete filled metal decks, concrete slabs having a span-depth ratio of two or less to be idealized

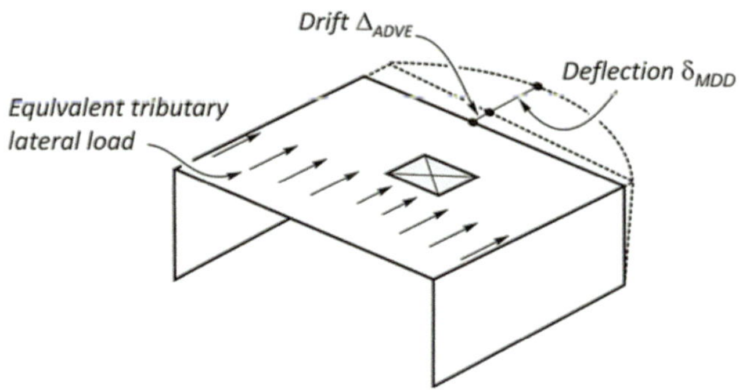

FIGURE 2.8 Extraction from figure 12.3-1 of ASCE7-22.

as rigid for consideration of wind loading. The requirement of no "diaphragm discontinuity irregularity" which is stated in 12.3.1.2 for seismic load is not included.

(ii) IBC (International Building Code 1604.4) – A diaphragm is rigid for the purpose of distribution of storey shear and torsional moment when the lateral deformation of the diaphragm is less than or equal to two times the average storey drift.

(iii) Eurocode BSEN 1998-1:2004+A1:2013 (4.3.1(4)) (the Eurocode for seismic design) – The diaphragm is taken as being rigid, if, when it is modelled with its actual in-plane flexibility, its horizontal displacements nowhere exceed those resulting from the rigid diaphragm assumption by more than 10% of the corresponding absolute horizontal displacements in the seismic design situation.

Comparatively the requirement in the Eurocode appears to be the most direct but yet the most stringent. Compliance with the requirement is difficult, especially when the deflection of a node is small.

Fundamentally, the rigid diaphragm assumption is invalid if the floor structure lacks sufficient in-plane stiffness to ensure all its nodes undergo nearly identical lateral movement (i.e., with differences within a specified tolerance, such as the 10% stipulated by Eurocode). This often occurs in floors with plan irregularities, such as the "bottle necks" illustrated in Figure 2.9(a) for a building. In such cases, unless the designer can demonstrate that areas on both sides of the bottleneck move as a single rigid body, the floor must be subdivided into multiple "split diaphragms," as shown in Figure 2.9(b). A split diaphragm is an idealized rigid unit covering a portion of the floor that is internally regular. In this model, the four split diaphragms are interconnected at the bottleneck regions, which should be modelled with their actual properties as semi-rigid elements. So, there are three split diaphragms joined by the two bottleneck structures simulated as "semi-rigid diaphragms." The lateral loads have to be split up in an appropriate manner onto each split diaphragm, say in

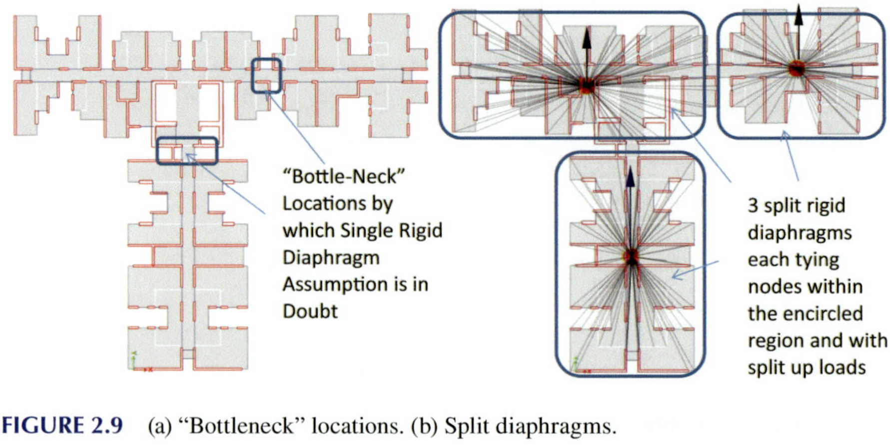

FIGURE 2.9 (a) "Bottleneck" locations. (b) Split diaphragms.

accordance with the exposed width for wind loads. Nevertheless, the bottle necks in Figure 2.9(a) actually fulfil the criterion ASCE7-22 as rigid diaphragms with their span to depth ratios.

A notable limitation of the rigid diaphragm assumption is that it precludes in-plane deformation, resulting in zero calculated in-plane forces. However, as these forces are rarely a controlling design factor, they are often neglected in structural design. In reinforced concrete structures, the low-level compression induced by diaphragm action can even enhance bending and shear capacity.

With the advances of computational methods and speeds, the rigid diaphragm idealization should be applied judiciously and only when justified by the floor's geometry and stiffness. Furthermore, the historical incentive for its use, efficient load input, is also diminishing, as modern software allows for easy application of distributed loads via spreadsheets.

2.2.5 Sub-frame Analysis

A building structure typically consists of multi-bay, multi-storey frames. Although modern computer software can analyze structures of many degrees of freedom, it is still uncommon to model the entire building for detailed gravity load design for floor beams of continuous spans. The major reason is the inefficiency of handling a full structural model for minor local revisions.

2.2.5.1 Sub-frame Analysis for Continuous Floor Beams

In addition to the inefficiency mentioned in the previous paragraph, another key reason for adopting sub-frame analysis is the impracticality of analyzing all possible loading combinations required by design codes, particularly the various "loading patterns" for continuous beams and slabs. The following are examples:

(i) Maximum loads on adjacent spans to determine peak support moments.
(ii) Maximum loads on alternate spans to determine peak sagging moments at mid-spans.

As a result, the structural analysis of a multi-storey frame is complicated by the need to consider numerous load cases, as gravity loads are factored differently based on whether their effect on a section is beneficial or adverse. To manage this complexity, the traditional approach is to use a "sub-frame" analysis. This method involves extracting a representative portion of the full structure for isolated analysis. The most common sub-frame, used for gravity load analysis of a continuous beam, includes the beam in question and the columns directly above and below it, as shown in Figure 2.10. In the model, the far ends of the columns are typically idealized as fixed supports, and the entire sub-frame is restrained against lateral translation. Consequently, the system is reduced to a continuous beam supported by rigid props, each with an associated rotational stiffness.

This rotational stiffness is provided by the columns. Under the fixed-far-end assumption, each column contributes a stiffness of $4EI / L$ to the joint, where E is

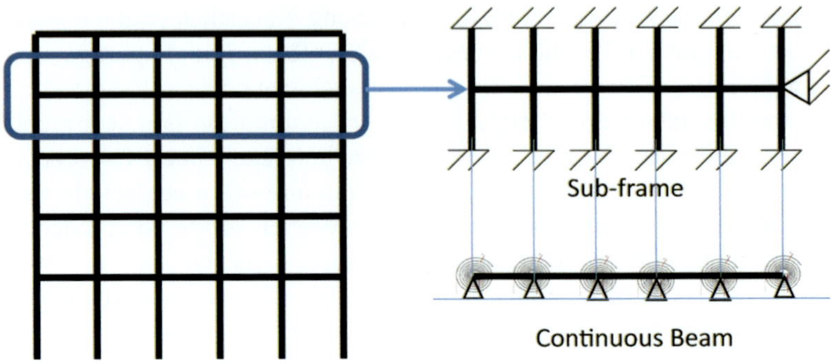

FIGURE 2.10 Formation of sub-frame.

the elastic modulus of the member, I is its second moment of area and L is the column height. This stiffness value represents the moment required to produce a unit rotation (1 radian) at the column end where the moment is applied. If a column's far end is pinned, its rotational stiffness is reduced to $3EI / L$.

The total rotational stiffness at a support is the sum of the stiffnesses from the column above and below: $\sum 4EI / L$ (or $\sum 3EI / L$ for pinned ends). This cumulative restraint is symbolized by the "coil" springs in the continuous beam model illustrated in Figure 2.10.

Sub-frame analysis is commonly employed to determine the internal forces in floor beams due to gravity loads for rebar design. This approach is significantly more efficient than analyzing the entire structure, as it simplifies the application of patterned loading for a limited number of load combinations. Furthermore, in buildings with typical floors, a single sub-frame model can represent all identical levels, drastically reducing the effort required to review results compared to a full global analysis. Any inaccuracies introduced by assuming fixed far ends for columns and preventing lateral sway are often accommodated through moment redistribution in the beam design.

2.2.5.2 Sub-frame Analysis at Interfaces of Stiff Horizontal Structures

While the sub-frame analysis described in Section 2.2.5.1 focuses on continuous beams, the practice of analyzing multi-storey sub-frames extracted from a larger building model remains common, even with modern computational power. This approach is often adopted to circumvent the complexity of analyzing the entire structure.

A critical limitation of this method is the transfer of reactions from one sub-frame to another (e.g., from a superstructure to a foundation). During this load transfer, deflection compatibility (ensuring that connected elements displace consistently) is rarely enforced. Consequently, the practice is only valid when the interface is a very stiff horizontal element, such as a thick pile cap or a transfer structure, where deformations are negligible as compared with the sub-frame.

This analytical inconsistency is typically tolerated for two reasons. Firstly, inherent safety factors in design codes provide a margin of cover. Secondly, since the total load is conserved, some internal redistribution can occur, allowing overstressed members to shed load to adjacent elements.

However, care must be taken when intermediate settlements are significant. In such cases, the sub-frame method should be replaced by a full global analysis to ensure deflection compatibility. Alternatively, the effects of estimated settlements may be explicitly prescribed to the sub-frames. For example, consider the system in Figure 2.11. The upper multi-storey frame is first analyzed assuming rigid supports at the transfer structure. The resulting reactions are then applied to the lower sub-frame (the transfer structure and its supports) for analysis. The settlements calculated from the lower frame are incompatible with the initial assumption of rigid supports for the upper frame, potentially leading to an underestimation of forces. To allow for these effects, the calculated settlements from the lower frame can be prescribed as displacement boundary conditions to the upper frame in a subsequent analysis. This iterative approach, while not perfectly compatible, allows designers to establish reasonable upper-bound estimates for forces and moments.

Nevertheless, even Figure 2.11(a) is not the most accurate analysis. The analysis is only true when the full frame has gained strength before the loads are applied, which is generally not achieved in multi-storey construction of buildings. This will be further discussed in Section 2.12 under "Staged Construction."

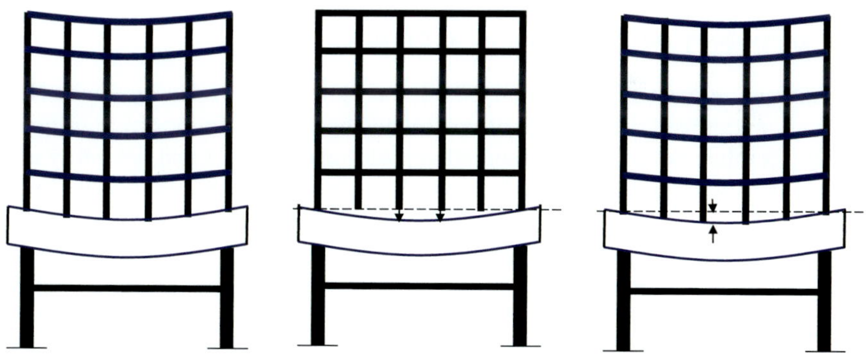

(a) Full frame analysed with full consistency in deflection.

(b) Analysis in split frames with inconsistency in deflection.

(c) Settlement in the transfer structure be prescribed to the frame above for additional analysis.

FIGURE 2.11 Sub-frame analysis and consistency in deflection.

2.3 TIMES BEFORE THE INTRODUCTION OF THE MOMENT DISTRIBUTION METHOD

The introduction of the Moment Distribution Method in 1930 marked a pivotal advancement in structural analysis for building design. Prior to this development, engineers primarily relied on two approaches for analyzing statically indeterminate structures, which are the Slope-Deflection Method and the Flexibility Method.

The Slope-Deflection Method can be considered as an early form of the stiffness method, with the axial deformation neglected in its calculations. The Flexibility Method, on the other hand, required a more involved process. First, redundant forces or moments were removed to create a statically determinate system. Then, these redundancies were reapplied while enforcing deformation compatibility conditions to solve for the unknown forces through systems of equations. Both methods required solving multiple simultaneous equations, which was a computationally intensive task in the pre-digital era.

To streamline these complex calculations, engineers developed comprehensive tables and charts for the determination of (i) fixed-end moments; (ii) structural deformations under various loading conditions; and (iii) moment coefficient tables for continuous beams with relatively unform spans.

The use of moment coefficients, in particular, became standardized in the 1934 publication of the Code of Practice for Reinforced Concrete of the UK by Scott and Glanville (1934), representing an important step in codifying structural analysis methodologies. The moment coefficients are still found in some of the modern codes of practice.

2.4 MOMENT DISTRIBUTION METHOD

The famous moment distribution method was developed by Hardy Cross (1930) which revolutionized the analysis of continuous beams and frames by significantly reducing computational effort compared to traditional stiffness and flexibility methods. Unlike these earlier approaches, which required solving systems of simultaneous equations to determine internal forces through displacement analysis – the moment distribution method employs an iterative, convergence-based process for calculating member moments. An overview of the method is outlined as follows:

(i) Initial Fixed-End Moments: The analysis begins with calculating fixed-end moments at each member end, representing moments induced by applied loads under the assumption of fully clamped joints (no rotation permitted). It should be noted that fixed-end moments are different for prismatic and non-prismatic members for the same loading on the same span.

(ii) Member Stiffness and Distribution Factor: The rotational stiffness of each member is calculated which is $4EI / L$ if the far end is fixed and $3EI / L$ when the far end is pinned. Distribution factors are also worked out among members meeting at a common joint.

(iii) Joint Release and Moment Distribution: Joints are sequentially released, allowing the initially unbalanced moments to redistribute according to member stiffness until both moment and rotational equilibrium are achieved.

(iv) Carry-Over Mechanism: For prismatic members (with a carry-over factor of 0.5), half of any distributed moment is carried over to the opposite end, creating new imbalances that require additional distribution cycles. The carry-over factor will not be 0.5 for non-prismatic members.

(v) Iterative Convergence: The process repeats with successively smaller carry-over moment adjustments until residual values become negligible. While theoretically requiring infinite iterations for exact solutions, practical analyses typically achieve sufficient accuracy within a few cycles.

(vi) Upon solution of moments, shears and axial loads can be readily solved.

Figure 2.12 shows a schematic representation of the method as extracted from Hardy Cross (1930).

A practical use of the moment distribution method is demonstrated in Table A3 of Appendix A for a continuous beam.

Theoretically, Hardy Cross's original method can achieve exact solutions under many cycles of distribution. For decades it has been taught in undergraduate courses, even up to now. But the method is still considered tedious as each process solves for only one load case and there can be many load cases if a continuous beam with many spans has to be analyzed. So, a simplified method has been developed which is outlined in Appendix A. By this simplified method, the moments induced by a span on all other spans of a continuous beam can be calculated in a single row of the moment distribution table. Selection of the hogging and sagging moments from the spans as appropriate can then be carried out to find the most critical hogging and sagging moments of a section. Comparing with the exact method of moment distribution, the

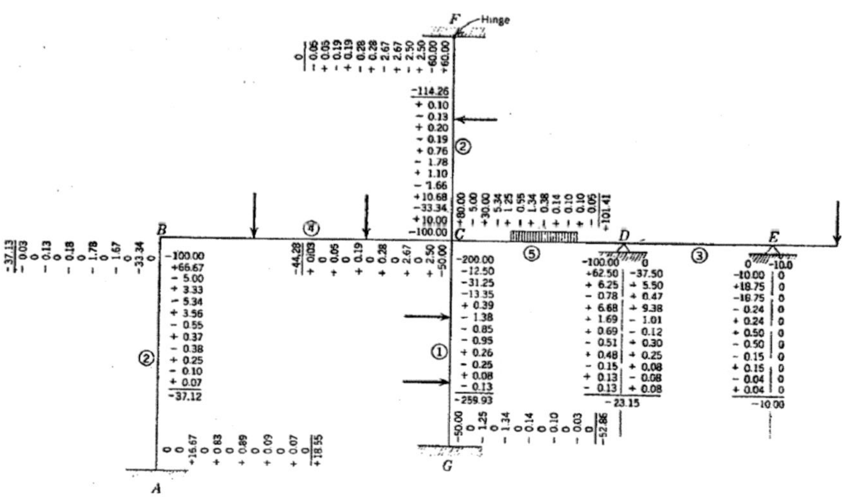

FIGURE 2.12 Extract from Hardy Cross showing the moment distribution procedure.

errors have been reported to be less than 10%. A worked example has been enclosed in Appendix A and the dead load case has been checked with Hardy Cross' exact method. The errors were within 10%.

To facilitate the calculation of the fixed-end moments due to partial uniformly distributed load, triangular load and trapezoidal load which involve lengthy formulae, design charts have been provided by references such as Reynolds and Steedman (1988) for estimating coefficients in the computations of the fixed-end moments. Demonstrations are enclosed in Appendix A.

Although the moment distribution method has been largely superseded by computer-based analysis in commercial design, it remains a valuable tool for structural engineers. It provides a fundamental understanding of structural behaviour that is essential for sound engineering judgment. Furthermore, practical knowledge of the method is still necessary for reviewing older designs and calculations that relied on this technique.

2.5 LATERAL LOAD ANALYSIS BY HAND CALCULATION

Building structures must be designed to resist lateral loads, such as those from wind and soil pressure. This section describes hand-calculation methods for analyzing this load distribution in the pre-computer era, which depend on whether the floor diaphragms are assumed to be rigid or flexible.

(i) Flexible Diaphragm Assumption: Lateral loads are distributed to lateral force-resisting elements (e.g., frames, walls) based on their tributary areas. This method assumes the diaphragm as being too flexible to affect the lateral displacement of these elements. So, it acts simply as a load-transferring membrane.

(ii) Rigid Diaphragm Assumption: The diaphragm is assumed to be infinitely rigid in-plane, causing all vertical elements to move together as a single unit. Consequently, lateral loads are distributed to frames and walls in proportion to their relative lateral stiffness, as illustrated in Figure 2.13.

In the rigid diaphragm load distribution method, the following simplified assumptions are often adopted:

(i) Walls are typically considered significantly stiffer than beam-column frames. Consequently, when shear walls or core walls are present, lateral loads are often assumed to be resisted entirely by these walls, with beam-column frames carrying no lateral load.

(ii) However, if the influence of these walls does not extend across the entire building, the beam-column frames in the unaffected areas will resist a portion of the wind load, as illustrated in Figure 2.14.

(iii) If the designer has adopted the assumption of a rigid diaphragm for distributing the lateral loads onto the walls and beam-column frames in accordance with their stiffnesses, a simplified approach is to assign the stiffness

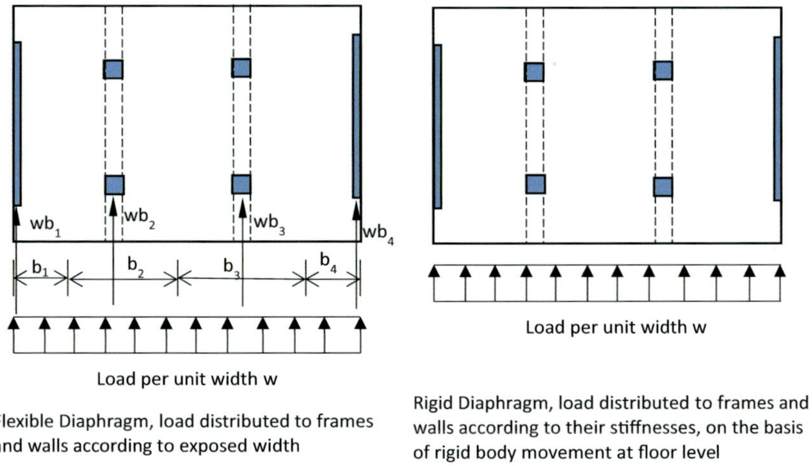

FIGURE 2.13 Lateral load distribution in accordance with "Rigid" and "Flexible" diaphragm.

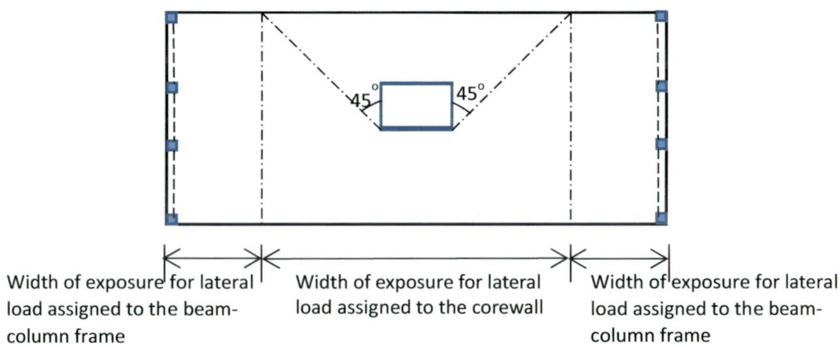

FIGURE 2.14 Lateral load distribution between walls and beam-column frames.

of a wall as $3EI / H^3$ where EI is the flexural rigidity of the wall and H is its height. This is under the assumption that the wall is fixed at its bottom level and is free to translate and rotate at its top level and lateral displacement. The assumption is justified by the comparatively large bending stiffness of the wall as compared with the comparatively small out-of-plane rotation stiffness of the floor structure generally. However, the stiffness assigned to a column is normally assumed to be $12EI / H^3$ which is based on the assumption that the column is restrained from rotation at its top and bottom levels. Furthermore, if the resultant of the applied load does not align with the shear centre (stiffness centre) of the system, there will be twisting of the system on plan which affects also the load distribution. The following procedures for analysis are adopted to achieve both force equilibrium and rigid body movement of the floor structure:

(a) Locate the "shear centre" of the system at (x_c, y_c) by $x_c = \frac{\sum K_{yi} x_i}{\sum K_{yi}}$; $y_c = \frac{\sum K_{xi} y_i}{\sum K_{xi}}$ where K_{xi}, K_{yi} are the stiffnesses of the wall or column i with the centroid at (x_i, y_i). The total shear stiffness in X and Y directions are then $K_X = \sum K_{xi}$ and $K_Y = \sum K_{yi}$

(b) Calculate the torsional rigidity of the system about the shear centre as $J = \sum \left[K_{yi} (x_i - x_c)^2 + K_{xi} (y_i - y_c)^2 \right]$;

(c) Taking S_X and S_Y as the total floor shear in X and Y directions and T as the torsion about the shear centre which is positive when it is anti-clockwise, calculate the column and wall shears in X and Y directions as

$$S_{xi} = S_X \frac{K_{xi}}{K_X} - T \frac{K_{xi}(y_i - y_c)}{J} \tag{2.1a}$$

$$S_{yi} = S_Y \frac{K_{yi}}{K_Y} + T \frac{K(x_i - x_c)}{J} \tag{2.1b}$$

(d) Calculate the end moments of a column i due to lateral shear by $0.5S_i \times H_i$ and that of wall i is $S_i \times H_i$ where H_i is the storey height, as in consistency with the fixity assumption of their ends.

2.6 DETAIL LATERAL LOAD ANALYSIS ON BEAM-COLUMN FRAME

Once the lateral shear distributed to a beam-column frame is determined, the frame can be easily analyzed using the stiffness method on a computer nowadays. However, this was impractical in the pre-computer era, as solving the resulting system of simultaneous equations, often numbering in the tens even for a simple frame, was highly laborious. Consequently, a simplified method was devised which assumed that floor beams had infinite stiffness relative to the columns. This assumption results in column end moments of 0.5S × H, where S is the column shear and H is its height. Simple moment distribution could then be used to calculate beam moments and shears, as well as column axial loads. This method is illustrated in Worked Example A2 in Appendix A. While historically important, it is generally not used today, as it can produce significant errors compared to modern computer analysis.

2.7 USE OF DESIGN CHARTS OR TABLES

In the pre-computer era, calculating parameters with lengthy formulae such as the fixed-end moments for partial uniformly distributed or triangular loads (as required

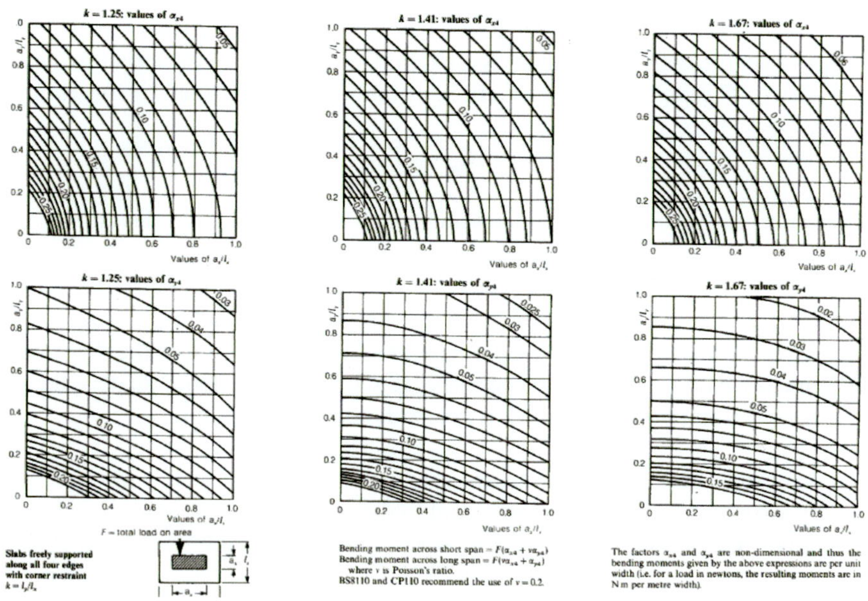

FIGURE 2.15 Design charts for 2-way slabs under partial uniformly distributed loads as extracted from Reynolds and Steedman (1988).

in Appendix A) was a tedious task. To avoid this laborious hand-calculation, engineers created design tables and charts. For example, Figure 2.15, extracted from Reynolds and Steedman (1988), shows a design chart for the moment coefficients in two-way slabs under partial loads, likely generated using the finite difference method. Such resources were indispensable tools in the pre-computer era.

These design charts are becoming unfamiliar to modern designers, who typically turn first to computer analysis. However, the authors maintain that these charts remain valuable for the fundamental engineering concepts they illustrate. Furthermore, they provide a useful way to quickly verify computer results. Since software often functions as a "black box," these charts allow a designer to check for potential errors in the digital analysis.

2.8 FIRST-ORDER LINEAR ANALYSIS

First-order linear structural analysis is a method that assumes a structure behaves linearly under load, meaning deflections and stresses are directly proportional to the applied forces and the proportionality is the stiffness of the structure. This approach simplifies calculations by neglecting the change in structural stiffness due to deformation, allowing for a fast analysis compared to non-linear methods and also a convenient analysis allowing linear superimposition of loads.

The analysis is often done by matrix analysis by which the stiffness matrix $\boldsymbol{K}_e$ of the structure, the deflection matrix Δ containing deflections of the structure and the force matrix $\boldsymbol{F}$ representing the externally applied load are related by Equation 2.2.

$$\boldsymbol{K}_e\Delta = \boldsymbol{F} \tag{2.2}$$

In Equation 2.2, $\boldsymbol{K}_e$ is built by assembling the stiffness of its individual members in accordance with their stiffness and geometry. $\boldsymbol{F}$ is also built by assigning the externally applied force at the various nodes of the structure. By Equation 2.2 Δ is solved by $\Delta = \boldsymbol{K}_e^{-1}\boldsymbol{F}$. The forces on the individual members i can then be calculated by $k_i\Delta_i$ where k_i is the stiffness matrix of the individual member and Δ_i is the deformation of members i extracted from Δ. These are included in the undergraduate syllabus.

The key assumptions embodied in the method are listed as follows:

(i) Linear Behaviour – The structure is assumed to behave linearly, i.e., its deformation is linearly proportional to the applied load.
(ii) Small Deformation – The deformation is assumed to be small so that the geometry of the structure is not affected.

Under (i) and (ii), the stiffness matrix $\boldsymbol{K}_e$ is taken as constant throughout the analysis process. Generally, there is also no upper bound to the deflection. The designer should therefore check if the deflection is so excessive that the method is no longer applicable to the analysis. In addition, if the secondary effects including buckling, P-Delta effects and effects due to stage construction are significant, the method is then not applicable or the errors will be significant.

2.9 BUCKLING ANALYSIS AND P-DELTA EFFECTS

Buckling is a well-documented phenomenon in structural mechanics, typically introduced in elementary textbooks as the instability of an axially loaded strut. Simply put, it is the sudden, significant lateral deflection of a compressed member, which can lead to catastrophic failure. The most fundamental model for analyzing buckling is a pin-ended strut (restrained from translation but free to rotate at both ends). The behaviour of this strut under an axial load P is governed by the following differential equation:

$$EI\frac{d^2v}{dx^2} + Pv = 0 \tag{2.3}$$

where EI is the flexural rigidity of the strut; v is the lateral deflection of the strut at coordinate x and P is the applied axial load and L is the length of the strut.

The solution of Equation 2.3 is

$$v = A\sin\left(\sqrt{\frac{P}{EI}}x\right) \tag{2.4}$$

The solution satisfies $v = 0$ when $x = 0$. However, to satisfy $v = 0$ when $x = L$, $\sqrt{\frac{P}{EI}}L = \pi, 2\pi, 3\pi$ etc., or $P = \frac{\pi^2 EI}{L^2}, \frac{4\pi^2 EI}{L^2}, \frac{9\pi^2 EI}{L^2}$ etc. They are the buckling loads of the strut with the lowest one at $\frac{\pi^2 EI}{L^2}$ which is also referred to as the Euler load.

Mathematically, Equation 2.4 implies that A can be any value if $P = \frac{\pi^2 EI}{L^2}$, $\frac{4\pi^2 EI}{L^2}, \frac{9\pi^2 EI}{L^2}$ etc. Conversely, if P is not equal to one of these discrete values, then A must be zero, implying that the strut remains straight. This result appears to contradict the common elementary textbook statement that buckling occurs when P exceeds $\frac{\pi^2 EI}{L^2}$ as the equation does not indicate instability for higher loads.

But when there is an "imperfection" in a strut which implies that the strut is not perfectly straight but has a "bowing" along its length, it will fundamentally change the strut's structural behaviour under axial compression by creating an eccentric moment. The further bowing v will then increase, together with the eccentric moment. Nevertheless, it can be proved that under an imperfection, a strut will have infinite deflection when the compressive axial load reaches the buckling load as indicated by Equation 2.5 for a pin-ended strut by solving $EI\frac{d^2 v}{dx^2} + P(v + v_0) = 0$ and the solution is

$$v = \frac{P / EI}{(\pi / L)^2 - (P / EI)} V \sin\frac{\pi x}{L} \tag{2.5}$$

where $v_0 = V \sin(\pi x / L)$ represents the bowing of the strut simulated as a half sine curve in which V is the maximum bowing, EI, P and L are the same symbols used in Equation 2.4.

From Equation 2.5, it can readily be seen that the denominator $(\pi / L)^2 - (P / EI)$ vanishes when $P = \frac{\pi^2 EI}{L^2}$ as the Euler load, implying an infinite value of the lateral deflection v. So, the strut fails at this load.

Therefore, the theoretical buckling load plays a critical role in a strut's stability. In practical terms, a strut really fails by buckling as its load approaches this critical value. However, this failure is ultimately triggered by the inevitable imperfections present in any real-world member.

The buckling load described by Equation 2.5 applies to a strut with pinned ends. For other support conditions such as fixed, free, or a combination, the buckling load is modified. A common practice is to express these loads using the concept of an "effective length" L_e, generalizing the formula to $P = \frac{\pi^2 EI}{L_e^2}$. So, for a strut with both ends fixed which has a theoretical buckling load of $\frac{4\pi^2 EI}{L^2}$, the effective length is $L_e = 0.5L$ whereas $L_e = 2L$ for a cantilever strut with buckling load $\frac{\pi^2 EI}{4L^2}$.

Because imperfections are inherent in real struts, checking slender compression members for buckling is a standard design requirement. This check is initially based on estimated end conditions, as structural instability occurs if the axial load in any member exceeds its buckling capacity.

In reality, however, the support conditions of a member within a building frame are seldom perfectly pinned or fixed, making the assignment of a precise effective length challenging. To overcome this limitation, a more rigorous method calculates the buckling load of a member within the context of the entire structure and its applied loading, as detailed in Appendix B.

This advanced approach accounts for the effect of axial load by incorporating a geometric stiffness matrix (also known as a stress stiffness matrix) for each member. This matrix, which is a function of the axial force (P), is added to the member's standard elastic stiffness matrix. The global stiffness matrix for the structure is then assembled conventionally, but from these modified member stiffness matrices, as shown:

$$\text{An individual stiffness matrix for member } i \quad \boldsymbol{k}_{ei} + P_i \boldsymbol{k}_{gi} \tag{2.6}$$

where $\boldsymbol{k}_{ei}$ and $\boldsymbol{k}_{gi}$ are the elastic stiffness and geometric matrices of the member and P_i is the axial load in the member.

After assembly, the full matrix of the structure becomes $\boldsymbol{K}_e + P\boldsymbol{K}_g$ and the relation with the deflection and externally applied load is

$$\left(\boldsymbol{K}_e + P\boldsymbol{K}_g\right)\Delta = \boldsymbol{P}_e \tag{2.7}$$

where $\boldsymbol{K}_e$ and $\boldsymbol{K}_g$ are the assembled elastic stiffness and geometric matrices of the full structure, P (scalars) are the axial loads of the struts, Δ is the deflection matrix, and $\boldsymbol{P}_e$ is the matrix of the externally applied loads.

However, as the axial loads are generally not known at the beginning, they can be first determined in the usual elastic stiffness analysis without the geometric matrices. Then the geometric matrices are included with the first determined axial loads. The structure is then re-analyzed to obtain a new set of axial loads. The analysis is then re-iterated until successive sets of axial loads converge to pre-determined limits of errors.

To find the least buckling load of the structure, the exercise is to find a factor λ applied to P such as the stiffness matrix $\left(\boldsymbol{K}_e + \lambda P\boldsymbol{K}_g\right) = 0$ or becomes singular, such that Δ becomes very large. So, we can list

$$\left(\boldsymbol{K}_e + \lambda P\boldsymbol{K}_g\right)\Delta = 0 \Rightarrow \boldsymbol{K}_e\Delta = -\lambda P\boldsymbol{K}_g\Delta \Rightarrow \boldsymbol{K}_e^{-1}P\boldsymbol{K}_g\Delta = \frac{-1}{\lambda}\Delta \tag{2.8}$$

Equation 2.8 is an eigenvalue problem by which $\frac{-1}{\lambda}$ is the eigenvalue of $\boldsymbol{K}_e^{-1}P\boldsymbol{K}_g$.

The eigenvalue is the buckling load factor by which the axial loads in the struts are increased to $-1/\lambda$ times such that at least one member reaches the buckling load, such that the structure becomes unstable.

Equation 2.8 shows an inconsistent stiffness matrix $\left(\boldsymbol{K}_e + \lambda P \boldsymbol{K}_g\right)$ as compared with Equation 2.2 with a constant stiffness matrix $\boldsymbol{K}_e$. So, the analysis is non-linear.

The foregoing can be demonstrated by the simple concrete structure shown in Figure 2.16 as Worked Example 2.2.

Worked Example 2.2

A four storey structure as shown in Figure 2.16 with storey heights 3 m, Young's modulus of 26,400,000 kN/m² is supported by a 400 mm thick shear wall on one side and 2 nos. of columns of 400 mm sq and 500 mm sq on the other side. All supports are assumed to be pinned. Uniformly distributed loads of 2 kN/m² for dead load and 5 kN/m² for live load are applied on the slabs.

As a result, the buckling load factor of the first mode is 11.150 for the load combination case of 1.4 × Dead + 1.6 × Live which should apply to the 400 mm sq column which shows significant buckling by the buckling mode shape as indicated on the right side of Figure 2.16. As the load in the column is 2803.28 kN analyzed by the software at the applied load, the buckling load is $P_{cr} = 11.15 \times 2803.28 = 31256.57\,\text{kN}$. The effective length of the column can be back-calculated as $L_e = \pi\sqrt{EI/P_{cr}} = \pi\sqrt{26400000 \times 0.4^4/12/31256.57} = 4.217$ m which is longer than the storey height of 3 m. So, the usual assumption of the

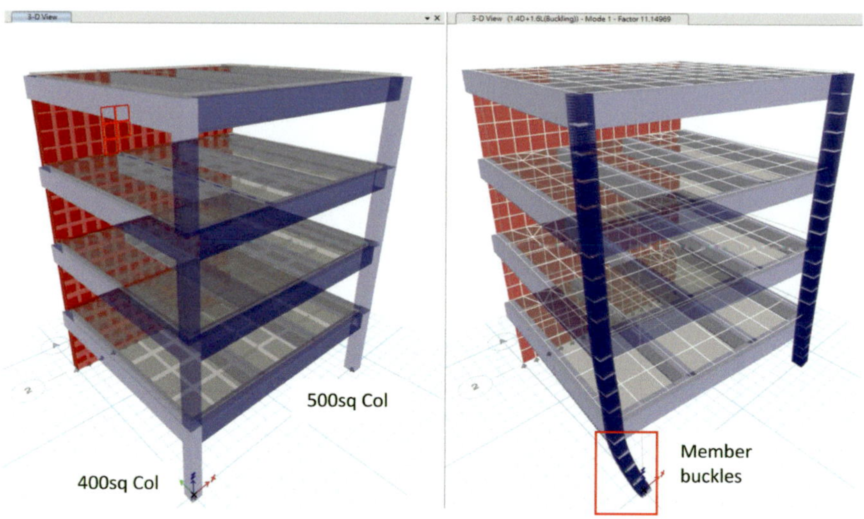

FIGURE 2.16 Buckling analysis of a simple structure (Worked Example 2.2).

physical length as the effective length is not conservative. This is likely due to lateral movement at the column upper end.

In addition, generally the buckling mode also varies with the applied load. Different buckling load factors will be arrived at different load cases. Theoretically, a constant effective length may not exist for a column member.

Higher buckling modes may be used for determination of effective lengths of other columns. But very often, the buckling modes are not distinct as to which columns are correspondingly failing.

Nevertheless, some codes of practice contain provisions for the effective length of compressive strut such as the Eurocode BSEN1992-1-1:2023. In the Eurocode, the effective length is the free length of the strut modified by a factor related to rotational stiffness (moment required to produce unit rotation) at its ends. Based on the effective length, slenderness ratio λ is determined by effective length divided by radius of gyration (uncracked section). The Code then restricts λ to some minimum values so as to protect the column from buckling failure.

2.10 P-DELTA ANALYSIS

The use of the geometric matrix with the stiffness matrix can be used to analyze the P-Δ-δ analysis.

P-Δ can be studied by reference to Figure 2.17 as a numerical example as Worked Example 2.3.

Worked Example 2.3
A cantilever strut is acted on by an axial load P and a lateral shear S at its free end as shown in Figure 2.17 with the following numerical values:

$P = 1500\text{kN}$; $S = 200\text{kN}$; $L = 4.8\text{m}$; $EI = 675840\text{kN/m}^2$.

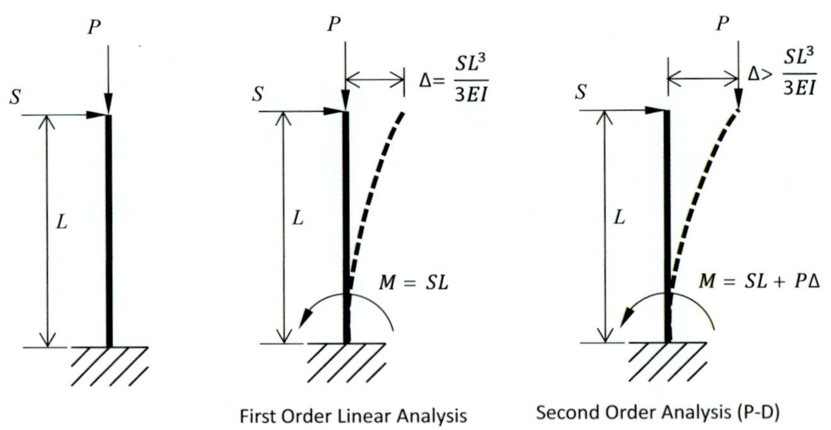

FIGURE 2.17 Illustration of P-Δ effect.

The lateral displacement obtained by first-order linear analysis is $\Delta = \frac{SL^3}{3EI} = \frac{200 \times 4.8^3}{3 \times 675840} = 0.0109\text{m}$. To calculate the additional effect due to P-Δ, Equation 2.7 is applied. The linear stiffness matrix k_e for a strut is

$$k_e = \begin{bmatrix} 12EI/L^3 & 6EI/L^2 & -12EI/L^3 & 6EI/L^2 \\ 6EI/L^2 & 4EI/L & -6EI/L^2 & 2EI/L \\ -12EI/L^3 & -6EI/L^2 & 12EI/L^3 & -6EI/L^2 \\ 6EI/L^2 & 2EI/L & -6EI/L^2 & 4EI/L \end{bmatrix}$$

The strut is divided into five members and the stiffness matrix is assembled in the usual manner. The first node is modified as the first support. A 12 × 12 stiffness matrix is then formulated in the following K_e, with the length of each segment $L = 4.8/5 = 0.96\text{m}$.

$$Ke = EI \begin{bmatrix} 1 & 0 & 0 & 0 & 0 & 0 & 0 & 0 & 0 & 0 & 0 & 0 \\ 0 & 1 & 0 & 0 & 0 & 0 & 0 & 0 & 0 & 0 & 0 & 0 \\ 0 & 0 & 24/L^3 & 0 & -12/L^3 & 6/L^2 & 0 & 0 & 0 & 0 & 0 & 0 \\ 0 & 0 & 0 & 8/L & -6/L^2 & 2/L & 0 & 0 & 0 & 0 & 0 & 0 \\ 0 & 0 & -12/L^3 & -6/L^2 & 24/L^3 & 0 & -12/L^3 & 6/L^2 & 0 & 0 & 0 & 0 \\ 0 & 0 & 6/L^2 & 2/L & 0 & 8/L & -6/L^2 & 2/L & 0 & 0 & 0 & 0 \\ 0 & 0 & 0 & 0 & -12/L^3 & -6/L^2 & 24/L^3 & 0 & -12/L^3 & 6/L^2 & 0 & 0 \\ 0 & 0 & 0 & 0 & 6/L^2 & 2/L & 0 & 8/L & -6/L^2 & 2/L & 0 & 0 \\ 0 & 0 & 0 & 0 & 0 & 0 & -12/L^3 & -6/L^2 & 24/L^3 & 0 & -12/L^3 & 6/L^2 \\ 0 & 0 & 0 & 0 & 0 & 0 & 6/L^2 & 2/L & 0 & 8/L & -6/L^2 & 2/L \\ 0 & 0 & 0 & 0 & 0 & 0 & 0 & 0 & -12/L^3 & -6/L^2 & 12/L^3 & -6/L^2 \\ 0 & 0 & 0 & 0 & 0 & 0 & 0 & 0 & 6/L^2 & 2/L & -6/L^2 & 4/L \end{bmatrix}$$

The geometric matrix for a single strut is

$$k_g = \frac{-1}{30L} \begin{bmatrix} 36 & 3L & -36 & 3L \\ 3L & 4L^2 & -3L & -L^2 \\ -36 & -3L & 36 & -3L \\ 3L & -L^2 & -3L & 4L^2 \end{bmatrix}$$

Similarly, the assembled geometric matrix for the whole strut divided into five segments is

$$K_g = \frac{-1}{30L}\begin{bmatrix} 1 & 0 & 0 & 0 & 0 & 0 & 0 & 0 & 0 & 0 & 0 & 0 \\ 0 & 1 & 0 & 0 & 0 & 0 & 0 & 0 & 0 & 0 & 0 & 0 \\ 0 & 0 & 72 & 0 & -36 & 3L & 0 & 0 & 0 & 0 & 0 & 0 \\ 0 & 0 & 0 & 8L^2 & -3L & -L^2 & 0 & 0 & 0 & 0 & 0 & 0 \\ 0 & 0 & -36 & -3L & 72 & 0 & -36 & 3L & 0 & 0 & 0 & 0 \\ 0 & 0 & 3L & -L^2 & 0 & 8L^2 & -3L & -L^2 & 0 & 0 & 0 & 0 \\ 0 & 0 & 0 & 0 & -36 & -3L & 72 & 0 & -36 & 3L & 0 & 0 \\ 0 & 0 & 0 & 0 & 3L & -L^2 & 0 & 8L^2 & -3L & -L^2 & 0 & 0 \\ 0 & 0 & 0 & 0 & 0 & 0 & -36 & -3L & 72 & 0 & -36 & 3L \\ 0 & 0 & 0 & 0 & 0 & 0 & 3L & -L^2 & 0 & 8L^2 & -3L & -L^2 \\ 0 & 0 & 0 & 0 & 0 & 0 & 0 & 0 & -36 & -3L & 36 & -3L \\ 0 & 0 & 0 & 0 & 0 & 0 & 0 & 0 & 3L & -L^2 & -3L & 4L^2 \end{bmatrix}$$

According to Equation 2.7 the stiffness matrix modified by the geometric matrix is $\boldsymbol{K_e} + P\boldsymbol{K_g}$ and the load matrix is $\boldsymbol{F} = (0 \quad 0 \quad 0 \quad 0 \quad 0 \quad 0 \quad 0 \quad 0 \quad 0 \quad 0 \quad S \quad 0)^T$, where $S = 200$.

Solving $(K_e + PK_g)\Delta = F$,

$$\Delta = 0.001\ (0 \quad 0 \quad 0.6222 \quad 1.2506 \quad 2.313 \quad 2.2259 \quad 4.8071 \quad 2.9239 \quad 7.8378 \quad 3.3434 \quad 11.1369 \quad 3.4833)^T$$

So, by the additional P-Δ effect, the deflection at the cantilever end is 11.1369 mm which is greater than 10.91 mm obtained previously.

A counter-check on the individual matrix multiplied by the displacement at the first two nodes as $k_c.\Delta_1$ where

$$K_c = \begin{bmatrix} 12EI/L^3 & 6EI/L^2 & -12EI/L^3 & 6EI/L^2 \\ 6EI/L^2 & 4EI/L & -6EI/L^2 & 2EI/L \\ -12EI/L^3 & -6EI/L^2 & 12EI/L^3 & -6EI/L^2 \\ 6EI/L^2 & 2EI/L & -6EI/L^2 & 4EI/L \end{bmatrix} + P\frac{-1}{30L}\begin{bmatrix} 36 & 3L & -36 & 3L \\ 3L & 4L^2 & -3L & -L^2 \\ -36 & -3L & 36 & -3L \\ 3L & -L^2 & -3L & 4L^2 \end{bmatrix}$$

$$\Delta_1 = \begin{bmatrix} 0 & 0 & 0.0006222 & 0.0012506 \end{bmatrix}^T$$

$$k_c.\Delta_1 = \begin{bmatrix} -200 & -976.7054 & 200 & 783.7721 \end{bmatrix}^T$$

So the base shear is 200 kN and the base moment is 976.7054 kNm.

As a counter-check by equilibrium, the base moment due to lateral shear is $200 \times 4.8 = 960\text{kNm}$ and that due to eccentric effect by the axial load is $1500 \times 0.0111369 = 16.7054\text{kNm}$. So the total moment $= 960 + 16.7054 = 976.7054$ kNm which is equal to that obtained by matrix multiplication above.

Analysis for more complicated structure for P-Δ follows the same principle by the use of the geometric matrix. Nevertheless, as in general the axial loads in the structural members are not known initially (as being different in the cantilever discussed in the Worked Example 2.3 where the axial load is known), usually they have to be found by the first-order analysis as a start. After the acquisition of the axial loads, analysis with the use of the Equation 2.7 will be carried out with the determination of the second set of axial loads P which will then be used in further analysis by Equation 2.7. The iterative process will be continued until convergence to a desired limit is reached. A Worked Example 2.4 by computer analysis is demonstrated as follows:

Worked Example 2.4

In the Worked Example, a computer model analysis is carried out for the symmetrical core wall frame building shown in Figure 2.18. The lateral displacements due to Dead + Live + wind load in the X-direction at the main roof level under first-order elastic analysis is 579 mm while that under inclusion of the P-Δ effect is 652 mm as shown in Figure 2.19. So the displacement is magnified by the P-Δ effect.

The additional deflection is due to the P-Δ effect in which the gravity load has a contribution. An additional overturning moment at the base of the building is also resulted. The reactions at the building base are summarized in Table 2.2.

FIGURE 2.18 Core wall building for study of P-Δ effects (Worked Example 2.4).

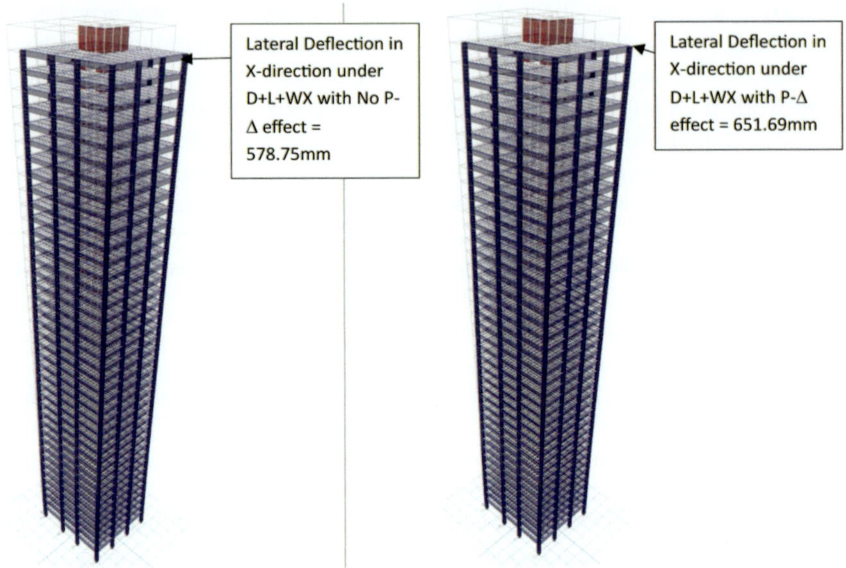

FIGURE 2.19 Deflections with and without P-Δ effects (Worked Example 2.4).

TABLE 2.2
Summary of Building Base Reaction with and without P-Δ Effect

	Dead Load	Live Load	D + L	Wind X (No P-Δ Effect)	D + L + WX (with P-Δ Effect)
Vertical Load (kN)	349997	126720	476717	0	476717
Overturning Moment (kNm)	–5080	0	–5080	–1354116	–1510864

From Table 2.2, an extra eccentric moment of $-1510864-(-1354116-5080)$ $=-151668\text{kNm}$ is found when comparing the P-Δ Effect and first-order analysis. As the lateral sway of the structure is 651.69 at the main roof level, assuming the average deflection of all floors is half of the maximum of 686.97 = 325.85 mm, the extra eccentric moment is $476717\times0.32585=155335\text{kNm}$ which is quite close to the extra moment of 151668 kNm by analysis (differing by 2.4%).

In addition to P-Δ effect, there is also a P-d effect which is the eccentric load effect due to "imperfection" in a strut under compression as illustrated in Figure 2.20 where a strut initially with a "bowing" shape of magnitude d which increases to d'. Accordingly, the eccentric moment within the strut is also magnified.

The P-δ (second-order) effect depends on the initial geometric imperfection of a compressive member, such as a column or wall, which is primarily caused

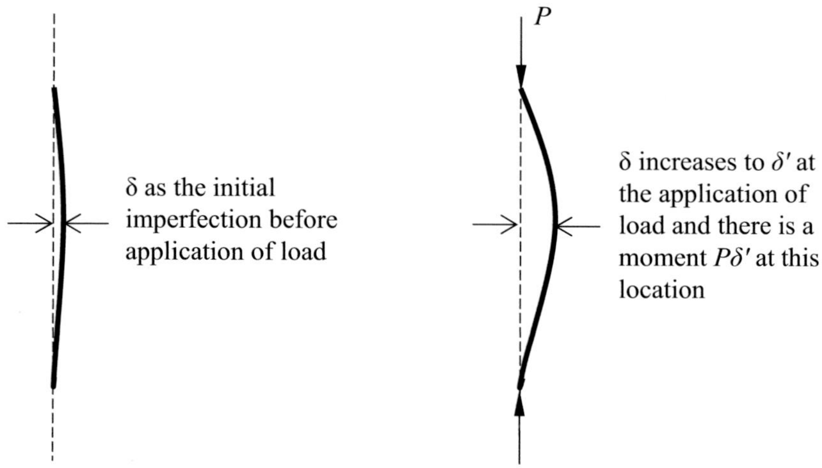

FIGURE 2.20 Illustration of P-d effect.

by construction tolerances. To account for this in design, building codes assign a pre-determined, standardized imperfection to members based on their length and cross-section; these values are typically derived from experimental data and past experience.

The additional eccentric moment induced by this imperfection, *V*, can be determined using Equation 2.5. This imperfection-based design approach is more prevalent in structural steel design. For instance, Eurocode EN1993-1 and the former British Standard BS5950 specify imperfection factors that vary with section type and bending direction to calculate this moment.

There are two primary methods to account for this effect:

1. Strength Reduction: The reduction in the member's compressive strength due to the P-δ effect is calculated and subtracted from its original axial capacity.
2. Equivalent Moment: The member is designed to resist an additional eccentric moment arising from the imperfection.

In both approaches, the resulting strength reduction and the eccentric moment increase with the member's length. Consequently, the key parameter governing this behaviour is the slenderness ratio, defined as the effective length (a concept identical to that used in buckling analysis, as discussed in Section 2.9) divided by the radius of gyration of the cross-section.

The same design concept, accounting for geometric imperfections and second-order effects, also applies to concrete compressive members (walls and columns). Eurocode EN1992-1-1 specifies minimum initial imperfections, which are a function of member length, and requires the inclusion of corresponding additional eccentric moments in the design.

A comprehensive analysis that incorporates both global structural sway (P-Δ effects) and local member imperfections (P-δ effects) is termed a P-Δ-δ analysis. A simplified, conservative approach is to perform a P-Δ analysis (considering global sway) and then design individual members for imperfection effects as stipulated by the design codes.

For a more accurate representation, a compressive member can be subdivided into multiple segments with introduced nodes, each assigned an appropriate initial imperfection (δ). A subsequent P-Δ analysis of this refined model effectively captures the combined P-Δ-δ behaviour.

A significant challenge with this detailed method is the determination of the direction of the initial "bowing" for each member. For a structure with many members, analyzing all possible bowing combinations becomes computationally impossible. A practical solution is to perform a buckling analysis and align the imperfection directions with the shape of the fundamental buckling mode, as this represents the most critical instability pattern.

2.11 LARGE DEFORMATION ANALYSIS

Large deformation analysis is an analytical method used when a structure undergoes significant deformations that alter its stiffness. Unlike first-order linear analysis, which assumes a constant stiffness matrix based on the original geometry, this method accounts for the change in structural geometry under load. Since the final deformed shape is unknown at the beginning, the analysis has to be performed iteratively so as to capture the change of deformation in stages. The analysis has to be P-Δ-δ analysis in each stage. The load is applied in incremental stages (e.g., 5 to 10 steps of 10–20% each). After each increment, the stiffness matrix is updated to reflect the current deformed geometry. To accurately model member curvature and internal forces, each member is subdivided into multiple segments with intermediate nodes. This process, illustrated in Figure 2.21, is also known as "large deformation analysis" which stands in contrast to the conventional first-order linear analysis based on the assumption of small displacements.

To accurately capture the effects of member imperfections, the structural model must explicitly incorporate them, as is done in a P-Δ-δ (P-Delta) analysis. The direction of these initial geometric imperfections (or bowing) should be aligned with the shape of the first buckling mode as discussed in Section 2.10.

The building shown in Figure 2.18, which has undergone P-Δ analysis is also analyzed by large deformation analysis. The lateral deflection at the same corner node at roof level, as shown in Figure 2.19(b), is 650.18 mm at the same load case of D + L + WX, which is very close to that by P-Δ analysis at 651.69 mm for this building.

2.12 STAGE CONSTRUCTION ANALYSIS

Conventional structural analysis models a building in its completed state, which fails to account for the sequential, floor-by-floor nature of construction. This simplified approach assumes the full building stiffness resists the entire dead load

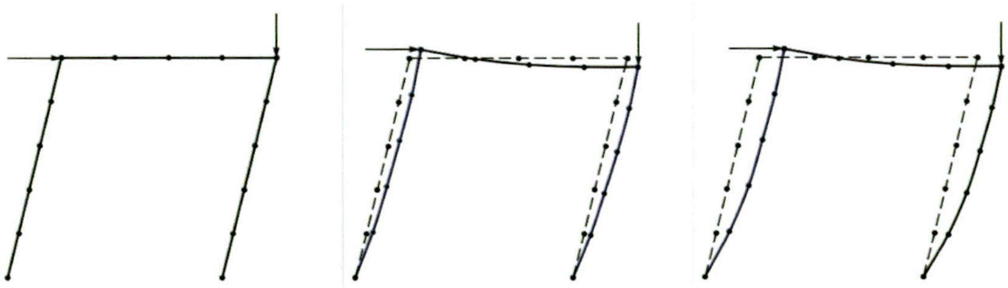

Initial structure (segmented by nodes) with defined "initial imperfections" and loads for 1st Analysis with a percentage of load.

After 1st Analysis, deformations taken and used to modify stiffness of the structure and load application for the next analysis with the next load increment.

After a number of analyses until full load, the final deformation and internal forces of the structure are obtained.

FIGURE 2.21 Illustration of the process of large deformation analysis.

simultaneously. In reality, both the application of dead load and the development of structural stiffness occur progressively.

This discrepancy is often negligible in structures with continuous vertical members. However, it can cause significant errors in buildings with transfer structures, as the case studies in subsequent sections will demonstrate.

The issue is illustrated in Figure 2.22. Essentially, the dead load of each new floor (including the "wet" weight of fresh concrete) is initially supported by the stiffness of the already constructed floors below it, not by the full building. The resulting deformations are "locked in" as the structure is being built, creating a stress state different from that of a full-model analysis.

The effect is now illustrated by the following Worked Example 2.5.

Worked Example 2.5

Worked Example 2.5, as shown in Figure 2.23, has been tried for demonstration, which is a 20 storey building with a 2 m thick transfer plate at the first storey level and the 4 interior columns omitted in the ground floor.

First-order linear analysis as per Model (a) and floor by floor staged analysis as per Model (b) are performed for the dead weight case (structure's own weight) by computer software. The deflection profile and axial loads of the frame above the transfer plate at grid two are shown in Figures 2.24 and 2.25 respectively. By the results of the analyses, the following can be summarized:

(i) Model (a) has a greater stiffening effect on the full structure than Model (b) as the deflections on the transfer plate are smaller, together with smaller axial loads by the interior columns on the transfer structure. As such, there can be under-design by Model (a).

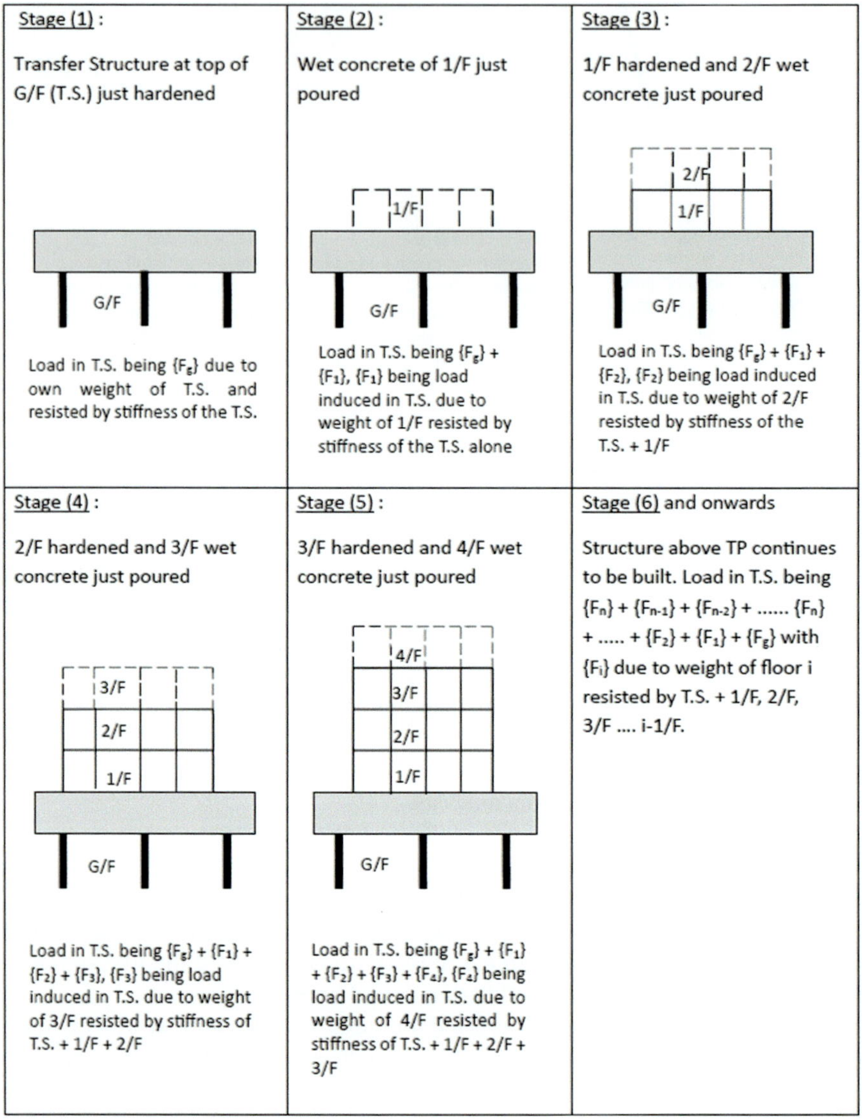

FIGURE 2.22 Diagrammatic illustration of the effects of construction sequence of loads induced on transfer structure.

(ii) The deflection at the roof level of Model (a) is however significantly greater than Model (b). This is due to the self-adjustment of the column height by the software during the construction stages, i.e., at each floor the columns

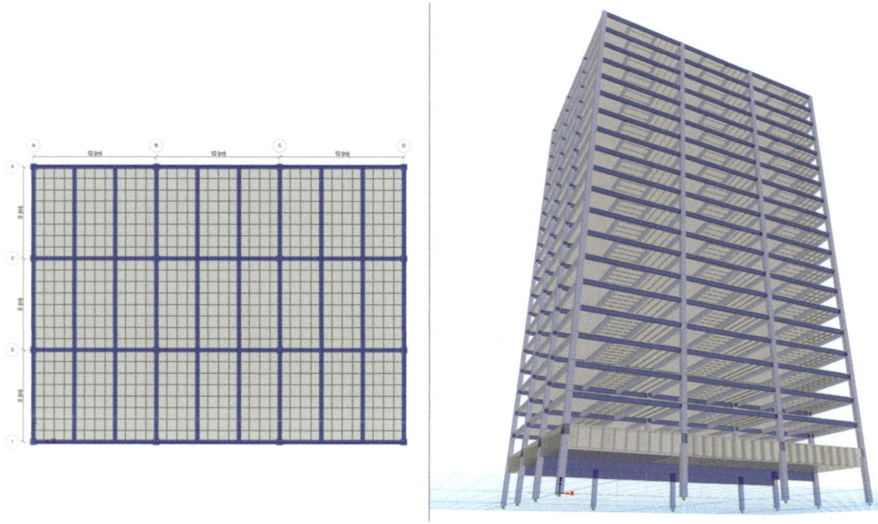

FIGURE 2.23 Twenty storey building with a transfer plate at first storey and the interior columns omitted at the ground floor (Worked Example 2.6).

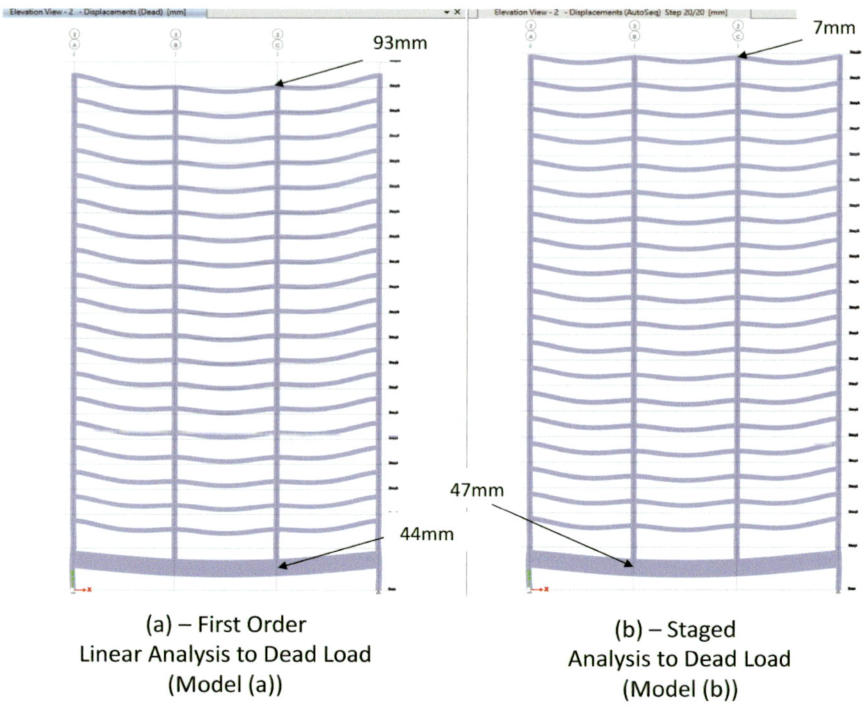

FIGURE 2.24 Comparison of deflection profiles of analysis due to dead load of Model (a) and Model (b) (Worked Example 2.5).

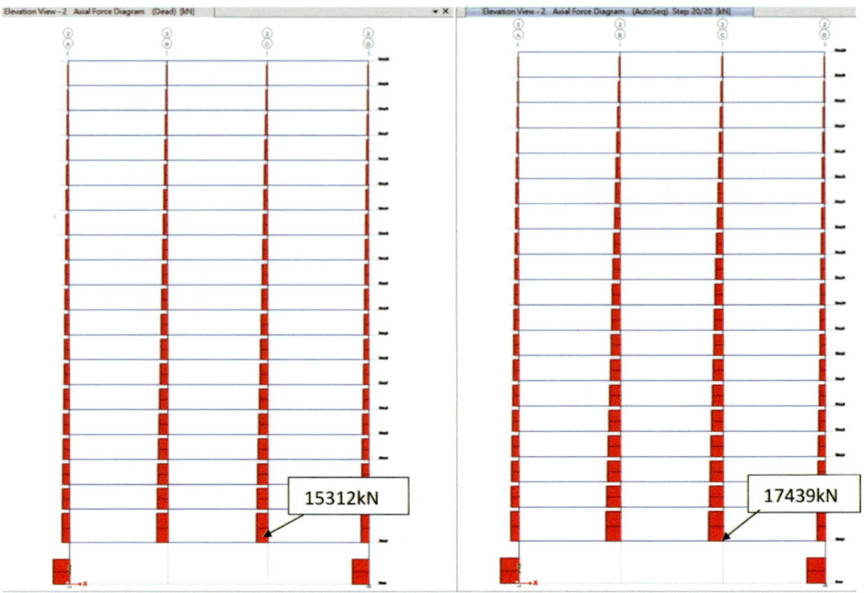

(a) – First Order
Linear Analysis to Dead Load
(Model (a))

(b) – First Order
Linear Analysis to Dead Load
(Model (b))

FIGURE 2.25 Comparison of column axial loads profiles of analysis due to dead load for Model (a) and Model (b) (Worked Example 2.5).

are modelled to the floor level in the stiffness matrix which will be longer than the floor height while the column heights of Model (a) remain unchanged at floor height.

The staged analysis by Model (b) is more accurate. The Worked Example demonstrates that errors can be significant when a full model is used to analyze dead weight of the structure when transfer structure is used.

Another interesting phenomenon is that demonstrated by Worked Example 2.6 for a building exhibiting building "lean."

Worked Example 2.6

In Worked Example 2.6, the building under consideration is 162 m high with 40 storeys. It possesses a "side core wall" such that its axial stiffness at one end on plan is significantly greater than at the other as shown in Figure 2.25. The analysis is first carried out by the first-order elastic analysis in which the full building stiffness is used for analysis against the full gravity loads. Significant lateral deflections in the Y-direction, i.e., towards the side with weaker axial stiffness, are obtained which are 201.9 mm under dead load and 313.5 mm under dead + live load at the main roof level as shown in Figure 2.26. The phenomenon can be justified as the large significant sway or the building lean.

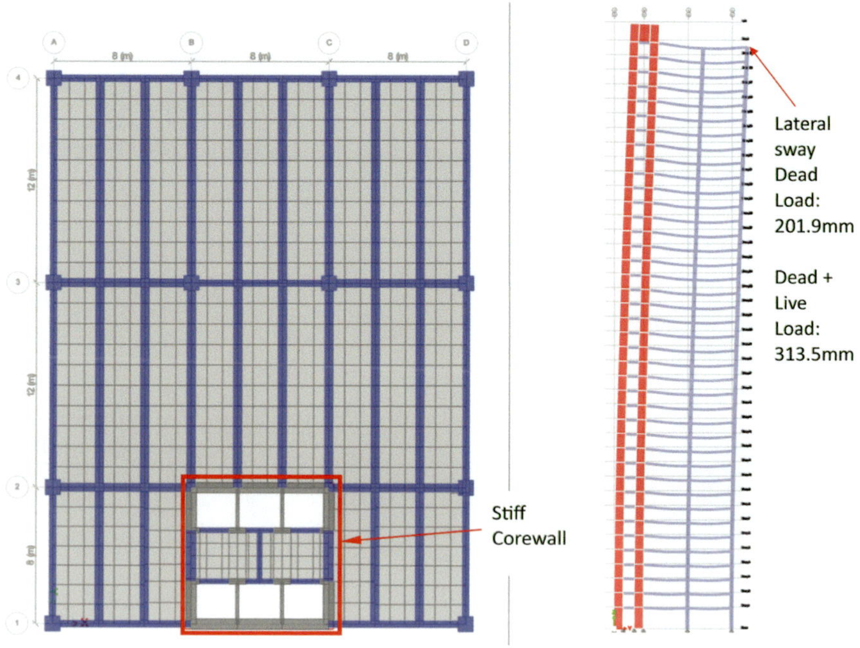

FIGURE 2.26 Gravity load analysis of an unsymmetrical building by first-order elastic analysis (Worked Example 2.6).

difference in axial stiffness on both ends of the building would give rise to the significant sway or the building lean.

However, the building lean can actually be "rectified" during the construction process. Although the floor sways laterally to some extent due to the additional weight of an upper floor on the unsymmetrical structural layout, the Contractor usually constructs the new floor to its correct position (if not for every floor, at least at every 3 to 4 floors), i.e., the newly constructed floor will be constructed to its original design position, i.e., the building will then lean in the reverse direction. As such, the building lean will be rectified floor by floor with the roof level having no sidesway at all due to dead load as shown in Figure 2.27. The computer software has calculated the building's deflection profile by "stage analysis" as shown. The maximum sidesway due to the structure's own weight alone finally takes place at the 28th storey of magnitude 46.44 mm instead of the roof floor and is significantly less than the previous exercise at 201.9 mm.

It should be emphasized that the foregoing analysis is based on the elastic behaviour of the structure, though plastic analysis gives similar behaviour but of different magnitudes.

Furthermore, by careful consideration in the adjustment of the column wall positions at floor levels during construction, theoretically, it is possible to achieve

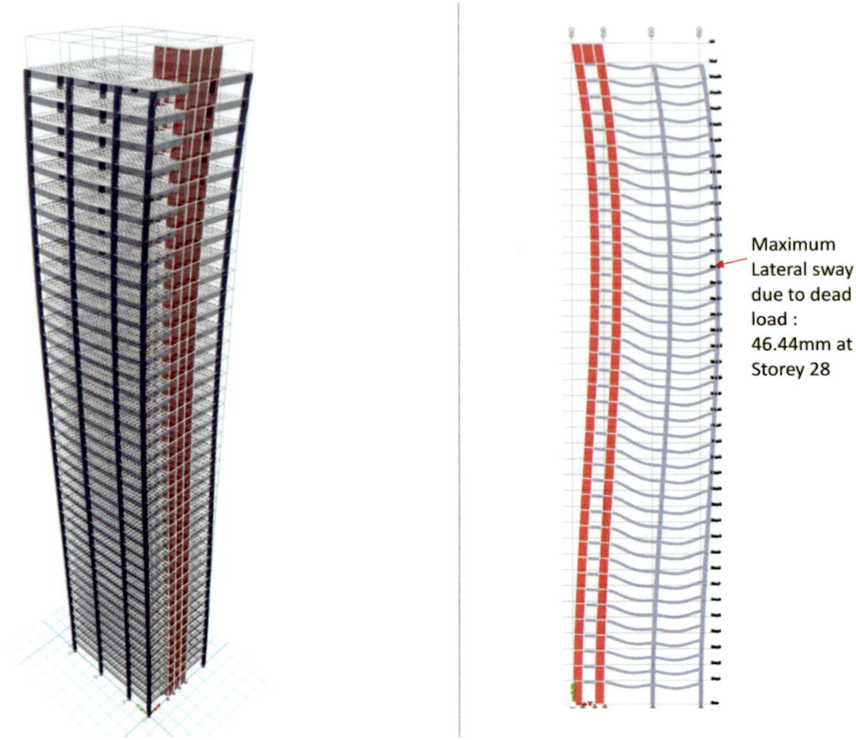

FIGURE 2.27 "Rectified" building lean by staged construction (Worked Example 2.5).

a perfectly straight profile for the dead weight at the end of the construction. But in the case of a concrete building where the deflection is also affected by shrinkage and creep, accurate deflection is almost impossible as these effects can hardly be determined accurately. In addition, by the application of live load and the post-construction creep of the structural material, especially concrete, lean of the building to the more flexible side is almost unavoidable. But anyhow, the effects will be comparatively small.

Summing up, analysis with consideration of construction sequence is sometimes required for structures with unusual layout, as otherwise significant errors will be incurred.

2.13 PLASTIC ANALYSIS

So far, the analytical methods discussed in this Chapter are based on the elastic theory where the deformation follows linearly with the applied load without limit of deformation, though non-linear analysis has been adopted in the P-Δ effect, large deformation and the staged analyses. However, if a part of the structure has reached the elastic limit beyond which the deformation will no longer follow the elastic behaviour, plastic analysis would be a better solution.

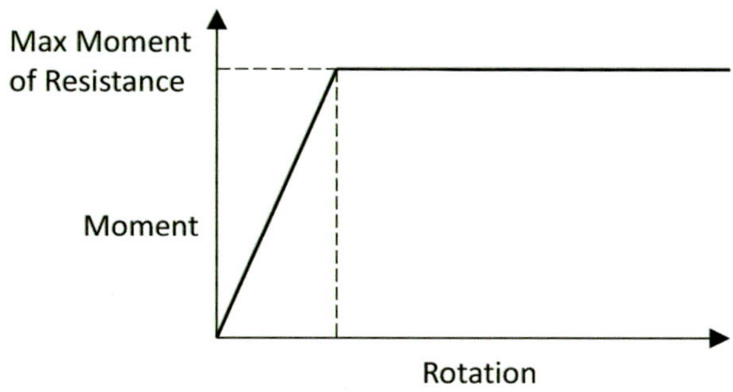

FIGURE 2.28 Formation of plastic hinge.

The most popular plastic analysis is that due to the formation of a "plastic hinge" for moment in a beam or column, or even a wall. It is formed when the entire section has reached its plastic stage. That is, the section has yielded and it cannot resist any additional moment even upon further rotation of the section. The threshold moment beyond which the section cannot resist is thus termed the "plastic moment." Figure 2.28 shows the simplest configuration of the plastic hinge with a constant value after the plastic moment is reached.

Plastic hinges would likely develop at locations with high moments which are mostly at the ends (or supports) of the members and locations under heavy transverse load. When one plastic hinge is formed, the load will be redistributed under further increase in external applied loads such that other parts will take up the additional loads. So, one redundancy is reduced. The structure will collapse if there are a sufficient number of plastic hinges or removal of redundancies.

Plastic hinge analysis is commonly performed in a stepwise manner where the stiffness matrix is updated each time when a new plastic hinge is formed. An approach by Aktaş E. and Korkmaz A (2007) can be used for analysis. A description of the approach with reference to Aktaş E. and Korkmaz A (2007) and a worked example is contained in Appendix C.

2.14 CONCLUSION

This chapter has presented a historical review of structural analysis methods for buildings. It began with the pre-computer era, when analysis relied on hand calculations and simplified assumptions. The advent of computers, however, has led to the continuous evolution of these methods which have progressed from simple first-order linear analysis to the advanced non-linear techniques. These modern methods can now simulate complex structural behaviours much more accurately, providing a far more realistic representation of a building's performance.

REFERENCES

Aktaş, E. and Korkmaz, A. (2007) *Plastic Hinge Analysis Using Reduced Basis Approximations*. International Symposium on Advances in Earthquake and Structural Engineering. Süleyman Demirel University, Isparta–Antalya, Turkey, 24–26 October.

ASCE Standard (2022) ASCE/SEI 7–22: Minimum Design Loads and Associated Criteria for Buildings and Other Structures. American Society of Civil Engineers.

BS EN 1992-1-1:2023 (2023) Eurocode 2: Design of Concrete Structures – Part 1-1: General Rules and Rules for Buildings, Bridges and Civil Engineering Structures. British Standards Institution.

BS EN 1993-1-8:2024 (2024) Eurocode 3: Design of Steel Structures – Part 1-8: Joints. British Standards Institution.

Eurocode 8 BS EN 1998-1:2004+A1:2013 (2004) Design of Structures for Earthquake Resistance – Part 1: General rules, seismic actions and rules for buildings.

Hardy Cross (1930) *Analysis of Continuous Frames by Distributing Fixed End Moments*. American Society of Civil Engineers, Papers and Discussions.

Reynolds, E.C. and Steedman, J.C. (1988) Reinforced Concrete Designer's Handbook, 10th Edition. E & FN Spon, Taylor & Francis Group.

Scott, W.L. and Glanville, W.H. (1934) Code of Practice for Reinforced Concrete. Concrete Publications Limited.

3 An Overview of the Finite Element Method and Its Application in Building Structure Analysis

3.1 INTRODUCTION

The finite element method (FEM) is a powerful computational tool widely used in building structure analysis to simulate 2-dimensional (2D) and 3-dimensional (3D) structures using plate and solid elements. Its integration with advanced computing has made FEM indispensable for structural designers. However, many designers lack a deep understanding of its theoretical underpinnings and optimal application strategies. This chapter aims to elucidate the core principles of FEM, its practical applications, and its limitations. For a deeper exploration, Appendix D provides formulations of basic elements, performance evaluations, and worked examples to equip readers with a clear and accurate understanding of FEM in building structure design.

3.2 OVERVIEW OF THE FINITE ELEMENT METHOD – SHAPE FUNCTIONS AND MESHING

Literally the finite element method refers to analysis based on elements of finite sizes. So, by the method in structural analysis, the structure is "discretized" into elements of finite sizes joined together as illustrated in Figure 3.1 extracted from Zienkiewicz et al. (2005).

The elements can actually be strut members modelled as line elements. So, a frame structure discretized into individual line members can also be regarded as being analyzed by the finite element method, though often in some structural engineers' view, the finite element method is restricted to analysis of plate or brick structures as continua.

The basic behaviour of a finite element is governed by the shape function (or also known as displacement function) initially assumed. The function dictates the deflected shape of the element. Unlike a strut member in stiffness analysis whose stiffness derived from the elastic theory is exact to theory, the shape functions of 2-dimensional "plate" structures and 3-dimensional "solid" structures are assumed

DOI: 10.1201/9781003636380-3

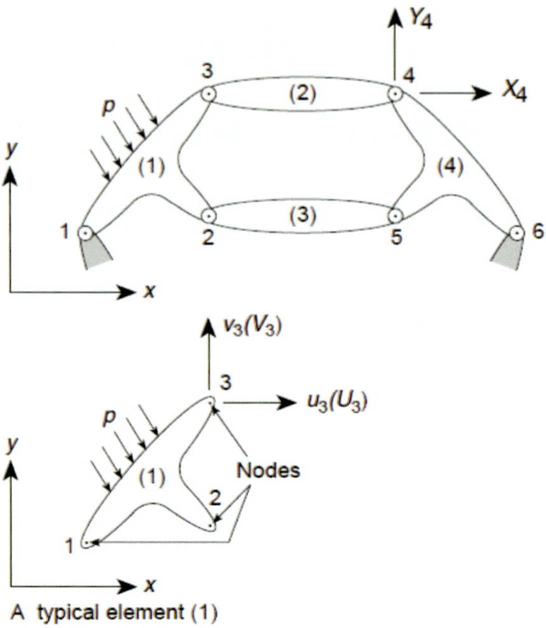

FIGURE 3.1 Extract from Zienkiewicz (2005) – structure discretized into elements connected together.

functions and exact solutions may not be obtained unless the shape function corresponds exactly to its actual behaviour under the specified load. For example, for a cantilever wall structure under point loads, the deflected shape follows a cubic parabolic profile. If the shape function allows cubic parabolic shape, exact solutions can be obtained. However, if the shape function is of lower order, errors will be incurred. But the error can be reduced under the use of greater number of elements. That is, the structure is subdivided into a number of elements joined together as in Figure 3.1 by a process called "meshing" in the finite element method. In general, the results from finite element method will converge to the elastic theory with increasing number of elements. For details of the finite element formulation, the readers can refer to the texts by Zienkiewicz et al. (2005) or Cheng et al. (2024). Cheng has developed many finite element programs for teaching, research, and engineering applications. Sample programs for 2D/3D truss, frame, plate/shell, seepage, consolidation, plane stress/strain, 3D brick element, pile driving, and others can be obtained from the authors for education purposes.

The shape functions are generally taken as polynomials for the ease of mathematical manipulation which involves differentiation and integration in calculus. So, if a linear function is chosen, the deflected shape of an element is restricted to linear shape while a quadratic or cubic function allows the member to achieve quadratic or cubic deflection profiles under the application of load. Figure 3.2 illustrates the deflection profiles of plane stress elements (elements undergoing in-plane

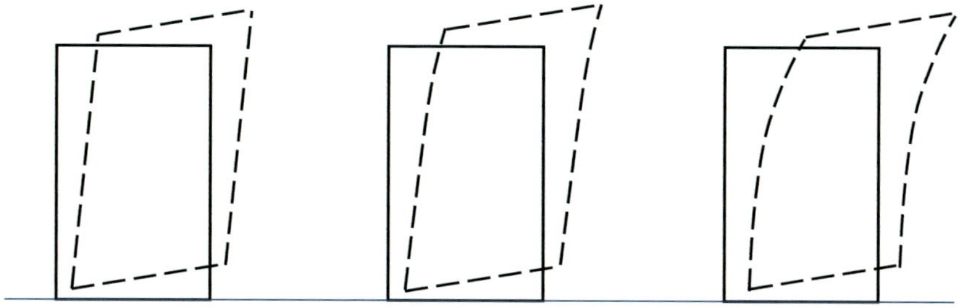

(a) Linear in X- and Y-Directions
$u = a_1 + a_2x + a_3y + a_4xy$
$v = b_1 + b_2x + b_3y + b_4xy$

(b) Quadratic in X-, linear in Y-direction
$u = a_1 + a_2x + a_3y + a_4xy + a_5y^2 + a_6xy^2$
$v = b_1 + b_2x + b_3y + b_4xy$

(c) Cubic in X-, linear in Y-direction
$u = a_1 + a_2x + a_3y + a_4xy + a_5y^2 + a_6xy^2 + a_7y^3 + a_8xy^3$
$v = b_1 + b_2x + b_3y + b_4xy$

FIGURE 3.2 Shape functions governing the element's deflected shape.

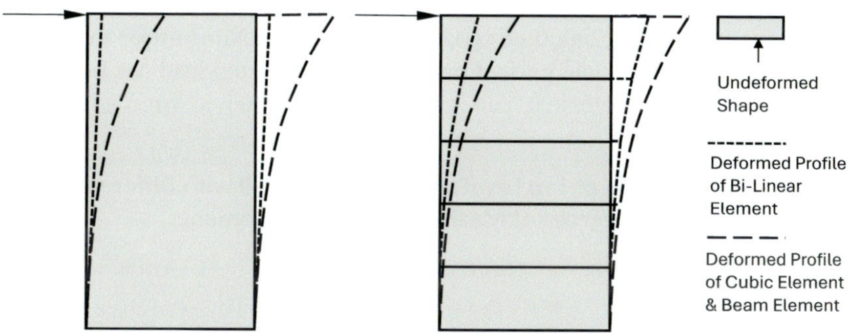

Deflected Shape of a Single Element using Bi-linear Shape Function

Deflected Shape of a "Meshed" Element Based on Bi-Linear Shape Function

FIGURE 3.3 Demonstration of the use of meshing in finite element analysis.

deformations and loads) with different shape functions. The dotted lines represent the deflected shapes, u and v are the deflections in the X and Y directions and the a_i and b_i are constants.

Obviously, different results will be obtained under different shape functions assumed for the same structural element. As discussed in the foregoing, if the shape function assumed in an element is of lower order than that of the actual deflected shape, error will be incurred. As an example, the cantilever wall shown in Figure 3.3 having a lateral load acting at its top edge will undergo a cubic parabolic shape as similar to a beam element. So, to achieve good accuracy, shape functions of order in x^3 has to be used or, if lower order is used, the structure has to be finely "meshed" allowing high degree of freedom for the structure to deflect under minimum energy to achieve accuracy. Figure 3.3 demonstrates how a bi-linear element simulates the actual behaviour better by "meshing." In this case, the meshing is by dividing the structure into elements stacking on one another. For comparison purposes, the deformed shape of an element with a cubic parabolic shape function which gives the exact solution to the elastic theory is also added.

A numerical example based on a concrete cantilever wall element of length 3 m, height 12 m and thickness 0.4 m under two lateral point loads each of 1000 kN at the top corners has been performed in Appendix D. The lateral deflections of the bi-linear element at "no meshing," and meshing of 2, 3, 4, 6, 10 numbers along the element height are plotted in Figure 3.4.

In Figure 3.4, the deflection profile of an element with shape function equal to a cubic parabola is also plotted. It can be seen that the deflection of an element with bi-linear shape function can only approach the element with cubic parabola at high degrees of meshing.

The vertical component of the stresses of the bottom level of the element is also examined. The one with cubic parabolic shape function shows triangular stress distribution due to the horizontal top level load with extreme value of ±40000 kN/m². Checking with the bending moment at the bottom level of the wall due to the external load which is $2000 \times 12 = 24000\text{kNm}$, the bending stress at the extreme fibre at the bottom level is $\pm 24000 \times 6 / \left(0.4 \times 3^2\right) = \pm 40000\text{kN/m}^2$ under the assumption of linear stress distribution, perfect balance to the external load has been achieved. However, a bi-linear element can only achieve ±2703 kN/m², ±11538 kN/m²,

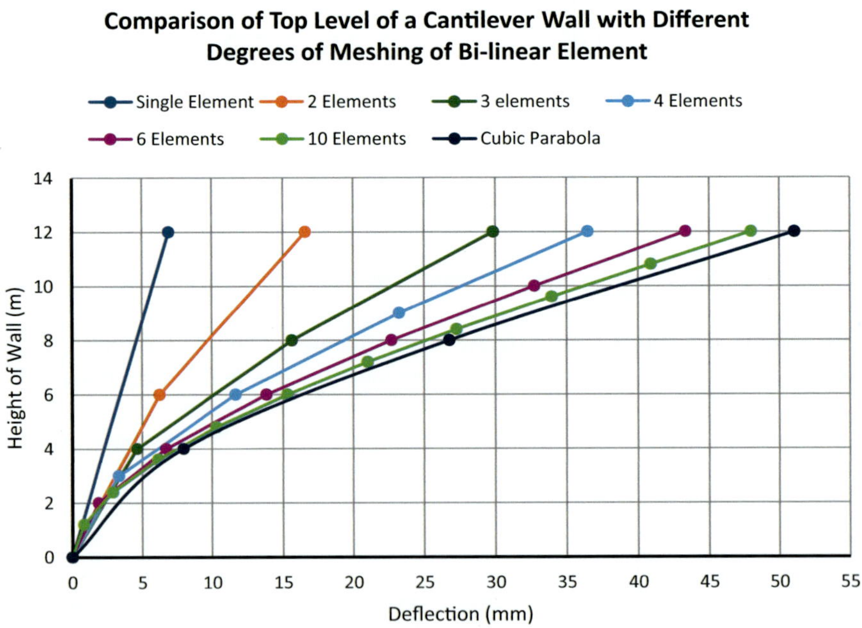

FIGURE 3.4 Variation of lateral deflections with different degrees of meshing of a bi-linear element.

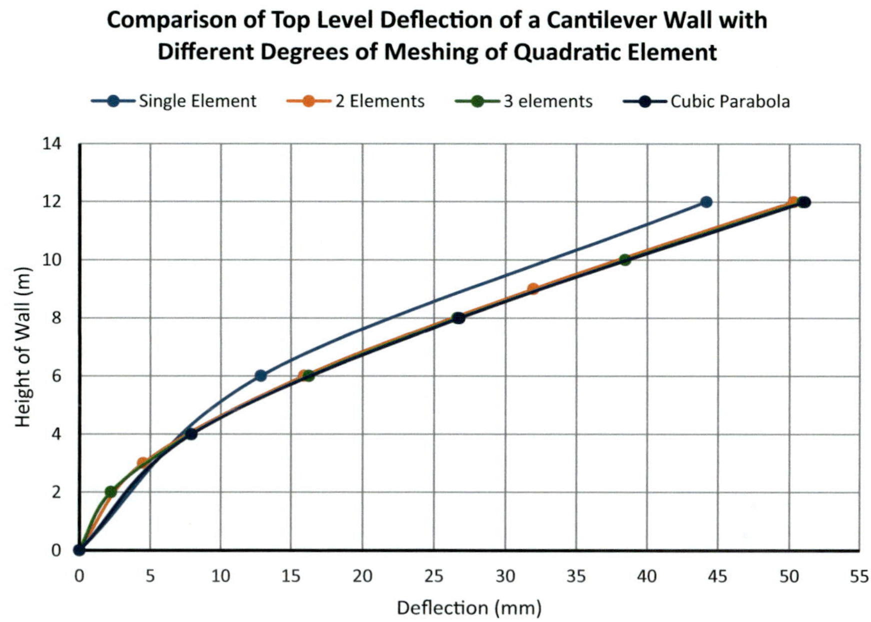

FIGURE 3.5 Variation of lateral deflections with different degrees of meshing of quadratic element.

±19180 kN/m², ±25000 kN/m², ±31132 kN/m², ±35714 kN/m², and ±38359 kN/m² at the models with no meshing and meshing with 2, 3, 4, 6, 10 and 20 elements respectively. As perfect balance to external load is achieved by the element with the cubic parabola shape function in the vertical direction, achievement of this stress or close to this stress is an indicator of convergence of the element under the degree of meshing used.

Details of the formulation of the stiffness matrix under the use of shape function of different degrees and the check on deflections and stresses with different degrees of meshing are contained in Appendix D.

A study of the quadratic element as similar to that of the bi-linear element is also carried out as discussed in Appendix D. Convergence is much faster as shown in Figure 3.5. From Figure 3.5, perfect convergence at the meshing of three is almost achieved in terms of deflection. As for the vertical stress at the bottom level of the wall, the bending stresses are 28772 kN/m², 37576 kN/m², and 39170 kN/m² at the models with no meshing and meshing with 2 and 3 elements respectively. So, good convergence at low degree of meshing can be achieved by the quadratic element, showing its high efficiency.

3.3 ANALYSIS PROCESSES AND NODE FORCES

The analysis process by the finite element method follows the stiffness method of analysis by

$$\boldsymbol{K}.\Delta = \boldsymbol{P} \tag{3.1}$$

where $\boldsymbol{K}$, Δ and $\boldsymbol{P}$ are respectively the stiffness, node displacement and force matrices.

As similar to the frame structures, $\boldsymbol{K}$ is assembled from the stiffness matrices of individual members at the respective nodes (where the members join) and the nodal displacements are analyzed by solving Δ in Equation 3.1 under the loads formed in $\boldsymbol{P}$. The stiffness matrix of individual element, when multiplied to its nodal displacements can recover its nodal forces (including moments). Also, as similar to the frame structure, the nodal forces of all elements joining at a node can achieve perfect balance.

3.4 STRESSES IN FINITE ELEMENT ANALYSIS

Stress as a term in finite element analysis refers to force per unit width or area of a plate or solid element. It may also refer to bending moment per unit width for plate bending structure. Stress can be obtained by multiplying the strain displacement matrix (the $\boldsymbol{B}$ matrix) to the nodal displacements as discussed in Appendix D. It should however be noted that these stresses may not converge to good answers unless the assumed shape function is of adequately high degree or the meshing is fine enough. The stresses as calculated by different elements joining at the same node even under no external load are generally not identical as demonstrated in the following example.

A one-way concrete slab of 500 mm thick and width = 20 m, span = 6 m as shown in Figure 3.6 is to be analyzed by a world-wide used commercial software. The large dimension of the width compared with its span is to ensure dominance of the one-way action of the slab for ease of comparison with that by hand calculation. The concrete Young's modulus is taken as 26,400,000 kN/m^2. The load applied comprises uniformly distributed dead load is equal to $24.5 \times 0.5 = 12.25$ kN/m^2 and live load = 10 kN/m^2 and the load case to be examined is 1.4 times dead load plus 1.6 times live load.

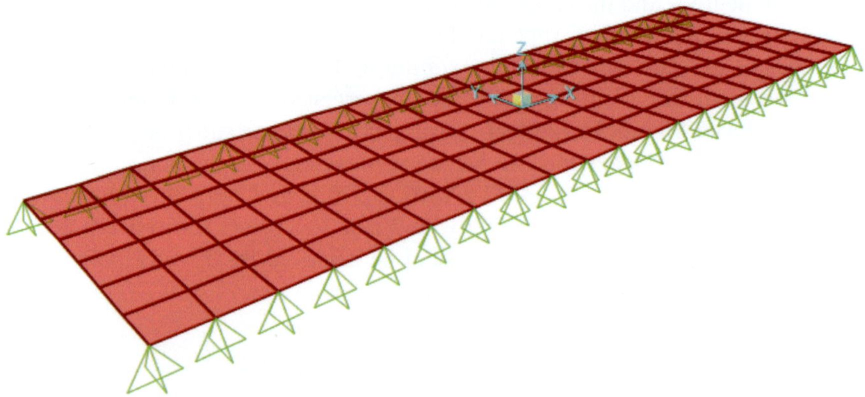

FIGURE 3.6 A one-way slab for examination of stress calculation.

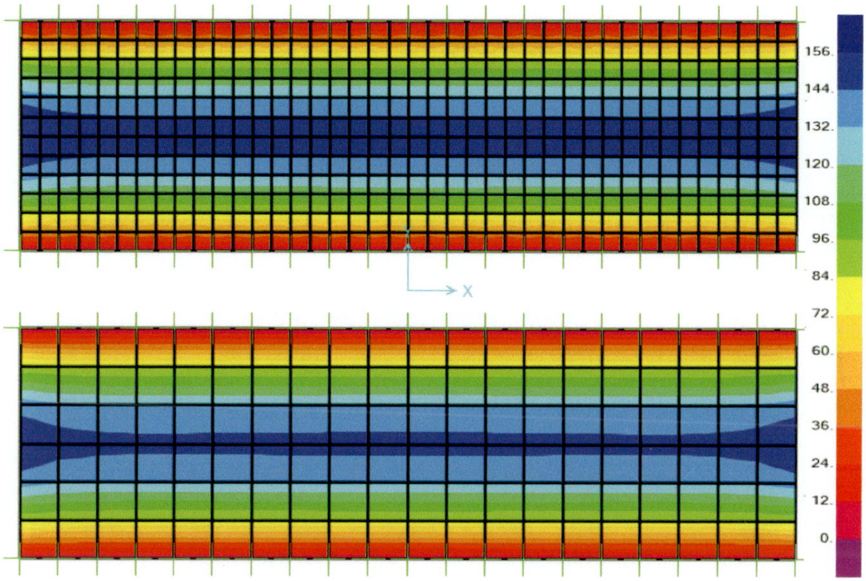

FIGURE 3.7 Contours of the bending moment along the short span under UDL.

Meshing of the slab into 20 × 6 (element size 1 m by 1 m) and 40 × 12 (element size 0.5 m by 0.5 m) are tried. The analytical results of the moments along the short span of the two meshings are shown in Figure 3.7. It can be shown that the bending moment at the centre in both models are 148.84 kNm/m, which is very close to that by hand calculation at $(1.4 \times 0.5 \times 24.5 + 1.6 \times 10) \times 6^2 / 8 = 149.18$ kNm/m. The plot in Figure 3.7 is that without stress averaging across the nodes. As such, the stresses on both sides of a node may not be identical, even though there are no external loads acting on the node, although physically the stresses should be "continuous" to achieve equilibrium. This application of stress averaging will be explained and demonstrated in the coming paragraphs.

The out-of-plane shear force is then examined. The plots for the out-of-plane shear for the two models of meshing are contained in Figure 3.8 where again no stress averaging across the nodes has been used. It can be seen that "distinctive" stresses are shown in the individual elements by the different colours of the plot. There would be a stress jump if a plot is carried out across the nodes.

From Figure 3.8, the out-of-plane shear in kN/m for the coarser meshing model is at almost constant values in each element of 16.5, 49.6, 82.7 kN/m from the centre to the edge. For the finer meshing model, they are 5.26, 24.79, 41.34, 59.73, 73.15, 138.87 kN/m. Nevertheless, the true shear is 99.45 kN/m for achievement of equilibrium at support. So, the coarser meshed model (with support shear at 82.7 kN/m) underestimates the reaction while the finer meshed model (with support shear 138.87 kN/m) over-estimates.

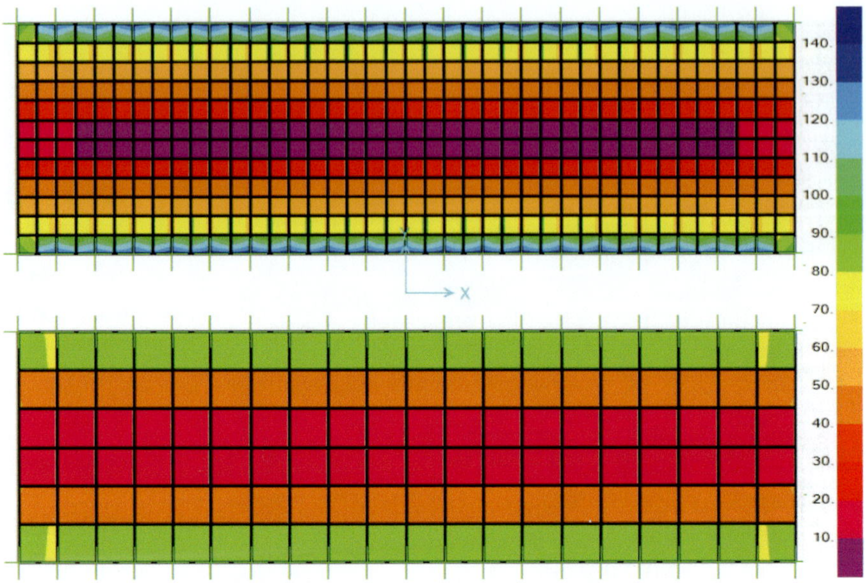

FIGURE 3.8 Contours of the out-of-plane shear within the slab (no stress averaging).

As pointed out previously and demonstrated in Appendix D, "stresses" of a node where adjoining elements meet are calculated separately in the adjoining elements through the dot product of the strain displacement matrix and the nodal displacement matrix. Generally, the stresses so arrived in the adjoining elements are not identical at the common node, although physically the stresses should be unique values unless there are external loads acting on the node. An element with good shape functions may achieve exact answer or answer close to the exact value. To solve the problem, an option by averaging the stresses of adjacent elements to a node is often adopted in software. The averaged stresses are often taken for design purposes. So, this option is adopted for the out-of-plane shear and the contour is shown in Figure 3.9.

Smoother contours are now obtained. However, the values of the support shears stay almost unchanged as compared with those without averaging, which remains as a source of error.

The use of stress averaging appears not too important in the one-way slab having simple loadings and layouts. However, as will be discussed in a later section, the use of averaging does make significant differences in real structures under general unsymmetrical arrangements of structural layouts and loads.

Another problem often encountered in finite element analysis is the problem of "peak stress" due to concentrated loads, which is even more severe under fine meshing. For demonstration, a heavy point load of 500 kN is exerted at the mid-point of the one-way slab as shown in Figure 3.10.

The bending moment at the centre of the finer mesh model where the point load applies is 216.24 kNm/m while that at the coarser model is only 183.36 kNm/m,

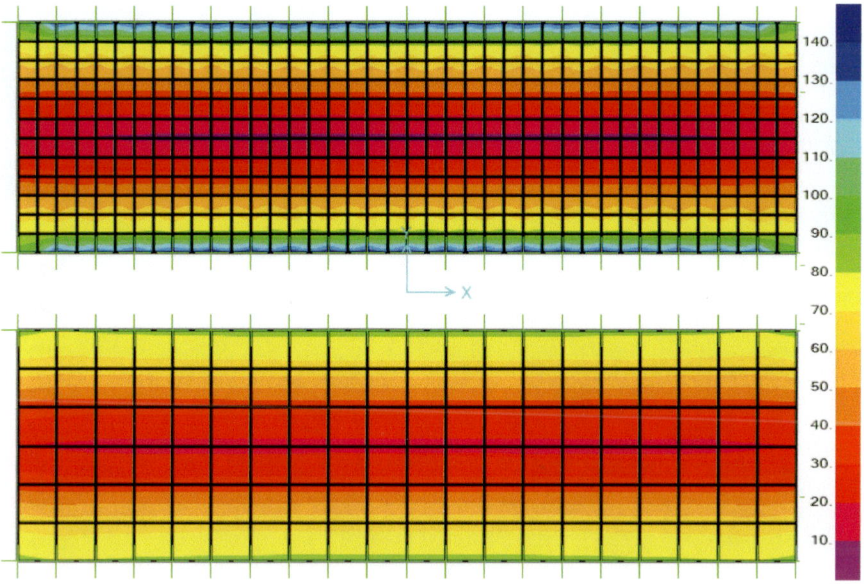

FIGURE 3.9 Contours of the out-of-plane shear within the slab (with stress averaging).

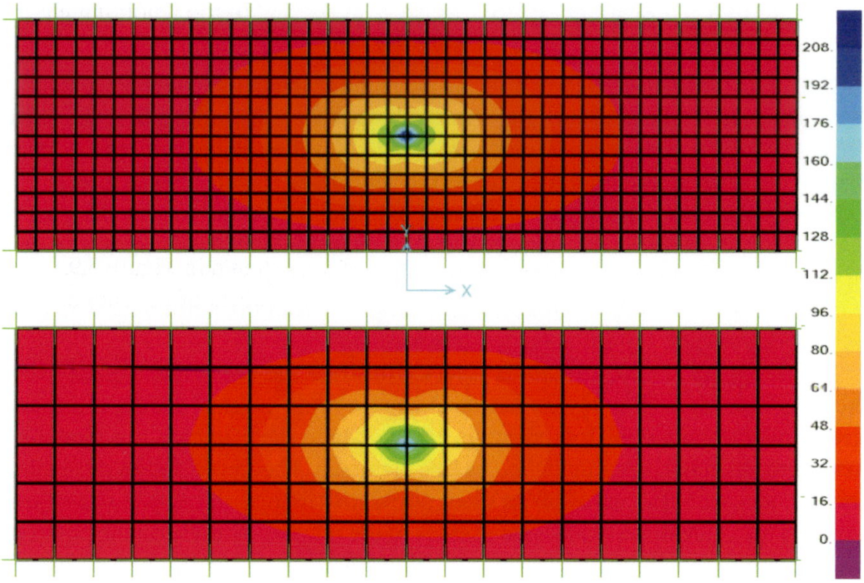

FIGURE 3.10 Contours of the bending moment per unit width along the short span under mid-span point load.

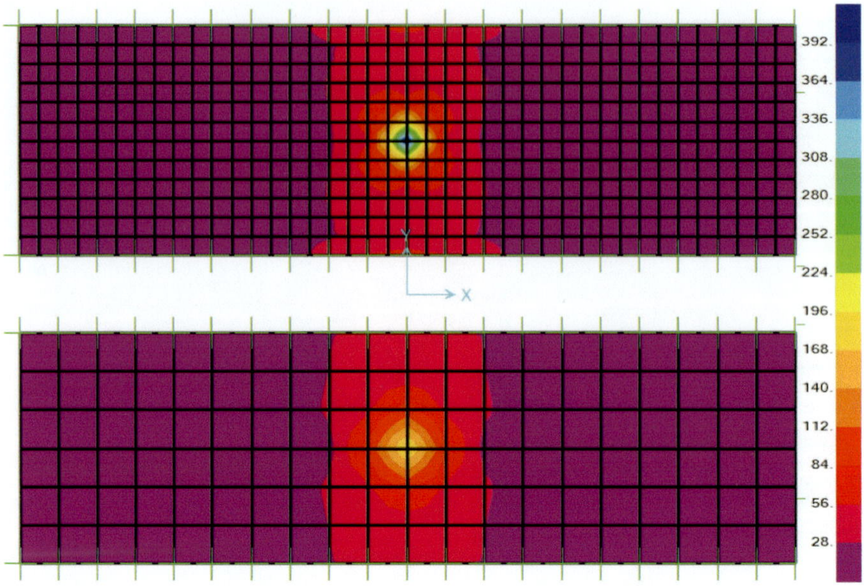

FIGURE 3.11 Contours of the out-of-plane shear under mid-span point load.

making some 18% difference. Higher values will be obtained if the meshing is finer. Nevertheless, the differences are diminishing at further distances from the point load.

The effect of meshing is even more pronounced in the out-of-plane shear as shown in Figure 3.11. At the centre of the figure where the point load applies, the shear is 398 kN/m for the finer meshed model and 199 kN/m for the coarser meshed model. Again, the differences are reduced at further distances from the point load.

Nevertheless, as the support or column exerting the point load resulting in high out-of-plane shear stress has finite physical dimensions, it is not necessary to design for the immediate high stress which is within the physical dimensions of the support or column. Generally, the high stress drops rapidly from the point of action. Design at the nodes outside can often be designable.

The foregoing simple structures help to illustrate the phenomena often encountered in finite element analysis. Figure 3.12 is a finite element model representing several piled foundations linked by tie beams where the pile caps are modelled as plate elements and beams modelled as line elements. The figure shows the 3-dimensional model with the out-of-plane shear stress distribution in the pile cap superimposed on them. Peak vertical shear stresses in the pile cap directly on the bored piles can be readily seen as shown by the dark blue colours.

Peak stresses in plate elements exist not only at the column supports. They can also be found at joints with beams being modelled as line elements, as shown in Figure 3.13 by the dark blue colours at the ends of the line elements.

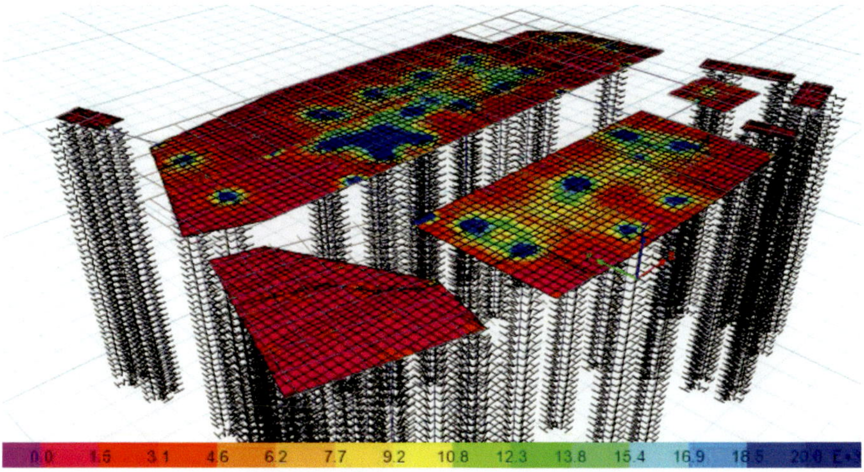

FIGURE 3.12 A finite element model for design of a piled foundation.

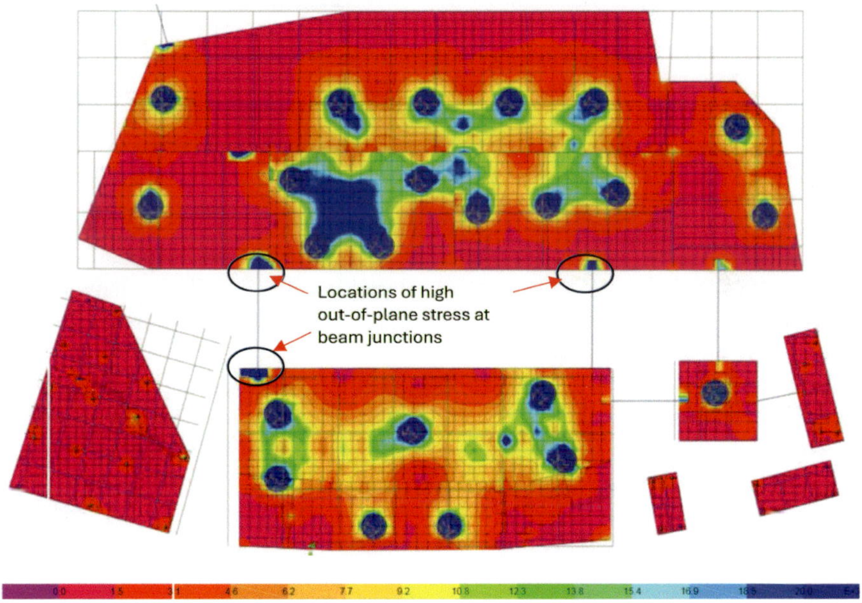

FIGURE 3.13 Peak out-of-plane shear stress in plate at junction with line element.

Nevertheless, the problem of high beam stresses can be alleviated by simulating the tie beams as an assembly of plate elements instead of using line elements. Figure 3.14 shows that the high stresses are significantly reduced.

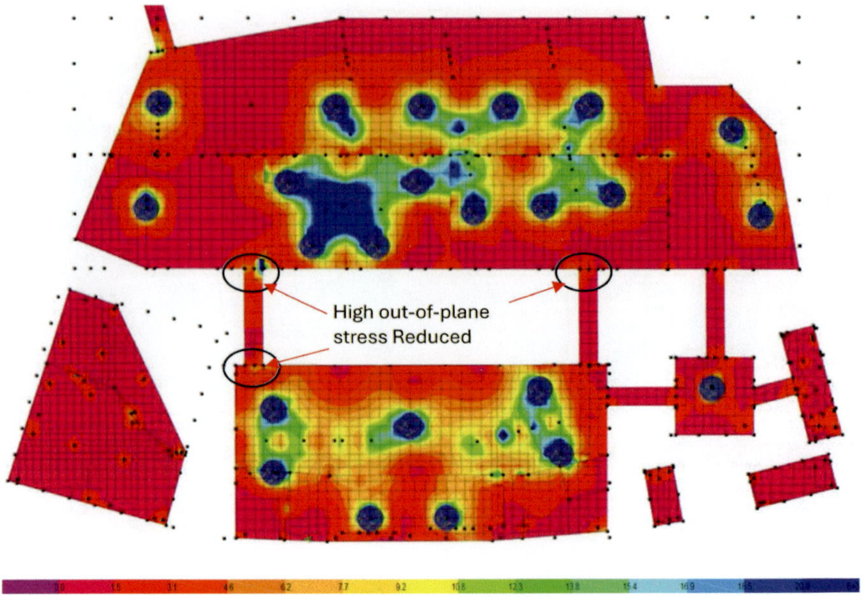

FIGURE 3.14 Peak stress significantly reduced at beam plate junction when beam is modelled as assembly of plate element in lieu of line element.

3.5 DISCONTINUITIES IN STRESSES

As discussed in the foregoing, while the stresses of adjoining elements meeting at a node are not identical even under no external load application, averaged values are often adopted in structural design. Such discontinuities are also indicators of the performance of the finite element method in terms of types of element chosen and the degree of meshing.

The structure shown in Figure 3.14 illustrates the "Discontinuities in Stresses" at a specific arrangement by this software. A closer look shows that the out-of-plane shear discontinues across the boundary where the meshing on both sides of the boundary do not match. The discontinuity is magnified and highlighted in Figure 3.15. This is a particular technique of the commercial software in using rectangular elements by which the deflections are ensured to be continuous even though the meshing on both sides of the nodes are not continuous.

However, an alternative meshing, as shown in Figure 3.16 can improve.

The discontinuity of stress, as shown in Figure 3.15 is due to the different meshing across a boundary where full stress averaging cannot be carried out on the nodes on the boundary. The alternative meshing in Figure 3.16 solves the problem by enabling stress averaging on all nodes to be carried out.

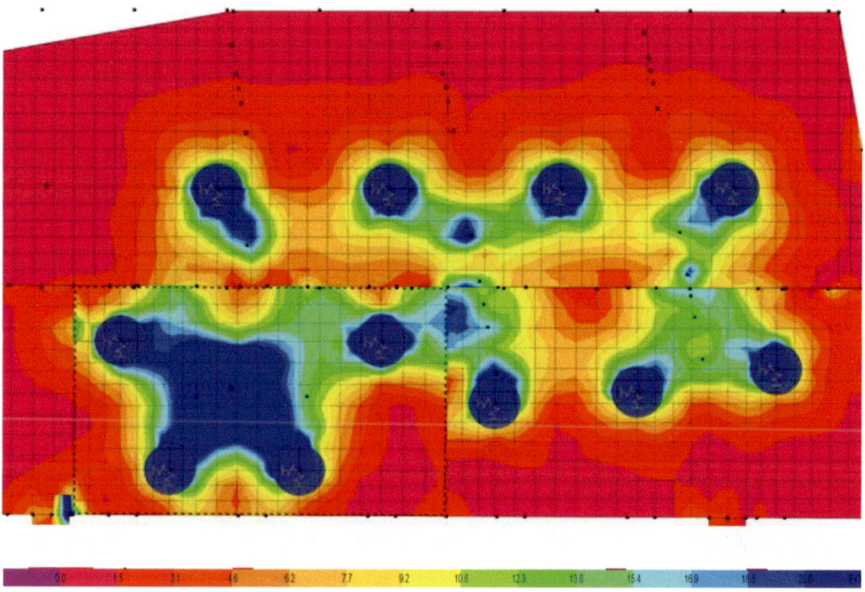

FIGURE 3.15 Magnified view on discontinuities of out-of-plane shear per unit width.

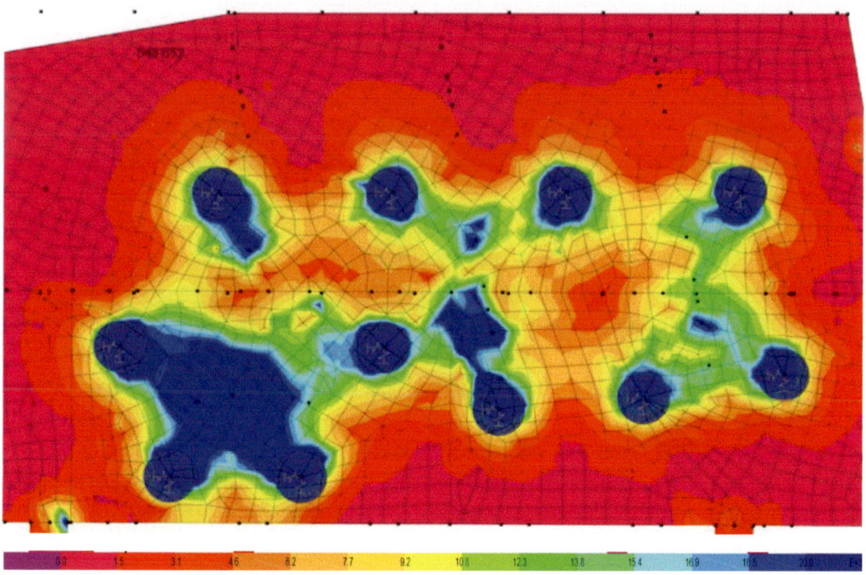

FIGURE 3.16 Improvement of stress discontinuities by alternative meshing.

3.6 APPLICATIONS AND LIMITATIONS

In the simulation of the building structure, various types of structural elements are available in the finite element software, including line elements, plate elements, and solid elements. The choice of the right types is essential both for the accuracy of the analytical results, efficiency in the analysis, and also the convenience in the structural design. The simulation of a structural element by 3-dimensional bricks at a finely meshed level is, of course, the most accurate and realistic approach. But the complexities of the structural model and the abundance of data produced will result in impracticality in analysis and design for concrete structures at a large scale. This is because the analytical results will involve 3-dimensional stresses of six components, requiring tedious design methods for reinforced concrete structures not commonly known to general designers, as will be discussed in Chapter 10. So, for walls and slab structures where the in-plane dimensions are great compared with their out-of-plane thickness, simulation by plate elements will be simpler. In addition, reinforcement design methods for the plate structures are more commonly known. Thus, simulation by plate elements is more practical. But the designer has to bear in mind that the plate structure is often formulated by the "plane remains plane" assumption for the out-of-plane deformation, which may not be valid if the structure is thick. In addition, the designer should also note if the element chosen can undergo substantial shear deformation in the out-of-plane direction, which may be important for thick structures.

Nevertheless, the 3-dimensional brick element is good to simulate a part of a structure where detailed examination of stress is required, e.g., in locations where a sharp change of stress is anticipated.

If the structural element has relatively small sectional dimensions as compared with its length, such as a beam or a column, so that the "plane remains plane" assumption is valid across its full cross-section, the simulation by strut elements is simpler, of which the design method for reinforced concrete structures is well-known to an average designer. However, as discussed in the foregoing, peak stresses may be incurred when line elements meet plate elements. Switching to plate element formulation for the beam or column may have to be adopted.

The choice of an element with the appropriate shape function is also important, as discussed in the foregoing and Appendix D. But some software do not disclose the shape functions involved in various elements. The designer may have to test by simple structures.

Most commercial software can automatically perform meshing on the plate and solid structures, upon default or specified degree of meshing expressed in number of elements or maximum sizes of individual elements. For example, the maximum plan size of an element may be specified as 1.2 m. As discussed in the foregoing, the appropriate degree of meshing at the balance of accuracy of the analytical results, time of analysis, and complexities of the reinforced concrete design work have to be well balanced. Figure 3.17 shows two storeys of a building modelled by plate elements for the slabs and the beams.

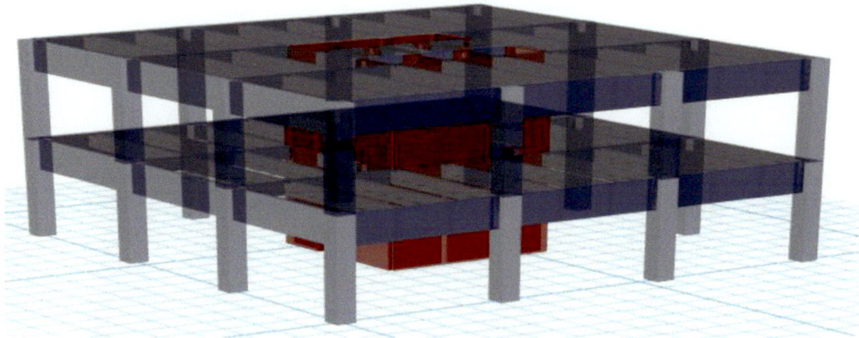

(a)

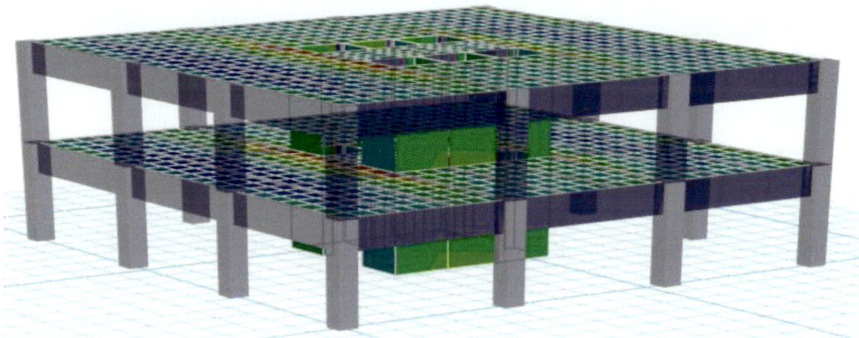

(b)

FIGURE 3.17 (a) Structural model of two storeys (a and b) of a building by finite element. (b) In-plane stress diagram of the model in (a).

The modelling inputs of the wall and slab panels by the designer are in large dimensions without meshing. The stress diagrams for in-plane forces show that auto-meshing has been done by the software. The slabs have been finely meshed but the walls are not. Obviously, there are missing common nodes at the boundaries between the walls and the slabs. The software has implemented some measures to equate the deflection profiles at the boundaries to ensure compatibility. However, stress compatibility can unlikely be ensured. But the error may not be significant. The slabs and the walls can be separately designed to the stresses induced on them.

In addition to modelling of the full structure, the finite element method is especially useful in the study of detailed stress distribution in a structure or a part of a structure. Consider a deep beam of 5 m structural depth and span 10 m supporting a heavy point load at the top level at its mid-span. The analysis by the finite element method with the beam being simulated as an assembly of plate elements shows the following principal stress diagram in Figure 3.18.

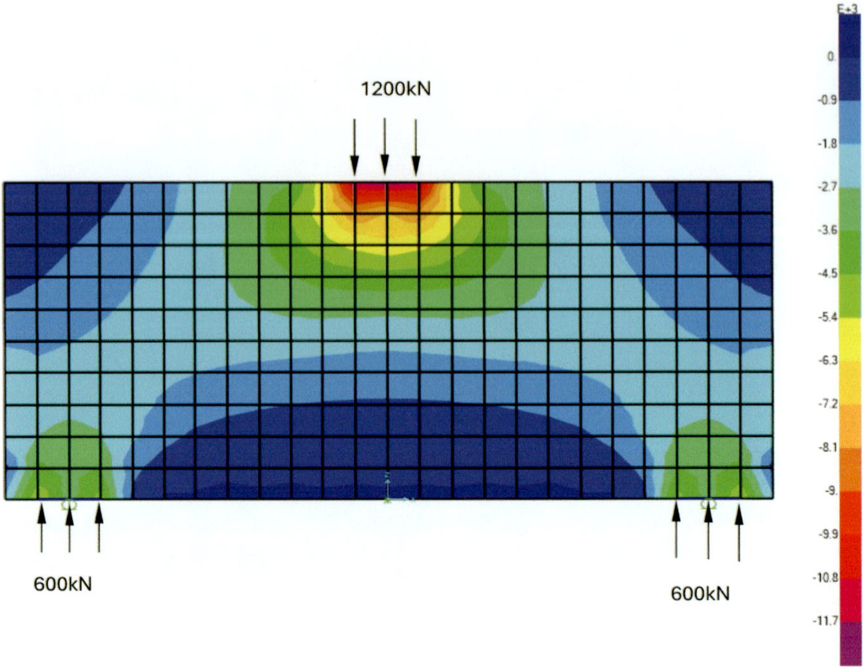

FIGURE 3.18 A 5 m deep beam.

The diagram clearly illustrates the strut-and-tie action with the heavy compressive stress along the "imaginary struts." Heavy local bearing stresses are also revealed.

The horizontal stresses at mid-span are also read and plotted in Figure 3.19. The non-linear stress profile reveals that the section does not fulfil the "plane remains plane" criterion.

3.7 SUMMARY AND CONCLUDING REMARKS

As pursuant to the discussions in the foregoing, the following can be summarized:

(i) The shape function is the basic parameter assumed in the finite element method which directly affects the accuracy of the element for the loading conditions under consideration.

(ii) Meshing can generally increase the accuracy of the analysis, except at locations under heavy point loads including supports where peak stresses will increase with finer meshes. However, as there are physical dimensions for columns exerting point loads, the peak stresses may be ignored in design. As for beams, if there are peak stresses, the designer may remodel the beams as assemblies of plate elements which may alleviate the peak stresses.

(iii) In a continuum formed by plate or solid elements, stresses of adjoining elements at the adjoining nodes are generally not identical even under no

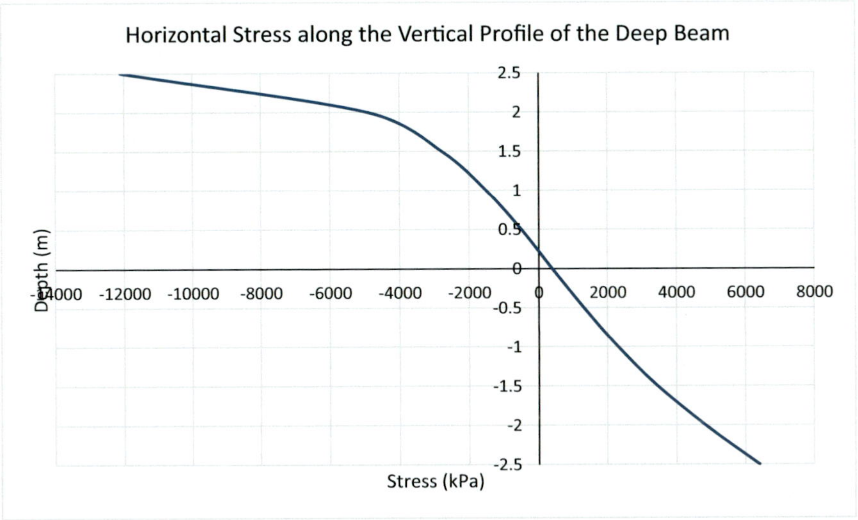

FIGURE 3.19 Horizontal stress at mid-span of the deep beam in Figure 3.18.

external loads. Average stresses may be taken as the accurate stresses which can be used for design purpose.

(iv) Simulation of the structural elements by various types of finite elements and the degree of meshing should depend on the structural dimensions and complexity of the structure. A fine mesh may be used in those parts of the structure where there are sharp changes in dimensions and stresses. Nevertheless, complexities of the problem incurred in analysis and design should also be considered.

REFERENCES

Cheng, Y.M., Law, C.W. and Liu, L.L. (2024) Analysis, Design and Construction of Foundations. CRC Press.

Zienkiewicz, O.C., Taylor, R.L. and Zhu, J.Z. (2005) The Finite Element Method: Its Basics and Fundamentals, 6th Edition. Elsevier.

4 Structural Systems of Building Superstructure

4.1 INTRODUCTION

A building's structural system is an integrated network of components designed to provide stability, integrity, and serviceability. The selection of this system is primarily dictated by the building's intended function, such as residential, commercial, or industrial, which dictates spatial requirements. For instance, a theatre necessitates a large, column-free space such that a long-span system like trusses may be an ideal solution.

The choice is further influenced by the building's height, plan dimensions, and external factors like ground conditions, wind, and seismic loads. Selecting the optimal system is critical, as an inappropriate choice can lead to functional failure, safety risks, and increased costs.

The following sections explore common structural systems and the key factors guiding their selection.

4.2 FACTORS AFFECTING THE CHOICE OF STRUCTURAL SYSTEMS FOR A BUILDING

The following factors are listed and discussed. They are actually interrelated in the final choice of the most appropriate structural system or combination of systems for a building (see Sakisian, 2012).

4.2.1 Building Use and Architectural Layout

Under a given architectural layout designed for the intended purpose of the building, the column wall layout may be more or less prescribed which would also dictate the span lengths of the floor structures and thus the structural floor system. The structural engineer will make the best use of the available columns and walls and check if they are adequate to carry the imposed loads comprising gravity loads and lateral loads. While the structural engineer must verify that these elements can support all gravity and lateral loads, he should also advise on optimizing the layout for constructability and cost-efficiency. Conventionally, all the columns and walls should be continuous throughout the floors down to the foundation. However, in case of discontinuity to suit change of the architectural layouts with floors, consideration of "transfer structure" and/or "hanging structure" may be required. The latter is less

DOI: 10.1201/9781003636380-4

preferred as the removal of the temporary work has to be deferred to the completion of the supporting structure at upper floors.

For the choice of the floor system, span, headroom and imposed loads are the determining factors. Ordinary beam-slab floor systems should be used for normal spans depending on the available headroom and imposed loads which are more economical with high structural efficiency. Otherwise, flat slab and voided slab may be used due to headroom or aesthetic reasons. In case of large span, prestressed concrete, composite (concrete and structural steel) construction or truss structures may have to be considered.

4.2.2 Ground Conditions

Ground geology primarily influences the selection of the foundation system. However, poor ground conditions, such as those that may lead to excessive or differential settlement, can also affect the choice of the superstructure's structural system. In such cases, the stiffness of the superstructure may need to be carefully adjusted to mitigate settlement-related issues. For instance, "hinged" connections between the floor system and supporting walls or columns may be introduced at specific locations to accommodate significant settlement or differential movement, preventing cracking or structural failure. Conversely, stiffness may need to be increased in other areas to minimize differential settlement. As a popular practice, transfer structures are often designed with high stiffness to reduce differential settlement in the superstructures they support, which is useful in ground of significantly varying geology.

4.2.3 Loads

Loads on buildings comprise gravity load, wind load, seismic action, soil load, etc. Gravity load is applicable in all buildings, while others depend on the building's locality. In reinforced concrete buildings, generally the building's own weight contributes to the predominant part of the gravity loads, while the live load has a lesser contribution. Heavy load sometimes affects the choice of a structural system, requiring prestressed concrete instead of reinforced concrete frames to resist the heavy gravity loads.

Wind loads and seismic loads are mostly lateral loads creating not only internal forces within the building structures but also lateral displacement and vibration, which are also design parameters for the building. These loads are dynamic in nature, and their magnitudes for design on the building also depend on the dynamic properties of the building that govern the building's response to them. For ease of design, these loads are mostly converted to lateral static (pseudo-static) loads acting on the building, though dynamic analyses (time history analyses) are sometimes required for very "slender" buildings for accurate and economical design. Thus, high wind loads generally require the building to possess adequate lateral stiffness to keep lateral deflection within pre-determined allowable limits. This criterion often governs the choice of the structural system by which frames comprising relatively stiff walls under good "coupling" are required instead of the column-beam frames with much lesser stiffness. The magnitude of the direct wind pressure on a building depends primarily on the locality of the building, i.e., whether it is in a high wind zone (tropical or sub-tropical). In addition to the magnitude of the direct wind pressure creating primarily the static response, the dynamic response of the building to wind loads depends on the closeness of its fundamental natural frequency to frequency ranges where the wind loading has significant energy. For Kaimal/Davenport-type spectra, the peak energy occurs at very low frequencies (periods of hundreds of seconds), but substantial energy extends into the range around 0.1 Hz to 1 Hz. The closeness of the fundamental natural frequency of the building to that of wind with high energy may cause resonance of the building, inducing magnified deflections and internal forces. So, generally high stiffness and low mass resulting in high natural frequency for a building are favoured in resisting wind loads. Furthermore, the damping of the building (often expressed in terms of damping ratio) also affects the magnitudes of the applied wind loads by reducing its dynamic responses. Damping can be enhanced by the use of dampers.

Unlike wind loads, seismic loads are not actually "real" loads acting on the structure. The seismic action created on a building structure is due to the relative movements among its upper floors and the ground during an earthquake. That is, when the ground is in motion (shaking), carrying the foundation of the building structure to move with it, the upper floors do not move simultaneously (or with delayed responses) with the ground motion due to their inertia. The relative displacements among the floors thus created induce internal forces on the structure by stiffness (elastic or plastic) multiplied to the displacement. Nevertheless, for convenience in design, the seismic effects are often converted to lateral static loads acting on the floor levels also. These lateral loads are often calculated as the product of mass of the floor structure and the estimated acceleration. However, dynamic analysis to predetermined ground vibration is sometimes carried out for some buildings as "time history analysis," which is a more direct but time-consuming analysis. Like wind loads, the magnitudes of seismic effects originating from ground motions depend on the locality of the building, i.e., the seismicity of the locality. But compared with wind load, seismic effects on a building depend much more on the dynamic properties of the building structure as it is the dynamic response that entirely governs the effects. Generally, there are more stringent requirements on the building system by seismic effect than that due to wind load, especially in regions of high seismicity. For example, a highly elevated mass is not preferred in seismic design, which would result in high inertial forces at the floor during seismic motion. Thus, the use of transfer structures is not favoured in seismic design. Some national codes of practice restrict the use of transfer structures if the building heights exceed certain limits. A high degree of structural regularity along the building height is also required for the same reason. In addition, structural regularity on plan is also a design requirement to minimize large relative motions (torsion) of the structural elements on plan. It is not uncommon for a building code to not allow the first fundamental vibration mode to be torsional, which indicates that the building is not weak in resisting torsion.

Often known as a "secondary effect," temperature change can create load on a building structure which may be significant if the building has a long plan length. The effect is due to the restraints by the supporting columns and walls on the floor structure which expands or contracts under temperature change. A long structure will have large plan elongation or shortening, which will result in large reaction forces from the columns and walls. For a reinforced concrete structure, some significant tension may take place to crack the concrete. It is therefore a practice to limit the plan length by the installation of movement joints.

Similar to the temperature effect, concrete shrinkage and creep in a concrete building will also create dimensional change in the floor structure and thus similarly stresses and cracks. By the shrinkage phenomenon, concrete contracts by losing water during its ageing. Under restraint from the supporting walls and columns, tensile stress will develop in the concrete floor. Coexisting with shrinkage, creep which acts in the opposing direction in the floor structure will reduce the tensile stress. The ultimate stress that will be developed in the long term can be up to 4 N/mm^2 to 5 N/mm^2, depending on the restraints. As the plan length has a significant effect on the magnitude of the shrinkage and creep stresses, limitation of the plan length of the floor structure by movement joints is a means to reduce the effect. In addition, the sequence of alternate bays of concrete pouring is also practised as the shrinkage rate is quick at the early age of concrete.

Some fuller discussions on these load effects are contained in Chapter 5.

4.2.4 Building Stiffness Required

A structure should possess adequate stiffness against imposed loads so as to achieve limited deflections both in the vertical and horizontal directions. While generally the axial stiffness of a structural member is high, settlement in the vertical direction would be a concern only when there is a discontinuity of the columns and walls upon the use of transfer structures or when the foundation is weak. A transfer structure is one which carries loads from columns and walls above terminating on it and spreads the loads onto the columns and walls beneath with different vertical alignments. The requirement on a transfer structure, which may be a plate structure or a beam system or a combination of both, is that it should be stiff and strong such that the settlement of it will be small under the relatively heavy imposed loads from the columns and walls carrying multi-storey loads. If the transfer structure is not stiff enough, considerable settlement will be incurred to the structures above, which may cause excessive cracking or even failures. Thus, the limitation of vertical settlement would dictate the use and sizing of the transfer structure system.

An important determining factor on the choice of structural system is the lateral stiffness of the building required, particularly for high-rise buildings where the lateral deflection and vibration under lateral wind or seismic loads are of key concern. For a building structure, the order of structural system with the increasing lateral load resistance may generally be (1) beam-column frame; (2) shear wall-beam-column frame; (3) braced frame; (4) outrigger structure; and (5) framed tube structure. (4) and (5) are generally adopted by super high-rise and/or very slender buildings. The choice of the structural system of a building depends very much on its height and slenderness. The height and slenderness together dictate the building's stiffness, which is the determining factor for design against lateral load. As an illustration, the super-high and slender International Finance Centre of Hong Kong, at 412 m high, is an outrigger structure possessing three levels of outriggers to stiffen itself against lateral deflection. Furthermore, for buildings designed to resist seismic loads, damping may be an important design parameter necessitating the use of "dampers." The Taipei 101 has a 660T skyscraper damper for damping building motions during strong wind and earthquakes. These structural systems will be discussed in more detail in the coming sections of this chapter.

4.3 DESCRIPTIONS OF VARIOUS HOLISTIC STRUCTURAL SYSTEMS OF BUILDINGS

Discussions related to the choice of the holistic system of the building are more focused on the resistance of lateral loads, as shown in the following sub-sections.

4.3.1 Beam-Column Frame Building

As schematically illustrated in Figure 4.1, this structural system utilizes a frame of beams and columns to form a loadbearing skeleton. This skeleton supports the floor structure and resists both gravity and lateral loads.

However, the system's load carrying capacity is limited due to the relatively small size of its columns compared to shear walls. Furthermore, even with rigid beam-column connections, the overall lateral stiffness of the frame is low. This can result in excessive lateral deflection under wind or seismic forces.

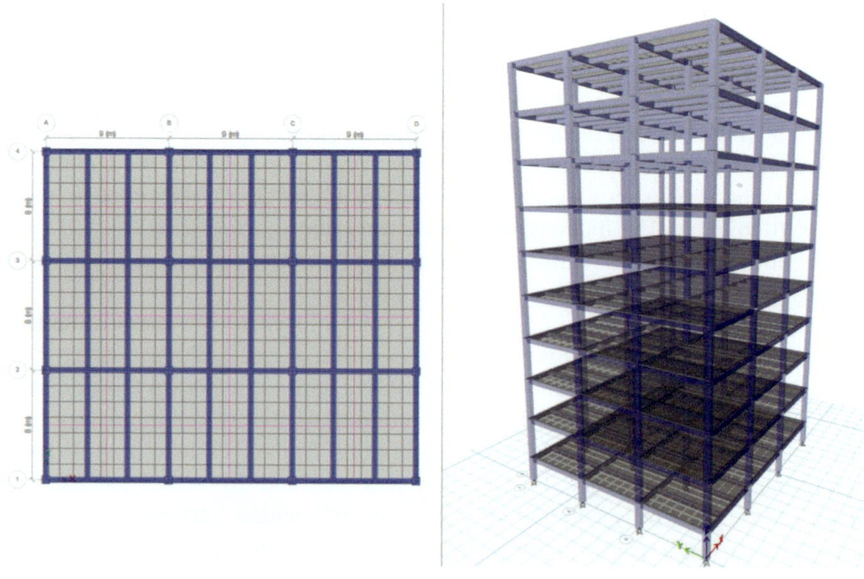

FIGURE 4.1 Beam column frame building.

Consequently, this frame system is unsuitable for high-rise construction. Nevertheless, it is generally easier and more economical to build. It also offers greater flexibility for future spatial reconfiguration, making it a favoured choice for low-rise reinforced concrete buildings.

4.3.2 Shear Wall/Core Wall Building

Concrete walls possess high in-plane stiffness, making them highly effective at resisting lateral loads acting along their plane and minimizing lateral deflection. Furthermore, their large cross-sectional area, compared to columns, provides significant strength against bending and axial loads.

When designed specifically to resist lateral forces from wind or earthquakes, these walls are termed "shear walls," as their primary function is to resist lateral shear. This is illustrated in Figure 4.2, which shows a 20-storey building (total height 58.2 m) supported by 250 mm thick shear walls and a core wall. The building features 200 mm thick floor slabs, and the concrete has an elastic modulus (E) of 26400000 kN/m^2.

The building is strong in resisting lateral loads even in the absence of "coupling" (which is the linking up of the shear wall by floor members). In commercial design, it is often a practice to ignore the out-of-plane bending stiffness of the floor slabs (in coupling of the shear walls) so that the slabs need not be designed for wind loads.

Wind pressures in accordance with the Hong Kong Wind Code (2019) (Hong Kong as a locality of high wind pressure) are applied to the building in Figure 4.2 in the X and Y directions. However, with the floor slabs assumed as membrane elements, the lateral deflection at the roof in the X-direction is 131.16 mm, resulting

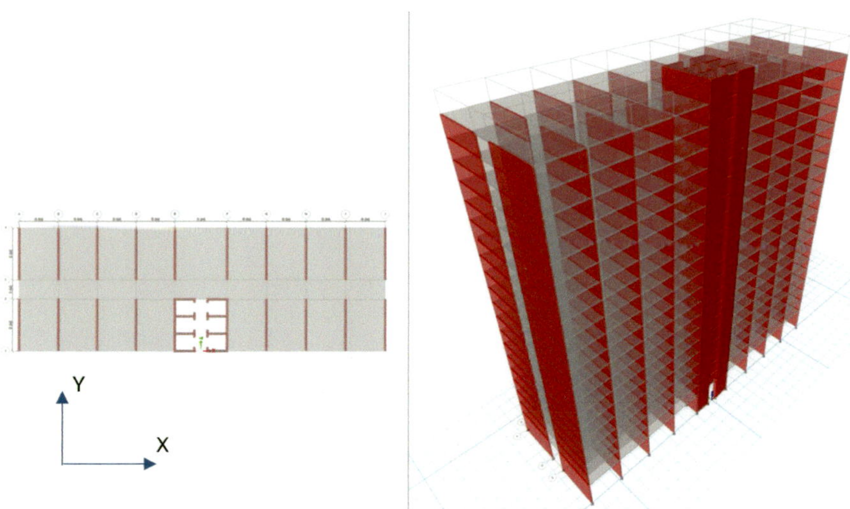

FIGURE 4.2 A shear wall/ore wall building.

in an overall deflection gradient $131.16/58600 = 1:447$ which exceeds the limit of $1:500$ imposed by the Hong Kong Concrete Code (2013). This is because the lateral stiffness of the shear walls are very weak when bending about their minor axes. Nevertheless, if the slab is taken as a shell element that can couple the walls, the deflection is reduced to 25.05 mm so that the deflection gradient is $25.05/58600 = 1:2339$. Anyhow, the deflection in the X-direction is ignored as a concern in the structural design of the building as the slabs can actually play the coupling role.

On the other hand, the deflection in the Y-direction is only 28.41 mm under the membrane assumption for the slabs, giving a deflection gradient of $28.41/58600 = 1:2063$ which is small and well within the allowable limit.

The shear walls along the same grid lines in the Y-direction are now "coupled" by linking opposite walls with beams of structural sizes 250 × 550. These beams are known as "coupling beams." The new configuration is shown at the top part of the plans in Figure 4.3. The deflection at the top storey becomes 10.79 mm versus 28.41 mm of the previous configuration without a coupling beam, achieving a 62% reduction.

Figure 4.3 illustrates the effectiveness of the coupling of shear walls by beams in enhancing the stiffness of the shear walls.

Nevertheless, if the floor slabs are modelled as "shell" elements which can resist out-of-plane bending, the lateral deflections of the "with coupling beams" model and "without coupling beams" model at the top storey become 9.41 mm and 14.44 mm respectively. So, the slabs do play a significant role in coupling in the "without coupling beams" model (reducing deflection from 28.41 mm to 14.44 mm by 49%) but not very significantly for the "with coupling beam" model (reducing deflection from 10.79 mm to 9.41 mm only by 13%).

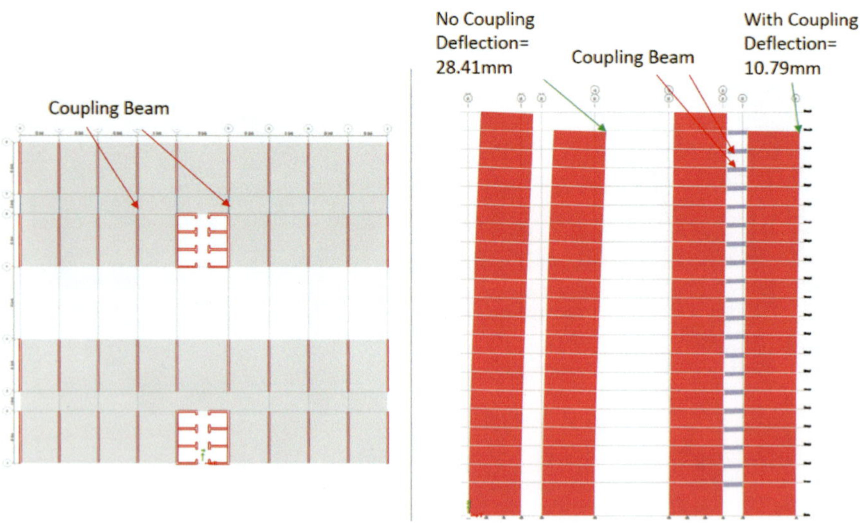

FIGURE 4.3 Comparison of deflection with and without coupling of shear walls.

4.3.3 Wall-Frame Building

To maximize lateral stiffness in a building that combines beam-column frames with shear walls or core walls, the most efficient approach is to couple these systems together. This integrated structure is classified as a "Wall-Frame Building," a common category for such designs. The combined action is demonstrated by the synchronized movement of the columns and walls at each floor level. The relatively rigid floor structures act as diaphragms, forcing the walls and columns to move as a single, cohesive unit. This "rigid body movement" allows for translation and rotation, but the relative positions of the components remain unchanged at each floor.

This behaviour is modelled analytically using the "rigid diaphragm" assumption, as detailed in Chapter 2. This assumption is valid due to the high in-plane stiffness of typical floor slabs. Furthermore, rigid moment-transfer connections between elements provide coupling, which significantly enhances the overall structural stiffness.

As a demonstration, Figure 4.4 shows a typical 30-storey building (with a standard storey height of 4 m) that utilizes this principle. A central core wall is directly coupled with the beam-column frames along Grids 2 and 3 in the X-direction and Grids B and D in the Y-direction, creating an effective wall-frame interaction. The concrete used is of elastic modulus $E = 26400000$ kN/m^2 for all members. The core wall is 500 mm thick. The columns are 900 mm × 900 mm, the perimeter beams are 400 mm(B) × 1200 mm(D) and the interior beams are 500 mm(B) × 500 mm(D). All slabs are 200 mm thick.

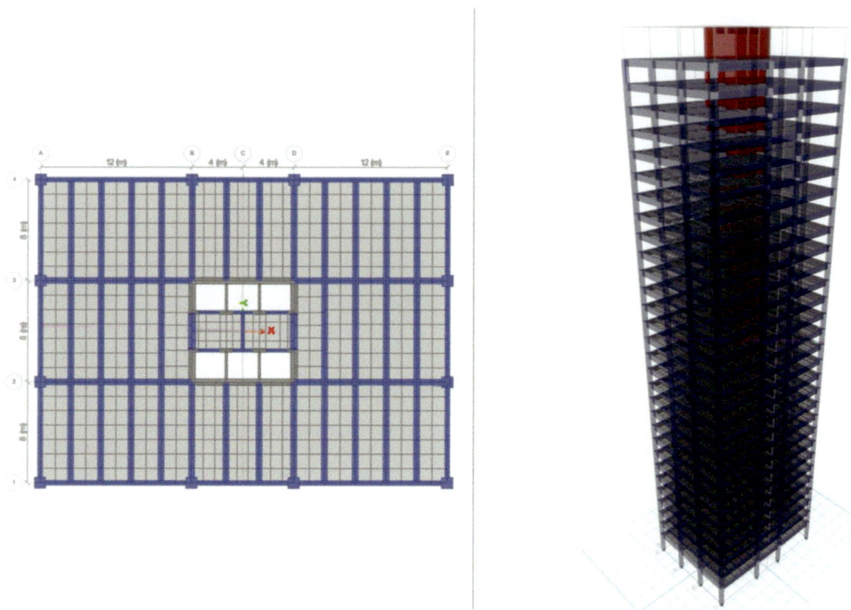

FIGURE 4.4 Wall-frame building.

Cross-sections showing the deflected shapes under wind loads to the Hong Kong Design Wind Code (2019) are shown in Figure 4.5. The deflections are quite significant. The greatest deflection at main roof level is that along the Y-axis at 244 mm, which is at the allowable limit of Height/Deflection ≥ 500 as specified in the Concrete Code of Hong Kong (2013). So, if the building is taller, other systems with higher lateral stiffness will be required for efficient control of lateral deflection.

4.3.4 Framed Tube Building

An improvement to the wall-frame building shown in Figure 4.4 is to strengthen (or stiffen) the peripheral structural frame by forming a "framed tube," which is a frame comprising closely spaced columns or walls structurally linked (coupled) by horizontal beams along the perimeter of the building, forming a "tube" structure. The high stiffness of the tube structure arises from the large plan dimensions of the framed tube and strong coupling of the walls or columns by the linking beams which are generally of short span and large structural sections, as they also act as the parapets of the building. In addition, the framed tube structure also allows lateral shear flow along it in a closed loop. So, a very high torsional stiffness can also be

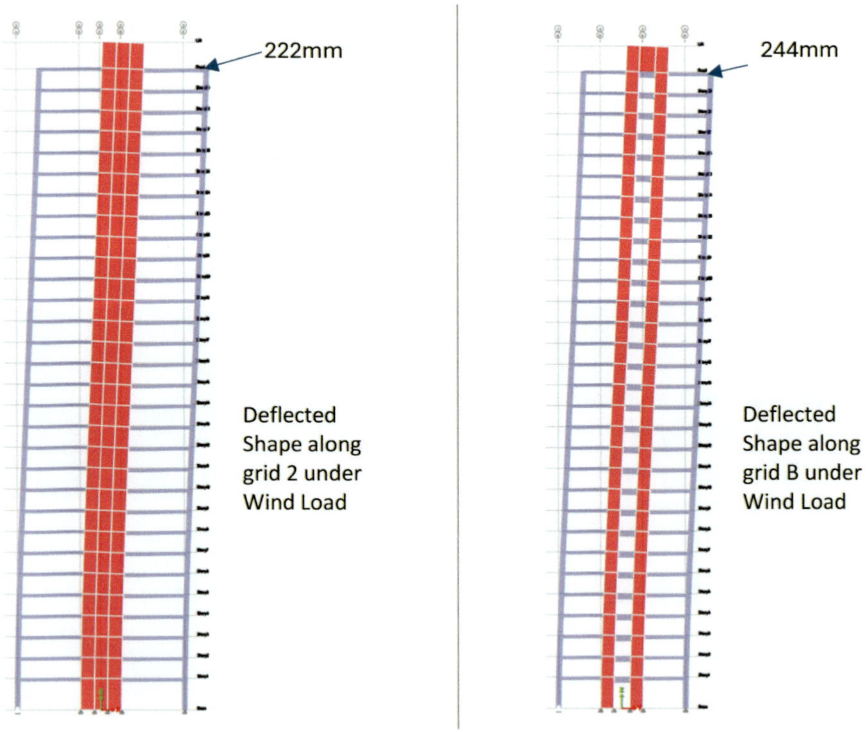

FIGURE 4.5 Deflected shapes of the core wall-frame building under wind load.

achieved which is favoured for or required by a high-rise building resisting eccentric high wind and/or seismic loads.

Furthermore, the inner lift core wall of the framed tube building can also be made use of to further enhance the bending stiffness of the building. By structural linking up the outer framed tube and the inner core wall structure, a "tube-in-tube" building structure is formed as shown in Figure 4.6, as modified from the wall and framed building shown in Figure 4.4. The computer model is shown in Figure 4.7.

Under the same wind load, the deflections have been significantly reduced from 222 mm to 43 mm in the X-direction and 244 mm to 74 mm in the Y-direction, showing the tube-in-tube configuration can achieve a substantially increase in stiffness.

To check the maximum height of the tube-in-tube building that the Hong Kong Concrete Code's (2013) criterion of deflection to building height not exceeding 1 in 500 is fulfilled or almost fulfilled, the number of storeys of the building structure in Figure 4.6 is increased to 50. The wind loads determined by the Hong Kong Wind Code (2019) are also increased as the height and natural frequencies of the building structure change. The greatest deflection is in the Y-direction at 426 mm as shown in Figure 4.8. The deflection to height ratio is 1 : 474, approximately equal to 1 : 500. So, the framed tube or the tube-in-tube system is very effective in enhancing lateral stiffness of buildings as compared with the wall-frame system.

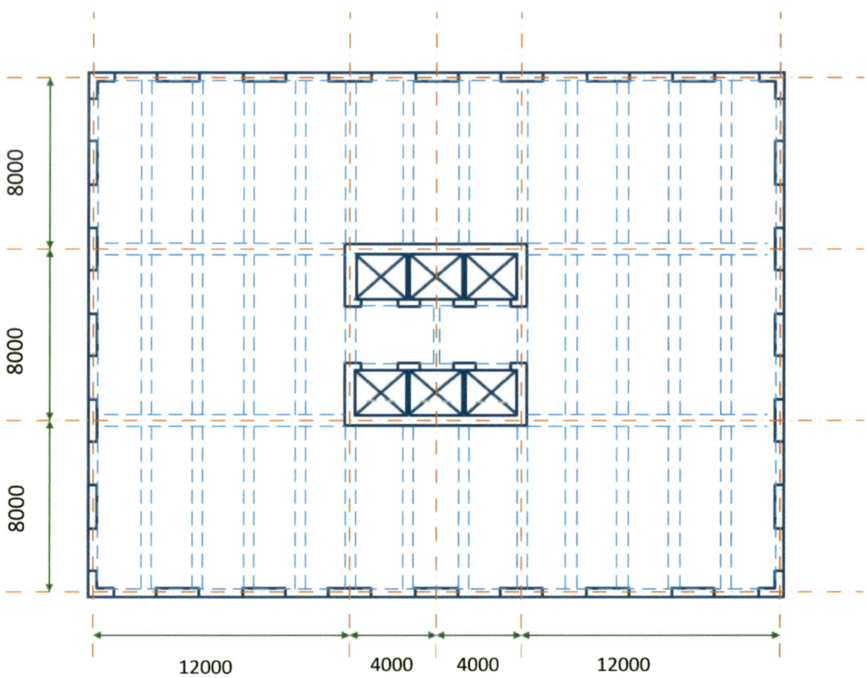

FIGURE 4.6 Tube-tube building – floor plan.

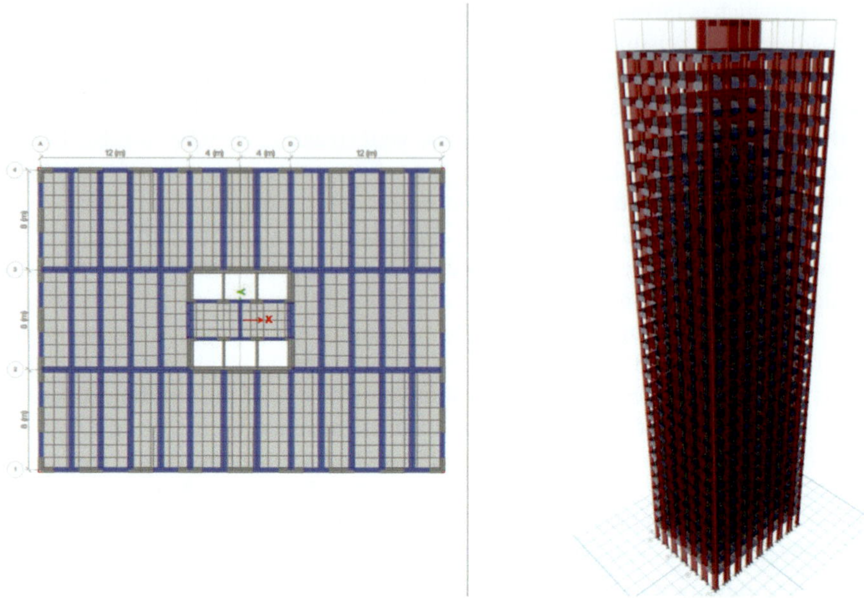

FIGURE 4.7 Tube-in-tube building (30 storeys).

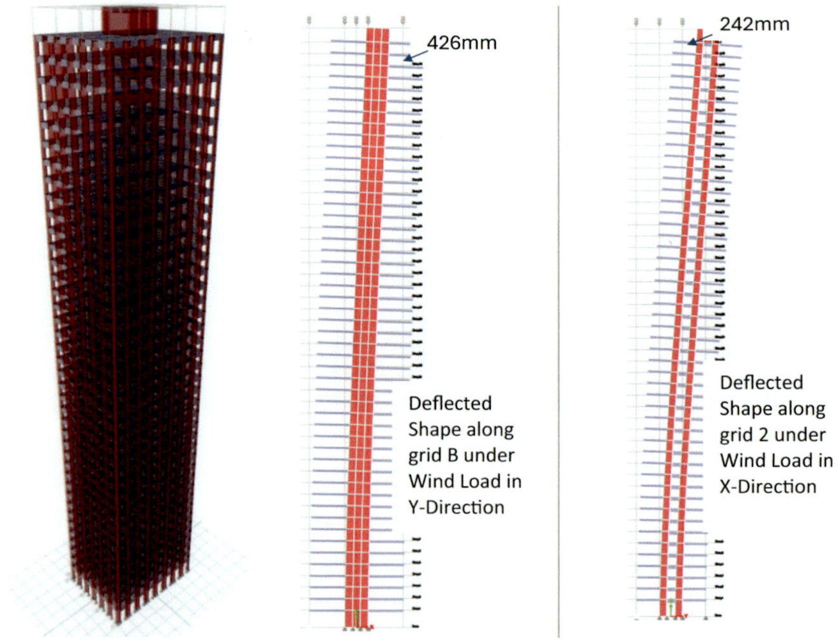

FIGURE 4.8 An example demonstrating the effectiveness of framed tube or tube-in-tube (50 storeys) in enhancing lateral stiffness of building.

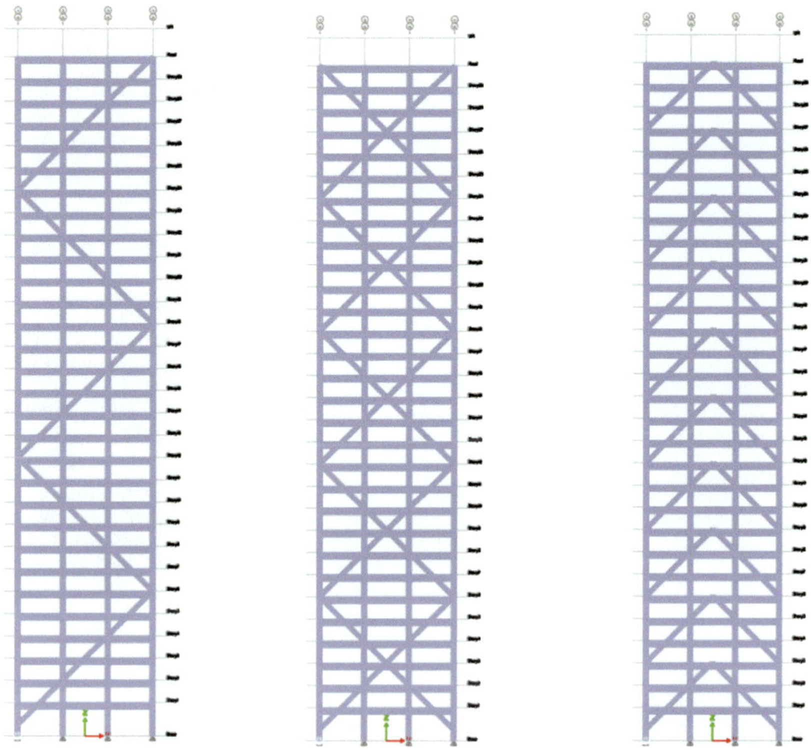

(a) K-Form (b) X-Form (c) A-Form

FIGURE 4.9 Different forms of external bracing – K-form, X-form, and A-form.

4.3.5 Externally Braced Building

A building's stiffness can be enhanced by using "bracing" across floors at its perimeter frames. The bracing can be of X-form, K-form, A-form etc. as shown in Figure 4.9. The choice of the type of bracing depends not only on the efficiency of the structural system but also on the aesthetic and architectural aspects.

Adding concrete bracing of 400 × 800 to the 30 storey Wall-Frame Building in Figure 4.4 at its 4 external faces in the K-, X-, and A-Forms and performing lateral load analysis with the same lateral loads, the following results are displayed in Figures 4.10, 4.11, and 4.12 respectively and summarized in Table 4.1.

From Table 4.1, it can be seen that for the wall-frame building which does not originally possess stiff frames along its perimeter, stiffness will be significantly increased by the bracing, even by the simplest K-frames. However, after the addition of K-frames, further increase by adding more bracing to form the X-frame does not substantially increase building stiffness. The A-frames, however, achieve the best performance among the three.

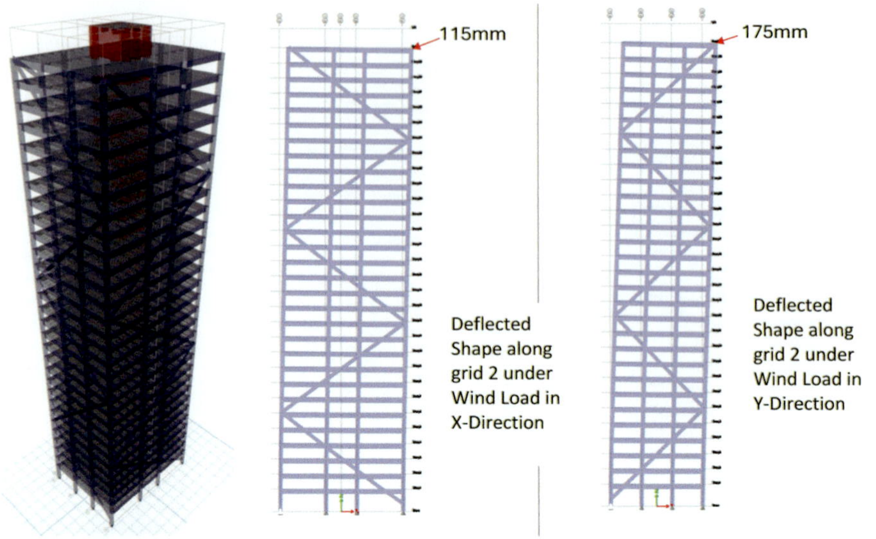

FIGURE 4.10 K-form bracing.

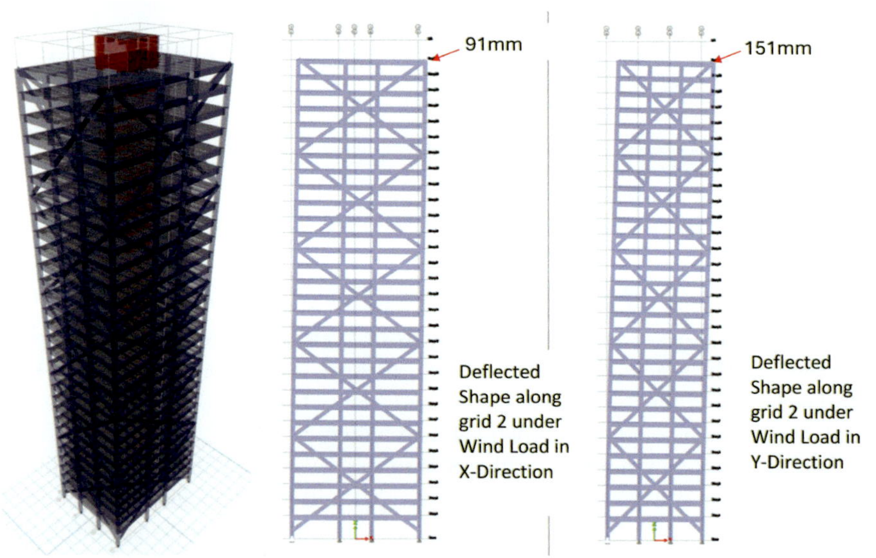

FIGURE 4.11 X-form bracing.

The external bracing of (size 400 mm by 1200 mm) is now added to the 50 storey tube-in-tube framed structure in Figure 4.7. Re-analysis based on the same lateral loads shows that the improvements are not significant as shown in Figure 4.13 where the reductions are 242 mm to 229 mm in the X-direction and 426 mm to 408 mm

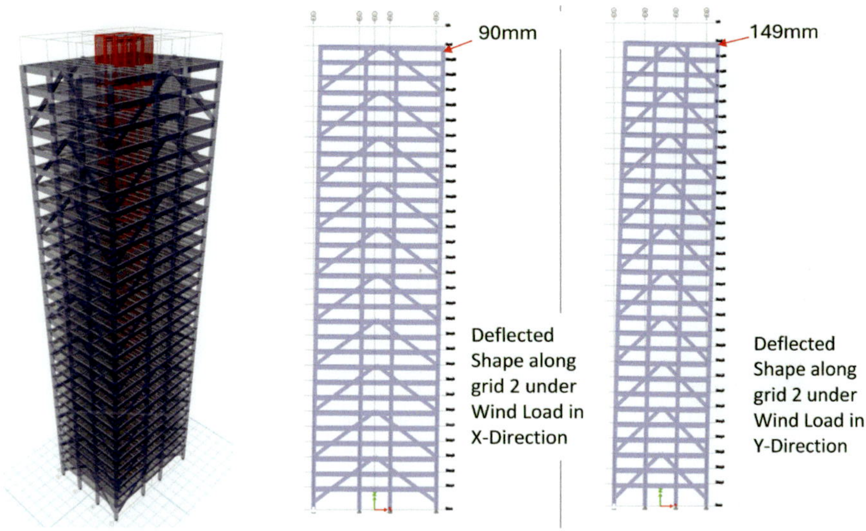

FIGURE 4.12 A-form bracing.

TABLE 4.1
Comparison of Deflections of the 30 storey Wall Frame Building shown in Figure 4.4 under Wind Load to Hong Kong Wind Code for the Different Forms of External Bracing

Bracing Type	Deflection in X-Direction (mm)	Deflection in Y-Direction (mm)
No bracing	222	244
K-frame	115	175
X-frame	91	151
A-frame	90	149

in the Y-direction, i.e., around 5% deflection reduction. This is because the peripheral frames are already very stiff. So, the increases brought by the bracing are not substantial.

Summing up, the external bracing contributes stiffness to the building significantly if the external frame is not stiff. If not, the contribution may be insignificant.

4.3.6 Outrigger Building Structure (Belt truss)

An outrigger building structure generally comprises a stiff core wall structure and a number of "giant columns" along its perimeter. Stiff horizontal members known as "outriggers" which are often in the form of steel trusses are used to structurally connect the core wall and the giant columns. Under lateral loads, these outrigger

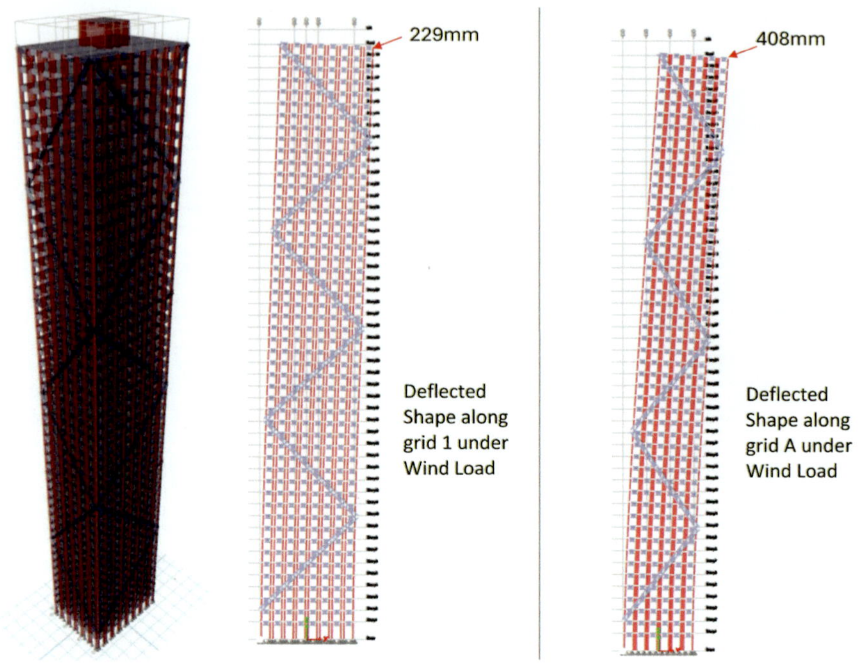

FIGURE 4.13 Further improvement by K-bracing to the tube-in-tube structure.

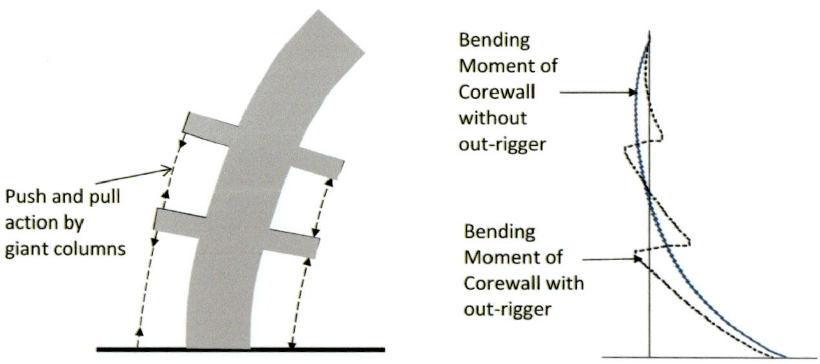

FIGURE 4.14 Working mechanism of outrigger structure.

structures will attract large up and down shears by their high stiffnesses which will transfer vertical push and pull loads onto the giant columns. By such actions, both the bending moments on the core wall and the lateral deflection of the building are reduced. The phenomenon is illustrated in Figure 4.14. As the outrigger building structure works by such shear transfer, the outrigger structure should therefore be placed at locations of considerably high moments in the core wall. By experience and in accordance with Taranath (2012), to achieve high efficiency of the intended

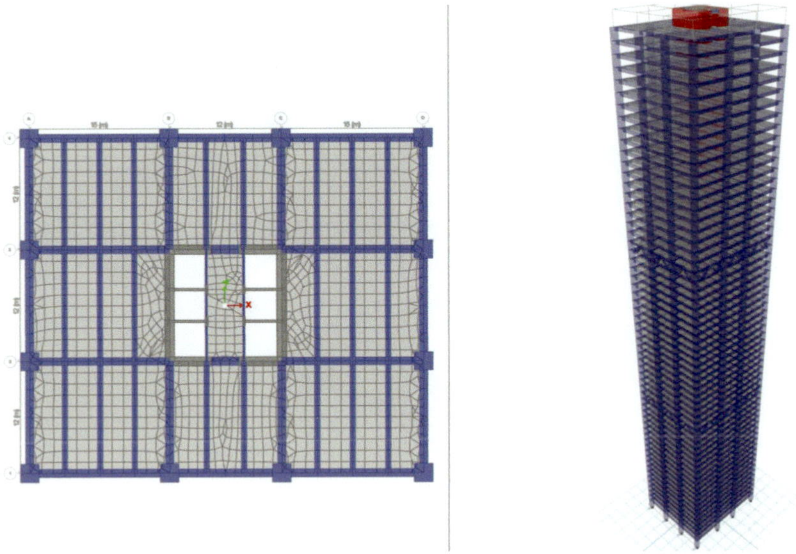

FIGURE 4.15 An outrigger building structure with belt trusses.

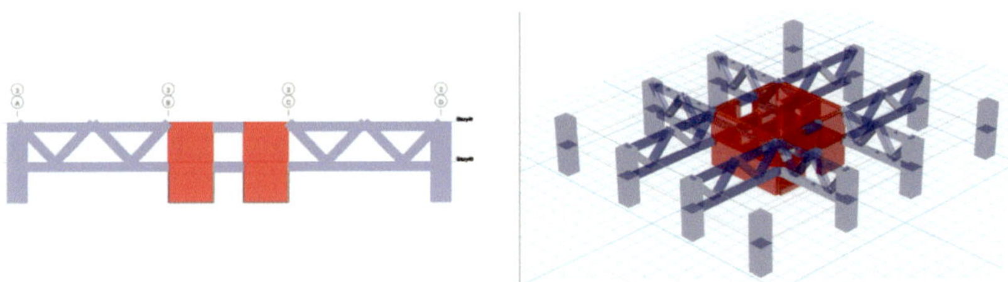

FIGURE 4.16 Layout of the outrigger structure.

purpose, these outriggers should be placed at one-third and two-thirds building heights if outriggers are used in two levels or be placed at half building height if one level outrigger is used.

For illustration of the efficiency of the "outrigger structure" with and without "belt truss" structure, a 60-storey building whose plan and perspective views are as shown in Figure 4.15. The building has plan dimensions of 40 m × 38 m. A central core wall and 8 no. of giant columns of size 2 m × 2 m connected by 8 pairs of outrigger structures in the form of steel trusses and belt truss structures at the 20th and 40th storeys.

Cut views of the outrigger structures are shown in Figure 4.16, with other floor structural members omitted for clarity. The steel members in the truss structures are heavy ones. The main girders of the outriggers are 900 mm square hollow section (SHS) with thickness of 100 mm all round while the inclined struts are 900 mm SHS with thickness of 75 mm.

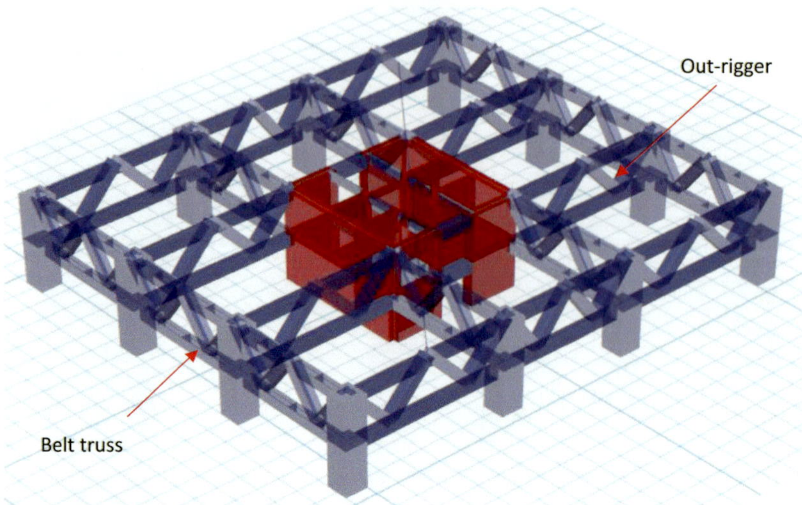

FIGURE 4.17 Outrigger truss and belt truss.

TABLE 4.2
Summary of Lateral Deflection due to Wind on the Building Structure Shown in Figure 4.15

	No Outrigger and No Belt Truss	With Outrigger Only	With Outrigger and Belt Truss
Wind in X-Direction	412 mm	310 mm (0.75)	295 mm (0.72)
Wind in Y-Direction	364 mm	304 mm (0.84)	288 mm (0.79)

The bracketed values are the ratios of the deflection to that of the structure without outrigger and belt truss.

Further strengthening or stiffening of the building can be carried out by the use of a "belt truss" which is a truss structure along the perimeter of the building. The belt truss serves to mobilize all the perimeter columns in resisting lateral loads, as shown in Figure 4.17 and, as a closed loop along the perimeter of the building, can enhance the torsional stiffness of the building significantly.

Three models which are ones (i) without outrigger and belt structures; (ii) with outrigger structures only; and (iii) with outrigger and belt truss structures are analysed by lateral wind loads determined by the Hong Kong Wind Code 2019 (2019). The deflections at the roof level are summarized in Table 4.2.

It can be seen that the reductions in deflections upon the use of outrigger structures alone are quite significant. But the further reductions by the belt trusses are not substantial, mainly because the belt trusses, being not directly linked to the core wall do not contribute additional stiffness to significant extents. The main function of the

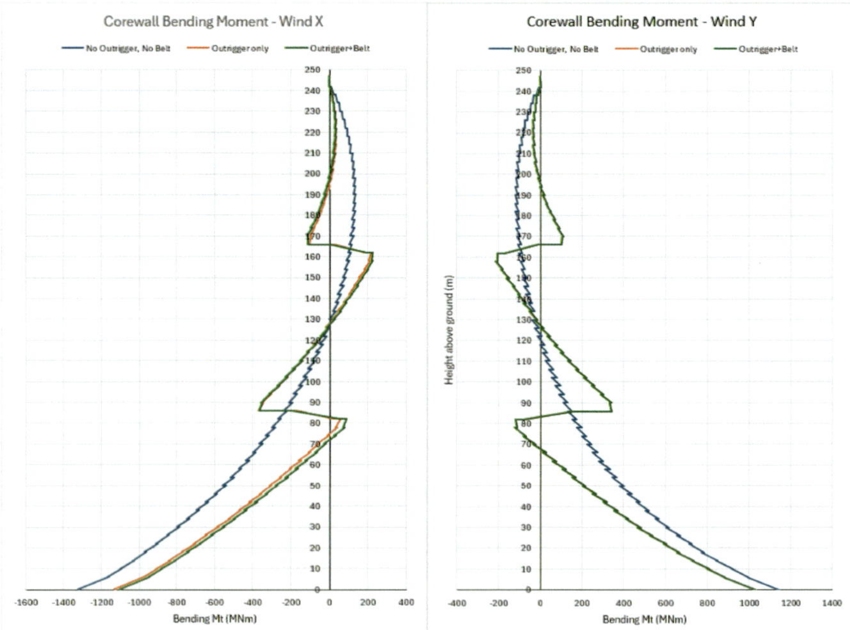

FIGURE 4.18 Effect on bending of the central core wall by the outrigger structure.

belt truss, however, lies in its provision of torsional stiffness which is not important in the building under consideration as the building is very symmetrical.

The bending moment of the core wall is also examined for the three models which are plotted in Figure 4.18 for wind load in the X and Y directions.

As shown by the figure, the outriggers can reduce and reverse the bending moments of the core wall. Generally, there can be substantial "savings" for the structural design of the core wall, especially at floors below the lower outrigger. The extent of savings depends on the comparative stiffness of the outrigger structures and the floor structures originally provided. In this particular example, the lateral stiffness provided by the floor structures other than the outrigger trusses is also substantial, as demonstrated by the moment reversals at the upper floors. In addition, the further reduction by the "belt truss" structure is not significant at all, as has been explained in the foregoing.

The Taipei 101 (509 m), Hong Kong International Finance Centre 2 (380 m), Hong Kong International Commerce Centre (450 m), the Henderson in Hong Kong (190 m), Qianhai CTF Tower (520 m) in Shenzhen, China, and Capital Gate in Abu Dhabi (165 m) are examples of buildings constructed using core wall and/or outrigger structural systems.

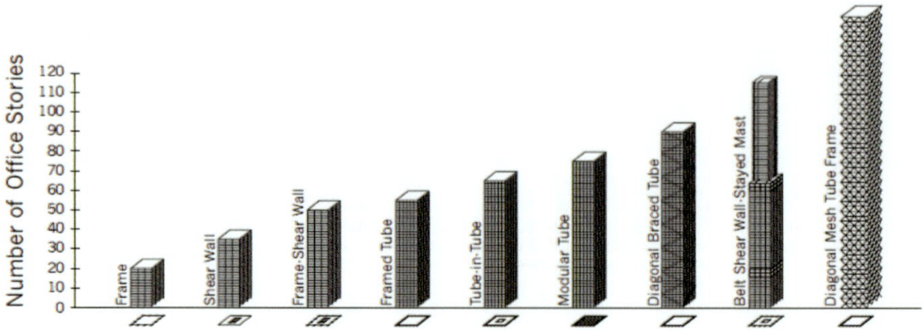

FIGURE 4.19 Building Structural System for Concrete Buildings (Extracted from Sarkisian (2012).

4.4 CHOICE OF HOLISTIC STRUCTURAL SYSTEM FOR BUILDING

It can be deduced from the foregoing that lateral load resistance governs the choice of the structural system in requiring an effective means to achieve lateral stiffness. There are guides given based on the number of storeys which are suggested by Sarkisian (2012) as Figure 4.19 for reinforced concrete buildings. Nevertheless, the suggestions are only rough guides as they have not accounted for specific factors such as plan dimensions, intensities of wind loads, and seismic loads which vary from one locality to another.

So, in the ascending order of strength and stiffness against lateral loads as discussed in the foregoing, a hierarchy of holistic structural systems can be listed from beam-column frame to framed tube and outrigger in terms of complexity in design and construction. The following Table 4.3 extracted from Cheng and Law (2024) briefly lists the comparison of the various systems in the order of the hierarchy for the readers' reference in their selection of structural systems.

Accordingly, to select the most appropriate structural system for a particular building, the following general principles can be applied:

(i) Develop an Efficient Structural Layout: Start by analyzing the given locations for walls and columns to create an efficient preliminary structural layout. This should be an iterative process done in collaboration with the architect and other disciplines to balance aesthetic, functional, economic, and structural requirements.
(ii) Estimate the lateral wind and/or seismic loads if applicable: Calculate the anticipated wind and seismic loads as mandated by local building codes. Wind loads are primarily influenced by the building's external shape and can often be estimated fairly accurately. Seismic loads, however, depend on the structure's dynamic properties, particularly its fundamental natural frequency. This can be approximated empirically (e.g., using the formula 46/H, where H is the building height in metres).

TABLE 4.3
Comparison of Different Holistic Structural System

	Beam-Column Frame	Shear/ Core Wall	Wall and Frame	Framed Tube	Braced Tube	Outrigger
Easy to construct	Easy	Easy	Easy	Easy unless techniques such as slipped form construction are used.	Difficult to construct as inclined bracing are involved	Difficult to construct as composite construction is often involved.
Flexibility in use	Flexible with greatest freedom in repartitioning	Less flexible as structural walls cannot be removed	Less flexible as structural walls cannot be removed	Less flexible in removing structural walls and frame tube. But generally large free spaces are given for flexibility in use.	Less flexible in removing structural walls and frame tube. But generally large free spaces are given for flexibility in use.	Less flexible in removing structural walls and columns. But generally large free spaces are given for flexibility in use. However, restricted use in the outrigger floors.
Economy in construction	Cheap and requiring the least construction technology	Cheap and requiring simple construction technology	Cheap and requiring simple construction technology	Cheap and requiring simple construction technology	More expensive as inclined struts are present	Expensive due to the presence of outriggers and the connections with the wall/column structure
Complexity in construction	Simplest	Simple	Simple	Simple	More difficult involving construction of inclined struts	Difficult in the construction of outrigger and connection to the wall/column structure
Gravity load carrying capacity	Least as the column load carrying capacities are relatively small	High as per the high cross-sectional area of walls	Moderate due to the presence of some relatively high cross-sectional area of walls	High as per the high cross-sectional area of walls and columns both internal and external	High as per the high cross-sectional area of walls and columns both internal and external	Moderate due to the presence of some relatively high cross-sectional area of walls
Lateral Stiffness	Least	Moderate	High	Very high	Very high	Very high

(iii) Evaluate Structural System: Assess various structural systems (e.g., moment frames, shear walls, outriggers, tube frames) against the calculated lateral loads. The goal is to ensure the system meets code requirements for deflection and vibration. Preliminary analyses can be based on simplified models or computer simulations that include only essential elements. Engineer's experience is valuable for eliminating infeasible or inefficient options early on.

(iv) Check for gravity load: Verify that the proposed systems can adequately support gravity loads. This is especially critical for designs incorporating large transfer structures, long spans, or hanging elements.

(v) Consider Constructability and Economy: Evaluate the economic implications and constructability of each system. The optimal solution is the simplest and most cost-effective option that meets all design and performance criteria.

(vi) Finalize and Refine: With the optimal system selected, perform a preliminary sizing of the main structural members. Be prepared to revisit the chosen system if this sizing reveals inefficiencies or layout conflicts with functioning.

4.5 FLOOR SYSTEM

The floor structure is a primary building system that supports and transmits loads from occupants, furniture, and machinery to the supporting structural elements. Its design must fulfil two critical requirements, which are strength to ensure safety against collapse (ultimate limit state) and stiffness to ensure serviceability by limiting deflection, vibration, and cracking (serviceability limit state). The following common floor systems are described and discussed.

4.5.1 Reinforced Concrete Beam and Slab

The reinforced concrete beam and slab system is the most prevalent floor system due to its design simplicity, high efficiency, and overall economy in both design and construction. This system comprises concrete slabs supported by a grid of beams.

Because slab thickness in buildings is typically limited to 100–200 mm, their effective span is generally restricted to 3–4 m, necessitating regular support from beams. These supporting beams are categorized as secondary beams and main beams. Secondary beams transfer loads to the main beams, which are ultimately supported by columns or walls. This beam grid effectively provides the necessary strength and stiffness to the entire floor system.

Slabs are also classified based on their spanning direction as either one-way or two-way.

Building codes often provide rough guides for design, such as limiting the span-to-depth ratio for slabs and beams to approximately 20–30. These ratios serve as practical starting points for designing standard building floors.

FIGURE 4.20 Flat slab construction with column drops (photo taken by the Authors).

A primary drawback of this system is that the beams require greater structural depth among other floor systems, resulting in less headroom beneath the beam soffits.

4.5.2 Flat Slab

The flat slab is a concrete floor system supported directly by columns, without the use of beams. Its primary advantages are a clean aesthetic appearance and the achievement of minimal headroom, making it ideal for buildings with low storey heights.

However, the absence of beams necessitates a greater slab thickness to span the wider column spacing. This results in a structurally less efficient system than a beam-slab alternative, as it incurs a higher self-weight.

A critical design consideration, in addition to bending moments, is punching shear at the column supports. The concentrated shear forces applied to the relatively thin slab require meticulous checks. To enhance shear resistance, an enlarged column head, known as a column drop or capital, is often incorporated into the design.

Figure 4.20 provides a view of the underside of a flat slab.

4.5.3 Voided Slab

A voided slab is a concrete floor system that incorporates internal voids to reduce its self-weight. These voids are strategically positioned at the mid-depth of the slab to minimize encroachment into the critical concrete compression zone, thereby preserving the slab's bending strength.

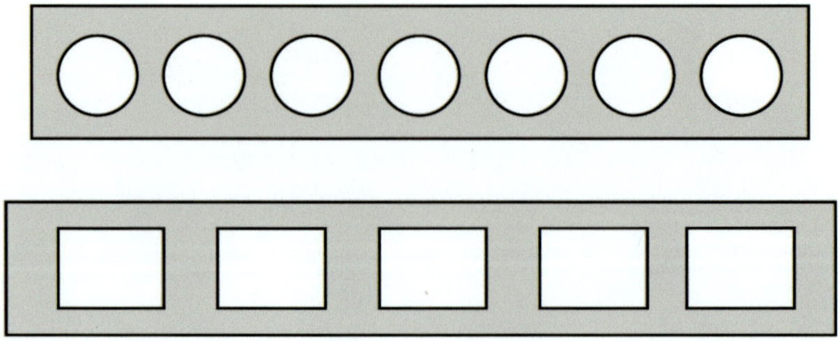

FIGURE 4.21 Voided slabs.

The voids are typically spherical or cuboidal and are created by using pre-manufactured "void formers," often made from plastic as shown in Figure 4.21.

This system is structurally efficient, achieving significant strength and stiffness while substantially reducing dead weight. Furthermore, the use of pre-fabricated void formers makes the construction process both practical and convenient.

4.5.4 Ribbed and Waffle Slabs

Ribbed and waffle slabs are reinforced concrete slab systems that achieve long spans with reduced weight. They consist of a thin topping slab supported by a grid of closely spaced, deep ribs. The primary distinction is their directionality between ribbed slabs and waffle slabs are their ribs arrangement. Ribbed slabs have ribs running in one direction while waffle slabs have ribs running in two directions, creating a grid pattern.

Structurally these ribs act as integral stiffeners, providing the system's strength and stiffness. Standard spacings between ribs typically range from 750 mm to 1200 mm. Construction employs reusable modular formwork, often called "moulds," to create the void spaces between the ribs.

To enhance punching shear resistance around columns, the voids are often filled with concrete to create solid caps. Beyond their structural efficiency – enabling long spans with minimal self-weight – these systems are also valued for their distinctive aesthetic, often leaving the grid pattern exposed as a ceiling feature. Figure 4.22 shows the underside of a waffle slab.

Computer modelling of a waffle slab may not require the explicit modelling of every individual rib and slab panel. Instead, the system can be idealized as an equivalent solid slab with a comparable bending stiffness in the relevant direction.

However, accurately representing the torsional (twisting) stiffness of the equivalent slab is more challenging. One approximate method is to define the effective torsional stiffness as the average of the bending stiffnesses in the two orthogonal directions.

FIGURE 4.22 Waffle lab (photo taken by the Authors).

4.5.5 Precast Slab and Semi-precast Slab

Precast construction is particularly suitable for slabs, as they are typically the lightest and most numerous structural components in a building. This method offers well-documented advantages, including faster construction, higher quality control, and a tidier site.

However, a key design consideration is the jointing between precast slabs and cast-in situ structural members. These connections require careful detailing to ensure adequate stiffness and overall structural robustness.

To address this, semi-precast construction is often preferred for slabs, as it facilitates a more integral connection with in situ members. In this system, a precast concrete plank acts as permanent formwork for a cast-in situ concrete "topping." Reinforcement is placed across both elements, allowing the precast and in situ components to act compositely as shown in Figure 4.23. A common requirement is to roughen the top surface of the precast plank to ensure a strong structural bond.

This semi-precast method is not only used for buildings but also for bridge decks.

4.5.6 Prestressing Slab

Post-tensioned slab construction involves installing cable ducts within the formwork prior to concreting. Once the concrete has gained sufficient strength, prestressing tendons are threaded through these ducts and tensioned.

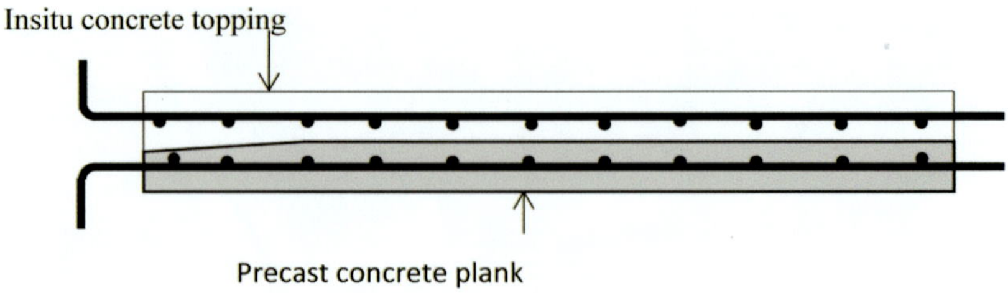

FIGURE 4.23 Semi-precast slabs.

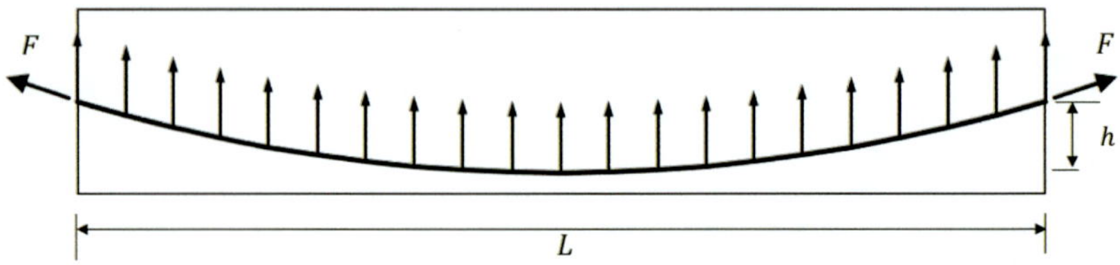

FIGURE 4.24 Prestressing cable exerting upward forces on the slab.

The most significant benefit of prestressing slab is its strength. The tendons are typically profiled along a parabolic curve. This curvature generates a beneficial upward thrust force on the slab, as illustrated in Figure 4.23. The magnitude of this equivalent upward load w can be calculated using the formula $w = 8Fh/L^2$, where F is the prestressing force, h is the tendon drape, and L is the span, as defined by T.Y. Lin (1961) and shown in Figure 4.24.

The prestressing force serves to reduce slab deflection and simultaneously counteracts (or "balances") a portion of the applied bending moments. This makes post-tensioned slabs particularly suitable for long-span applications where minimizing structural thickness is critical.

It is important to note that conventional reinforcement is still required within the slab.

A related technique, pre-tensioning, can also be employed. In this method, concrete is cast around tendons that are already under tension. After the concrete has hardened and gained sufficient strength, the tendons are released. The force is transferred to the concrete through bond (or "grip"), inducing beneficial compressive stresses.

4.5.7 Composite Beam-Slab (Profiled Metal Deck Floor)

This floor system utilizes a profiled metal deck as a permanent formwork and tensile reinforcement for a cast-in-place concrete slab. The concrete bonds to the deck, and

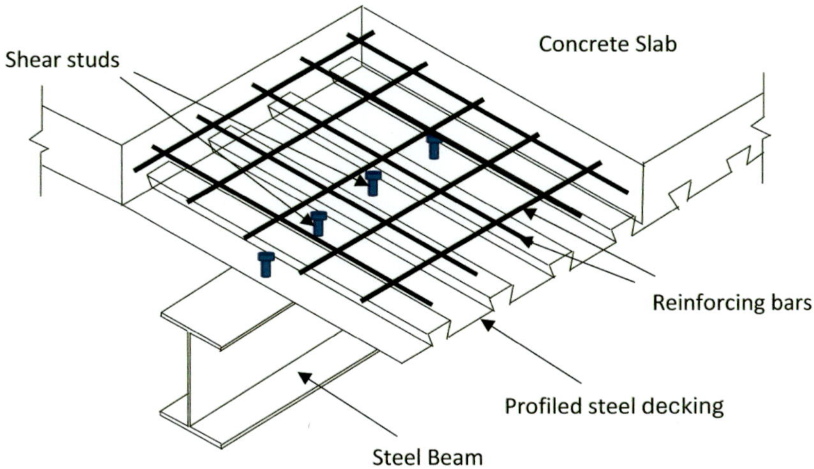

FIGURE 4.25 Composite beam-slab construction (Extracted from Cheng and Law, 2024).

the assembly is connected to the supporting structural steel members via shear studs welded through the deck, as illustrated in Figure 4.25.

This composite construction combines the principal advantages of both materials: the high durability, mass, and stability of concrete with the high strength-to-weight ratio of structural steel. Consequently, it is an ideal solution for long-span structures and applications where minimizing weight is a critical design consideration (Figure 4.24).

4.5.8 Comparison of Different Structural Floor Systems

Summing up the foregoing discussion, a comparison of different floor systems can be listed in Table 4.4, extracted from Cheng and Law (2024).

4.6 SUMMARY AND CONCLUDING REMARKS

This foregoing has provided an overview of holistic building and floor systems, including their working principles, comparative advantages, and disadvantages. Worked examples have also been provided for illustration with quantification in terms of strength and deflection. The discussions in Sections 4.4 and 4.5.8 aim to furnish readers with a fundamental understanding of the common structural forms used in modern construction practice.

TABLE 4.4
Comparison of structural floor systems

	Beam and Slab	Flat Slab	Voided Slab	Ribbed and Waffle Slab	Precast and Semi-precast Slab	Prestressing Slab	Composite Beam-Slab
Easy to construct	Easy	Easy, least formwork	Require some technique	Require some technique	Require relatively higher level of technique	Require high level of technique	Require relatively high level of technique
Load carrying efficiency	Moderate	Low	Moderate	Moderate	Moderate	High	High
Headroom provision	Beams are deep in comparison, giving least headroom	Least structural thickness, greatest headroom	Moderate thickness to house the void formers	Moderate thickness, greater headroom	Beams are deep in comparison, giving least headroom	Lesser slab thickness and lighter weights, Beams can then be shallower.	Lesser slab and beam thickness, so greater headroom
Cost effectiveness	Moderate in construction cost	Not effective though least formwork but heaviest concrete	Effective, require proprietary void former	Effective, require proprietary formwork	Effective if produced in cheap territories	Moderate as high construction cost	Moderate as structural steel is not cheap
Sustainability	Moderate as per the quantities of construction materials used	Low as quantity of concrete is the highest.	Relatively high as quantities of concrete are relatively low	Relatively high as quantities of concrete are relatively low	High if the precast components are factory made by efficient process with material and energy saving.	High as quantities of concrete is low	Relatively high as quantities of concrete is relatively low

REFERENCES

Buildings Department, Hong Kong SAR Government (2013) *Code of Practice for the Structural Use of Concrete 2013.*

Buildings Department, Hong Kong SAR Government (2019) *Code of Practice on Wind Effects of Hong Kong 2019.*

Cheng, Y.M. and Law, C.W. (2024) Design and Construction of Buildings and Foundations with Illustrative Examples. Bentham Books.

Lin, T.Y. (1961) A New Concept for Prestressed Concrete. Australian Constructional Review.

Sakisian, M. (2012) Designing Tall Buildings. TJ International Ltd.

Taranath, B.S. (2012) Structural Analysis and Design of Tall Buildings. CRC Press, Taylor & Francis Group.

5 Loads on Buildings

5.1 INTRODUCTION

A building which may be residential, commercial, or industrial inevitably has to be designed to carry the loads due to its intended purpose(s). These loads are mostly gravitational in nature, generated from human occupants, furniture, machinery, equipment, etc. They are termed live loads as they are movable and the magnitudes may vary. The building's own weight due to its construction materials, together with loads such as finishes and fixed partitions also constitutes another gravitational load known as the "dead load" because it is relatively constant and stationary throughout the life of the building.

In addition to the gravitational load, the building has to be designed to resist other load types that it may experience during its life time including wind load, seismic load, soil load, snow load, and flood load, etc. These loads are generally not constant and are less certain in magnitude. Generally, they are "random processes," and their determination for design often has to be based on statistical approaches.

5.2 DEAD LOAD AND LIVE LOAD

Dead loads (permanent actions) of a structure are those which are constant or stationary throughout the life of the structure. They normally refer to the own weight of the structure and the permanent attachments such as finishes, partitions, services, etc. The determination of dead load can be of high certainty as it is based on the densities of the materials which can generally be measured precisely.

Live loads (variable actions) as loads acting on a building refer to those which are not constant and vary with time. They normally include weights of occupants, furniture, machinery and the like which are referred to as "live loading elements" in this Chapter thereon. Although these weights are not constant, they however vary within some ranges which can be determined by statistical approaches based on surveyed data for structural design purposes. Generally, raw data in relation to the weights of the live loading elements are collected by survey including site measurement and/or survey by means such as questionnaires from the occupants. Statistical approaches applying probability distribution functions (e.g., normal, lognormal, gamma) for analysing the collected data are then carried out. The "fractile" or "percentile" of a sample or population of data is finally obtained as a value greater than a high percentage of the data, e.g., 95 percentile which is a value not less than 95% of the collected or analysed data fitted to a probability distribution function. This value, known as the characteristic value, may be used as the design value of the live load intensity for the building structure as specified in a number of design codes

DOI: 10.1201/9781003636380-5

including Eurocode BSEN 1990:2023 (2023), the Australian Load Code ASNZS 1170.1 (2002), the Hong Kong Load Code 2011 (2011). The American load code ASCE 7-16 (2018) however terms the design load as the "nominal loads" which are the loads given in the standard for design similar to the characteristic load in other codes. The characteristic load concept can also be applied to dead load where the standard deviations are generally small so that they are close to the mean values.

For convenience in design, the values presented in the design codes are generally uniformly distributed loads (UDL). For example, 2 kN/m^2 UDL for domestic floors and 3 kN/m^2 UDL for office floors. However, during the survey, the loads from the furniture and people are not uniformly distributed. So, a design approach of "equivalent uniformly distributed load" (EUDL) is adopted in some codes. The greatest bending moment due to the actually located live load components is first determined on a slab. The EUDL which is a UDL producing the same bending moment is calculated. In the exercise of determining live load intensity of a floor of a particular usage, e.g., residential or industrial, many similar EUDL statistical data may be acquired through survey and used for further processing to determine a percentile (or fractile) value, say 95 percentile which may be used as the characteristic design value for floors of that usage. While statistically the live load components are movable in a sample, some design codes such as the China Load Code GB50009 (2012) requires the components to be so located that the most adverse scenario is arrived at in determining EUDL.

As the chance of full live load on a floor slab is less in a slab of large area than that of a smaller area, many codes of practice allow reduction of live load with increase of area. The Eurocode BSEN1991-1-1:2025 (2025) introduces a α_A factor for floors not carrying heavy live load such as storage as

$$0.6 \le \alpha_A = 0.5 + 10 / A \le 1.0 \tag{5.1}$$

And the Australian/New Zealand Standard AS/NZS1170.1:2002 (2002) gives a factor ψ_A, again with restriction to heavily loaded floors and 2-way slabs

$$0.5 \le \psi_A = 0.3 + 3 / \sqrt{A} \le 1.0 \tag{5.2}$$

where A is the tributary area for the structure in m^2 for both codes.

In addition, live load reduction is also allowed for columns, walls and foundations carrying multi-floors, also on the grounds that all floors being fully loaded is small. The Eurocode BSEN1991-1-1:2025 (2025) provides formulae as follows :

$$\alpha_n = 0.7 + 0.6 / n \le 1.0 \tag{5.3}$$

The Hong Kong Load Code (2011) provides a table for live load reduction percentages as related to the number of floors.

While dead loads are unequivocally static, live loads from sources like moving occupants or operating machinery are inherently dynamic. Nevertheless, in standard building design, these live loads, along with environmental forces like wind and seismic loads, are typically modelled as equivalent static loads. This simplified approach uses amplification factors to account for dynamic effects and uncertainties of magnitudes and occurrences of the loads. Generally, only when the dynamic nature of a load is particularly pronounced is a full dynamic analysis required to address the structural effects of vibration.

5.3 DYNAMIC LOADS DUE TO LIVE LOAD

Although design for live load is mostly static for building structures, dynamic analysis is sometimes required in case the dynamic nature of the load is very pronounced. The nature of the vibratory load and the effects on strength design are discussed as follows.

5.3.1 Vibration

A primary concern regarding building vibration is occupant discomfort. While discomfort is a subjective sensation that varies among individuals, several national and international standards have established objective threshold values for acceleration, typically based on vibration frequency.

A widely referenced standard is the chart from AISC Steel Design Guide 11 (1997) extracted in Figure 5.1, which recommends tolerance limits for floor acceleration. This chart indicates that humans are most sensitive to vibrations in the 4–8 Hz range and that higher frequencies are generally less perceptible, thereby reducing discomfort.

Vibration of a floor structure is its response to dynamic loads acting on it as per its dynamic properties. The basic principles of structural dynamics are discussed in detail in Chapter 7, including the response to dynamic loads by structures. Briefly speaking, a structure under dynamic loads will essentially vibrate in accordance with the frequencies of the external excitation. In addition, it will exhibit a phenomenon of "resonance" if its natural frequencies are close to that of the external. Damping, which is inherent in the structure, can help reduce vibrations. To achieve better damping, external dampers may be used. This section of the chapter will focus only on the dynamic loads on structural floors due to human activities.

Human activities that would create dynamic loads on floors are those due to walking, jumping, or dancing. Based on the structural dynamic theories, empirical formulae have been developed to estimate accelerations of some simple structures due to human walking with the key parameters including weight of the walking person, natural frequency of the structure and the damping (mostly expressed in damping ratio). The weight of the person to be adopted in the analysis is in the range of 0.7 kN (approximately 70 kg).

However, with the use of computer methods nowadays, accurate analysis can be performed by "footfall analysis," by which the effects varying with locations and time on the structure, including acceleration due to a walking person along a designated path can be determined. The working principles can be based on the

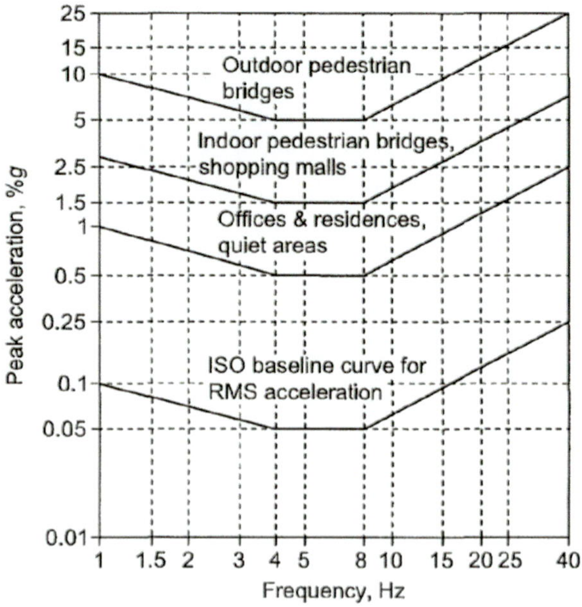

FIGURE 5.1 Recommended tolerance limits for human comfort extracted from AISC Steel Guide 11 (1997).

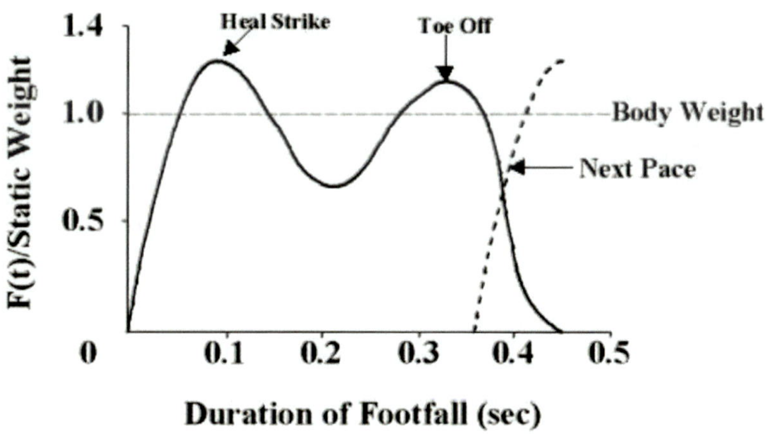

FIGURE 5.2 Force versus time function of a single footfall extracted from SEB guidelines of ASD of Hong Kong (2009).

modal method or the numerical integration described in Chapter 7. The load from the walking person is taken as an impulsive force lasting for a short duration on the floor structure. A popular simulation of the force is as shown in Figure 5.2 which is a force/body weight function with time as extracted from SEB Guidelines SEBGL-OTH4 of the Architectural Services Department (ASD) of Hong Kong (2009).

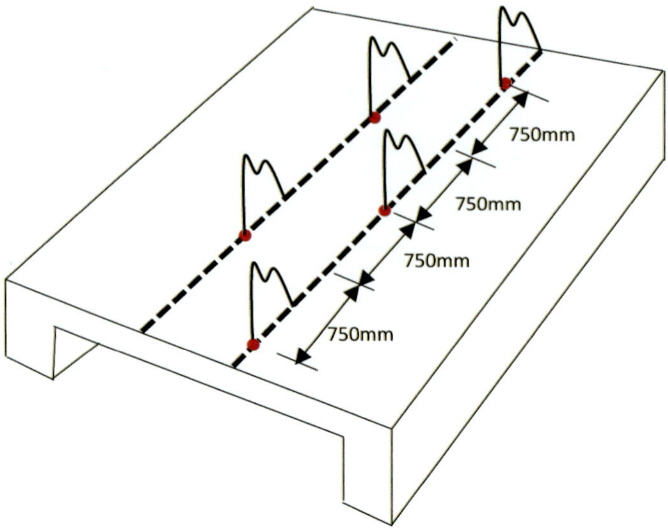

FIGURE 5.3 Footfall analysis of a single person.

The simulation of footfalls by humans is illustrated in Figure 5.3 in which a person walking on two legs along a deck with a footfall function shown in Figure 5.2. Generally, the following parameters can be assumed for analysis which are (i) Weight of a single person = 0.7 kN; (ii) Walking Speed = 1.5 m/sec; (iii) Stride = 1.5 m.

By Figure 5.3, the brown dots which are footfalls by the left and right feet of the walking person alternate with time-varying points in accordance with Figure 5.2. For simplicity in analysis, the left and right footfalls can be on the same line of action as they are generally very close laterally for a walking person. The successive footfalls are separated by a time equal to 0.75 m divided by the walking speed of the person.

A worked example illustrates a footfall analysis on a concrete beam is enclosed in Chapter 7 as Worked Example 7.6.

Nevertheless, for design, crowd loading must be considered. Ellis (2000) found that crowd-induced peak accelerations are similar to those from a single person, likely due to a lack of synchronization. However, the possibility of synchronized walking in confined spaces remains a concern for some researchers. Later, Ellis (2003) concluded that while peak acceleration increases with crowd size, it does not exceed double that of an individual. Synthesizing these findings, Hong Kong's SEBGL-OTH4 (2009) proposes a simplified magnification factor, m = 2 for floors with a fundamental natural frequency $f_n > 3$ Hz, and m = 3 for floors where f_n is between 1 Hz and 3 Hz.

Floor structures in areas with "rhythmic activities," such as dance halls or gymnasiums, are susceptible to excessive vibration. This risk is highest when activities become synchronized and the floor has large spans with low natural frequencies. Vibration analysis for these structures typically assumes uniformly distributed loads

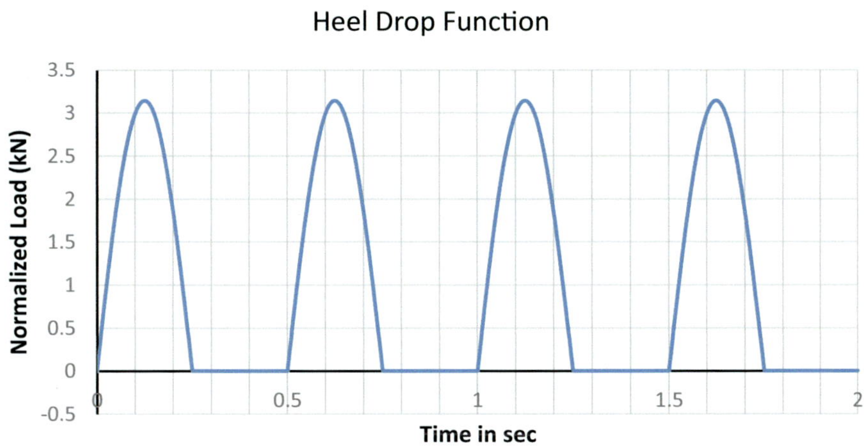

FIGURE 5.4 Typical heel drop force function for a single person with time.

vibrating at complex frequencies generated by "heel drops." Ellis and Ji (1997) modelled the contact force from a single heel drop as a series of half-sinusoidal pulses, given by Equation 5.4 for $F(t)$. This force-time relationship, illustrated in Figure 5.4, differs significantly from the standard footfall function shown in Figure 5.2.

$$F(t) = \begin{cases} K_p G_s \sin(\pi t / t_p) \text{ for } 0 \le t \le t_p \\ 0, \text{for } t_p \le t \le T_p \end{cases} \tag{5.4}$$

where G_s is the static weight of a single person normally taken as 0.7 kN

K_p is the impact factor such that the peak dynamic load $F_{max} = K_p G_s$

t_p is the contact duration

T_p is the period of the jumping load by which $1/T_p$ is the frequency of the jumping

A factor $\alpha = t_p / T_p$ is defined for ease of mathematical manipulation. Ellis and Ji (1997) recommends values of α in Table 5.1.

ASD (2009) quoted Ji and Ellis which indicates that the mean value of the time history of a vertical load corresponding to rhythmic jumping is always equal to the weight of the people such that $\frac{1}{T_p}\int_0^{T_p} K_p G \sin(\pi t / t_p) dt = G$, as per experimental observation. K_p is then reduced to

$$K_p = \frac{\pi}{2\alpha} \tag{5.5}$$

For a single person jumping with a certain frequency (number of heel drops per second), Equation 5.4 can be extended to cover a number of T_p. As an illustration, consider a gymnasium with rhythmic exercise of high impact aerobics such that $\alpha = 0.5$,

TABLE 5.1
Typical Values of the Contract Ratio α for Various Activities

Activity	α
Pedestrian movements, low impact aerobics	2/3
Rhythmic exercises, high impact aerobics	1/2
Normal jumping	1/3
High Jumping	1/4

$K_p = \frac{\pi}{(2\times 0.5)} = \pi$. If the frequency is 2 Hz such that $T_p = 0.5\,\text{sec}$ and $t_p = 0.25\,\text{sec}$, the time history function with normalized load (Weight of person = 1 kN) versus time can be plotted as in Figure 5.4.

While Equation 5.4 is a piecewise function, it has been a practice to model it by a Fourier series for ease of mathematical manipulation as follows

$$\mathrm{F}(t) = G_s\left(a_0 + \sum_{n=1}^{\infty} a_n \cos\frac{2n\pi}{T_p}t + \sum_{n=1}^{\infty} b_n \sin\frac{2n\pi}{T_p}t\right) \tag{5.6}$$

where $a_0 = \frac{2K_p\alpha}{\pi} = 1$ (mean value of a vertical load)

$$a_n = \frac{K_p\alpha}{\pi}\left[\frac{\cos(2n\alpha-1)\pi - 1}{2n\alpha-1} - \frac{\cos(2n\alpha+1)\pi - 1}{2n\alpha+1}\right]$$

$$b_n = \frac{K_p\alpha}{\pi}\left[\frac{\sin(2n\alpha-1)\pi}{2n\alpha-1} - \frac{\sin(2n\alpha+1)\pi}{2n\alpha+1}\right]$$

Equation 5.6 is used to re-calculate the case represented by Figure 5.4 with n up to eight. The results are plotted in Figure 5.5. Very close simulation to the original heel drop function can be achieved.

The periodic loads Equation 5.1 or Equation 5.3 can then be applied to a floor structure as a time-varying point load for analysis.

Crowd rhythmic activity in a synchronized manner will lead to more severe effects on vibrations than that by an individual. Ellis and Ji (1997) recommend the following equation for crowd rhythmic activities as similar to Equations 5.4 and 5.6 as

$$F(t) = G\left[1.0 + C_e\sum_{n=1}^{\infty} r_n \sin(2\pi n f_p t + \phi_n)\right] \tag{5.7}$$

where G replaces G_s and G is the load intensity of the crowd. The value of G has been widely discussed in various literature including Bachmann and Ammann

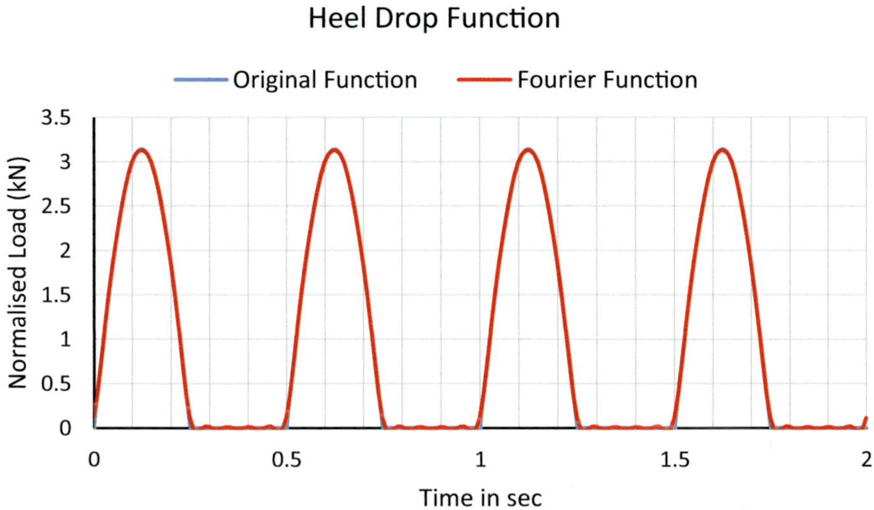

FIGURE 5.5 Typical heel drop functions under simulated Fourier function for a single person with time.

(1987), Murray et al. (1997), and Smith et al. (2009). Smith et al. (2009) adopted $G = 0.2\text{kN/m}^2$ for aerobic or sports events (i.e., 0.25 person/m^2) and 1.5 kN/m^2 for social dancing (i.e., 2 person/m^2) after review. C_e is the dynamic crowd effect which accounts for the lack of perfect synchronization within a crowd. A value of 2/3 has been recommended by Ellis and Ji (1997) for C_e. f_p is the frequency of rhythmic activity. In addition, the values of r_n and ϕ_n for the first six terms in the Fourier Series are tabulated in Table 5.2.

TABLE 5.2

Typical r_n and ϕ_n Values for Crowd Rhythmic Activity

Activity	α	Coeff	$n = 1$	$n = 2$	$n = 3$	$n = 4$	$n = 5$	$n = 6$
Pedestrian movements, low impact aerobics	2/3	r_n	1.286	0.164	0.133	0.036	0.023	0.032
		ϕ_n	−π/6	−5π/6	−π/2	−π/6	−5π/6	−π/2
Rhythmic exercises, high impact aerobics	1/2	r_n	1.570	0.667	0	0.133	0	0.057
		ϕ_n	0	−π/2	0	−π/2	0	−π/2
Normal jumping	1/3	r_n	1.800	1.286	0.667	0.164	0.099	0.133
		ϕ_n	π/6	−π/6	−π/2	−5π/6	−π/6	−π/2
High Jumping	1/4	r_n	1.886	1.571	1.132	0.667	0.269	0
		ϕ_n	π/4	0	−π/4	−π/2	π/4	0

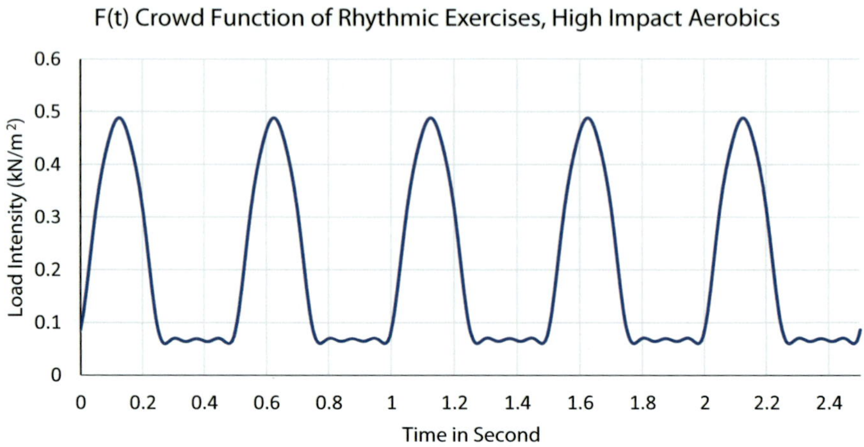

FIGURE 5.6 Plot of F(t) for rhythmic exercises, high impact aerobics.

Taking "Rhythmic exercises, high impact aerobics" as an example, $F(t)$ is plotted in Figure 5.6.

For vibration created by agents other than walking or jumping people, analysis can be directly done to the magnitude and frequency of the vibration source (if known), such as a working machine. These vibrations can be minimized by isolating mounts of motion armrest pads such as those installed for washing machines.

Vibration control is a critical design consideration for ensuring occupant comfort. To mitigate vibrations from human activities like walking or jogging (typically 1.5–2.5 Hz), a structure should have a high natural frequency, ideally exceeding 6 Hz, which can hardly be approached by human walking or jogging activities. This prevents resonance, where vibration amplitudes are amplified.

If analysis predicts excessive vibration, the most effective solution is to increase the stiffness of the structure to raise its natural frequency. However, reducing the mass is often counterproductive; although it raises the natural frequency, the lower mass can increase the acceleration amplitude for a given force, as defined by the principle *acceleration = force/mass.*

For example, a flexible system like a flat slab may be replaced with a stiffer alternative, such as a waffle slab or composite deck system.

Although less common in buildings, tuned mass dampers (TMDs) offer an effective solution. A TMD is a device comprising a mass, spring, and dashpot that dissipates vibrational energy. For effective damping, the TMD is tuned to the natural frequency of the problematic structural mode. While frequently used on flexible, long-span bridges with low frequencies, TMDs can also be applied to building floors.

5.3.2 Strength

Vibrations in a floor can cause instantaneous deflections that exceed those from static loads alone. These deflections become particularly significant when vibrations approach resonance, increasing internal structural forces and demanding greater strength. A detailed analysis can determine the specific strength requirements. Although this increased demand is transient, it allows for the use of reduced load and strength factors in design.

In practical design, especially when the specifics of vibration sources are unknown, building codes often provide conservative guidelines by applying dynamic amplification factors to static loads. For example:

(i) ASCE 7-22 (US) (2022): Specifies a 20% increase for impacts from light machinery and a 50% increase for reciprocating machinery, noting that manufacturer specifications may require higher values.
(ii) AS/NZS 1170.1 (Australia/New Zealand) (2002): Requires factors ranging from 1.1 (for hand-operated cranes) to 1.5 (for reciprocating machines), with factors for lifts reaching up to 2.0.
(iii) National Building Code of Canada (2010): Similarly specifies dynamic factors between 1.1 and 1.5.
(iv) Hong Kong Load Code (2011): Takes a different approach by mandating a blanket additional load of 2.5 kPa for beams and slabs in industrial buildings without specific machinery data. It also requires design for a concurrent horizontal force – not to be considered with wind load – equivalent to 10% of the vertical imposed load acting on a critical set of N floors, where N is at least 0.2 times the total number of floors subject to dynamic effects.

5.4 DESIGN LOAD FOR ROBUSTNESS

Robustness is defined in the Eurocode BSEN1991-1-7:2006 (2000) as the ability of the structure to withstand events like fire, explosions, impact or the consequences of human error, without being damaged to an extent disproportionate to the original cause. Disproportionate collapse is a collapse which is considered to be disproportionate to the cause initiating it. While there is no strict definition for "disproportionate," it is generally assessed on the scale of damage and its consequences. To mitigate this risk, design codes stipulate specific performance criteria. For instance, the former British Standard BS 5950-1 (2000) required that upon the notional removal of a column, pier, or wall, the subsequent collapse should not exceed the lesser of 15% of the floor area or 70 m². The Eurocode BSEN1991-1-7:2006 (2000) maintains a similar provision, with the absolute area limit increased to 100 m².

Many design codes incorporate principles and rules to ensure structures can withstand accidental actions, such as impact, explosion, and progressive collapse. These provisions generally include the following:

(i) Application of a Notional Minimum Lateral Load
A minimum lateral load, often termed a "notional horizontal load," is applied to the building structure. This load is typically distributed across floor levels and is calculated as a small percentage of the building's gravity load. Satisfying this check helps ensure a basic standard of robustness. There are examples as follows:

(a) The former British Standard BS 8110-1:1997 (1998) specified 1.5% of the characteristic dead load for concrete structures.
(b) AS/NZS 1170.0:2002 (2002) (Australia/New Zealand) recommends 1% to 1.5%.
(c) The former British Steel Code BS5950-1:2000 (2000) required 0.5% of the factored vertical dead and imposed load.

Nevertheless, this requirement is considered satisfied if other design lateral loads (e.g., wind or seismic) that exceed the notional load have already been considered.

(ii) Design of key elements

Certain critical components, known as "key elements," must be designed to withstand specific accidental loads. A key element is defined as a structural member upon which the stability of the remainder of the structure depends (e.g., BS EN 1991-1-7 (2006)). Such elements are designed for a defined accidental pressure (e.g., 34 kPa in BS EN 1991-1-7 (2006)) and for impact loads, particularly on columns.

(iii) Provision of Structural Ties

Codes specify tensile forces for which structural "ties" must be designed. These forces are based on the building's gravity loads, floor dimensions, and number of storeys (e.g., BS EN 1992-1-1:2023 (2023)). Ties create continuous linkages throughout the structure and include:

(a) Vertical Ties: Between columns.
(b) Column Ties: Between columns and floor structures.
(c) Internal Ties: Between floor elements.
(d) Peripheral Ties: Within elements along the building's perimeter. In reinforced concrete structures, these ties are provided by tensile reinforcement.

(iv) Ensuring Alternative Load Path

A favoured principle for ensuring robustness is the provision of alternative load paths. This design strategy aims to localize damage and prevent disproportionate collapse by allowing loads to be redistributed if a key structural element is removed.

Fuller discussions on robustness design requirements in terms of structural provisions are given in Chapter 8.

5.5 BRIEF REVIEW OF THE DYNAMIC PROPERTIES OF A BUILDING IN ITS FULL STRUCTURE

Before proceeding to discussions on the application of wind loads and seismic loads on buildings, a brief review of the dynamic properties of buildings is first summarized as follows, as both the wind load and seismic load on a building are very much related to its dynamic properties as responses to the dynamic part of the loads. Fuller discussions with illustrative examples are contained in Chapter 7.

(i) A natural frequency is that with which the building will naturally vibrate when it is excited and left to move freely.
(ii) A building structure comprises many degrees of freedom. Theoretically, each degree of freedom is associated with a natural frequency. The overall free natural movement is the sum of all modes, with the contribution of each mode in accordance with the modal participating mass ratio, i.e., the percentage of mass of the structure moving in the mode.
(iii) Generally, the lowest frequency associated with the mode in the movement is in the direction with the least stiffness and it has the greatest modal participating mass ratio. Reference can be made to Table 7.2 in Chapter 7 for a tube-in-tube building in which the first mode is the movement in the Y-direction, where the building possesses the least stiffness and the modal participating mass ratio is 0.6678. So, although the number of natural frequencies in a building structure is huge, examinations can be confined to the first 20 or so, depending on the complexity of the structure. As a rule of thumb, it would be adequate to select the number of modes up to where the cumulative modal participating mass ratios exceed 95%.
(iv) Under forced vibration due to an external source, the building will vibrate with the frequency of the excitation approximately. But when the external exciting frequency is close to the lowest natural frequencies of the structure, which possess large modal participating mass ratios, resonance will occur, leading to large displacements and subsequently large internal forces.

So, the main theme in the study of the responses of a building structure to dynamic loads due to wind and earthquake lies in the responses of the structure to the external excitation due to the closeness of the structural natural frequencies and that of the external excitation.

5.6 WIND LOAD

Wind load is the force exerted on a structure by moving air that experiences a change in velocity or momentum upon impact. This load is a critical design consideration, particularly for high-rise buildings in regions prone to high-speed winds. Its magnitude is dependent on geographical location, season, and height above ground.

While the specific parameters and methods for calculating wind loads vary by country and are detailed in local design codes, the underlying physical principles are universally consistent. The following sub-sections discuss these fundamental theories and illustrate their applications through examples from various international design codes.

5.6.1 Basic Wind Pressure

By the Bernoulli Theorem, the application to flowing air takes the following form

$$\rho gh + \frac{1}{2}\rho v^2 + p = \text{constant} \tag{5.8}$$

where ρ is the density of air (taken as 1.2 kg/m^3);

g is the acceleration due to gravity;
v is the velocity of the flow;
p is the pressure

ρgh can generally be ignored as wind is considered as a horizontal flow so that a streamline has no variation in h and therefore it can be absorbed in the constant. So Equation 5.8 can be simplified as

$$\frac{1}{2}\rho v^2 + p = constant \tag{5.9}$$

When the wind hits a body, its velocity will change as

$$\frac{1}{2}\rho v_1^2 + p_1 = \frac{1}{2}\rho v_2^2 + p_2 \Rightarrow (p_1 - p_2) = \frac{1}{2}\rho\left(v_2^2 - v_1^2\right) \tag{5.10}$$

At a point where velocity is 0 (stagnation point) on the face of a building where the wind blows on, the change in pressure is

$$p_1 - p_2 = \frac{1}{2}\rho v^2 \tag{5.11}$$

So, it follows that $p_1 - p_2$ is the change of pressure due to wind blowing on a surface. The term $\frac{1}{2}\rho v^2$ can therefore be taken as the "basic wind pressure."

However, the expression $\frac{1}{2}\rho v^2$ can only be considered a "nominal" or "reference" wind pressure. This is because the equation assumes a uniform, laminar airflow and does not account for turbulence or the building's shape, which causes wind velocity to vary across its surface. The design wind pressure is typically calculated by multiplying this reference value by several factors that account for effects including shape, topography, sheltering, size, and dynamics. This final pressure is then multiplied by the building's exposed surface area to determine the total wind load. So far the wind load is taken as a static load for design purposes, although wind load is fundamentally a dynamic pressure.

5.6.2 Determination of Wind Speed

The design wind pressure for a structure is primarily based on the highest expected wind speed for its location, whether caused by typhoons, monsoons, or other storm systems. As detailed in Section 5.6.3, wind speed increases with height; therefore, the design wind speed specified in building codes must be referenced to a standard height. Common reference heights include 10 metres (as in Eurocode BS EN 1991-1-4:2005 (2005)), the height of a meteorological station, or the "gradient height," the altitude at which wind speed ceases to increase.

National codes typically determine design wind speeds through a statistical analysis of historical wind data. The results of this analysis, often presented as design wind speeds or pressures, are then stipulated in the code for engineering use.

Wind speed is a random process, and its reported value is highly sensitive to the averaging period used in its measurement. The "mean wind speed" is commonly defined as the maximum average over consecutive one-hour intervals, yielding an "hourly mean speed." Other standard averaging periods include 10-minute means and shorter durations like a "3-second gust speed." Unsurprisingly, a 3-second gust is significantly higher than an hourly mean speed. It is important to note that gust speeds are sometimes derived statistically from longer-duration mean speeds rather than from raw, high-frequency data.

For such a statistical determination of the gust wind speed, it should be the mean wind speed plus speed fluctuations, which are related to the standard deviation σ by an expected peak factor g_v as

$$v_{3sec-gust} = v_{mean} + g_v \sigma \tag{5.12}$$

A parameter known as "turbulence intensity," defined as $I_z = \frac{\sigma}{v_{mean}}$ is introduced.

$$v_{3sec-gust} = v_{mean}\left(1 + g_v I_z\right) = v_{mean} G \tag{5.13}$$

The turbulence intensity I_z generally decreases with height above ground. As detailed in Section 5.6.3, this occurs because ground friction slows down the wind speed and generates greater turbulence closer to the ground surface. So, the G factor also depends on height. g_v is a statistical factor used to convert mean wind speed to 3-second gust, with a value ranging between 3.4 and 3.7. This methodology has been adopted by the Australian/New Zealand Wind Code AS/NZS 1170.2:2021.

ASCE 7-22 (2022), nevertheless, contains a chart by Durst (1960) for the "gust factor" for direct conversion of hourly mean wind speed to gust wind speed of various gust durations which is reproduced in Figure 5.7. From the curve, the G value for the 3-second gust is about 1.53.

In contrast, ASCE 7-22 (2022) provides a different approach based on Durst's (1960) chart for a "gust factor." This chart, reproduced in Figure 5.7, enables the direct conversion of an hourly mean wind speed to gust wind speeds of various durations. According to the curve, the G value for a 3-second gust is approximately 1.53.

National wind codes use different averaging periods to define their basic wind speed, such as the hourly mean (ASCE 7-16 (2018)), the 10-minute mean (Eurocode BSEN1991-1-4 (2005)), or the 3-second gust (AS/NZS 1170.2:2021). While the hourly and 10-minute mean speeds are relatively similar, both are substantially lower than the 3-second gust speed. The final design wind pressures are derived by applying

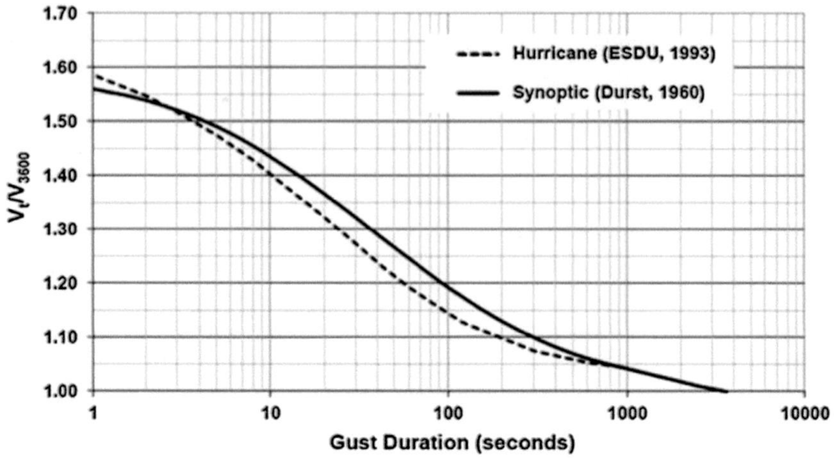

FIGURE 5.7 Durst curve, maximum wind speed averaged over *t* second to hourly mean speed extracted from ASCE 7-16 (2018).

various factors, such as those for size and dynamics (detailed in Section 5.6.8), to these base values, ensuring a realistic wind load is used for design.

Another critical temporal consideration is the "return period." Since wind speed data is collected over many years, a statistical analysis is required to determine a speed with a low probability of being exceeded during a building's design life. This probability is expressed as a return period. For a building with a 50-year design life, a wind speed with an N-year return period has a 1/N annual probability of being exceeded. Therefore, the probability that this speed will be exceeded at least once during the building's 50-year lifespan is calculated as follows:

$$p = 1 - \left(1 - 1/N\right)^{50} \tag{5.14}$$

So, if the design return period is 50 years, the wind speed is with probability of being exceeded at $p = 1 - \left(1 - 1/50\right)^{50} = 0.635$.

An approach for calculating wind speed at return period of N years is outlined in Simiu (1996) by a formula

$$\widehat{v_N} = \overline{X} + 0.78\left(\ln\left(N\right) - 0.577\right)s \tag{5.15}$$

where $\widehat{v_N}$ is the peak wind speed at return period of N years;

$\overline{X}$ is the mean value of the highest yearly wind speeds of the past years;

s is the standard deviation of the highest yearly wind speeds of the past years.

The determination process is an "extreme value" analysis, by which the wind speeds are considered to be falling into the Gumbel distribution. The values are more critical than those of normal distributions.

For the limiting 10% and 2% of exceedance, the return periods are respectively 475 years and 2475 years.

Consider 38 years of mean wind speeds in km/hr

102, 108, 128, 143, 135, 115, 144, 139, 131, 133, 135, 111, 127, 144, 135, 102, 105, 106, 133, 131, 116, 129, 111, 150, 104, 132, 147, 129, 144, 107, 139, 132, 145, 98, 134, 107, 124, 149

By statistical analysis of the above value, the mean value is $\overline{X} = 126.421\text{km/hr}$ and the standard deviation is $s = 15.623\text{km/hr}$.

So, by Equation 5.12, the wind speed at 50, 475, and 2475 years are respectively

$$\widehat{v_{50}} = 126.421 + 42.75 = 167.062\text{km/hr};$$

$$\widehat{v_{475}} = 118.29 + 71.62 = 194.497\text{km/hr};$$

$$\widehat{v_{2475}} = 118.29 + 92.78 = 214.613\text{km/hr}.$$

ASCE 7-10 (2010), taking 50 years as the nominal return period under its C26.5.1, lists the following formula (from Peterka and Shahid (1998) for converting wind speed of return period T from that at 50 years as

$$\frac{V_T}{V_{50}} = 0.36 + 0.1\ln(12T) \tag{5.16}$$

Equation 5.16 produces results slightly different from that by Equation 5.13.

Similarly, the Eurocode EN1991-1-4:2005 (2010) the NBCC (2010) and the former Hong Kong Wind Code 2004 (2004) adopt 50 years return period. The adoption of the return period depends on the risk to be allowed.

5.6.3 Wind Speed Variation with Height

As wind blows close to the ground surface, the ground roughness retards the wind speed. This retarding effect, however, diminishes with height, leading to increases in wind speed. While wind speed is 0 at the ground surface, there is a certain height where the influence by the ground roughness becomes negligible. Beyond this height, which is termed "gradient height" (z_g), the wind speed no longer increases and remains constant at "gradient wind speed" (v_g).

A common mathematical method has been used to simulate the wind speed profile with height, which is the power law:

$$v_z = v_g\left(\frac{z}{z_g}\right)^{\alpha} \quad 0 < z < z_g; \quad 0 < \pm < 1 \tag{5.17}$$

TABLE 5.3
Relevant Parameters Related to Terrain Categories by BSEN1991-1-4 (2010)

Terrain Category	0	I	II	III	IV
z_0 (m)	0.003	0.01	0.05	0.3	1.0
z_{min} (m)	1	1	2	5	10

Another mathematical simulation is by the logarithm law as

$$v_z = \frac{1}{k} v_* \ln\left(\frac{z}{z_0}\right) \tag{5.18}$$

where v_* is the friction velocity,

k is termed the "Von Karman constant" approximately equal to 0.41;
z_0 is the surface roughness in metre.

The Eurocode BSEN1991-1-4 (2010): introduces a "roughness factor" $c_r(z)$ to account for the variability of wind speed due to height above ground and the ground roughness defined by "Terrain Category." $c_r(z)$ follows logarithm law as $c_r(z) = k_r \ln(z / z_0)$ for $z_{min} \le z \le z_{max}$ and $c_r(z) = c_r(z_{min})$ for $z \le z_{min}$ where z_{max} is taken as 200 m and $k_r = 0.19(z_0 / z_{0,\text{II}})^{0.07}$. The parameters are related to "Terrain Categories" defined in table 4.1 of the Eurocode, summarized as in Table 5.3. Terrain category in the Eurocode is with the smallest values for Category "0" for open sea and "IV" for terrain well covered (at least 15%) by buildings, as described in table 4.1 of the Eurocode. So, the terrain categories 0 to IV are used to measure the increasing ground roughness as provided to reduce the wind speed and subsequently wind pressure.

The wind mean speed at height z is taken as $v_m(z) = c_r(z) c_o(z) v_b$ where $c_o(z)$ is taken as 1 unless there are significant topography effects discussed in Section 5.6.5 and v_b is the basic wind velocity defined as the fundamental basic wind velocity modified to account for the direction of the wind being considered and the season (if required). v_b will be specified by the National Annexes of individual nations in Europe. An illustration of the wind speed profile is given in Section 5.6.5 with $c_o(z) = 1$ as compared with another $c_o(z) > 1$ under topography effect.

Another example is the Hong Kong Wind Code 2004 (2004) which has used the power law to simulate the hourly mean wind speed with $v_g = 59.5\text{m/sec}$; $z_g = 500\text{m}$ and $\alpha = 0.11$ as $v_{mean} = v_g (z / z_g)^{0.11}$, which is explained in the Code's Explanatory Materials (2004). The effect of different ground roughness of different terrains is not considered, and in fact, the unique profile is based on open sea. If the wind speed profile is, however, simulated by the logarithm law, then $v_* = 24.6\text{m/sec}$, $z_0 = 0.0246\text{m}$.

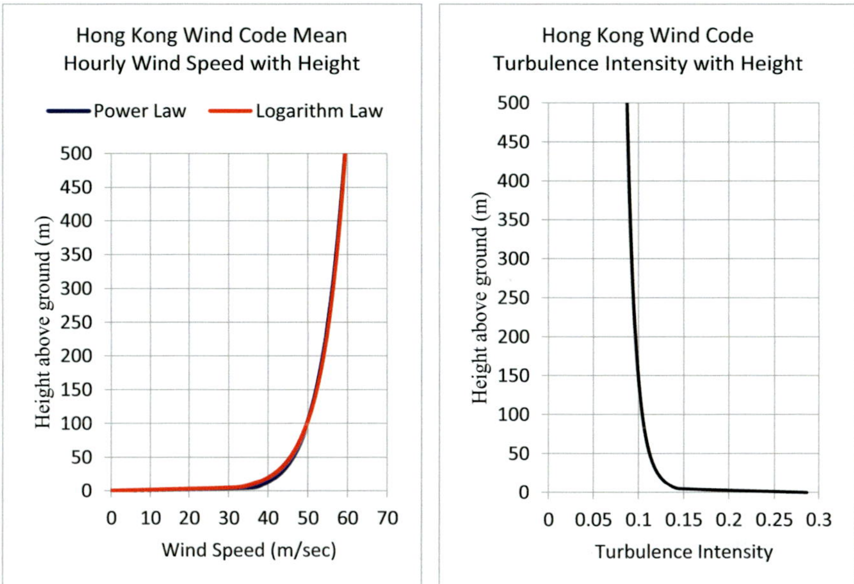

FIGURE 5.8 Variations of mean hourly wind speed and turbulence intensity with height for the former Hong Kong Wind Code 2004.

As illustration of the applicability of the power law and logarithm law, the wind speed profiles based on the power law and simulation by the logarithm law for the Hong Kong Wind Code are shown in Figure 5.8. The turbulence intensity profile is simulated by the equation, $I_z = I_g \left(z / z_g\right)^{-0.11}$ where $I_g = 0.087$ which is the turbulence intensity is at the gradient height. It should be noted that although the current Hong Kong Wind Code 2019 (2019) and its explanatory notes (2019) have not explicitly stated the wind speed used for design purposes, their basic wind pressure profile which is in 3-sec gust pressures follows very closely to that of Explanatory Materials (2004) to the 2004 Wind Code which is $v_{gust} = v_{mean}\left(1 + 3.7I_z\right)$, implying that the wind speeds of the codes are very similar.

5.6.4 Force coefficients

The shape of a building significantly influences how wind speed and direction change around it, resulting in varying surface pressures, as discussed in Section 5.6.1. For practical design, national codes of practice provide standardized force coefficients C_f for different building shapes. These coefficients generally increase with the building's aspect ratio (its height-to-depth ratio in the wind direction). They also increase with the plan ratio (the length-to-depth ratio, where length is perpendicular to the wind direction). All codes of practice for assessment wind loads have provisions of force coefficients for common shapes of buildings and structural elements including

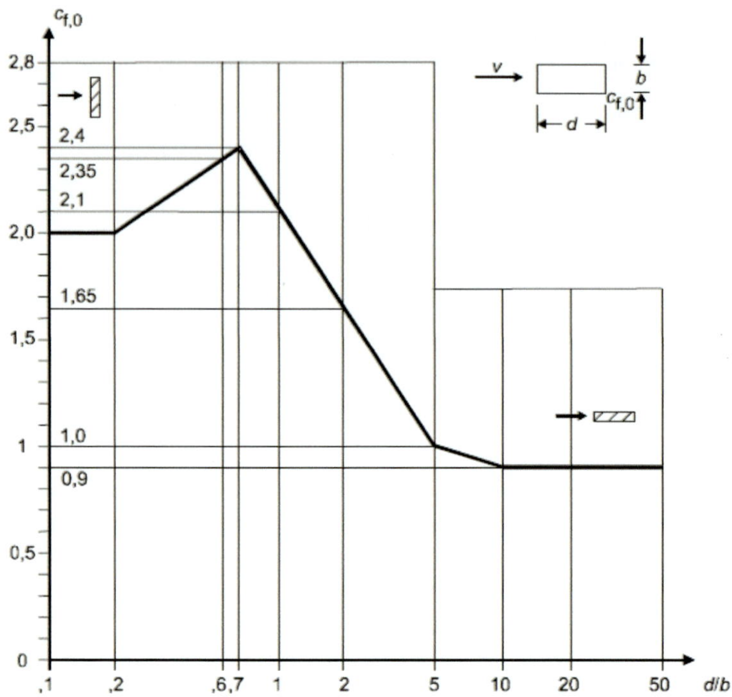

FIGURE 5.9 Extraction from Figure 7.23 of BSEN1991-1-4:2005 showing force coefficients $C_{f,0}$ of rectangular sections with sharp corners and without free end flow.

the Eurocode BSEN1991-1-4:2005 2010), the American Code ASCE 7-22 (2022) and the Hong Kong Wind Code 2019 (2019).

The Eurocode BSEN1991-1-4:2005 (2010) figure 7.23 shows the force coefficient $C_{f,0}$ for buildings with a rectangular section extracted in Figure 5.9 as follows. The chart contains a condition without free end flow.

"Free end flow" in wind engineering refers to the complex, 3-dimensional flow patterns at the edges or corners of a building at which the wind is no longer constrained to flow in a predictable path parallel to the surface and can instead wrap around the edge, forming powerful vortices (swirling motions). The Eurocode BSEN1991-1-4:2005 uses ψ_r and ψ_λ to account for the corner and free end effects so that $C_f = C_{f,0}\psi_r\psi_\lambda$. Readers may refer to the Eurocode for details.

In addition, BSEN1991-1-4:2005 (2010) table 7.11 shows the force coefficient $C_{f,0}$ for buildings with regular polygonal section and 7.9.2/7.9.3 for circular section under Reynold's number defined in 7.9.1 of the Eurocode. Readers may refer to the Eurocode for details.

In case of irregular shape, a designer may, based on the pressure coefficients defined in section 7, calculate the pressures on various parts of the exposed areas of the buildings. The total force coefficient can be worked out by integrating the forces summed by pressures times the respective tributary areas.

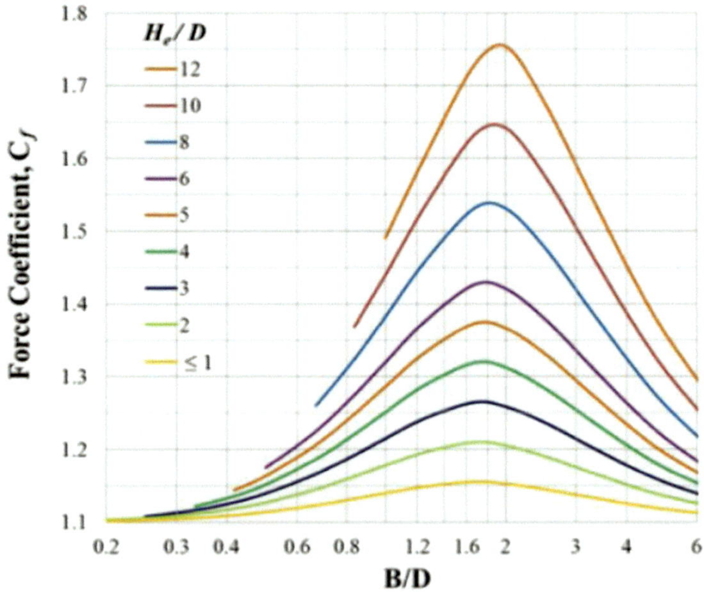

FIGURE 5.10 Force coefficient variation with He/D and B/D ratios in accordance with Hong Kong Wind Code 2019.

The Hong Kong Wind Code 2019 (2019) has outlined a precise approach for determining force coefficient C_f for some common plan shapes for buildings including rectangular, U-, X-, L-, Y-, double Y-plan shapes. Each of these plans is reduced to two dimensions which are B as the width exposed and normal to the applied wind load and D in the direction of the wind. D is taken as either as the average depth of the building plan or$B/1.8$. With B and D and the effective height of the building H_e , C_f can be determined in accordance with a chart or a detailed formula extracted from (2019) in Figure 5.10 and Equation 5.19. Furthermore, the Explanatory Notes of the Wind Code (2019) extend the concept for irregular plan shapes by circumcising the plan by the smallest rectangle of size $B \times D$ with $D \leq B/1.8$.

$$C_f = \frac{0.055H_e/D}{\exp\left\{\left|\log_e\left[(0.6B/D)(1-0.011H_e/D)\right]\right|^{\left[1.7-0.0013(H_e/D)^2\right]}\right\}} \tag{5.19}$$

5.6.5 Topographical Factor

The local topography surrounding a building can significantly influence the wind loads acting upon it. While accurate assessments can be obtained through wind tunnel testing or computational fluid dynamics (CFD) software, some codes provide empirical methods for buildings situated on hills, ridges, or escarpments. A general principle is that wind speed increases as it flows up a slope due to a reduction in

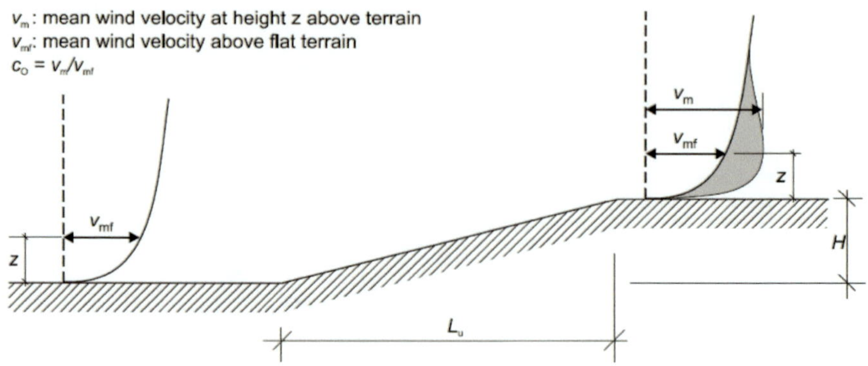

FIGURE 5.11 Topographical effect of wind (extract from figure A.1 of BSEN1991-1-4:2005).

the effective cross-sectional area, known as topographic acceleration, leading to an increase in wind pressure, as illustrated in Figure 5.11 (extracted from BSEN1991-1-4:2005 (2010)).

The Eurocode EN1991-1-4:2005 (2010) uses the term "orographic" effects and provides formulae and charts to account for the topographic effects, as similar to the former British Code BS6399-2:1995 (1995). Other codes have similar formulae or charts and include provisions for exemption of topographic effects for locations (below mid-height of the slope) termed "topographic insignificant zone" where the consideration of topographical effect can be ignored. A chart for determination the orographic effect and the defining parameters are extracted in Figure 5.12 for illustration.

A Worked Example 5.1 is used to demonstrate the determination of the topographical effect on a building at the crest of an escarpment as follows. Reference should be made to Section 5.6.3 for the determination of basic wind speeds and pressures.

Worked Example 5.1

A Building of 100 m tall stands on the crest of an escarpment of Terrain Category II as shown in Figure 5.12. Other parameters are

$L_u = 100\text{m}$; $H = 30\text{m}$; $\Phi = 30/100 = 0.3$; $z_0 = 0.05\text{m}$; $z_{min} = 2\text{m}$

The effective length of the upwind slope is $L_e = L_u = 100\text{m}$,

The basic wind pressure v_b is taken as 50 m/sec.

As Terrain Category II, the terrain factor $k_r = 0.19\left(z_0 / z_{0,\text{II}}\right)^{0.07} = 0.19$

With $\Phi = 0.4 > 0.3$, the orography factor, $c_o = 1 + 0.6s$. At various heights of z / L_e, s is read from Figure 5.12 and c_o , wind speed (in m/sec) and pressure (in kN/m^2 determined by $0.5\rho v^2$ where ρ is the density of air at 1.2 kg/m^3) are summarized in the following Table 5.4. Wind speeds and pressures are plotted in Figure 5.13.

Figure 5.13 shows that the topography effect increases wind speed and pressure, but at a decreasing rate with an increase in level. Nevertheless, the Hong Kong Wind Code (2019) has simplified the provision by applying a single factor to increase wind

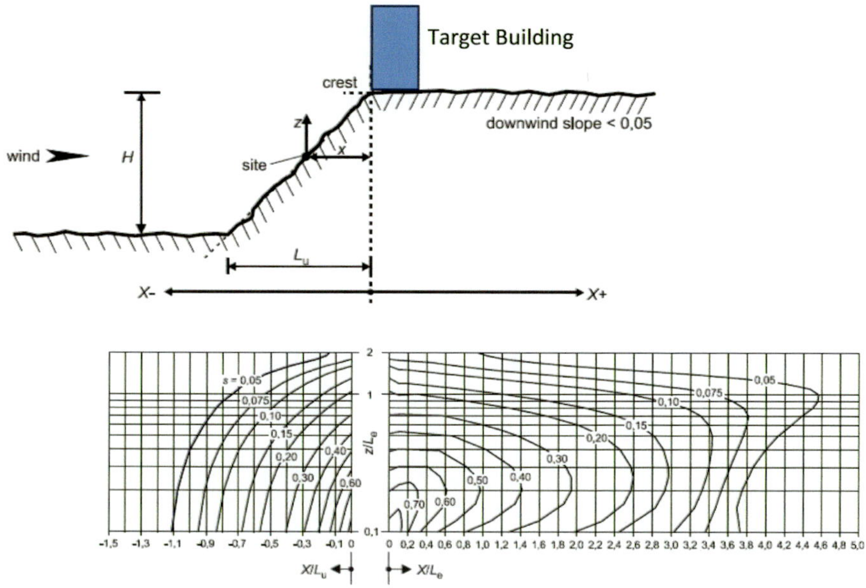

FIGURE 5.12 Factors for cliff and escarpment (extract from figure A.2 of BSEN1991-1-4:2010).

speed and pressures at all levels of a building, which is similar to the former British Standard BS6399-2:1997 (2002).

5.6.6 Directionality of Wind Load

Statistically, wind speeds are different in different directions for the same geographic locality due to a combination of factors including local topography, terrain, temperature differences, and large-scale weather systems such as fronts and storms. In addition, there is a reduced probability by which the direction of the maximum wind aligns with the building structure's most unfavourable direction in resisting wind loads.

The Eurocode BSEN1991-1-4 (2010) provides a direction factor c_{dir} to modify the basic wind speed. It leaves it to individual nations to decide in their national annexes. The UKNA of the Eurocode (2010) lists the factors from 0.73 at 30°, 60°, 120° (measuring clockwise from north) to 1 at 240°. AS/NZS1170.2 (2002) allows the use of wind direction multiplier M_d for 8 directions (N, NE, E, SE, S, SW, W, NW) in different geographic regions of Australia and New Zealand, again with values from 0.75 to 1. ASCE 7-22 (2022) includes a "directionality factor" k_d with tabulated values for different structure types with a nominal value of 0.85. In ASCE 7-22 (2022), k_d only depends on the structure type and not the geographic locality. According to ASCE 7-22 (2022), the factor accounts for two effects: (1) the reduced probability of maximum winds coming from any given direction; and (2) the reduced probability of the

TABLE 5.4
Summary of Results in Worked Example 5.1

z(m)	0	2	10	20	30	40	50	60	70	80	90	100
z/L_e	0	0.02	0.1	0.2	0.3	0.4	0.5	0.6	0.7	0.8	0.9	1
s	1.2	0.85	0.7	0.59	0.49	0.41	0.35	0.3	0.26	0.23	0.205	1.2
$c_r(z)$	0.701	0.701	1.007	1.138	1.215	1.270	1.312	1.347	1.376	1.402	1.424	1.444
$v_m(z)$	35.044	35.044	50.334	56.919	60.771	63.504	65.624	67.356	68.820	70.089	71.208	72.209
$c_o(z)$	1.510	1.720	1.510	1.420	1.354	1.294	1.246	1.210	1.180	1.156	1.138	1.123
$v_m(z)c_o(z)$	60.276	60.276	76.004	80.825	82.284	82.174	81.767	81.500	81.208	81.023	81.034	81.090
p_m	0.737	0.737	1.520	1.944	2.216	2.420	2.584	2.722	2.842	2.947	3.042	3.128
p_o	2.180	2.180	3.466	3.920	4.062	4.052	4.012	3.985	3.957	3.939	3.940	3.945

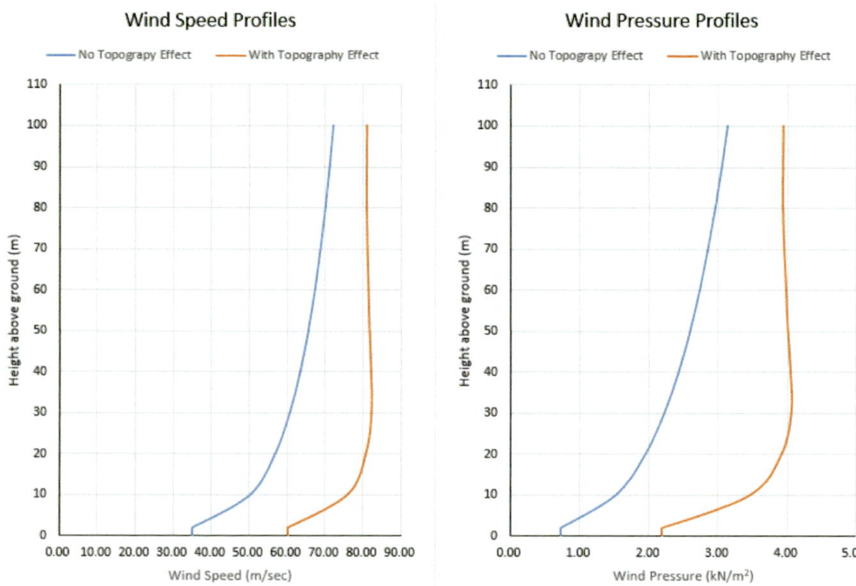

FIGURE 5.13 Wind speed and wind pressure profiles in Worked Example 5.1.

maximum pressure coefficient (related to the structure) occurring for any given wind direction. The Hong Kong Wind Code 2019 (2019) similarly adopts a directionality factor that is different in 8 directions (N, NE, E, SE, S, SW, W, NW) with the maximum value also at 0.85.

These codes, nevertheless, require the factors to be at unity for circular buildings which are axisymmetrical or almost axisymmetrical such as the octagonal shapes (ASCE 7-16 (2018)) where the scenario of maximum wind not aligning with the building structure's most unfavourable direction is not applicable.

5.6.7 Sheltering Effect

Wind on a building can be effectively shielded off or reduced by adjacent buildings, which is considered as "sheltering effect." The Eurocode BSEN1991-1-4:2005 (2010) and the former British Wind Code BS6399-2:1997 (2000) adopt the "displacement height" concept, by which the wind is considered to behave as if the ground is raised to a height (known as displacement height) just below the average height of the roofs of the surrounding buildings. The determination of displacement height for the target building depends on the plan density and average surrounding building height of the adjacent buildings. The Hong Kong Wind Code 2019 (2019) comprises a more precise procedure to determine the height of reduction (displacement height) and subsequently the reduced wind pressure on the target building. Basically, it considers various directions with sheltering effects of the sector and accounts for the effects in a statistical manner. The reader may refer to the Code for further information.

5.6.8 Size and Dynamic Factor

As wind blowing is a random action, the chance of having the greatest wind pressure on the entire exposed area decreases with the increase of the exposed area. It follows that the application of design wind pressure can be reduced if the exposed area is large with the application of a "size factor." The Eurocode BSEN1991-1-4:2005 (2010) gives a formula for the size factor as

$$c_s = \frac{1+7I_v(z_s)\sqrt{B^2}}{1+7I_v(z_s)} \quad (5.20)$$

where $I_v(z_s)$ is the turbulence intensity at the height z_s; and

$$B^2 = \frac{1}{1+0.9\left(\dfrac{b+h}{L(z_s)}\right)^{0.63}} \quad (5.21)$$

with b and h as the breadth and height of the exposed surface and $L(z_s)$ as the turbulence length scale defined in the Eurocode. It follows B^2 and subsequent c_s decrease with increase of the building surface dimension b and h.

In addition, the structure may respond to the wind turbulence in resonance. So, the Eurocode introduces another "dynamic factor" which takes into account the increasing effect in resonance of the structure with the natural wind as

$$c_d = \frac{1+2k_pI_v(z_s)\sqrt{B^2+R^2}}{1+7I_v(z_s)\sqrt{B^2}} \quad (5.22)$$

Readers may refer to the Code for detailed determination of k_p and R^2. Briefly speaking, k_p is a statistical factor close to Equation 3.7 and R^2 involves "power spectrum density function" of the wind, which is the distribution of wind energy over frequencies and the fundamental frequency of the building.

The Eurocode combines the two factors into c_sc_d known as the structural factor as

$$c_sc_d = \frac{1+2k_pI_v(z_s)\sqrt{B^2+R^2}}{1+7I_v(z_s)} \quad (5.23)$$

Readers may refer to the Eurocode for the details in the determination of the c_sc_d factor. Nevertheless, a demonstration of the determination of the factor is included in Chapter 13, in which the design of a hypothetical building is found. In addition to the recommendations by the Eurocode, modifications by the National Annex are allowed.

There are similar factors given in other wind codes.

The Hong Kong Wind Code 2019 (2019) gives the following formula to account for the size and dynamic effects together:

$$S_{q,h} = 0.5 + \sqrt{\left(S_{s(L_{0.5p=B})} - 0.5\right)^2 + \frac{0.25}{B^{0.5} H N_x^{\,2} \xi_x}} \tag{5.24}$$

where $S_{s(L_{0.5p=B})}$ is the size factor defined by the Code with half perimeter equal to the width B of the area under wind load; H is the height of the building, N_x is the fundamental natural frequency of the building aligned with the along wind direction and ξ_x is the damping ratio of the building ranging from 0.08 to 0.3 (depending on the aspect ratio of the building and the construction material of the building). The reader may refer to the Code for detailed application. So, it follows that the size effect by the first term inside the square root and the dynamic involving natural frequency of the building in the second term inside the square root have been taken into account.

The third example raised is the Australian/New Zealand Wind Code AS/NZS1170.2:2021 (2002) which gives a "dynamic response factor" as in Equation 5.25

$$C_{dyn} = \frac{1 + 2I_h \sqrt{g_v^{\,2} B_S + \dfrac{H_s g_R^{\,2} S E_t}{\zeta}}}{\left(1 + 2 g_v I_h\right)} \tag{5.25}$$

In Equation 5.25, $g_v^{\,2} B_S$ measures the effect due to the size of the exposed area by the product of the factor B_S involving the size of the exposed face under the modification by the statistical factorg_v which decreases with the increase of area of the exposed face. The term $\frac{H_s g_R^{\,2} S E_t}{\zeta}$ measures the resonant response with E_t as a measure of the spectrum of turbulence of the approaching wind and S involves the natural frequency of the building. Again, the readers may refer to this code for the details of the equation.

E_t in Equation 5.25 is the spectrum of turbulence which describes how turbulent wind energy is distributed across different frequencies given by $E_t = \pi N / \left(1 + 70.8N^2\right)^{5/6}$, where $N = n_a L_h \left(1 + g_v I_h\right) / V_{des,\theta}$ with n_a as the natural frequency of the building, $L_h = 85\left(h/10\right)^{0.25}$ ($h$is the height of the building), $V_{des,\theta}$ is the design wind speed at the orientation θ, $I_h$as the turbulence intensity. For a building with height at $h = 120$m and design wind speed at building height $V_{des,\theta} = 50.8$m/sec under a turbulence intensity $I_h = 0.1022$ and $L_h = 85\left(h/10\right)^{0.25} = 158.2$m, a plot of E_t with n_a is given as follows in Figure 5.14.

The plot shows a maximum E_t when the building natural frequency is at 0.034 Hz and the factor decreases more sharply at higher frequencies. As most buildings have natural frequencies greater than 0.1 Hz, the chart demonstrates that generally flexible buildings with low natural frequencies will be more dynamically susceptible to resonance by wind loads and thus the design wind loads are higher.

Other codes give similar formulation. The factor is one of the factors multiplied to the basic wind pressures to account for the size and resonant effects.

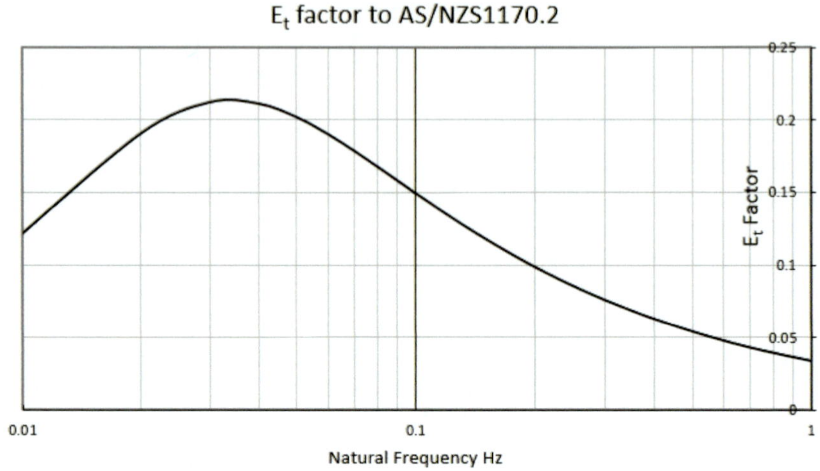

FIGURE 5.14 Plot of variation of E_t factor with natural frequency of building at selected parameters.

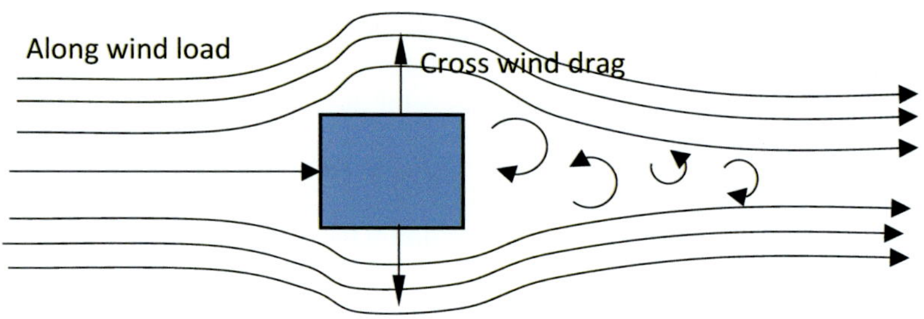

FIGURE 5.15 The phenomenon of cross-wind effects.

5.6.9 Cross-Wind Effect

When wind flows across a building, alternating vortices may be formed behind the building as shown in Figure 5.15. The vortices would create a lateral push and pull on the building inducing both lateral forces and vibrations.

Various empirical formulae for assessing cross-wind effects, primarily concerning moment and acceleration, are provided in building codes. Key design parameters include building dimensions, wind speed, turbulence intensity, and the structure's natural frequency. For very slender buildings, cross-wind effects can exceed those in the along-wind direction, making them the critical design case. A detailed account of these methods can be found in the Hong Kong Wind Code 2019 (2019).

5.6.10 Torsional Effect

Wind load is a random process that acts unevenly on a structure's exposed surfaces. Though the uneven application reduces the net load, torsional moments are induced.

Even under these reduced loads, torsion can be a critical design consideration for buildings that are not torsionally strong.

Consequently, most wind codes mandate a torsional analysis for buildings exceeding certain heights or lacking torsional strength. The specific methodologies, however, vary:

(i) ASCE 7-22 (2022): Requires analysis for torsion caused by an eccentricity of 0.15 times the building width, applied concurrently with a 25% reduction in the direct wind load (figure 27.3-8 of ASCE7-22). This check can be exempted for "torsionally regular" structures. According to its 26.2, a building qualifies as torsionally regular if, under the reduced load with the specified eccentricity, the deflection at any point on any storey does not exceed the deflection caused by the full, unreduced wind load without torsion.

(ii) Hong Kong Wind Code 2019 (2019): Requires checking for torsion due to eccentricities ranging from 0.05 to 0.2 times the building width, depending on the plan's aspect ratio. Exemptions are granted for single-storey buildings under 10 m, buildings under 70 m with a peripheral lateral system, or buildings where the inter-storey drift under torsion is less than 25% of that under full lateral load.

(iii) AS/NZS 1170.2:2021 (2021): For buildings taller than 70 m (Section 2.5.4), requires design for a torsional moment due to a 0.2 times width eccentricity with no reduction in lateral load and offers no exemptions for torsional regularity. Furthermore, it cautions that for buildings with a plan aspect ratio exceeding 1.5, torsion is primarily generated by cross-wind forces, necessitating specialist advice.

5.6.11 Design Wind Load Directions

Since wind can originate from any direction and a building's structural response varies accordingly, the design must account for loads from critical wind directions. For practical purposes, analysis is typically performed for winds aligned with the building's principal axes. As illustrated in Figure 5.16 for rectangular and L-shaped

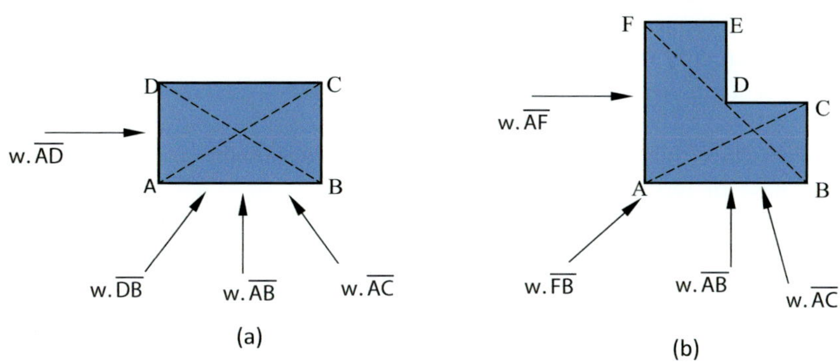

FIGURE 5.16 Examples of design wind load in the principal directions.

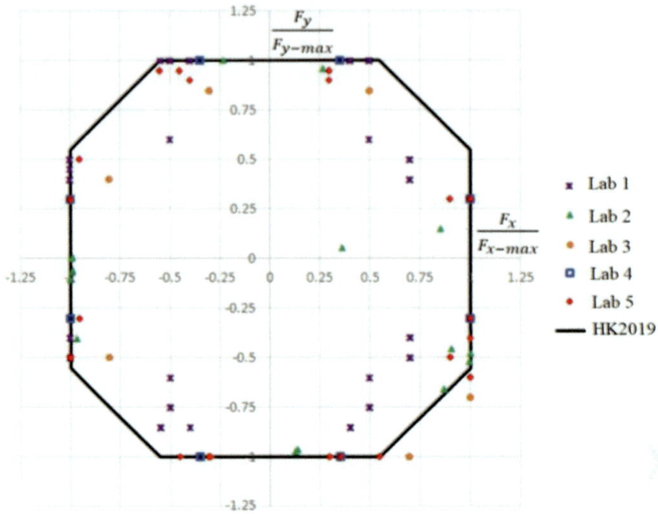

FIGURE 5.17 Translational wind load cases by the Hong Kong Wind Code 2019.

buildings, these directions are considered reversible (e.g., north-south also implies south-north). In the figure, w denotes the wind load per unit width, and dimensions like $\overline{AD}$ represent the exposed length of the building face. Therefore, the total wind load on a face is calculated as the product w.$\overline{AD}$.

The Australian/New Zealand Standard AS/NZS1170.2:2021 (2002) requires wind actions to be considered for no fewer than four orthogonal directions aligned with the structure's principal axes. Consequently, in Figure 5.16(a), wind loads such as w.$\overline{DB}$ and w.$\overline{AC}$ may be eliminated, as they are not aligned with these required orthogonal directions.

In contrast, the Hong Kong Wind Code 2019 (2019) stipulates a more comprehensive analysis, requiring wind loads to be checked in eight directions. These directions are derived from two orthogonal components, WX and WY, and are defined by the combinations ±WX±0.55WY and ±0.55WX±WY. This approach, based on extensive wind tunnel testing, defines an octagonal envelope of load cases. As illustrated in Figure 5.17 (extracted from the Code's Explanatory Handbook (2019)), this octagon effectively encompasses the critical wind load results for complex structural responses.

Under further combinations of torsions in opposite directions and with a further consideration of torsion (ΔTZ) dominating cases as ±0.55WX±0.55WY±ΔT, the maximum number of wind load cases can rise to 24.

5.6.12 Acceleration

The dynamically resonant component of wind loading can induce periodic lateral or torsional accelerations in a building. These accelerations may cause discomfort to occupants, particularly on the upper floors. The principle for estimating this along-wind acceleration is analogous to that of simple harmonic motion, given by:

$$a = \omega^2 \Delta \tag{5.26}$$

where $\omega = 2\pi f$ and f is the natural frequency of the building and Δ is the "resonant part" of the displacement.

This principle forms the basis of the acceleration expression provided in the Australian/New Zealand Wind Code AS/NZS1170.2:2021 (2002). Consider a cantilever of height h and with an end mass m under a lateral point load at the top and that $\omega^2 = K / M$ where K is the stiffness. As $\Delta = F / K$ where F is the lateral force, the base moment is $M = Fh$, the acceleration is then $a = \omega^2 \Delta = \frac{K}{m}\frac{F}{K} = \frac{K}{m}\frac{M}{Kh} = \frac{M}{mh}$.

But the building is with a continuous mass per unit height m_0 where $m = m_0 h$. So, the Code assumes only 1/3 of the mass is accounted for acceleration in the top floor; the Code lists the equation in its E.2 as follows :

$$a = \frac{M}{\frac{1}{3}m_0 h \times h} = \frac{3M}{m_0 h^2} \tag{5.27}$$

But, as it is the resonant part of the acceleration under concern, the moment M as the total base moment of the building due to wind should be taken from the part of the resonant component due to wind load only. So, following Equation 5.25, it should be the part of $2I_h\sqrt{\frac{g_R{}^2 SE_t}{\zeta}} = 2I_h g_R\sqrt{\frac{SE_t}{\zeta}}$ out of the total of $1 + 2I_h\sqrt{g_v{}^2 B_S + \frac{H_s g_R{}^2 SE_t}{\zeta}}$,

in which 1 is the static component and $2I_h\sqrt{g_v{}^2 B_S}$ is the background component.

Beyond the along-wind acceleration associated with lateral displacement and bending, buildings can also experience significant cross-wind acceleration. This orthogonal effect, discussed in Section 5.6.9, arises from mechanisms like vortex shedding. Major international wind codes, including AS/NZS 1170.2:2021 (Australia/New Zealand) (2021), NBCC (Canada) (2010), and the Hong Kong Wind Code 2019 (2019), provide specific formulae for its estimation. Similar to the along-wind case, cross-wind acceleration is derived from the cross-wind moment. For detailed methodologies, designers should consult the relevant sections of these codes.

5.6.13 Design Wind Loads Application on Buildings

Once all relevant factors have been determined, the final wind pressure profiles can be established. For convenience in structural analysis, these pressures are often converted into equivalent loads on the structural frame. This typically involves calculating point loads and moments at each floor level, or line loads (uniformly distributed or linearly varying) applied to beams and slabs, as illustrated in Figure 5.18. This conversion is achieved by multiplying the wind pressure by the corresponding tributary area for point loads or the tributary width for line loads.

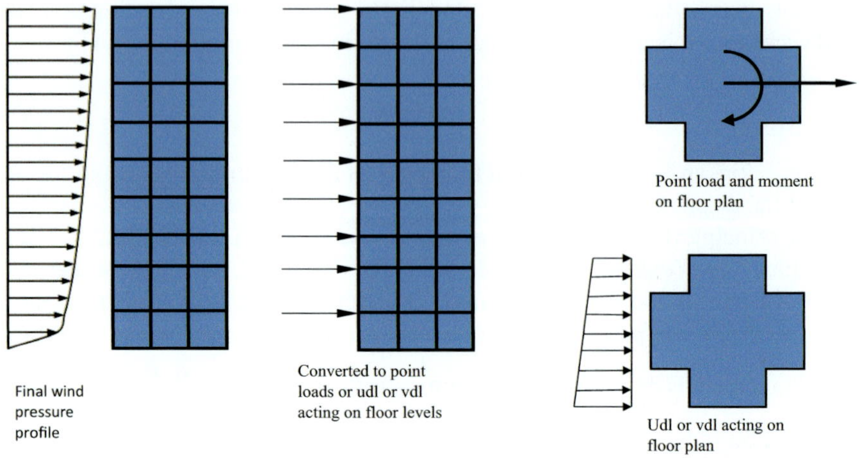

FIGURE 5.18 Conversion of design wind pressure into point or line loads on buildings.

5.6.14 Wind Loads on Building Elements

Elements in a building directly exposed to wind, including facades, signposts, canopies, and roofs, require specific wind load calculations in their design. These loads are primarily a function of the element's shape and location. Wind codes often simplify this process by providing "pressure coefficients." The design pressure is found by multiplying these coefficients by an adjusted wind pressure (accounting for sheltering, topography, and direction). Conservatively, the pressure at roof height may be used. This calculation also typically includes a size factor to consider the scale of the exposed area.

Wind loads on buildings can also be obtained by wind tunnel tests as explained in the next section.

5.6.15 Wind Tunnel Test

While wind codes incorporate various factors such as building shape (force coefficient), sheltering and topography (topography factor), it should be well understood that they are only rough guides and can unlikely account for the interaction of these factors. For a more precise assessment, direct measurement through wind tunnel testing is recommended. This technique involves constructing a scaled model (typically between 1 : 5000 and 1 : 400) of the building and its surrounding topography on a turntable, which is then subjected to controlled wind flows within the tunnel. A typical wind tunnel setup is shown in Figure 5.19.

Wind tunnel testing involves placing a scaled model of the target building on a turntable and subjecting it to scaled wind conditions that simulate the design wind, including its speed, turbulence intensity, and profile as typically required by codes. These conditions are created by adjusting devices like splines and rougheners within

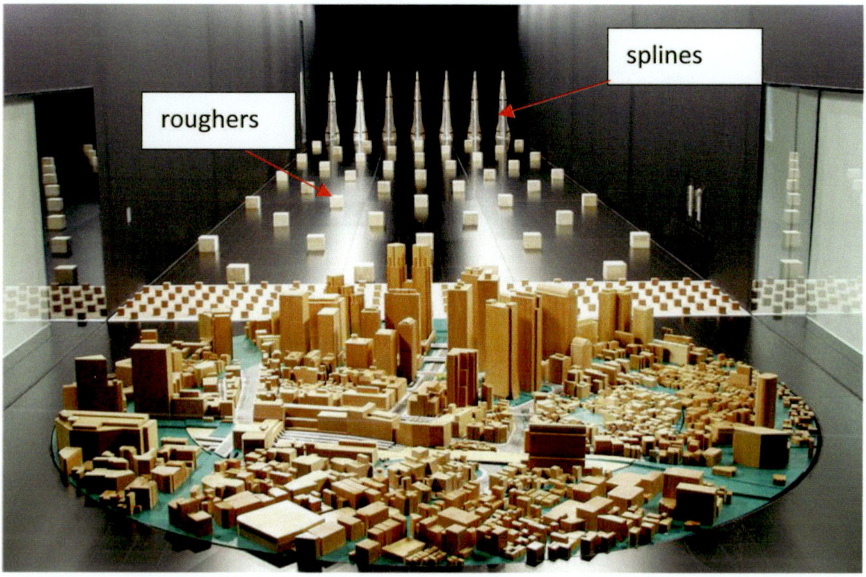

FIGURE 5.19 Photographic view of a wind tunnel (source: Shimz Corporation Wind Tunnel Testing Laboratory).

the tunnel. The model is extensively instrumented to measure its response to these winds. Key measurements include:

(i) Local Pressures: Captured at selected points on the model's surface.
(ii) Global Forces: Measured as base reactions (shear and moments) using a high-frequency force balance installed at the model's base.

The measured data is then scaled up to represent the full-scale prototype structure. The scaling laws are derived from the ratios of the model's dimensions and wind speed to those of the real building. If the length scale is λ_L and the velocity ratio is λ_v, then the scaling factors are applied as follows:

(i) Pressures are scaled by $(1/\lambda_v)^2$
(ii) Forces have an additional length component, scaling by $(1/\lambda_v)^2(1/\lambda_L)^2$;
(iii) Moments scale by $(1/\lambda_v)^2(1/\lambda_L)^3$

Data is acquired at a high frequency (e.g., 200 Hz), providing a large dataset for robust statistical analysis. From this data, the mean response is calculated. The extreme maximum and minimum design values, such as bending moments, are then determined by adding or subtracting a peak factor (g_p) multiplied by the standard deviation from the mean (i.e., $Mean \pm g_p \times \sigma$).

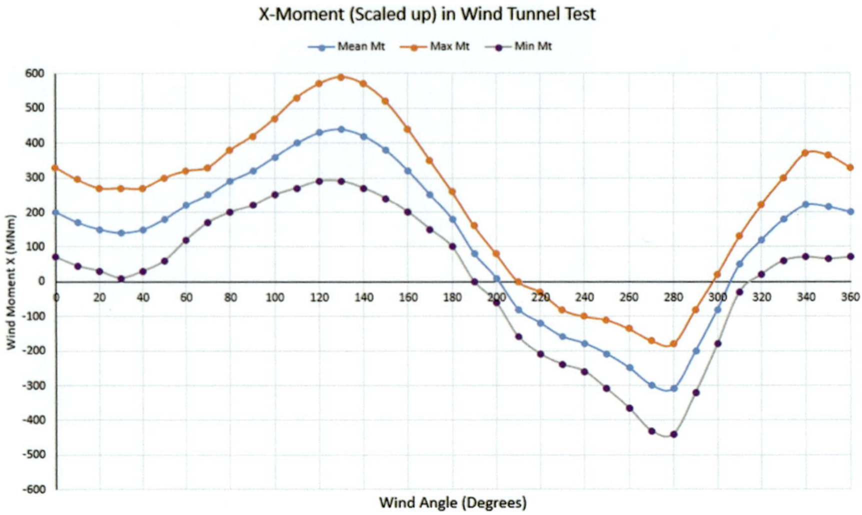

FIGURE 5.20 Mean and design moments from wind tunnel test.

The peak factor g_p is a statistical value, approximately 3.7 for typical applications, calculated by the formula $\sqrt{2\ln(3600f_0)} + 0.5772/\sqrt{2\ln(3600f_0)}$ where f_0 is the fundamental frequency of the structure and hourly mean speed is used. In case of 10 minutes wind speed is used, the figure 3600 will become 600.

This process of deriving extreme values from measured data is illustrated in Figure 5.20.

Generally, the physical models in the wind tunnel are taken as rigid bodies, except for very slender and flexible buildings. So, the dynamic responses of the structure can only be calculated by structural dynamic theories based on the measured data. Such models are known as aerodynamic models. The total response is the sum of static and dynamic responses where the static response is that due to the mean value. The dynamic analysis is mostly carried out in the frequency domain instead of direct dynamic analysis to a time history of wind loads. Briefly speaking, the dynamic part of the wind load is converted to a frequency domain with energies versus frequency as similar to that shown in Figure 5.14 known as spectral power density $S_p(f)$. The variance of the dynamic load which is the square of the standard deviation of the structural response will be the summation of $|H(f)|^2 S_p(f)$ ($|H(f)|$ is called the mechanical admittance):

$$\sigma_{\mathbf{P}}^2 = \int_0^\infty S_{\mathbf{P}}(f)df = \int_0^\infty |H(f)|^2 S_p(f)df \tag{5.28}$$

where $\left|H(f)\right|^2 = \dfrac{1}{\left(1 - f/f_0\right)^2 + \left(2\zeta f/f_0\right)^2}$ with f as the frequency of the external excitation, f_0 as the natural frequency of the structural system and ζ as the damping ratio.

Equation 5.28 is a general equation applicable to shear and moment. When applied to overturning moment for design, the standard deviation will then be $\sigma_{\mathbf{M}}{}^2 = \int_0^{\infty} S_{\mathbf{M}}(f)\,df = \int_0^{\infty} \left|H(f)\right|^2 S_M(f)\,df$ and the extreme design peak values are

$$\hat{\mathbf{M}} = \bar{\mathbf{M}} \pm g_p \sigma_{\mathbf{M}} \tag{5.29}$$

So extreme values of $M_{max} = \bar{\mathbf{M}} + g_p\sigma_{\mathbf{M}}$ and $M_{min} = \bar{\mathbf{M}} - g_p\sigma_{\mathbf{M}}$ can be obtained.

If, however for very flexible buildings, it may be desirable to model its stiffness and natural frequencies physically for more accurate results. They are known as aero-elastic models. More sophisticated techniques in analysis are required in such models.

To identify critical wind load cases for design, an approach is to adopt Equation 5.30 to calculate the overturning about X, Y axes and the torsion (denoted respectively as M_X, M_Y and M_Z) so that eight combinations can be formed for a wind load direction as

$$M_{Xmax} + M_{Ymax} + M_{Zmax}; \quad M_{Xmax} + M_{Ymax} - M_{Zmin};$$
$$M_{Xmax} - M_{Ymax} + M_{Zmax}; \quad M_{Xmax} - M_{Ymax} - M_{Zmin};$$
$$M_{Xmin} + M_{Ymax} + M_{Zmax}; \quad M_{Xmin} + M_{Ymax} - M_{Zmin};$$
$$M_{Xmin} - M_{Ymax} + M_{Zmax}; \quad M_{Xmin} - M_{Ymax} - M_{Zmin};$$

As the building movement under wind load is more or less elliptical on plan instead of following the outline of a rectangle, so it should be more reasonable to adopt extreme values as $\sqrt{M_{Xmax}{}^2 + M_{Ymax}{}^2}$. Considering only M_{Xmax} and M_{Ymax}, the eight values adopted for design can be as in Figure 5.21 as:

The two cases with extreme values of M_Z may be added for design so that ten critical cases can be summed for a wind load direction.

Generally, there are 24 or 36 wind directions in a wind tunnel test, making up 240 to 360 wind load cases for design by such an approach. However, visual inspection of the plots of these elliptical presentations, as in Figure 5.21, would allow a large number of load cases which are not critical to be eliminated in design. By such process, generally 10 to 20 numbers of load cases are left for design.

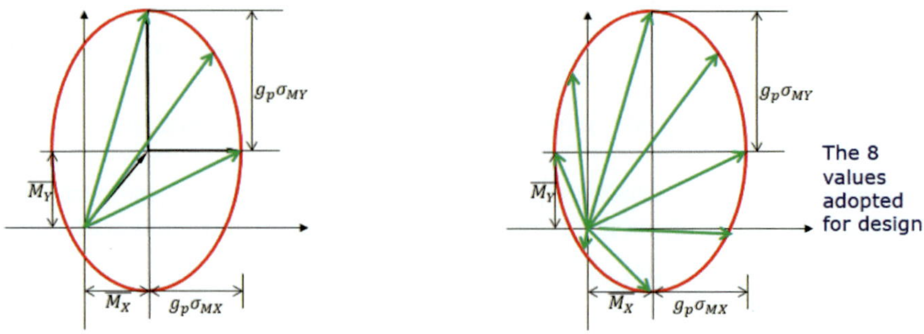

FIGURE 5.21 Critical case for design by wind tunnel test.

5.7 SEISMIC LOAD

Buildings are susceptible to damage or collapse during earthquakes. Seismic intensity, like wind, varies by geographic location, requiring structures in active seismic zones to be designed for specific seismic actions. However, the design philosophy for earthquakes differs significantly from that for wind. While wind design aims to prevent damage to ensure serviceability after a storm, seismic design acknowledges that preventing all damage in a major earthquake is impractical. Therefore, the objective is to manage and control damage to ensure it occurs in a predictable, ductile manner, and preserving the building's overall stability to facilitate occupant escape and prevent catastrophic collapse.

5.7.1 Nature of the Seismic Action (Load)

Unlike wind load, which is an actual force applied to a structure, seismic effects originate from ground motion during an earthquake. As the ground moves, the building's inertia and flexibility cause its floors to experience relative displacements, both with respect to the ground and among themselves. These relative movements manifest as structural deflections, which in turn induce internal forces. These internal forces can generally be calculated as the product of the structural stiffness and the deflections, as illustrated in Figure 5.22.

The differential movements within a structure responding to seismic action depend on the ground motion's magnitude and the building's dynamic properties. If the building's natural frequency is close to the dominant frequency of the ground motion, resonance can significantly amplify these movements. Consequently, unlike wind load, which has a significant static component, seismic effects are purely dynamic.

5.7.2 Return Period of Seismic Actions

Similar to wind load, seismic action is a random event which can be defined by a return period. Common return periods used in design are 50, 475, and 2475 years.

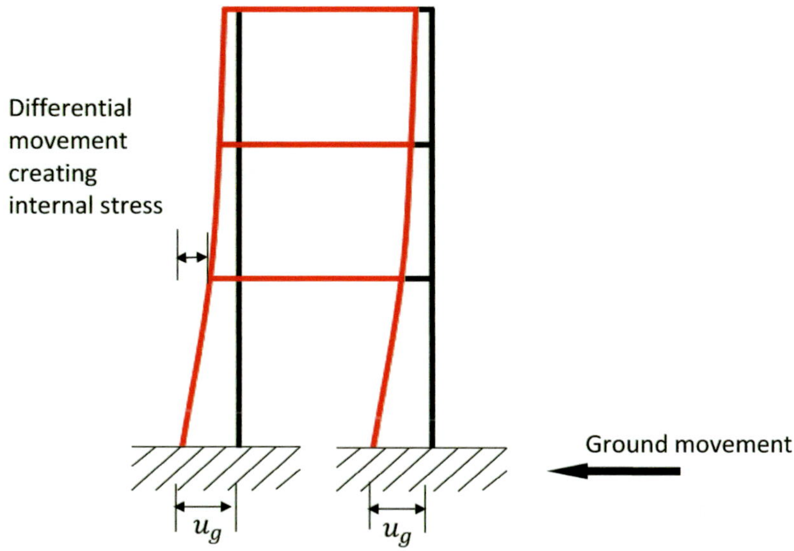

FIGURE 5.22 Seismic effect creating internal stress in structure.

These correspond to annual exceedance probabilities of 1/50, 1/475, and 1/2475, respectively. Over a 50-year structure lifespan, these probabilities translate to a 63.6%, 10%, and 2% chance of being exceeded.

5.7.3 Territorial Seismicity, Seismic Records and Time History Analysis

Earthquakes result from ground movements, often caused by tectonic or volcanic activity. As this activity varies geographically, seismicity differs from one region to another.

Determining a territory's seismicity for design purposes relies on two key factors which are its proximity to active seismic features (e.g., faults) and its historical earthquake records. Since predicting earthquakes remains impossible, these historical records are a vital reference.

Seismologists use this data to construct "time histories" which are recordings of ground motion over time for use in seismic design. Well-known examples include the El Centro (1940), Kobe (1995), and Tang Shan (1976) records. For design purposes, scaled and combined versions of these time histories are often used to model potential future ground motions. Time histories of the earthquake El Centro (1940) and Kobe (1995) are extracted in Figure 5.23 for illustration.

A structure being excited by ground acceleration can be analysed by equivalent forces in accordance with Figure 5.24 which are derived in the next paragraphs. The figure illustrates a 3-storey structure under an underground acceleration of $\ddot{u}_g$.

The following equation can be used to describe the motion of a mass in the structure:

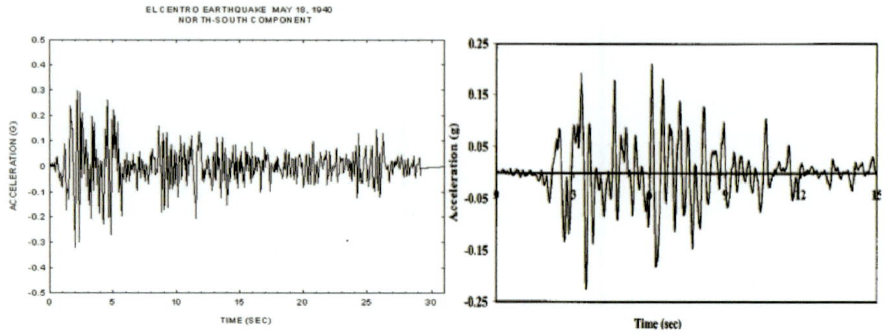

(a) From Vibration data El Centro Earthquake Page (b) From Mortezaei and Zaharai (2009)

FIGURE 5.23 Seismic records of Vibrationdata (2006) and Mortezaei and Zaharai (2009).

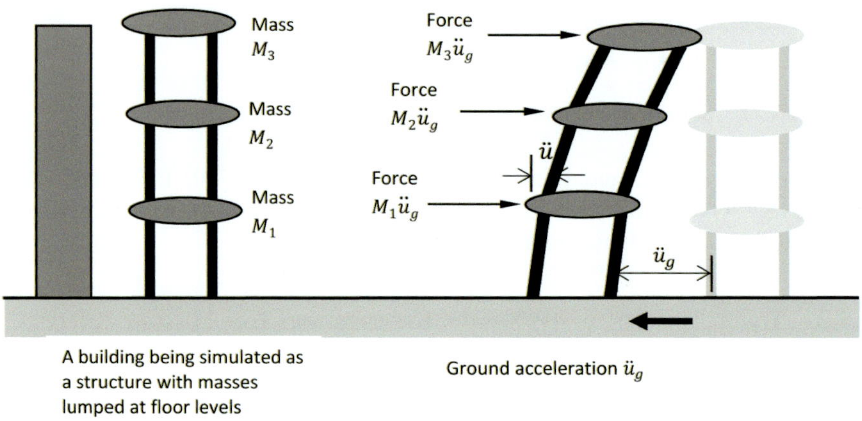

FIGURE 5.24 A building under ground acceleration.

$$M\left(\ddot{u}+\ddot{u}_g\right)+C\dot{u}+Ku=0 \tag{5.30}$$

where $\ddot{u}$ is the acceleration of the mass relative to the ground support; $\ddot{u}_g$ is the ground acceleration; $\ddot{u}$, $\dot{u}$ and u are respectively the acceleration, velocity and displacement of the mass relative to the ground; M is the mass of the structure; C is the damping coefficient representing an opposing force originated from the internal resistance of the structure being considered as proportional to the velocity and K is the stiffness providing an opposing force proportional to the displacement as an elastic structure to the mass. A fuller explanation of the equation can be found in Chapter 7. It should be noted that while the mass is undergoing total acceleration of $\ddot{u}+\ddot{u}_g$, while the restoring force by K and C depend only on the relative movement within the structure.

Re-arranging Equation 5.30

$$M\ddot{u} + C\dot{u} + Ku = -M\ddot{u}_g \tag{5.31}$$

So, the motion of the mass is equivalent to that being acted upon by an external force of magnitude $M\ddot{u}_g$ and in a direction opposing the displacement u as shown in Figure 5.24. The full structure can then be considered as being discretized into a number of masses each with its motion governed by Equation 5.31 for analysis against seismic action with known $\ddot{u}_g$. During structural analysis, the masses are generally added as point masses in the various nodes of the structural model as illustrated in Figure 5.25. which can be used in seismic analysis of a structure.

The structural mass is discretized into point masses, and the dynamic analysis is governed by Equation 5.31, where $\ddot{u}_g$ represents the time-varying ground acceleration. This acceleration induces equivalent time-varying forces at each story level, as illustrated in Figure 5.24. As these forces are dynamic, the analysis must account for their variation over time.

Equation 5.31 is a second-order differential equation that can be solved using either the "modal analysis method" or "numerical integration method"; both approaches are detailed in Chapter 7 with applications.

A critical input for this analysis is the ground acceleration time history, a record of acceleration versus time. The selection of an appropriate $\ddot{u}_g$ is based on local seismicity, which includes factors such as proximity to active faults, local geology, and historical seismic activity. Common practice involves using a combination of scaled historical seismic records or functions derived from them, factored to match

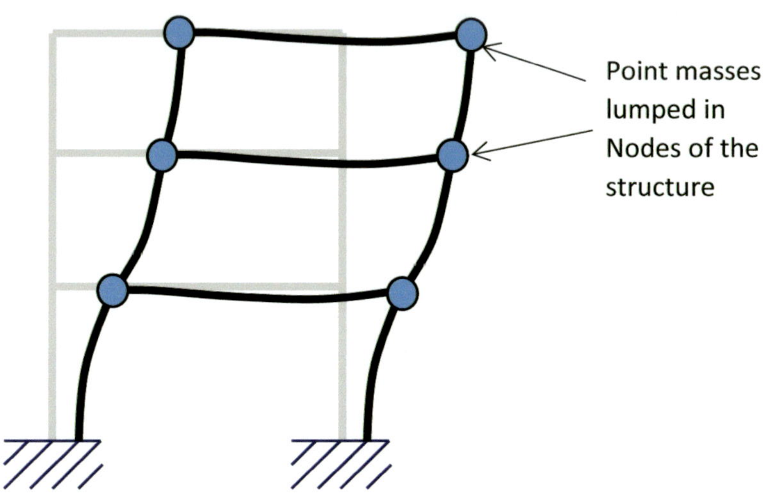

FIGURE 5.25 Mass of structure being lumped at joints of floor for dynamic analysis.

the assumed intensity of the seismic event. As per Eurocode BS EN 1998-1:2024 (2024), time history analysis generally requires a minimum of seven ground motion records. This requirement may be reduced to three for structures in low or very low seismicity classes.

5.7.4 Response Spectrum Method

Though powerful, time history analysis is computationally intensive and time-consuming. Its requirement to be performed for multiple seismic directions further adds to the laborious process. Consequently, it is not commonly used in general practice. Building codes typically mandate this method only for exceptional structures, such as very tall buildings, those with significant vertical irregularities (e.g., transfer structures or outriggers), or buildings on sites of high seismicity.

The more prevalent alternative is the response spectrum method. This approach uses an elastic response spectrum that plots the peak acceleration response of a single degree of freedom (SDOF) system against its natural period for a given level of damping. As illustrated in Figure 5.26, this spectrum is generated by calculating the maximum responses of numerous SDOF systems (each a simple mass-spring model with a unique natural period) to a specific ground motion record. The resulting data points are then idealized into a smooth curve of straight lines and simple curves for design use. A key feature of response spectra, also shown in Figure 5.26, is that the structural response decreases as damping increases.

A response spectrum constructed from the response to a time history record representative of the local seismicity provides a fundamental basis for seismic design. Consequently, building codes in seismic regions often include standardized response spectra.

Since seismic ground motion and structural response are significantly influenced by local soil conditions, with weaker soils amplifying shaking more than hard rock, design codes typically provide multiple spectra tailored to different ground types. For example, Eurocode 8 (BS EN 1998-1-1:2004) (2004) classifies ground into types

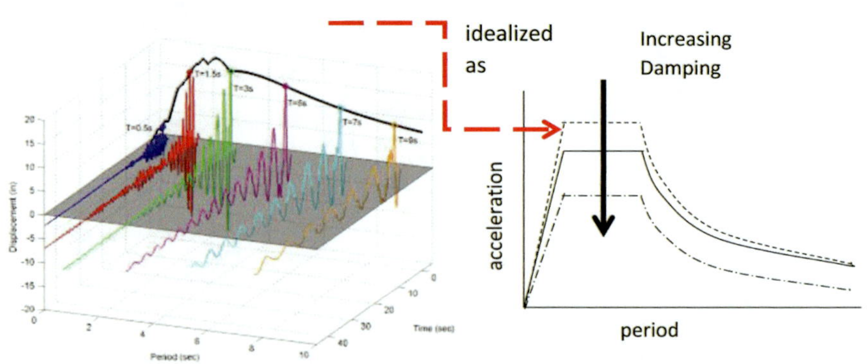

FIGURE 5.26 Construction of a response spectrum (from QuakeManager Wiki).

A (rock) through E (soft alluvium), with types B, C, and D representing soils of intermediate stiffness; each type has a corresponding response spectrum with unique critical design values.

A response spectrum plots the peak spectral acceleration response of a single degree of freedom (SDOF) system against its natural period. For a real building structure, which has multiple natural periods (or modes) as discussed in Section 5.5, the peak acceleration for each mode can be read directly from the spectrum. The total seismic response of the building is then calculated by summing the contributions from each significant mode, with each contribution weighted by its respective modal mass participation factor.

This concept is illustrated in Figure 5.27, which shows the peak accelerations corresponding to the first three fundamental periods (T1, T2, T3) of a structure. In this case, the acceleration value for period T3 has reached the spectrum's maximum (peak) acceleration.

Upon determination of the acceleration from the respective natural period of the structure, a force can be assumed to apply to account for the effect which is equivalent to mass × acceleration in accordance with Newton's Law. With application of the force, the analysis becomes a static problem which can easily be handled.

The characteristics of a response spectrum can be summarized as follows, with reference to a typical one extracted from figure 11.4-1 of ASCE 7-22 (2022) as Figure 5.28:

- (i) Response spectrum is normally designated by an authority in a nation or a territory for seismic design. It is normally specified in a design code.
- (ii) As a random action similar to wind load, the seismic load has to be defined by a return period as discussed in Section 7.2. The Eurocode BS EN 1998-1:2024 (2024) and BSEN1998-1: 2004 (2004) recommend 475 years. The ASCE 7-16 (2018) requires a mean return period to be 2475 years as Maximum Considered Earthquake (MCE), but the design is typically taken

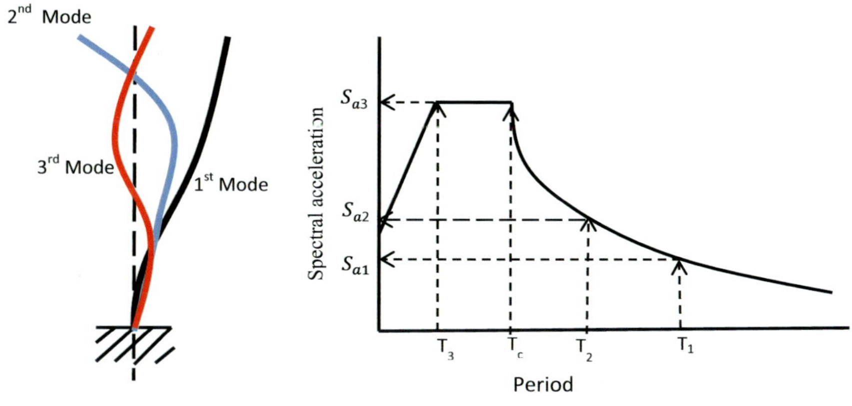

FIGURE 5.27 Use of response spectrum.

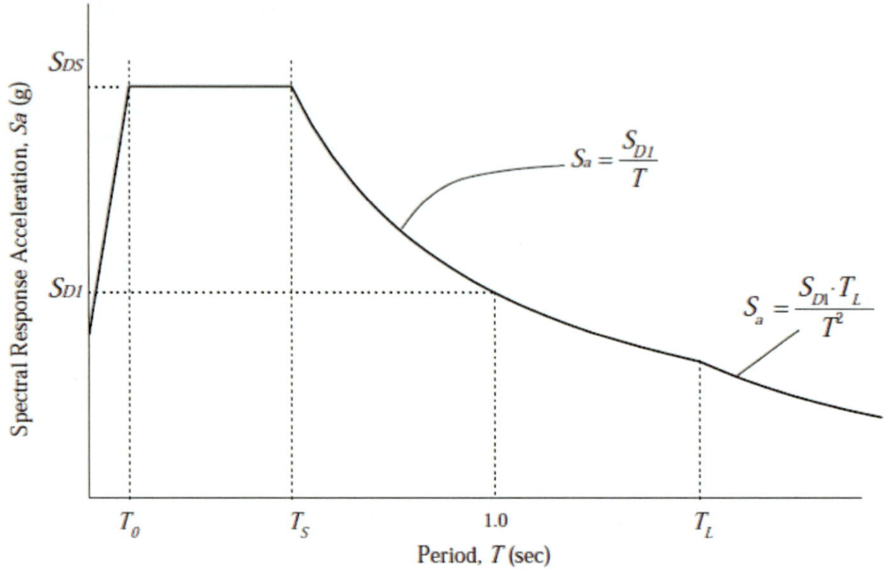

FIGURE 5.28 Typical response spectrum extracted from ASCE7-16.

as 2/3 of MCE. The Australian/New Zealand Standard AS/NZS1170.0:2002 (2002) utilizes a return period of 500 years for normal structures (by its classification of Importance Level IL = 2) and requires other return periods for other IL.

(iii) A typical response spectrum is often defined by equations in different intervals simulating the designed spectral response acceleration to structural periods.

(iv) The spectral acceleration at structural period = 0 is the peak ground acceleration as it refers to the accelerations of a structure of infinite stiffness whose movements are identical to the ground motion.

(v) The greatest acceleration which is the plateau of the response spectrum is generally at short period. It is known as "design spectral acceleration at short periods" denoted by the symbol S_{DS}. It is typically set at 0.15 to 0.2 sec.

(vi) S_{D1} is spectral acceleration at a structural period of 1 sec. It is also a key value for defining the shape of the response spectrum in the intermediate period range.

So, the aim of the response spectrum method is to determine static lateral shears applied to the building at its storeys that can simulate the seismic effect. The general procedure is as follows:

(i) The seismic analysis begins by determining the building's natural periods. The number of periods required depends on the building's complexity. According to Eurocode BS EN 1998-1:2004 (2004), the analysis must include enough modes so that the sum of their effective masses is at least 90% of the total mass. Furthermore, any single mode with an effective mass greater than 5% must be included. Natural periods and their effective masses can be calculated using structural dynamics principles or standard computer software;

(ii) Reference is then made to a response spectrum appropriate to the Site which is often provided by a local authority or prepared by a specialist. The accelerations as appropriate to the respective periods are read as in Figure 5.27, i.e., accelerations a_1, a_2, …. a_i …. a_n as corresponding periods T_1, T_2, …. T_i …. T_n are taken;

(iii) The mode shape for each natural period, which defines the deformed structure's vibration pattern at each story level (as illustrated in figure 7.7), is also determined. For a building, the mode shape value at the j-th floor for the i-th mode is denoted as ϕ_{ji}. Although these values are pure ratios representing relative displacements, they are often normalized (typically by setting the maximum absolute value to 1) to simplify subsequent calculations;

(iv) The mass of each storey effective for the translational movement denoted by M_j for the j-th floor, which is generally taken as dead load + 30 ~ 40% of live load;

(v) The modal participation ratio of a period i is calculated by

$$\Gamma_i = \frac{\Sigma M_j \times \phi_{ji}}{\Sigma M_j \times \phi_{ji}^2} \tag{5.32}$$

(vi) The maximum response acceleration at the j-th floor for the period i is then calculated as $\Gamma_i \times \phi_{ji} \times a_i$. With the mass as M_j, the storey shear is

$$S_{ji} = M_j \times \Gamma_i \times \phi_{ji} \times a_i \tag{5.33}$$

(vii) The total storey shear at the j-th floor S_j should be the summation of all modes. However, it would be too conservative to add their absolute values together as the maximum of them does not occur at the same time. So they may generally be summed by SRSS (square root of sum of squares) as

$$S_j = \sqrt{\Sigma S_{ji}^2} \tag{5.34}$$

(viii) Or if the periods of any two modes are too close, CQC (complete quadratic combination) is more applicable than SRSS. The Eurocode BSEN1998-1-1:2024 (2024) (6.4.3.2) requires CQC to be used unless any pair of periods T_i and T_j satisfy the following where ξ_i and ξ_j are the damping ratios of the periods

$$\left|T_i - T_j\right| / \left(T_i + T_j\right) > \xi_i + \xi_j \tag{5.35}$$

If Equation 5.35 is not satisfied, the summed effect E (including shear) by the method is

$$E_E = \sqrt{\Sigma E_{Ei} r_{ij} E_{Ej}} \tag{5.36}$$

where $r_{ij} = \dfrac{1}{1+\left(\alpha_{ij}/\xi\right)^2}$ with $\alpha_{ij} = \dfrac{T_i - T_j}{T_i + T_j}$ if $\xi_i = \xi_j = \xi$

$$r_{ij} = \frac{8\sqrt{\xi_i \xi_j}\left(\xi_i + \rho_{ij}\xi_j\right)\rho_{ij}^{3/2}}{\left(1-\rho_{ij}^2\right)^2 + 4\xi_i\xi_j\left(1+\rho_{ij}^2\right)\lambda_T + 4\left(\xi_i^2 + \xi_j^2\right)\rho_{ij}^2} \quad \text{with} \quad \rho_{ij} = \frac{T_i}{T_j} \tag{5.37}$$

A worked example for determining lateral shear by the response spectrum method of a simple 3-storey frame is enclosed in Appendix E.

5.7.5 Base Shear Method

For buildings in regions of low seismicity or for structures less susceptible to seismic effects, the simplified base shear method may be used as an alternative to the full response spectrum analysis. Eurocode 8 (BS EN 1998-1:2004) (2004) permits this method for buildings that meet the following criteria:

(i) Their fundamental natural period of vibration is less than 2.0 seconds and less than $4T_C$ in both principal directions, where T_C is the upper limit of the constant spectral acceleration branch (see Figure 5.27);
(ii) They possess sufficient regularity in elevation, as defined by the code.

The base shear method simplifies the design by considering only the response of the fundamental vibration mode in each principal direction. The total design lateral seismic force, or base shear (F_b), is calculated for each direction as:

$$F_b = \alpha M_{eff} \lambda \tag{5.38}$$

where α is the acceleration due to the fundamental period T_1

M_{eff} is the effective total mass, typically taken as 85% to 100% of the total mass (calculated from dead load plus 30–40% of the live load).
$\lambda = 0.85$ to 1. The Eurocode EC8 recommends λ to be 0.85 if $T_1 \le 2T_C$ and the building has more than two storeys, or $\lambda = 1$ otherwise.

The total load is then distributed onto the building floor in accordance with the effective mass of the floor times the fundamental mode shape lateral deflection as shown in Equation 5.39. As the fundamental mode shape is almost linear, it may be simpler to distribute the load onto the respective floor in accordance with floor mass times floor height as per Equation 5.40.

$$F_j = F_b \frac{M_j \phi_{j1}}{\sum M_j \phi_{j1}} \tag{5.39}$$

Or

$$F_j = F_b \frac{M_j h_j}{\sum M_j h_j} \tag{5.40}$$

where F_j is the distributed floor shear on the floor j, M_j is the effective floor mass, ϕ_{j1} is the fundamental mode shape of the floor j and h_j is the height of the floor j above ground.

Similar provisions for the base shear methods are found in other codes including the China Code for seismic design GB50011-2010 (2010).

5.7.6 Importance Factor

Building codes often assign an importance factor to increase design loads based on a structure's societal role and the consequences of its failure. This classification system may apply to various hazards, including seismic, wind, and tsunami loads.

For instance, ASCE 7-22 (2022) defines Risk Categories I through IV, with corresponding importance factors ranging from 1.0 to 1.5 (table 1.5-2). A building's category depends on whether it provides essential functions or if its failure would pose a significant hazard. The required seismic design level is then determined by combining the Risk Category with the Seismic Design Category.

Similarly, Eurocode BSEN1998-1:2004 (2004) employs an "importance class" system with associated factors.

In contrast, the China seismic code (GB50011-2010) (2010) uses a different approach. Instead of applying an importance factor, it requires buildings of high importance to be designed to the specifications of the next higher seismic precautionary intensity level, thereby directly enhancing their seismic resilience.

5.7.7 Behaviour Factor

The Eurocode BSEN1998-1:2004 (2004) includes the use of a behaviour factor denoted by q. The factor allows elastic analysis to be carried out under reduced load by dividing the loads obtained from the elastic analysis by the factor $(q \geq 1)$. The factor accounts for the ability of the structure to dissipate energy through inelastic behaviour by deforming beyond the elastic limit, such as plastic hinge formation during an earthquake. The Eurocode lists the minimum behaviour factor to be used under different categories of ductility limit design.

5.8 SOIL LOAD

Well-established theories exist for determining design soil loads on structures, though a detailed examination is beyond the scope of this chapter. For building structures, which are typically much stiffer than dedicated earth-retaining structures, soil pressures are usually designed for the "at-rest" condition $\left(K_o\right)$.

Soil loading is often accompanied by lateral hydrostatic pressure from groundwater, which is typically significantly greater than the soil pressure itself. However, the effective weight of the soil is reduced by buoyancy when submerged, which in turn reduces the lateral soil pressure.

5.9 SNOW LOAD

Snow load is the weight of accumulated snow and ice on a building's roof. The roof and its supporting structure must be designed to withstand this load.

Buildings in regions that experience snow, including frigid zones, must be designed for these loads. Like wind and seismic loads, snowfall is a random action. Consequently, the design snow load is a statistical value determined by a specified "return period," as well as the site's location and altitude.

According to Eurocode BS EN 1991-1-3:2003 (2003), the "characteristic" snow load is defined as the value with an annual probability of exceedance of 0.02, equivalent to a 50-year return period. For the UK and Republic of Ireland, the characteristic snow load on the ground s_k is provided in table C.1 of the code:

$$s_k = 0.14\mathrm{Z} - 0.1 + \mathrm{A} / 501 \tag{5.41}$$

where Z is the zone number, A is the altitude.

So accordingly, s_k on ground of Inverness of Zone number 4.5 as per figure C.9 of BSEN1991-1-3:2003 (2003) at an altitude 5m is $s_k = 0.14 \times 4.5 - 0.1 + 5 / 501 = 0.53\mathrm{kN/m}^2$.

The snow load on the roof is modified by a number of factors:

(i) construction materials of the roof;
(ii) detailing and construction of the roof;
(iii) pitch and thermal conditions of the roof;
(iv) roof layout and geometry;
(v) sun and wind exposures.

The factors can be found from relevant codes of practice. For example, Eurocode BSEN1991-1-3:2003 (2003) specifies the following equation for the persistent/transient design situations

$$\mathrm{s} = \mu_i C_e C_t s_k \tag{5.42}$$

where μ_i $(i = 1 \text{ or } 2)$ is the slope load shape coefficient depending on the slope of the roof, C_e is the exposure coefficient taking into account development of the Site which should normally be taken as 1 (or 0.8 for windswept, 1.2 for sheltered); and C_t is the thermal coefficient taking into account the thermal transmittance of the roof (>1 W/m^{2}K) and should also be taken as 1 normally.

For a flat roof where $\mu_i = \mu_1 = 0.8$ and $C_e = C_t = 1$, s by Equation 5.42 for Inverness at an altitude 5 m will be

$$s = 0.8 \times 1 \times 1 \times 0.53 = 0.424 \text{kN/m}^2$$

5.10 TEMPERATURE LOAD

Thermal expansion or contraction causes a structure's dimensions to change. If this movement is restrained, internal forces develop, creating a "temperature load." This is illustrated in Figure 5.29 in which a floor slab expands as its temperature rises. This expansion induces lateral sway and shear in the supporting columns. The columns resist this movement, generating compressive reactions in the floor slab in addition to the shears and moments on the columns. These forces are reversed if the slab contracts due to a temperature drop.

The temperature load illustrated in Figure 5.29 can be significant in buildings with large plan dimensions and rigid floor restraints. A large span will result in greater lateral deflection under restraint, imposing larger forces on the supporting columns and larger compression on the floor. Similarly, stronger restraints generate greater reactive forces in the floor slab.

Theoretically, no internal forces are induced if a floor structure is free to expand and contract. Conversely, a perfectly rigid restraint would generate the maximum theoretical thermal stress, σ_{Tmax}, in the floor structure.

$$\sigma_{Tmax} = E \times \alpha \times \Delta\theta \tag{5.43}$$

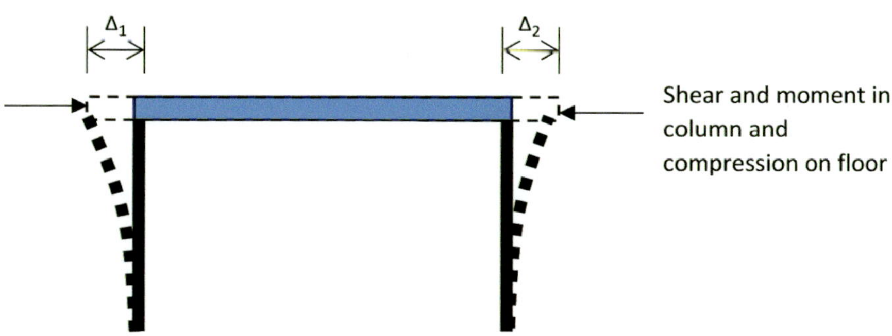

FIGURE 5.29 Forces Induced on building structure due to temperature rise.

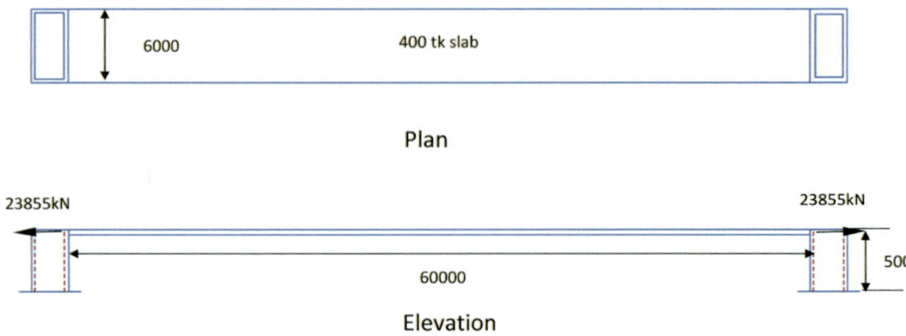

FIGURE 5.30 Portal frame under temperature rise in Worked Example 5.2.

where E is the Young's modulus of the material, α is the coefficient of thermal expansion and $\Delta\theta$ is the temperature change.

For analysis, forces equal to σ_{Tmax} times the cross-sectional area of the structural member under temperature change are applied to its two ends. The applied forces as external loads are subtracted from the structure after analysis. This approach is well incorporated in many software. Worked Example 5.2 illustrates the design principle.

Worked Example 5.2

A slab of 400 mm thick with clear length 60 m under lateral restraints by two core walls at its end as shown in Figure 5.30. The design parameters are as follows:

	Cross-sectional Area (m²)	Second Moment of Area (m⁴)
Left core wall	5.04	7.279
Right core wall	5.81	8.126
Slab	2.4	0.032

Young's Modulus of Concrete for the Core wall are 27700 N/mm² and that of the slab is 25100 N/mm². The coefficient of linear expansion of concrete is 0.0000099/°C.

By Equation 5.30 $\sigma_{Tmax} = E \times \alpha \times \Delta\theta = 25100 \times 0.0000099 \times 40 = 9.94\text{N/mm}^2$.

Tension force to be applied to the slab is $9.94 \times 6000 \times 400 / 1000 = 23855.04\text{kN}$

The structure can be solved as a portal frame by the stiffness method with the sectional properties given under the tension forces at 23855.04 kN.

Solving, the tension in the slab is 6731.705 kN tension with displacement on the left end at 0.003537 m to the left and 0.0031684 m at the right end to the right. After subtraction of the applied load of 23855.04 kN, the net compression in the

$6731.705 - 23855.04 = 17123.34\text{kN}$, creating $17123.34 / (6 \times 0.4) = 7135\text{kN/m}^2$.

The axial load in the slab is also equal to the shear on both core walls obtained by analysis.

If the length of the slab is increased to 100 m, the compressive stress in the slab will rise to 8042 kN/m².

The design temperature change for a structure depends on the local climate. In tropical regions, roof surfaces exposed to direct sunlight can experience temperature increases exceeding 40°C.

For thermal analysis, a uniform temperature change can be applied to thin roof slabs. Thicker slabs, however, require consideration of a differential temperature gradient to accurately assess thermal stresses.

In many national design codes, thermal loading is considered a secondary effect. Special design for thermal expansion is often only mandated for structures with large plan dimensions, typically those exceeding 60 m.

The effects on structure due to temperature are mainly due to movement of the floor structure which induce forces on the supporting columns and walls. The stresses in the floor structures if in compression can generally be ignored as they are small and can in effect enhance the bending capacities of the floors. Tensile stresses, however, pose a different challenge. It is generally impractical and uneconomical to resist them by additional reinforcements. A more efficient approach is to permit the concrete to crack. Once the concrete is cracked, the stresses will be released. Formulae to calculate crack widths based on the products of tensile strain and the crack spacing are found in codes like the Eurocode BSEN 1991-1-1:2023 (2023). The crack widths so obtained are generally small and well below the allowable values in the range of 0.2 mm to 0.3 mm. In fact, they are much narrower than those due to structural loads.

5.11 SHRINKAGE AND CREEP LOAD

Shrinkage of concrete is the loss of water (or moisture) contents to the exterior, which leads to a decrease in the dimensions of the concrete structure. Shrinkage of concrete will take place once it has hardened unless it is fully immersed in water.

As similar to temperature load, the change of dimension of the concrete structure will induce internal loads in itself if it is under restraints to movement. However, shrinkage will only lead to contraction. Shrinkage depends on the following:

(i) Time: the shrinkage rate will be fast at the early age of the concrete and slow down with time.
(ii) Humidity of the surrounding: obviously, a humid climate will lead to a slower shrinkage rate.
(iii) Composition of the concrete: the higher the cement content and water-cement ratio, the higher will be shrinkage rate.
(iv) Thickness of the structure: a thicker structure exhibits a slower shrinkage rate and the ratio of the surface to volume for losing water is smaller.

The Eurocode BSEN1992-1-1:2023 (2023) lists in its table 5.3(ND) the "nominal total shrinkage strain" in 50 years for concrete structures (denoted as $\varepsilon_{sh,50y}$) in terms of (i) concrete strength development CS (slow strength development), CN (normal strength development) and CR (rapid strength development); (ii) concrete strength;

(iii) environmental relative humidity; (iv) thickness. Other codes have similar provisions, by charts or formulae.

Accompanied with concrete shrinkage is the concrete "creep." Creep refers to the continuous change of dimension of the concrete structure in the direction of the applied load even when the load remains constant over time. So, this phenomenon is different from the elastic deformation, which can be recovered upon removal of the applied load. Again, creep depends on the factors similar to those of shrinkage, including humidity, composition, effective thickness, and the composition listed above. In addition, creep also depends on the age of the concrete when the load is applied and the duration of the load. Thus, in analysis, which generally has to be numerical, the process generally has to be broken into load steps by which loads are progressively applied at various times and durations with interactions with shrinkage, and the effects are finally summed. Like other codes, the Eurocode BSEN1992-1-1:2023 (2023) calculates creep strain in its Equation 5.2, which is reproduced as Equation 5.44:

$$\varepsilon_{cc}\left(t,t_0\right)=\varphi\left(t,t_0\right)\left(\sigma_c / E_{28}\right) \tag{5.44}$$

In Equation 5.44, t is the age of concrete at which the creep strain is to be assessed and t_0 is the age of concrete at which a load applies, σ_c is the stress induced by the load and E_{28} is the tangent modulus of elasticity of concrete (may be taken as $1.05E_{cm}$ where E_{cm} is the secant modulus of elasticity). The Eurocode tabulates values ($\varphi\left(50y,t_0\right)$in its Table 5.4) and formulae (in its Annex B) for the determination of $\varphi\left(t,t_0\right)$.

Take an example of a column at the ground floor of a high-rise building in which the load acting on it increases incrementally due to the load from each of the above floors, t_0 is different from each floor load, so is $\varphi\left(t,t_0\right)$. The creep strain due to the load from a floor can then be calculated by Equation 5.44 and the total creep strain is the summation of all the $\varepsilon_{cc}\left(t,t_0\right)$ due to all the floors. Shrinkage and creep strains are in the direction of a column, which should be added to the elastic strain for the calculation of the total shortening of the column at time t due to a load starting to act at the concrete's age, t_0 as shown in Equation 5.45

$$\varepsilon_{total,t_0}=\varepsilon_{sh,t_0}+\varepsilon_{elastic}+\varepsilon_{cc}\left(t,t_0\right) \tag{5.45}$$

As $\varepsilon_{elastic}=\sigma_c / E_{28}$, Equation 5.45 can be re-written as

$$\varepsilon_{total}=\varepsilon_{sh,t_0}+\left(\sigma_c / E_{28}\right)\left[1+\varphi\left(t,t_0\right)\right] \tag{5.46}$$

where $\varphi\left(t,t_0\right)$ is sometimes known as "creep coefficient."

In a concrete floor structure, creep, however, acts oppositely to shrinkage. As concrete contracts due to shrinkage, tension will be created within the floor structure if it is under lateral restraint. The tension as load in the opposite direction of the shrinkage will therefore reduce the tension in the floor by "relaxing" the concrete.

Creep in the floor structure therefore reduces shrinkage strain. So Equation 5.46 can be re-written as $\varepsilon_{total} = \varepsilon_{sh,t_0} - \varepsilon_{elastic} - \varepsilon_{cc}(t,t_0)$ implying

$$\varepsilon_{total} = \varepsilon_{sh,t_0} - (\sigma_c / E_{28})[1 + \varphi(t,t_0)] \tag{5.47}$$

$\varepsilon_{total} L = \Delta$ is the total shortening of the floor where L is the length of the floor slab. If the floor slab is supported on two columns or walls of lateral stiffness K_1 and K_2, the lateral force on them is $\sigma_c A_f$ where A_f is the cross-sectional area of the floor slab, then we may write $\sigma_c A_f / K_1 = \Delta_1$ and $\sigma_c A_f / K_2 = \Delta_2$ as the lateral movements of the two supports. With $\Delta = \Delta_1 + \Delta_2$, we may finally set

$$\sigma_c \left(1 + \varphi(t,t_0) + \frac{1/K_1 + 1/K_2}{L / A_f E_{28}} \right) = E_{28} \varepsilon_{sh,t_0} \tag{5.48}$$

So Equation 5.48 can be used to calculate the stress σ_c induced. However, it should be noted that σ_c in Equation 5.48 actually refers to shrinkage and creep effects starting from t_0 to t and σ_c is a continuous function. Thus, the analysis should be a numerical one by dividing time into a finite number of steps, each with a small interval, say 3 or 7 days at the beginning (with $t_0 = 0$) for summing up stepwise to t. The method has been discussed by Bazant (1988).

Nevertheless, the estimation of shrinkage and creep is derived from experimental data, and many mathematical models for their prediction are provided in various codes of practice, such as ACI-209 and CEB-FIP. However, these models can vary significantly, as the effects of shrinkage and creep are highly dependent on specific material properties and local environmental conditions. Consequently, the designer must exercise careful judgment in selecting the most appropriate model.

5.12 LOAD COMBINATIONS

Structural members must be designed to withstand the combined effect of all simultaneous loads, such as gravity, wind, and seismic forces. This is achieved through "load combinations," where factored loads are summed.

In these combinations, characteristic loads F_{ki} are multiplied by partial safety factors γ_{fi}. These factors account for the uncertainty in load magnitude and the reduced probability of all loads acting at their maximum levels simultaneously. The result is a set of design load cases $\sum \gamma_{fi} F_{ki}$ that ensure the structure has adequate strength and stability under adverse conditions.

The following outlines the load factors and combinations specified in Eurocode (BS EN 1990:2023), ASCE/SEI 7-22, and BS 8110-1:1997.

5.12.1 Load combinations in Accordance with the Eurocode BSEN1990:2023 (2023)

The Eurocode distinguishes between (1) permanent action (load); (2) variable action; (3) accidental action, and (4) seismic action. Permanent action basically refers to dead load which acts throughout the building life with small variation. Variable action refers to live load, wind load, snow load etc. Accidental actions are those due to accidents. Seismic actions are those due to earthquakes. In combinations of actions, the permanent action will be included. Variable actions which may occur concurrently, such as live load with wind load, will also be included under use of different partial load factors γ_Q. In a combination of actions, one variable action will be taken as the "leading variable action" and others are termed "accompanying variable actions." In addition to partial factors for actions with symbol prefixed γ, there are "combination factors" with symbols prefixed ψ. While the partial factors are directly multiplied to the characteristic load to account for variations in load values and uncertainties in their magnitudes, the combination factors are used with these partial factors to further reduce the overall loads when multiple variable actions are present. Some detailed explanations for the ψ factors will be given in the latter part of this section.

The Eurocode includes the following combinations of actions for ultimate limit states

(i) persistent and transient (fundamental) design situations;
(ii) accidental design situations;
(iii) seismic design situations;
(iv) fatigue design situations.

For combinations of actions for persistent and transient (fundamental) design situations, the following equations 8.12 to 8.14 in BSEN1990:2023 (2023): are reproduced as Equations 5.49 to 5.51 as follows. In design, any one of them can be used. In Equations 5.50 and 5.51, the most adverse expressions of the two in each of them shall be used.

$$\Sigma F_d = \Sigma \gamma_{G,i} G_{k,i} + \gamma_{Q,1} Q_{k,1} + \sum\nolimits_{j>1} \gamma_{Q,j} \psi_{0,j} Q_{k,j} + \gamma_P P_k \tag{5.49}$$

$$\Sigma F_d = \begin{cases} \Sigma \gamma_{G,i} G_{k,i} + \gamma_{Q,1} \psi_{0,1} Q_{k,1} + \sum\nolimits_{j>1} \gamma_{Q,j} \psi_{0,j} Q_{k,j} + \gamma_P P_k \\ \Sigma \xi_i \gamma_{G,i} G_{k,i} + \gamma_{Q,1} Q_{k,1} + \sum\nolimits_{j>1} \gamma_{Q,j} \psi_{0,j} Q_{k,j} + \gamma_P P_k \end{cases} \tag{5.50}$$

$$\Sigma F_d = \begin{cases} \Sigma \gamma_{G,i} G_{k,i} + \gamma_P P_k \\ \Sigma \xi_i \gamma_{G,i} G_{k,i} + \gamma_{Q,1} Q_{k,1} + \sum\nolimits_{j>1} \gamma_{Q,j} \psi_{0,j} Q_{k,j} + \gamma_P P_k \end{cases} \tag{5.51}$$

where

F_d represents the design value of an action;
$\gamma_{G,i}$ is the partial factor for permanent action i;
$G_{k,i}$ is the characteristic value of permanent action i;
$\gamma_{Q,1}$ is the partial factor for the leading variable action 1;
$\psi_{0,1}$ is the combination factor for the leading variable action 1 (if applied);

$Q_{k,1}$ is the characteristic value of the leading action 1;
$Q_{k,j}$ is the characteristic value of an accompanying variable action j;
$\psi_{0,j}$ is the combination factor for the variable action j;
$\gamma_{Q,j}$ is the partial factor for the variable action j;
P_k is the characteristic value of a prestressing force;
γ_P is the partial factor for the prestressing force;
ξ is a reduction factor applied to the unfavourable permanent action only.

Although it is stipulated in 8.3.4.2 (2) of BSEN1990:2023 (2023) that any of equations 8.12 to 8.13 (reproduced as Equations 5.49 to 5.51) above can be used, Note 1 of the end of 8.3.4.2 states that the formula to be used is equation 8.12 (i.e., Equation 5.49 above), unless the National Annex gives a different choice. For ξ, Note 2 suggests $\xi = 0.85$, unless National Annex gives a different value.

For combination of actions for accidental design situations A_d, equation 8.15 of BSEN1990:2023 (2023) should be used which is reproduced as Equation 5.52 as follows:

$$\Sigma F_d = \Sigma G_{k,i} + A_d + \left(\psi_{1,1}\ or\, \psi_{2,1}\right)Q_{k,1} + \sum\nolimits_{j>1}\psi_{2,j}Q_{k,j} + P_k \tag{5.52}$$

where $\psi_{1,1}$, $\psi_{2,1}$ and $\psi_{2,j}$ are combination factors

For combination of actions for seismic design situations $A_{Ed,ULS}$, equation 8.16 of BSEN1990:2023 (2023) should be used which is reproduced as Equation 5.53 as follows:

$$\Sigma F_d = \Sigma G_{k,i} + A_{Ed,ULS} + \sum\nolimits_{j}\psi_{2,j}Q_{k,j} + P_k \tag{5.53}$$

While the γ factors that are the partial load factors to cater for uncertainty of load to achieve safe designs, the combination factors in the Eurocode with suffixes 0, 1, and 2 account for the following:

ψ_0: combination factor to account for reduced probability of simultaneous occurrence of maximum values of more than one variable actions.

ψ_1: combination factor to determine the frequent value of the variable action, i.e., the action is likely to be present for a significant portion of time. In building, the fraction of time chosen to be exceeded is 1% by the Eurocode.

ψ_2: combination factor to determine the quasi-permanent value of the variable action, i.e., the action is likely to be present for most of the time. The quasi-permanent value is chosen so that the fraction of time it is exceeded is 50%.

For structural and geotechnical design (under Verification Case VC1), the Eurocode BSEN1990:2023 (2023) gives values for γ_G and γ_Q respectively as $1.35K_F$ and $1.5K_F$ for permanent and variable actions resulting in an unfavourable/destabilising effect (K_F will be explained later). The same factors will be used in Verification Case VC2, which is for combined verification of strength and static equilibrium when they are applicable and unfavourable to equilibrium. These factors will, however, be changed to 1 (mainly for the equilibrium check) if they are favourable. There are Verification

Cases VC3 and VC4 for geotechnical design. Readers may refer to table A.1.8 (NDP) in BSEN1990:2023 (2023) for details.

K_F as mentioned in the previous paragraph is termed the "consequence factor" which are 0.9, 1.0, and 1.1 for Consequence Class CC1, CC2, and CC3 respectively. They aim at increasing the loads for structures whose failure bears high consequence. Table A.1.1 of BSEN1990:2023 (2023) gives examples of how buildings are classified to CC0, CC1, CC2, CC3, and CC4. The Eurocode does not give K_F for CC0 and CC4. High-rise buildings are classified as CC3. But high-rise building is not precisely defined in the Eurocode.

Table A.1.7 defines the combination factors as reproduced in Table 5.5 as follows.

For illustration, the following combinations are generated for the ultimate limit state in accordance with Equation 5.49 involving dead load, live load, and wind load on a domestic floor. In the following, G_k is the characteristic load due to permanent

TABLE 5.5

ψ Factors in Table A.1.7 of the Eurocode BSEN1990:2023 (2023)

Action	ψ_0	ψ_1	ψ_2
Imposed loads in buildings, category (see EN 1991-1-1)			
Category A: domestic, residential areas	0.7	0.5	0.3
Category B: office areas	0.7	0.5	0.3
Category C: congregation areas	0.7	0.7	0.6
Category D: congregation areas	0.7	0.7	0.6
Category E: storage areas	1.0	0.9	0.8
Category F: traffic area vehicle weight ≤ 30kN	0.7	0.7	0.6
Category G: traffic area, 30kN < vehicle weight ≤ 160kN	0.7	0.5	0.3
Category H: roofs accessible for normal maintenance and repair only	0.7	0	0
Construction loads (see EN 1991-1-6)	1.0	–	–
Snow loads on Buildings (see EN 1991-1-3)	0.7	0.5	0.2
– Finland, Iceland, Norway, Sweden	0.7	0.5	0.2
– remainder of CEN Member States, for sites located at altitude H > 1000 m a.s.l.;	0.5	0.2	–
– remainder of CEN Member States, for sites located at altitude H ≤ 1000 m a.s.l.			
Remainder of CEN Member States, for sites located at H > 1000 m.s.l.	0.7	0.5	0.2
Remainder of CEN Member States, for sites located at H ≤ 1000 m.s.l.	0.5	0.2	0
Wind loads on building (see EN 1991-1-4)	0.6	0.2	0
Temperature (non-fire) in buildings (see EN 1991-1-5)	0.6	0.5	0
Icing (see EN 1991-1-9)	0.5	0.2	0

NOTE: The combination value for water actions can be based on a 10% probability that it is exceeded during a one-year reference period.

action, Q_k and Q_w are respectively the characteristic live load and wind load on the floor:

(i) $\gamma_g G_k$: $1.35G_k$ (No variable action);
(ii) $\gamma_g G_k + \gamma_q Q_k$: $1.35G_k + 1.5Q_k$ (live load as the only variable action);
(iii) $\gamma_g G_k + \gamma_q Q_w$: $1.35G_k + 1.5Q_w$ (wind load as the only variable action);
(iv) $\gamma_g G_k + \gamma_q Q_k + \psi_{0,w}\gamma_q Q_w$:
$1.35G_k + 1.5Q_k + 0.6 \times 1.5Q_w = 1.35G_k + 1.5Q_k + 0.9Q_w$ (live load as the leading variable action + wind load);
(v) $\gamma_g G_k + \psi_{0,q}\gamma_q Q_k + \gamma_q Q_w$:
$1.35G_k + 0.7 \times 1.5Q_k + 1.5Q_w = 1.35G_k + 1.05Q_k + 1.5Q_w =$ (wind load as the leading variable action + live load);
(vi) $\gamma_{g,inf} G_k + \gamma_q Q_w$: $1.0G_k + 1.5Q_w$ (lower bound of permanent action and wind load as the only variable action).

If there is soil load as an additional variable action, the load combinations are:

(i) $\gamma_g G_k$: $1.35G_k$ (No variable action);
(ii) $\gamma_g G_k + \gamma_q Q_k$: $1.35G_k + 1.5Q_k$ (live load as the only variable action);
(iii) $\gamma_g G_k + \gamma_q Q_s$: $1.35G_k + 1.5Q_s$ (soil load as the only variable action)
(iv) $\gamma_g G_k + \gamma_q Q_w$: $1.35G_k + 1.5Q_w$ (wind load as the only variable action);
(v) $\gamma_g G_k + \gamma_q Q_k + \psi_{0,s}\gamma_q Q_s$:
$1.35G_k + 1.5Q_k + 0.7 \times 1.5Q_s = 1.35G_k + 1.5Q_k + 1.05Q_s$ (live load as the leading variable action + soil load);
(vi) $\gamma_g G_k + \psi_{0,q}\gamma_q Q_k + \gamma_q Q_s$:
$1.35G_k + 1.5Q_k + 0.7 \times 1.5Q_s = 1.35G_k + 1.5Q_k + 1.05Q_s$ (soil load as the leading variable action + live load);
(vii) $\gamma_g G_k + \gamma_q Q_k + \psi_{0,w}\gamma_q Q_w$:
$1.35G_k + 1.5Q_k + 0.6 \times 1.5Q_w = 1.35G_k + 1.5Q_k + 0.9Q_w$ (live load as the leading variable action + wind load);
(viii) $\gamma_g G_k + \psi_{0,q}\gamma_q Q_s + \gamma_q Q_w$:
$1.35G_k + 0.7 \times 1.5Q_s + 1.5Q_w = 1.35G_k + 1.05Q_s + Q_w$ (wind load as the leading variable action + live load);
(ix) $\gamma_g G_k + \psi_{0,s}\gamma_q Q_s + \gamma_q Q_w$:
$1.35G_k + 0.7 \times 1.5Q_s + 1.5Q_w = 1.35G_k + 1.05Q_s + 1.5Q_w$ (wind load as the leading variable action + soil load);
(x) $\gamma_g G_k + \gamma_q Q_s + \psi_{0,w}\gamma_q Q_w$:
$1.35G_k + 1.5Q_s + 0.6 \times 1.5Q_w = 1.35G_k + 1.5Q_s + 0.9Q_w$ (soil load as the leading variable action + wind load);

(xi) $\gamma_g G_k + \gamma_q Q_k + \psi_{0,s}\gamma_q Q_s + \psi_{0,w}\gamma_q Q_w$:

$1.35G_k + 1.5Q_k + 0.7 \times 1.5Q_s + 0.6 \times 1.5Q_w = 1.35G_k + 1.5Q_k + 1.05Q_s + 0.9Q_w$ (live load as the leading variable action + soil load + wind load);

(xii) $\gamma_g G_k + \psi_{0,q}\gamma_q Q_k + \gamma_q Q_s + \psi_{0,w}\gamma_q Q_w$:

$1.35G_k + 0 \times 1.5Q_k + 1.5Q_s + 0.6 \times 1.5Q_w = 1.35G_k + 0Q_k + 1.5Q_s + 0.9Q_w$ (soil load as the leading variable action + live load + wind load);

(xiii) $\gamma_g G_k + \psi_{0,q}\gamma_q Q_k + \psi_{0,s}\gamma_q Q_s + \gamma_q Q_w$:

$1.35G_k + 0 \times 1.5Q_k + 0.7 \times 1.5Q_s + 1.5Q_w = 1.35G_k + 0Q_k + 1.05Q_s + 1.5Q_w$ (wind load as the leading variable action + live load + wind load);

(xiv) $\gamma_{g,inf} G_k + \gamma_q Q_w$: $1.0G_k + 1.5Q_w$ (lower bound of permanent action and wind load as the only variable action).

The load combinations for the serviceability limit state are similar with the same application of the ψ factors but the partial factors $\gamma = 1$.

Load combinations involving accident or seismic actions should be in accordance with Equations 5.52 and 5.53.

For the combination of actions for the serviceability limit state, the same principles as those of the ultimate limit state are applied except that the partial factors γ are removed.

The foregoing is in accordance with BSEN1990:2023 (2023). However, the older version of the Eurocode BSEN1990:2002 (2002) can still be used until the end of March 2028. Though with some differences in terminology and arrangement, the two are more or less identical in application.

5.12.2 ASCE Standard 7-22

The ASCE 07 (2022) recommends the following basic load combinations with load factors for "Load and Resistance Factor Design" (LRFD) structural member strength design

(i) 1.4D
(ii) 1.2D + 1.6L + 0.5(L_r or S or R)
(iii) 1.2D + 1.6(L_r or S or R) + (L or 0.5W)
(iv) 1.2D + 1.0W + L + 0.5(L_r or S or R)
(v) 1.2D + 1.0E + L + 0.2S
(vi) 0.9D + 1.0W
(vii) 0.9D + 1.0E

where D stands for dead load; L stands for live load; L_r stands for roof live load, S stands for snow load; W stands for wind load; E stands for earthquake load.

Depending on the specification of the material strengths, i.e., whether ultimate or the allowable, the ASCE 7-16 (2018) also lists combinations for "Allowable Stress Design" (ASD) as

(i) D
(ii) D + L
(iii) D + (L_r or S or R)
(iv) D + 0.75L + 0.75(L_r or S or R)
(v) D + (0.6W or 0.7E)
(vi) (a) D + 0.75L + 0.75(0.6W) + 0.75(L_r or S or R; (b) D + 0.75L + 0.75(0.7E) + 0.75S
(vii) 0.6D + 0.6W
(viii) 0.6D + 0.7E

5.12.3 BS8110

The load factors used in the load combination are listed in table 2.1 of the BS, which is reproduced for ease of reference in Table 5.6.

The factors are still in use in Hong Kong whose Concrete and Steel Codes are largely based on the British Standards.

TABLE 5.6
Load Factors for Load Combinations in BS8110-1:1997 (1997)

	Load Type					
	Dead		Imposed			
Load Combination	Adverse	Beneficial	Adverse	Beneficial	Earth[a] and Water[b] Pressure	Wind
1. Dead and imposed (and earth and water pressure)	1.4	1.0	1.6	0	1.2[c] 1.0[d]	–
2. Dead and wind (and earth and water pressure)	1.4	1.0	–	–	1.2[c] 1.0[d]	1.4
3. Dead and imposed (and earth and water pressure)	1.2	1.2	1.2	1.2	1.2[c] 1.0[d]	1.2

a The earth pressure is that obtained from BS8002 including an appropriate mobilization factor. The more onerous of the two factored conditions should be taken.

b. The value of 1.2 may be used where the maximum credible level of the water can be clearly defined. If this is not feasible, a factor of 1.4 should be used.

c. Unplanned excavation in accordance with BS8002, 3.2.2.2 not included in the calculation.

d. Unplanned excavation in accordance with BS8002, 3.2.2.2 included in the calculation.

5.13 LOAD PATTERNS

On a continuous beam or slab with several spans, it is well recognized that internal forces on a span (bending moments and shears) will be affected by loads of adjacent spans. For strength design by Eurocode, while the variable action (live load) may be present or absent on a span by which the partial factor γ_q varies between 0 and 1.5, the effect of the permanent action (dead load) also varies under the application of different partial factors γ_g which would be 1.0 and 1.35. The former British Code BS8110-1:1997 (1997) adopts the same principle but with different factors.

5.14 CONCLUSION

This chapter has presented a comprehensive overview of the various loads acting on building structures, which are fundamental parameters in structural design. The nature of these loads, including their inherent randomness, has been examined in depth. The discussion has also highlighted statistical methods for load determination and outlined the methodologies and requirements prescribed by major international design codes for quantifying load magnitudes.

REFERENCES

American Concrete Institute (2008) *ACI 209.2R-08 Guide for Modelling and Calculating Shrinkage and Creep in Hardened Concrete*. American Concrete Institute.

American Society of Civil Engineer (2010) *ASCE 7-10 Minimum Design Loads for Buildings and other Structures*. American Society of Civil Engineer.

American Society of Civil Engineer (2018) *ASCE 7-16 Minimum Design Loads for Buildings and other Structures*. American Society of Civil Engineer.

American Society of Civil Engineer (2022) *ASCE 7-22 Minimum Design Loads for Buildings and other Structures*. American Society of Civil Engineer.

Architectural Services Department of Hong Kong Special Administration Region Government (2008) *SEB Guidelines SEB-GL-OTH2, Guidelines on the Design for Floor Vibrations due to Human Actions Part II – Floor Vibrations due Rhythmic Activities*. Architectural Services Department, Hong Kong SAR Government.

Architectural Services Department of Hong Kong Special Administration Region Government, SEB Guidelines (2009) *SEB-GL-OTH4, Guidelines on the Design for Floor Vibrations due to Human Actions Part II – Floor Vibrations due to Walking Loads*. Architectural Services Department, Hong Kong SAR Government.

Australian / New Zealand Standard Joint Technical Committee – BD006 (2002) *AS/NZS 1170.0:2002 Part 0 Structural Design Actions General Part 0*. Standards Australia / Standards New Zealand.

Australian / New Zealand Standard Joint Technical Committee – BD006 (2002) *AS/NZS 1170.1:2002 Structural Design Actions Part 1: Permanent,* Imposed and Other Actions. Standards Australia / Standards New Zealand.

Australian / New Zealand Standard Joint Technical Committee – BD006 (2002) *AS/NZS 1170.2:2021 Part 2 Wind Actions*. Standards Australia / Standards New Zealand.

Bachmann, H. and Ammann, W. (1987) *Structural Engineering Document 3e : Vibrations in Structures Induced by Man and Machines*. International Association for Bridge and Structural Engineering.

Bazant, Z.P. (1988) Material Models for Structural Creep Analysis. In: Bazant, Z.P. (ed.) Mathematical Modeling of Creep and Shrinkage of Concrete,. John Wiley and Sons, pp. 99–215.

British Standard Institute (1995) *BS6399-2:1995 Loading for Buildings – Part 2: Code of Practice for Wind Loads*. British Standard Institution.

British Standard Institution (1997) *BS8110-1:1997 Structural Use of Concrete Part 1 – Code of Practice for Design and Construction*. British Standard Institution.

British Standard Institute (2000) *BS5950-1:2000 Structural Use of Steelwork in Building – Part 1: Code of Practice for Design – Rolled and Welded Sections*. British Standard Institution.

British Standard Institute (2000) *BS8110-1:1997 Structural Use of Concrete – Part 1: Code of Practice for Design and Construction*. British Standard Institution.

British Standard Institute (2000) *BSEN1991-1-7:2006 Structural Use of Steelwork in Building – Part 1: General Actions Accidental Actions*. British Standard Institution.

British Standard Institute (2002) *BS6399-2:1997 Loading for Buildings – Part 2: Code of Practice for Wind Loads*. British Standard Institution.

British Standard Institution (2002) BSEN 1990:2002 Eurocode – Basis of Design. British Standard Institution.

British Standard Institute (2010) *NA to BSEN1991-1-4:2005+A1:2010 UK National Annex to Eurocode 1 – Actions on Structures Part 1-4: General Actions – Wind Actions*. British Standard Institution.

British Standard Institution (2023) *BS EN 1990:2023 Eurocode Basis of Structural Design*. British Standard Institution.

British Standard Institution (2025) *BS EN 1991-1-1:2025 Eurocode 1 Actions on Structures*. British Standard Institution.

Buildings Department of Hong Kong (2011) *Code of Practice for Dead and Imposed Load 2011*. Buildings Department of Hong Kong.

Buildings Department, Hong Kong SAR Government (2004) *Code of Practice on Wind Effects of Hong Kong 2004*. Buildings Department, Hong Kong SAR Government.

Buildings Department, Hong Kong SAR Government (2004) *Explanatory Materials to the Code of Practice on Wind Effects 2004*. Buildings Department, Hong Kong SAR Government.

Buildings Department, Hong Kong SAR Government (2019) *Code of Practice on Wind Effects of Hong Kong 2019*. Buildings Department, Hong Kong SAR Government.

Buildings Department, Hong Kong SAR Government (2019) *Explanatory Notes to the Code of Practice on Wind Effects of Hong Kong 2019*. Buildings Department, Hong Kong SAR Government.

Durst, C.S. (1960) Wind speeds over short period of time. *Meteorol. Mag.* 89, 181–187.

Ellis, B.R. (2000) On the Response of Long-Span Floors to Walking Loads Generated by Individuals and Crowds. *The Structural Engineer*, 78(10), 17–25.

Ellis, B.R. (2003) The Influence of Crowd Size on Floor Vibrations Induced by Walking. *The Structural Engineer*, 81(6), 20–27.

Ellis, B.R. and Ji, T. (1997) BRE Digest 426: Response of Structures Subject to Dynamic Crowd Loads. London: BRE Centre for Structural Engineering.

Eurocode (1998) *BSEN 1998-1* Design of Structures Against Earthquake Resistance – Part 1: General Rules, Seismic Actions and Rules for Buildings. British Standard Institution.

Eurocode (2003) *EN 1991-1-3* Actions on Structures – General Actions Snow Loads. European Committee for Standardization.

Eurocode (2005) *EN 1991-1-4:* Actions on Structures – General Actions Snow Loads. European Committee for Standardization.

Eurocode (2023) *BSEN 1991-1:2023 Design of Concrete Structures Part 1-1: General Rules and Rules for Buildings, Bridges and Civil Engineering Structures.*

Government of Peoples' Republic of China (2012) *GB50009-2012 Load Code for the Design of Building Structures.* Ministry of Housing and Urban-Rural Development.

International Federation for Structural Concrete (*fib*) (2012) Model Code 2010. fib – Fédération Internationale du Béton.

Joint Technical Committee BD-006 (n.d.) General Design Requirements and Loading on Structures, AS/NZS 1170.1. Standards Australia / Standards New Zealand.

Mortezaei, A. and Zaharai, S.M. (2009) Seismic Response of Reinforced Concrete Building with Viscoelastic Damper under Near Field Earthquake. *Asian Journal of Civil Engineering* 10(3), 347–359.

Murray, T.M., Allen, D.E. and Ungar, E.E. (1997) AISC Steel Design Guide 11 – Vibrations of Structural Systems Due to Human Activity. American Institute of Steel Construction.

National Research Council of Canada (2010) National Building Code of Canada. National Research Council.

Newmark, N.M. (1959) A Method of Computation for Structural Dynamics. ASCE Journal of the Engineering Mechanics Division, 85(EM3), 67–94.

Peterka, J.A. and Shahid, S. (1998) Design Gust Wind Speeds in the United States. Journal of Structural Engineering, 1247(2), 207–214.

Simiu, E. (1996) *Wind Effects on Structures*, 3rd Edition. John Wiley & Sons.

Smith, A.L., Hicks, S.J. and Devine, P. (2009) Design of Floors for Vibration: A New Approach. Steel Construction Institute.

The Ministry of Housing and Urban-rural Development (MOHUD) and the General Administration of Quality Supervision, Inspection and Guarantine (AQSIQ) of the Peoples' Republic of China (2015) Code for Seismic Design of Buildings GB50010-2010 (2010 Edition). MOHUD/AQSIQ.

Vibrationdata. (2016) Vibrationdata El Centro Earthquake Homepage.

6 Approaches in Structural Analysis of Building Structures under Computer Modelling

6.1 INTRODUCTION

Structural analysis of building structures has evolved significantly with advancements in computational methods and software tools since mid-1900s. Nowadays, computer method and modelling have become an indispensable part of the analysis and design of building structures. The extensive use of computer method enables engineers to simulate complex structural behaviours, optimize designs with high efficiency and economy, while achieving structural safety. In the history of evolution, there are many levels in application in the use of computer methods, ranging from simplified linear methods to advance non-linear and dynamic analyses. These computer-aided techniques allow for more accurate predictions of structural responses under different loading conditions, including static, dynamic, seismic, and wind effects. However, the authors have also noticed that due to inadequacy in basic theory and limitations of computer programs, some computer results have to be assessed with care but are seldom noticed until some technical problems are found. For example, the shear reinforcement required in a raft foundation design is so excessive that it is impossible to put the steel on site. Further refined study by Cheng and Law (2008) revealed the limitations of commercial program where nodal incompatibility can generate very high local shear. Since a good and nice mesh design can be difficult in some cases, if an engineer simply relies on the generation of a commercial program without careful consideration, serious errors can be incurred which are found from many projects in Hong Kong. Towards this, Cheng has given a short course to the engineers (ppt available on request) about the over-reliance on computer programs, background for the error in analysis, post-processing, and others.

In this chapter, simulation of structural members in the computer models of structure will first be discussed, which will be followed by approaches in analyses by computer modelling.

DOI: 10.1201/9781003636380-6

6.2 SIMULATION OF STRUCTURAL MEMBERS

In computer modelling of structures, different types of finite elements are used to simulate structural members based on their behaviour and the computational efficiency being aimed at. Broadly speaking, finite element includes also "line members" formed by the conventional stiffness matrix. The three primary categories of elements are:

1. Line (1-dimensional) Elements – Generally used for beam, column, and truss members, and sometimes simulation of wall and slab structures by wide column analogy or grillage beam system.
2. Planar (2-dimensional) Elements – Used for slabs, walls, and domes.
3. 3D-mensional Brick (or Solid) Elements – Used for detailed modelling of structures with complex geometries and considerable thickness.

Each element type has specific applications, advantages, and limitations in structural simulation, which are discussed in some detail as follows:

1. A line element is generally used to simulate shallow beams and columns where the sectional dimensions are small compared with the lengths of the structural members. In the computer model, a line element carries the section's full sectional properties including area, moments of inertia, and torsional stiffness. Its stiffness, which is incorporated in the stiffness matrix of the full structure for analysis is well defined and "exact" to the elastic theory. Unlike planar and solid elements, as discussed in Chapter 3 and Appendix D, the analytical results due to different degrees of segmentation (or meshing, which also refers to the division of the line element into segments along its length) are identical. A line element in a 2-dimensional model has 2 or 3 degrees of freedom per node, which increases to 3 or 6 respectively in a space truss or space frame model. Generally, shear deformations in the lateral dimensions of the line member are ignored in the basic formulation of a line member, though more sophisticated elements are available to incorporate them. For reference, the 3-dimensional stiffness matrix for a beam with shear deformation is listed (by Przemieniecki, 1968) and Ghali and Neville (2017) as follows. For the beam as listed, x is the direction along its axis while y and z axes are the orthogonal axes perpendicular to the x-axis.

$$K = \begin{bmatrix}
\frac{EA}{L} & 0 & 0 & 0 & 0 & 0 & \frac{-EA}{L} & 0 & 0 & 0 & 0 & 0 \\
0 & \frac{12EI_z}{(1+\Phi_y)L^3} & 0 & 0 & 0 & \frac{6EI_z}{(1+\Phi_y)L^2} & 0 & \frac{-12EI_z}{(1+\Phi_y)L^3} & 0 & 0 & 0 & \frac{6EI_z}{(1+\Phi_y)L^2} \\
0 & 0 & \frac{12EI_y}{(1+\Phi_z)L^3} & 0 & \frac{-6EI_y}{(1+\Phi_z)L^2} & 0 & 0 & 0 & \frac{-12EI_y}{(1+\Phi_z)L^3} & 0 & \frac{-6EI_y}{(1+\Phi_z)L^2} & 0 \\
0 & 0 & 0 & \frac{GJ}{L} & 0 & 0 & 0 & 0 & 0 & \frac{-GJ}{L} & 0 & 0 \\
0 & 0 & \frac{-6EI_y}{(1+\Phi_z)L^2} & 0 & \frac{(4+\Phi_z)EI_y}{(1+\Phi_z)L} & 0 & 0 & 0 & \frac{-6EI_z}{(1+\Phi_y)L^2} & 0 & \frac{(2+\Phi_z)EI_y}{(1+\Phi_z)L} & 0 \\
0 & \frac{6EI_z}{(1+\Phi_y)L^2} & 0 & 0 & 0 & \frac{(4+\Phi_z)EI_z}{(1+\Phi_y)L} & 0 & \frac{-6EI_z}{(1+\Phi_y)L^2} & 0 & 0 & 0 & \frac{(2-\Phi_y)EI_z}{(1+\Phi_y)L} \\
\frac{-EA}{L} & 0 & 0 & 0 & 0 & 0 & \frac{EA}{L} & 0 & 0 & 0 & 0 & 0 \\
0 & \frac{-12EI_z}{(1+\Phi_y)L^3} & 0 & 0 & 0 & \frac{-6EI_z}{(1+\Phi_y)L^2} & 0 & \frac{12EI_z}{(1+\Phi_y)L^3} & 0 & 0 & 0 & \frac{-6EI_z}{(1+\Phi_y)L^2} \\
0 & 0 & \frac{-12EI_y}{(1+\Phi_z)L^3} & 0 & \frac{6EI_y}{(1+\Phi_z)L^2} & 0 & 0 & 0 & \frac{12EI_y}{(1+\Phi_z)L^3} & 0 & \frac{6EI_y}{(1+\Phi_z)L^2} & 0 \\
0 & 0 & 0 & \frac{-GJ}{L} & 0 & 0 & 0 & 0 & 0 & \frac{GJ}{L} & 0 & 0 \\
0 & 0 & \frac{-6EI_y}{(1+\Phi_z)L^2} & 0 & \frac{(2-\Phi_z)EI_y}{(1+\Phi_z)L} & 0 & 0 & 0 & \frac{6EI_y}{(1+\Phi_z)L^2} & 0 & \frac{(4+\Phi_z)EI_y}{(1+\Phi_z)L} & 0 \\
0 & \frac{6EI_z}{(1+\Phi_y)L^2} & 0 & 0 & 0 & \frac{(2-\Phi_y)EI_z}{(1+\Phi_z)L} & 0 & \frac{-6EI_z}{(1+\Phi_y)L^2} & 0 & 0 & 0 & \frac{(4+\Phi_y)EI_y}{(1+\Phi_y)L}
\end{bmatrix} \tag{6.1}$$

E, G, A and I_y, I_z in Equation 6.1 are the usual symbols for the Young's modulus, shear modulus, cross-sectional area and moments of inertia of the section about Y and Z axes. $\Phi_y = \dfrac{12EI_z}{L^2GA_{sy}}$ and A_{sy} is the effective shear area in the Y-direction as related to the shape of the section which can be taken as $A/1.2$ for rectangular section and $A/1.11$ for circular section, similarly for $\Phi_z = \dfrac{12EI_z}{L^2GA_{sz}}$ in the Z-direction. K will revert to the ordinary stiffness matrix without consideration of shear deformation where Φ_y and Φ_z are set to zero.

Equation 6.1 can be reduced to the 2-dimensional plane frame beam or grillage beam matrix by selecting only the relevant parameters. For the plane frame, the movement in the X and Z directions together with rotation about the Y-axis are chosen while for that of the grillage, the parameters chosen are movement in the Z-direction, twisting about the X-direction and rotation about the Y-axis.

As the elastic theory is used, the line element's deformation is limited by the "plane remains plane" assumption, even with shear deformation being taken into consideration. So, the use of the line element to simulate a deep beam behaviour which does not follow the "plane remains plane" assumption may not be valid. This is demonstrated in Figure 6.1 for real examples which show the horizontal stress of a shallow and a deep beam under uniformly distributed load on top. While the horizontal stress follows a linear

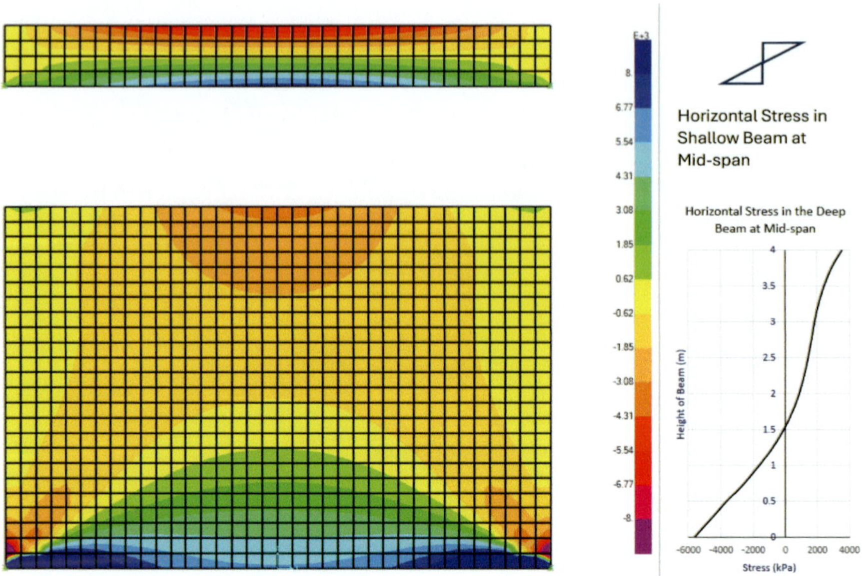

FIGURE 6.1 Horizontal stress distributions along beam depth for shallow and deep beams.

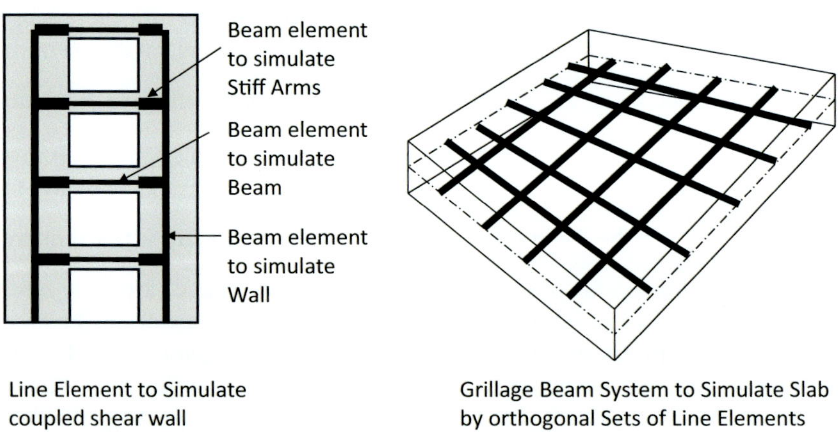

FIGURE 6.2 Structure simulation by line elements.

variation for the shallow beam, that of the deep beam is not linear as shown in Figure 6.1.

Nevertheless, before the wide application of planar members including the “membrane” or “shell” members as discussed in Chapter 3, the line members have been used to simulate shear walls by “wide column analogy” (in which the walls are simulated as "wide columns") or grillage beams to simulate slab as shown in Figure 6.2.

The simulation of coupled shear walls as shown in Figure 6.2 is generally good. However, that of grillage beam simulation of floor slab is generally not ideal. In assigning stiffness to the individual beams, duplication of stiffness may result if the same section is used for the beams in both directions. The correct distribution can hardly be determined, especially for a slab of irregular shape. In addition, the torsional stiffness has to be well accounted for which is difficult. With the use of the finite element method, the approach of simulation by grillage beam is currently seldom adopted. An in-house check by Cheng has demonstrated that the use of the grillage model can overestimate the design moment and shear as compared with the use of plate elements for slab type design.

2. A planar element is generally 2-dimensional resisting in-plane action or out-of-plane action or both as discussed in Chapter 3 and Appendix D. It is suitable for modelling wall and slab structures where the in-plane dimensions are significantly greater than their thicknesses. Being generally simulated by the finite element method as discussed in Chapter 3 and Appendix D, it does not require the "plane remains plane" assumption and is thus more versatile in the analysis of different shapes of structures at the expense of more complicated formulation.

   In the finite element method, out-of-plane shear deformation can typically be either neglected or incorporated into the formulation of planar structures. Appendix D provides formulations for both approaches. For thin plates, shear deformation is usually negligible, whereas for thick plates where shear deformation can be substantial, ignoring it may lead to significant errors. Nevertheless, plate elements that account for shear deformation generally yield more accurate results and should be preferred when available.

   In addition, a finely meshed planar structure can capture local stresses more correctly, which is important in locations where rapid change of stresses is anticipated. Nevertheless, care needs to be exercised in problems related to degrees of meshing, connection with line members, etc. Readers may refer to Chapter 3 and Appendix D for more details. Cheng et al. (2024) provided some examples where mesh refinement can greatly degrade the accuracy of the solution for a special case. An engineer must always exercise the knowledge and judgment in the analysis and should be able to assess the acceptability of the computer results by some simple methods. For some problems such as the action of a point load on a slab, the local results will tend to infinity with mesh refinement. It appears that some engineers are not aware of this phenomenon.

   Reinforced concrete design of planar elements also requires more sophisticated approaches which may not be familiar to general designers. There are, however, well-established approaches for reinforced concrete design for in-plane and out-of-plane loads (and combinations of both) as contained in Chapter 10 for the readers' reference.

3. A "solid" element in finite element analysis is one with 3-dimensional geometries as discussed in Chapter 3. In contrast to the line and planar elements, it is best suited to simulate structures where the three spatial dimensions (length, width and depth) are comparable. The element is also particularly useful in locations where stress gradients are great in 3-dimensions which cannot be accurately captured by a planar element, bearing in mind that a planar element allows only linearly varying stress distribution across its thickness. Consequently, thick structures such as pile caps, footings and transfer plates should be modelled as assemblies of "solid" elements. Law et al. (2007) has demonstrated the simulation of a pile cap by solid elements in which the stresses are studied, which are followed by reinforced concrete design.

   However, due to the complexity of modelling with solid elements and the tediousness in the subsequent reinforced concrete design (covered in Chapter 10), the solid element is seldom used in modelling of reinforced concrete building structures up to the present. The thick pile caps, footings, and transfer plates structures are often modelled as an assembly of plate elements instead. Nevertheless, this simplification ignores the violation of the "plane remains plane" assumption which may lead to inaccuracies, particularly at locations with closely spaced loads and supports where "truss action" dominates as reflected in Figure 3.18 in Chapter 3. As discussed in Chapter 5, the nodal stresses across a section can be used to recover the axial force, bending, and shear which has been used by Cheng for a complicated problem with more than one million solid elements.
4. Meshing is the division of a continuum structural element, say a wall or a slab, into elements of small sizes but interconnected together at the nodes. The importance of meshing has been discussed in Chapter 3 and Appendix D. While designers had to perform manual meshing by themselves in the early stage of the computer era, meshing is now mostly done automatically by the software. However, the auto-meshing by software sometimes can result in poor performance such as discontinuities of stresses, as discussed in Chapter 3, due to the nodal incompatibility problem. Users are therefore advised to examine the model after meshing and accept a generation mesh with knowledge and judgment. Such problem is also mentioned in the ppt to the engineers by Cheng.
5. Summing up, in the current practice, the beams (shallow beams) and columns are generally simulated by line elements, while the slabs and walls are simulated by planar elements. While currently most of the thick plate structures are modelled by plate elements, the validity in the use of plate elements should be carefully examined in the case of thick plate structures with closely spaced supports and loading elements. With the advancement in computer software and analysis/design techniques, solid elements should be more commonly in use in near future.

6.3 ANALYSIS METHOD

6.3.1 Linear Analysis

Under computer modelling, linear analysis for buildings is still the most popular method of analysis being employed in the industry currently due to its simplicity in application. In addition, linear load combinations can be adopted which greatly simplify the analysis, especially when there are a large number of load cases to solve. Although the modern computer can manipulate non-linear analysis with short computer time, onerous efforts in manual handling of data in the pre- and post- computer analysis stage are often required. So, for general analysis where no significant errors are anticipated from sources including the P-delta effect, construction sequence etc. as discussed in Chapter 2, linear analysis is still the preferred approach for the engineers.

6.3.2 Non-linear Analysis

Non-linear analysis is becoming more popular with the advancement of computer methods. The various methods and approaches of analysis including P-Delta effects, staged analysis, large deformation analysis, and elasto-plastic analysis forming plastic hinge have been discussed in Chapter 2. These methods are originally time-consuming analyses using primitive computers. But nowadays they can be analyzed within a reasonable time. More accurate results can then be acquired.

Designers should therefore exercise their judgment in choosing the method of analysis. For instance, the P-delta effect, which may be significant for tall and slender buildings, should be included in the design of these buildings. Similarly, a large span transfer structure supporting multi-floors above, staged analysis should be carried out to determine both locked-in stresses in the transfer structure and effects on the frames above due to successive settlement of the transfer structure. The modern computer method can now predict these effects more accurately instead of making wide assumptions before the computer era.

6.4 OPTIMIZATION

Computer method is an extremely useful and efficient tool to perform optimization in structural design. Before the computer era, optimization for an objective (minimization of cost, optimization of structural efficiency etc.) had to be performed by trial and error processes which were very time-consuming without computer. Often due to limitation of time, experience and resources, not all possible options can be examined, leading to omission of the most optimized solutions.

In the optimization process, some specific goals (or objective functions) have to be defined. These objective functions in structural design of buildings typically comprise minimization of material and optimization of stiffness, natural frequency and safety margins. These objectives, however, have to be achieved under "constraints." The constraints are generally limits imposed by the design codes or local authorities or the client which may include material strength limits, deflection tolerance, margins of safety, budget control, etc. An example for illustration purpose is the minimization of structural sizes under the constraint of achieving certain stiffness of the building to minimize deflection.

With input of other parameters for other objectives under concern, optimization can also be extended to minimization of construction cost and time, embodied carbon and some others.

Under the use of computer, optimization can best be performed by building a basic model and then formulating an optimization mathematical algorithm, say varying the related variables systematically through mathematical programming to find out the structure's performance in the objective function. For example, if the objective is to minimize structural sizes and the main constraint is the stiffness of the structure attaining certain threshold value so that the structure's deflection will not exceed a prescribed limit, the structural sizes in the basic model can be reduced systematically until the required performance can still be achieved.

There are two major types of optimization methods used by the engineers: continuity based gradient type method and heuristic type optimization methods. A detailed discussion about optimization theory is given in Cheng et al. (2020). In general, the gradient of the objective function is formed by a finite difference approach for the gradient type method, which is fast in computation. The main limitation in this method is the requirement of a good initial trial solution if multiple local minimum (maximum) exist in the solution, which is the common case for many practical solutions. In the heuristic type optimization methods which target towards global optimization, the gradient of the objective function and a good initial trial are not required. The main limitation of this approach is the relatively large number of computations. The choice of the optimization method depends heavily on the acceptability of the optimum design.

Summing up, by computer-based optimization, engineers can improve the design of structures that can achieve the required objectives in a fast and efficient manner. As such, safer, lighter, cheaper structures with high performance and sustainability can be built.

6.5 BUILDING INFORMATION MODELLING

Building Information Modelling (BIM) is the process for generating and managing intelligent 3-dimensional models of buildings throughout the construction life cycle, from planning and design to construction, and even operation and maintenance. Unlike the traditional 2-dimensional CAD drawings, BIM integrates multi-dimensional data (3-dimensional geometry + time, cost, sustainability etc.) into a centralized model accessible to all stakeholders including the architects, engineers, contractors and clients.

With the visualization aid provided by BIM, stakeholders can communicate efficiently. Conflicts or clashes can be resolved quickly. So, design can be performed in an accurate and efficient manner.

In recent years, efforts have been made to convert the BIM model into a structural engineering model for analysis and design. As such, the efforts spent by structural engineer in analysis and design can be significantly reduced. Besides, more accurate modelling can be achieved due to the minimization of human errors. However, such modelling is not yet mature for accurate application. There is room for further development in this aspect.

6.6 CONCLUSION

Computer modelling has become an indispensable tool in structural analysis and design. Structural design can be performed in a fast and accurate manner, provided the input parameters are correct. In fact, advanced computer software provides automation in input to minimize input errors.

REFERENCES

Cheng, Y.M. and Law, C.W. (2008) Development of a New and Efficient Thick Plate Element. Structural Engineering and Mechanics, 29(3), 327–354.

Cheng, Y.M., Law, C.W. and Liu, L.L. (2024) Foundation Analysis, Design and Construction, 2nd Edition. CRC Press.

Cheng, Y.M., Wang, J.H., Li, L. and Fung, W.H. (2020) Frontier in Civil Engineering, Vol. 4: Numerical Methods and Implementation in Geotechnical Engineering. Benjamin Press.

Ghali, A. and Neville, A.M. (2017) Structural Analysis: A Unified Classical and Matrix Approach, 7th Edition. CRC Press.

Law, C.W., Cheng, Y.M. and Su, R.K.L. (2007) Approach for Reinforcement Design in Reinforced Concrete Structures Based on 3-Dimensional Stress Field. The HKIE Transactions, 14(2), 9–18.

Przemieniecki, J.S. (1968) Theory of Matrix Structural Analysis. New York: Dover Publications.

7 Dynamic Analysis of Buildings

7.1 INTRODUCTION

Buildings, depending on their location and usage, are often subjected to dynamic loads such as those caused by wind, earthquakes, human activities and machinery vibrations throughout their service life. These dynamic loads typically produce more severe effects on structures compared to static loads. For simplicity, dynamic loads are frequently approximated as static loads using amplification factors, as discussed in Chapter 5. However, such approximations may fail to account for unforeseen dynamic structural behaviours, potentially compromising structural safety and human comfort. Conversely, these approximations may intentionally overestimate loads to address uncertainties not captured by static load analysis.

Dynamic analysis is inherently more complex than static analysis. However, advancements in theoretical frameworks and computational tools have made it increasingly feasible to perform dynamic analyses on building structures even of irregular shapes. In general, dynamic analysis involves studying a structure's response to time-varying loads. This can include time history analysis, which examines the structure's behaviour over a specific time period. Additionally, the modal method requires the evaluation of the structure's dynamic characteristics, such as natural frequencies and mode shapes, before proceeding with further analysis. Another widely used approach is the numerical integration method, where the structure's response is analyzed step-by-step under time-varying external loads.

This chapter begins with an overview of the fundamental principles of structural dynamic analysis. It then delves into the modal method and the numerical integration method, with the latter illustrated in detail using the Newmark Method. In practice, the choice of an appropriate analysis method depends on the complexity of the problem, as well as the availability of expertise and computational software.

7.2 BASIC PRINCIPLES

Illustration by a structure of single degree of freedom

The discussion of structural dynamics starts with the analysis of an idealized plane frame as shown in Figure 7.1 with a beam of infinite stiffness so that its motion is only horizontal which is a system of "single degree of freedom." The columns are weightless with lateral stiffness K which can be calculated by $12EI / L^3$ for each

 DOI: 10.1201/9781003636380-7

FIGURE 7.1 A portal frame with single degree of freedom in lateral motion.

where EI is the flexural rigidity of the column and L is its height, and the beam carries a mass M.

Let the lateral displacement be u. Under the condition of no damping and no external load, the following equation can be listed:

$$M\ddot{u} + ku = 0 \tag{7.1}$$

If the system is given an initial displacement u_0 and then allowed to sway freely, the solution of Equation 7.1 will be Equation 7.2 where t is the time and $\omega = \sqrt{K / M}$

$$u = u_0 \cos(\omega t) \tag{7.2}$$

By Equation 7.2, the lateral mass will oscillate with magnitude in accordance with a sine curve with time. ω is the circular frequency which is equal to $2\pi f$ where f is the frequency of the lateral motion by which the system will make f oscillations in one second. The frequency is also known as the natural frequency of the system. The motion described by Equation 7.2 is a simple harmonic motion.

However, there is inherently "damping" within the system which is the dissipation of energy as the structure oscillates. Damping originates from internal friction of the structure material which transforms energy to heat. Without new input of energy, the amplitude of the motion will diminish with time. Generally, it is assumed that the damping will incur a force opposing the motion which is assumed to be proportional to the velocity of the motion as shown by Equation 7.3

$$M\ddot{u} + c\dot{u} + ku = 0 \tag{7.3}$$

where c is the damping constant.

For light damping, the solution of Equation 7.3 is found to be

$$u = u_0 e^{-\frac{c}{2M}t} \cos(\omega_D t) \tag{7.4}$$

where u_0 is the initial displacement of the system and $\omega_D = \omega\sqrt{1-(c/2M\omega)^2}$.

By light damping as formed for the solution in Equation 7.4, it refers to the damping less than the "critical damping." Critical damping is that which will just bring the system to rest when the system reaches its original at rest position which can be proved to be

$$c_c = 2M\omega \tag{7.5}$$

Damping is however, more commonly expressed by a ratio of the damping constant c to the critical damping which is termed "damping ratio" ζ. Therefore, $c = 2\zeta M\omega$. Substituting into Equation 7.5, the equation can be written as

$$u = u_0 e^{-\zeta\omega t} \cos(\omega_D t) \tag{7.6}$$

Figure 7.2 shows the motions of the system without damping, with underdamping and the critical damping.

Equation 7.6 can be written more generally as:

$$u = e^{-\zeta\omega t}\left[A\cos(\omega_D t) + B\sin(\omega_D t)\right] \tag{7.7}$$

where the constants A and B are constants to cater for the initial magnitude of the vibration and the "phase angle" denoted by θ_i so that the initial magnitude need not

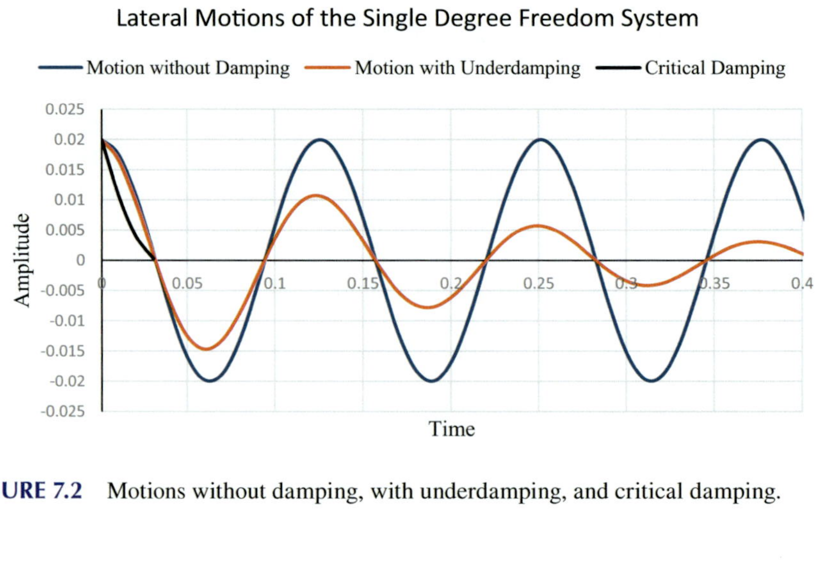

FIGURE 7.2 Motions without damping, with underdamping, and critical damping.

be 0 or the greatest magnitude, noting that $A\cos(\omega_D t)+B\sin(\omega_D t)=C\sin(\omega_D t+\theta_i)$ where C is the magnitude of vibration at $t=0$.

Next if there is an external lateral periodic force with angular frequency Ω acting on the beam of the portal frame in Figure 7.1 which can be expressed as $F=F_e\sin\Omega t$, the equation of motion can be written as:

$$M\ddot{u}+c\dot{u}+ku=F_e\sin\Omega t \tag{7.8}$$

The general solution to Equation 7.8, in accordance with Mario (2004) is

$$u=e^{-\zeta\omega t}\left[A\cos(\omega_D t)+B\sin(\omega_D t)\right]+\frac{F_e\sin(\Omega t-\theta)}{k\sqrt{\left(1-r^2\right)^2+\left(2r\zeta\right)^2}} \tag{7.9}$$

where $\theta=\tan^{-1}\left(\dfrac{2r\zeta}{1-r^2}\right)$ with $r=\dfrac{\Omega}{\omega}$.

The first term of Equation 7.9 is the transient state solution which persists at the early stage of motion of the system. Generally, this term will diminish when t increases so that only the second term remains as the steady stage solution. Nevertheless, when $\Omega=\omega$ such that $r=1$, i.e., the external excitation is of the same frequency as the natural frequency of the system, the amplitude of the motion will become very large. It becomes infinity when $\zeta=0$ (undamped motion). This is a state of "resonance" by which the structure will fail at displacements so large that it cannot tolerate.

A numerical analysis of the portal frame as shown in Figure 7.1 has been carried out to demonstrate the use of Equation 7.9 as Worked Example 7.1.

Worked Example 7.1 – Single Degree of Freedom

The columns of the frame have been assigned the structural sizes of 500 mm × 1200 mm with a height of 5 m, giving a total stiffness to the frame of $k=25000000\times2\times12\times\left(0.5\times1.2^3/12\right)/5^3=345600\,\text{kN/m}$ ($12EI/L^3$ for one column). The stiffness has to be converted to kg/sec² for ease of analysis which is $k=345600\times1000=345600000$kg/sec². The mass at the storey is taken as $m=44100\text{kg}$.

The circular frequency of the system is $\omega=\sqrt{k/m}=88.5253/\sec(14.089\text{Hz})$

Taking the damping ratio $\zeta=0.05$. So $\omega_D=\omega\left(1-\zeta^2\right)=88.30402/\text{sec}$.

Let the column be given a continuous peak ground horizontal acceleration of $a_g=1$ m/sec² with a circular frequency $\Omega=88$/sec, so that by the derivation in Chapter 5, it is equivalent to a horizontal force of $F_e\sin(\Omega t)=ma_g\sin(\Omega t)=44100\times1\sin(88t)\text{N}$.

$$r=\Omega/\omega=0.99656$$

in Equation 7.9.

The phase angle $\theta = \tan^{-1}\left(2\zeta r / \left(1 - r^2\right)\right) = 1.4523141$.

The initial conditions where time $t = 0$ are that both the displacement and velocity of the system are zero. These conditions are used to determine A and B in Equation 7.9. While the displacement is given by Equation 7.9, the velocity can be obtained by differentiating Equation 7.9 with respect to t.

$$\text{So } A = \frac{-F_e \sin(-\theta)}{k\sqrt{\left(1-r^2\right)^2 + (2r\zeta)^2}} = 0.001265723$$

$$B = \left[\zeta \omega A - \frac{F_e \cos\theta\Omega}{k\sqrt{\left(1-r^2\right)^2 + (2r\zeta)^2}}\right] / \omega_D = -0.000086708$$

Plugging in all the determined numerical values into Equation 7.9, the displacement versus time graph of the horizontal beam is plotted in Figure 7.3, together with the ground displacement which is $u_g = \left(1/\Omega^2\right)\sin\Omega t$.

From Figure 7.3, the vibration increases from zero and attains a steady state at around $t = 0.8\,\text{sec}$. The phase difference between the ground acceleration and the beam displacement is also obvious at $\theta = 1.452\,\text{rad}$ or 83.19°. The very high amplitude of oscillation of the horizontal beam is due to the closeness of the frame's natural frequency to the external excitation ($\omega = 88.525/\text{sec}$ versus $\Omega = 88/\text{sec}$ or 14.089 Hz versus 14.006 Hz) leading to vibration very close to "resonance."

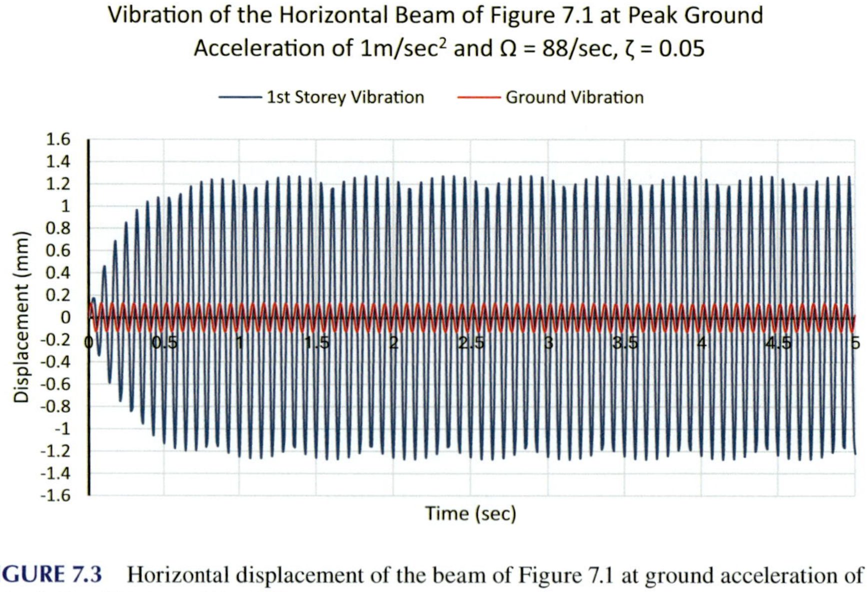

FIGURE 7.3 Horizontal displacement of the beam of Figure 7.1 at ground acceleration of 1 m/sec^2, $\Omega = 88/\text{sec}$, and $\zeta = 0.05$.

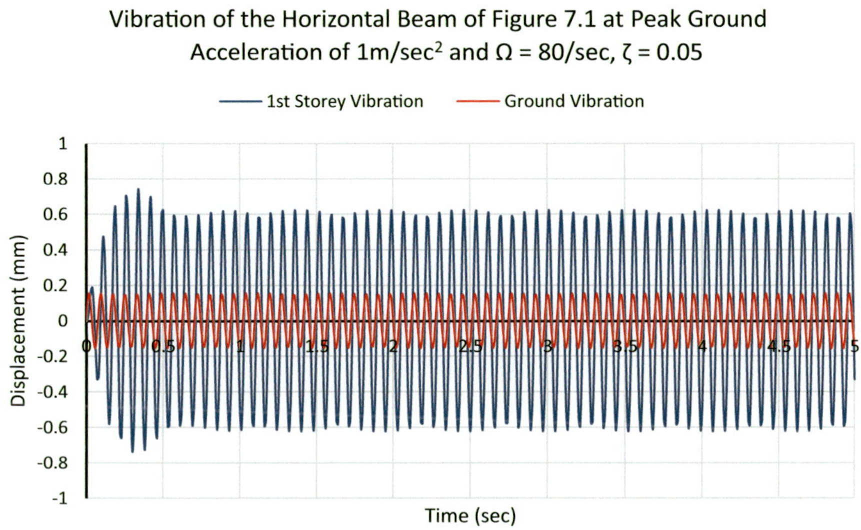

FIGURE 7.4 Horizontal displacement of the beam of Figure 7.1 at ground acceleration of 1 m/sec², $\Omega = 80$/sec, and $\zeta = 0.05$.

While keeping the peak ground acceleration at 1 m/sec² and decreasing the angular frequency Ω to 80/sec (or frequency at 12.732 Hz), the vibration of the beam is significantly reduced as plotted in Figure 7.4. It should be noted that the amplitude of the ground displacement increases as it is inversely proportional to Ω^2.

The phase angle is smaller at $\theta = 0.458\text{rad}$ or 26.24°.

If the damping ratio is set to zero $(=0)$ under the case with ground acceleration $\Omega = 88$/sec, excitation close to resonance takes place under no damping. The horizontal deflection of the beam increases with time to very large amplitudes as shown in Figure 7.5. Steady state is not found even at $t = 5$sec beyond which the amplitude continues to increase indefinitely until failure.

Upon determination of the displacement, the loads induced in the structure can be checked by multiplying the stiffness of the structure by the displacement. For the case of the single storey frame with $\Omega = 88$/sec and $\zeta = 0.05$ (Figure 7.3), the maximum deflection is 1.27 mm. So, the maximum shear on a column is $\left(12EI\Delta / L^3\right)$ $12 \times 25000000 \times 0.072 \times 0.00127 / 5^3 = 219.456\text{kN}$ and a moment of $219.456 \times 5 / 2 = 548.64\text{kNm}$.

The external excitation, however, may not follow a simple sine function as used for demonstration in the foregoing. Nevertheless, the excitation can be simulated as a superimposition of many sine functions by Fourier transform, and the motion can be obtained as a superimposition of the solutions for the respective sine functions.

An actual building structure is generally far more complicated than the single degree freedom structure shown in Figure 7.1 by possessing many degrees of freedom. By "discretizing" the mass of the building into the respective joints, the equations of motion of these masses can be listed. But unfortunately, these equations are

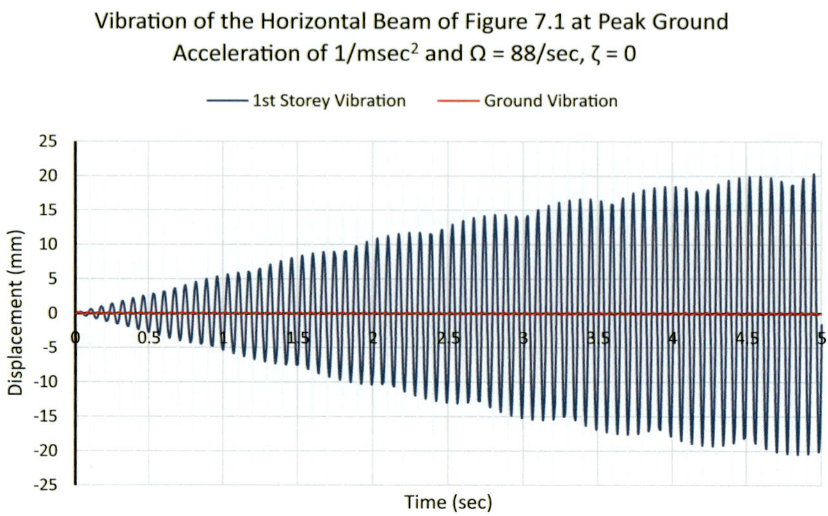

FIGURE 7.5 Horizontal displacement of the beam of Figure 7.1 at ground acceleration of 1 m/sec², Ω = 88/sec, and ζ = 0.

"coupled" by which each of them involves not only one variable with displacement, velocity, and acceleration. An example of a "shear building" (floor with infinite stiffness) of one Bay and three storeys as shown in Figure 7.6 will be used for illustration as Worked Example 7.2. In the figure, numerical values for the dimensions are also added which will be used for demonstration of analysis by the worked example.

7.2.1 Illustration by a One-Bay Three-Storey Shear Building

The shear building is first given some initial lateral displacements and left to vibrate freely without external loads. The equations of motion of the three storeys are

$$m_1\ddot{u}_1 + c_1\dot{u}_1 - c_2\left(\dot{u}_2 - \dot{u}_1\right) + k_1u_1 - k_2\left(u_2 - u_1\right) = 0 \tag{7.10}$$

$$m_2\ddot{u}_2 + c_2\left(\dot{u}_2 - \dot{u}_1\right) - c_3\left(\dot{u}_3 - \dot{u}_2\right) + k_2\left(u_2 - u_1\right) - k_3\left(u_3 - u_2\right) = 0 \tag{7.11}$$

$$m_3\ddot{u}_3 - c_3\dot{u}_2 + c_3\dot{u}_3 - k_3u_2 + k_3u_3 = 0 \tag{7.12}$$

In Equation 7.10, $u_2 - u_1$ and $\dot{u}_2 - \dot{u}_1$ are the displacement and velocity of the mass at the first storey that would produce the restoring force and damping force on the mass. So, it involves two variables and cannot be readily solved as a single differential equation as in Equation 7.8, similarly for Equations 7.11 and 7.12. So, the modal method that can solve the problem will be discussed in fuller detail in Section 7.3. As will be seen, there will be three modes of vibrations instead of a single mode for

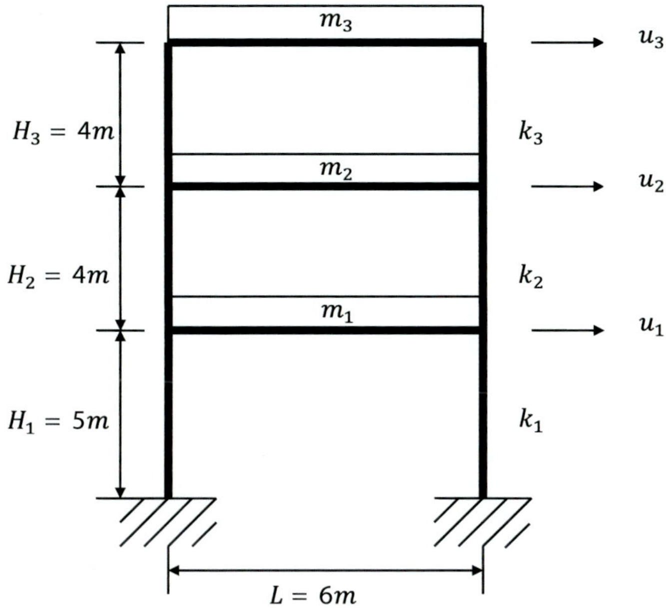

FIGURE 7.6 A one-bay three-storey shear building.

the single degree of freedom structure. Generally, the number of modes of vibrations is identical to the degree of freedom of the structure.

7.3 THE MODAL METHOD

The modal method in structural dynamic analysis involves the determination of the natural frequencies and modes of vibrations of the structure, with which the motions of the structure under dynamic loads are determined with appropriate superimposition of the modes. The analysis for the single degree of freedom structure, as described in the foregoing, is also by the modal method, which is simple as it involves only one variable. However, for a structure with multi-degree of freedom which involves many natural frequencies with each associated with a mode shape, the technique of "decoupling" has to be exercised for analysis.

7.3.1 One-Bay Three-Storey Shear Building

The method for multi-degree freedom of structure is described and demonstrated with reference to a one-bay three-storey shear building as shown in Figure 7.3.

Equations from 7.10 to 7.12 are re-arranged as in Equations 7.13 to 7.15

$$m_1\ddot{u}_1 + (c_1 + c_2)\dot{u}_1 - c_2\dot{u}_2 + (k_1 + k_2)u_1 - k_2u_2 = 0 \tag{7.13}$$

$$m_2\ddot{u}_2 - c_2\dot{u}_1 + (c_2 + c_3)\dot{u}_2 - c_3\dot{u}_3 - k_2u_1 + (k_2 + k_3)u_2 - k_3u_3 = 0 \tag{7.14}$$

$$m_3\ddot{u}_3 - c_3\dot{u}_2 + c_3\dot{u}_3 - k_3u_2 + k_3u_3 \tag{7.15}$$

In matrix form

$$\begin{bmatrix} m_1 & 0 & 0 \\ 0 & m_2 & 0 \\ 0 & 0 & m_3 \end{bmatrix}\begin{bmatrix} \ddot{u}_1 \\ \ddot{u}_2 \\ \ddot{u}_2 \end{bmatrix} = \begin{bmatrix} c_1 + c_2 & -c_2 & 0 \\ -c_2 & c_2 + c_3 & -c_3 \\ 0 & -c_3 & c_3 \end{bmatrix}\begin{bmatrix} \dot{u}_1 \\ \dot{u}_2 \\ \dot{u}_3 \end{bmatrix} + \begin{bmatrix} k_1 + k_2 & -k_2 & 0 \\ -k_2 & k_2 + k_3 & -k_3 \\ 0 & -k_3 & k_3 \end{bmatrix}\begin{bmatrix} u_1 \\ u_2 \\ u_3 \end{bmatrix} = \begin{bmatrix} 0 \\ 0 \\ 0 \end{bmatrix} \tag{7.16}$$

Or

$$\boldsymbol{m}\ddot{\boldsymbol{u}} + \boldsymbol{c}\dot{\boldsymbol{u}} + \boldsymbol{k}\boldsymbol{u} = 0 \tag{7.17}$$

The $\boldsymbol{k}$ matrix is actually identical to the stiffness matrix for static analysis. Under the static condition, the static displacement $\boldsymbol{u}_s$ due to external load $\boldsymbol{P}_s$ can be solved by $\boldsymbol{k}\boldsymbol{u}_s = \boldsymbol{P}_s$.

Ignoring the damping at the start, let the solution be
$u_1 = A_1 \sin(\omega t - \alpha)$; $u_2 = A_2 \sin(\omega t - \alpha)$; $u_3 = A_3 \sin(\omega t - \alpha)$; then
$\ddot{u}_1 = -\omega^2 A_1 \sin(\omega t - \alpha)$; $\ddot{u}_2 = -\omega^2 A_2 \sin(\omega t - \alpha)$; $\ddot{u}_3 = -\omega^2 A_3 \sin(\omega t - \alpha)$
Back substituting into three equations of motions

$$(7.13) \Rightarrow -\omega^2 m_1 A_1 + (k_1 + k_2)A_1 - k_2A_2 = 0$$

$$\Rightarrow (k_1 + k_2)A_1 - k_2A_2 = \omega^2 m_1 A_1 \tag{7.18}$$

$$(7.14) \Rightarrow -\omega^2 m_2 A_2 - k_2A_1 + (k_2 + k_3)A_2 - k_3A_3 = 0$$

$$\Rightarrow -k_2A_1 + (k_2 + k_3)A_2 - k_3A_3 = \omega^2 m_2 A_2 \tag{7.19}$$

$$(7.15) \Rightarrow -\omega^2 m_3 A_3 - k_3A_2 + k_3A_3 = 0$$

$$\Rightarrow -k_3A_2 + k_3A_3 = -\omega^2 m_3 A_3 \tag{7.20}$$

Putting into matrix form

$$\begin{bmatrix} k_1 + k_2 & -k_2 & 0 \\ -k_2 & k_2 + k_3 & -k_3 \\ 0 & -k_3 & k_3 \end{bmatrix}\begin{bmatrix} A_1 \\ A_2 \\ A_3 \end{bmatrix} = \omega^2 \begin{bmatrix} m_1 & 0 & 0 \\ 0 & m_2 & 0 \\ 0 & 0 & m_3 \end{bmatrix}\begin{bmatrix} A_1 \\ A_2 \\ A_3 \end{bmatrix} \tag{7.21}$$

Putting $\phi = \begin{bmatrix} A_1 \\ A_2 \\ A_3 \end{bmatrix}$;

Equation 7.21 can be written as

$$\boldsymbol{k}\phi = \omega^2 \boldsymbol{m}\phi \tag{7.22}$$

$$\Rightarrow \boldsymbol{k}^{-1}\boldsymbol{m}\phi = \frac{1}{\omega^2}\phi$$

$$\Rightarrow \boldsymbol{d}\phi = \lambda\phi \tag{7.23}$$

where $\boldsymbol{d} = \boldsymbol{k}^{-1}\boldsymbol{m}$ and $\lambda = 1/\omega^2$

Equation 7.23 is an eigenvalue problem for the matrix $\boldsymbol{k}^{-1}\boldsymbol{m}$ which is the inverse of the stiffness matrix times (dot product) the mass matrix by which ϕ and λ as eigenvectors and eigenvalues can be solved. Physically λ corresponds to the inverse of the square of the natural circular frequencies and ϕ corresponds to the mode shapes under the λ they are arrived at.

Put $\Phi = \begin{bmatrix} \phi_{11} & \phi_{12} & \phi_{13} \\ \phi_{21} & \phi_{22} & \phi_{23} \\ \phi_{31} & \phi_{32} & \phi_{33} \end{bmatrix}$, the first column is the mode shape of the first mode, and the second column is the second mode etc.

It should also be noted that the column vectors of Φ are pure ratios, though they are conveniently given by so assigned that $\phi_{11}^2 + \phi_{21}^2 + \phi_{31}^2 = 1$; $\phi_{12}^2 + \phi_{22}^2 + \phi_{32}^2 = 1$; $\phi_{13}^2 + \phi_{23}^2 + \phi_{33}^2 = 1$.

By the orthogonality property of Φ, it can be shown that

$\boldsymbol{\Phi}^T\boldsymbol{m}\boldsymbol{\Phi}$; $\boldsymbol{\Phi}^T\boldsymbol{k}\boldsymbol{\Phi}$ are diagonal matrices, i.e.,

$$\boldsymbol{M} = \boldsymbol{\Phi}^T\boldsymbol{m}\boldsymbol{\Phi} = \begin{bmatrix} M_1 & 0 & 0 \\ 0 & M_2 & 0 \\ 0 & 0 & M_3 \end{bmatrix} \tag{7.24}$$

$$\boldsymbol{K} = \boldsymbol{\Phi}^T\boldsymbol{k}\boldsymbol{\Phi} = \begin{bmatrix} K_1 & 0 & 0 \\ 0 & K_2 & 0 \\ 0 & 0 & K_3 \end{bmatrix} \tag{7.25}$$

The assignment of M_1, M_2, M_3 and K_1, K_2, K_3 will be in accordance with the mode shape where the storey has attained the maximum value in the mode. For example, if the three modes are as follows,

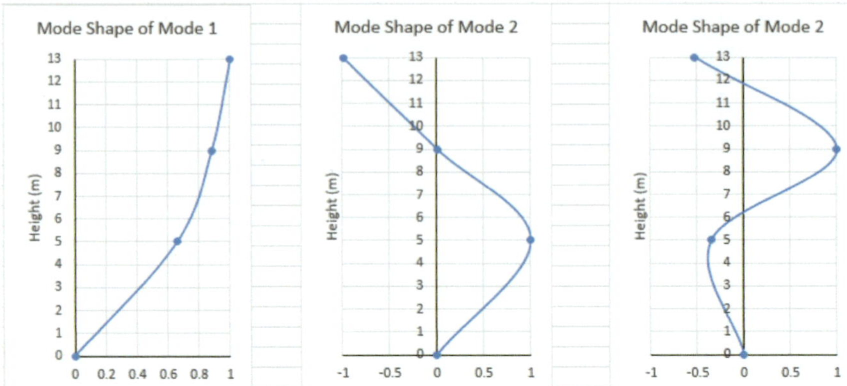

then M_1 and K_1 will be assigned to the third storey, M_2 and K_2 will be assigned to the first storey and M_3 and K_3 will be assigned to the first storey in the formulation of the equations of motion as

$$M_i\ddot{q}_i + K_i q_i = 0 \tag{7.26}$$

where $i = 1, 2 \text{ and } 3$ and q_i is the amplitude vector related to the original displacement vector u_i determined by

$$\boldsymbol{u} = \begin{bmatrix} u_1 \\ u_2 \\ u_3 \end{bmatrix} = \begin{bmatrix} \phi_{11} & \phi_{12} & \phi_{13} \\ \phi_{21} & \phi_{22} & \phi_{23} \\ \phi_{31} & \phi_{32} & \phi_{33} \end{bmatrix} \begin{bmatrix} q_1 \\ q_2 \\ q_3 \end{bmatrix} = \boldsymbol{\Phi q} \tag{7.27}$$

If $\boldsymbol{\Phi}$ is so scaled that $M_1 = M_2 = M_3 = 1$, the vectors of $\boldsymbol{\Phi}$ are called mass-normalized mode shape vectors.

$\boldsymbol{c}$ in Equation 7.17 represents "damping" in the system, which is a result of a complicated friction and energy dissipation mechanism that occurs in every component throughout the structure. It is difficult to account for its value. Furthermore, it is unlikely that $\boldsymbol{C} = \boldsymbol{\Phi}^T \boldsymbol{c} \boldsymbol{\Phi}$ will also become a diagonal matrix. However, an approximation for it to be a diagonal matrix for the ease of solution of Equation 7.17 should be appropriate. Bearing in mind that the critical damping is $2M\omega$, let

$$\boldsymbol{C} = \begin{bmatrix} 2M_1\omega_1\zeta_1 & 0 & 0 \\ 0 & 2M_2\omega_2\zeta_2 & 0 \\ 0 & 0 & 2M_3\omega_3\zeta_3 \end{bmatrix} \tag{7.28}$$

where $\zeta_1 = \dfrac{C_1}{2M_1\omega_1}$; $\zeta_2 = \dfrac{C_2}{2M_2\omega_2}$; $\zeta_3 = \dfrac{C_3}{2M_3\omega_3}$, known as "damping ratio" of the respective mode.

Following Equation 7.27 $\boldsymbol{u} = \begin{bmatrix} u_1 \\ u_2 \\ u_3 \end{bmatrix} = \begin{bmatrix} \phi_{11} & \phi_{12} & \phi_{13} \\ \phi_{21} & \phi_{22} & \phi_{23} \\ \phi_{31} & \phi_{32} & \phi_{33} \end{bmatrix} \begin{bmatrix} q_1 \\ q_2 \\ q_3 \end{bmatrix} = \boldsymbol{\Phi q}$ where $\{q\} = \begin{bmatrix} q_1 \\ q_2 \\ q_3 \end{bmatrix}$ is the modal amplitude vector. So the actual displacement is treated as a superposition of the modal contributions, i.e., $u_1 = q_1\phi_{11} + q_2\phi_{12} + q_3\phi_{13}$; $u_2 = q_1\phi_{21} + q_2\phi_{22} + q_3\phi_{23}$ and $u_3 = q_1\phi_{31} + q_2\phi_{32} + q_3\phi_{33}$

Substituting Equation 7.27 into Equation 7.17 with an external force $\boldsymbol{P}$

$$\boldsymbol{m\ddot{u}} + \boldsymbol{c\dot{u}} + \boldsymbol{ku} = \boldsymbol{P} \Rightarrow \boldsymbol{m\Phi\ddot{q}} + \boldsymbol{c\Phi\dot{q}} + \boldsymbol{k\Phi q} = \boldsymbol{P} \tag{7.29}$$

Multiplying $\boldsymbol{\Phi}^T$ on both sides of Equation 7.29

$$\boldsymbol{\Phi}^T \boldsymbol{m\Phi\ddot{q}} + \boldsymbol{\Phi}^T \boldsymbol{c\Phi\dot{q}} + \boldsymbol{\Phi}^T \boldsymbol{k\Phi q} = \boldsymbol{\Phi}^T \boldsymbol{P} \tag{7.30}$$

By Equations 7.24, 7.25, and 7.28, Equation 7.30 becomes

$$\boldsymbol{M\ddot{q}} + \boldsymbol{C\dot{q}} + \boldsymbol{Kq} = \boldsymbol{\Phi}^T \boldsymbol{P} \tag{7.31}$$

Or $\begin{cases} M_1\ddot{q}_1 + C_1\dot{q}_1 + K_1 q_1 = \phi_{11}P_1 + \phi_{21}P_2 + \phi_{31}P_3 \\ M_2\ddot{q}_2 + C_2\dot{q}_2 + K_2 q_2 = \phi_{12}P_1 + \phi_{22}P_2 + \phi_{32}P_3 \\ M_3\ddot{q}_3 + C_3\dot{q}_3 + K_3 q_3 = \phi_{13}P_1 + \phi_{23}P_2 + \phi_{33}P_3 \end{cases}$

So, the three equations can be solved independently. By the above approach, Equation 7.17 has been "decoupled" to Equation 7.31.

If the external load is zero, Equation 7.31 becomes

$$\begin{cases} M_1\ddot{q}_1 + C_1\dot{q}_1 + K_1 q_1 = 0 \\ M_2\ddot{q}_2 + C_2\dot{q}_2 + K_2 q_2 = 0 \\ M_3\ddot{q}_3 + C_3\dot{q}_3 + K_3 q_3 = 0 \end{cases} \tag{7.32}$$

The solution is

$$\begin{cases} q_1 = e^{-(C_1/2M_1)t} A_1 \sin\left(\omega_{D1}t + \theta_1\right) \\ q_2 = e^{-(C_2/2M_2)t} A_2 \sin\left(\omega_{D2}t + \theta_2\right) \\ q_3 = e^{-(C_3/2M_3)t} A_3 \sin\left(\omega_{D3}t + \theta_3\right) \end{cases} \Rightarrow \begin{matrix} q_1 = e^{-\zeta_1\omega_1 t} A_1 \sin\left(\omega_{D1}t + \theta_1\right) \\ q_2 = e^{-\zeta_2\omega_2 t} A_2 \sin\left(\omega_{D2}t + \theta_2\right) \\ q_3 = e^{-\zeta_3\omega_3 t} A_3 \sin\left(\omega_{D3}t + \theta_3\right) \end{matrix} \tag{7.33}$$

where $\zeta_1 = C_1/(2M_1\omega_1)$; $\zeta_2 = C_2/(2M_2\omega_2)$; $\zeta_3 = C_3/(2M_3\omega_3)$ and θ_1, θ_2, θ_3 are the initial phase angles denoting the displacements at $t = 0$.

$$\omega_{D1} = \omega_1\sqrt{1-\zeta_1^{\ 2}} \ ; \omega_{D2} = \omega_2\sqrt{1-\zeta_2^{\ 2}} \ ; \omega_{D3} = \omega_3\sqrt{1-\zeta_3^{\ 2}}$$

By Equation 7.27 where $\boldsymbol{u} = \begin{bmatrix} u_1 \\ u_2 \\ u_3 \end{bmatrix} = \begin{bmatrix} \phi_{11} & \phi_{12} & \phi_{13} \\ \phi_{21} & \phi_{22} & \phi_{23} \\ \phi_{31} & \phi_{32} & \phi_{33} \end{bmatrix} \begin{bmatrix} q_1 \\ q_2 \\ q_3 \end{bmatrix} = \boldsymbol{\Phi q}$

$$\boldsymbol{u} = \begin{bmatrix} u_1 \\ u_2 \\ u_3 \end{bmatrix} = \begin{bmatrix} \phi_{11} & \phi_{12} & \phi_{13} \\ \phi_{21} & \phi_{22} & \phi_{23} \\ \phi_{31} & \phi_{32} & \phi_{33} \end{bmatrix} \begin{bmatrix} e^{-\zeta_1\omega_1 t} A_1 \sin(\omega_{D1}t + \theta_1) \\ e^{-\zeta_2\omega_2 t} A_2 \sin(\omega_{D2}t + \theta_2) \\ e^{-\zeta_3\omega_3 t} A_3 \sin(\omega_{D3}t + \theta_3) \end{bmatrix}$$

For light damping where ζ_1, ζ_2 and ζ_3 are small, takes $\omega_{D1} = \omega_1$; $\omega_{D2} = \omega_2$ and $\omega_{D3} = \omega_3$

$$\begin{cases} u_1 = \phi_{11}e^{-\zeta_1\omega_1 t} A_1 \sin(\omega_1 t + \theta_1) + \phi_{12}e^{-\zeta_2\omega_2 t} A_2 \sin(\omega_2 t + \theta_2) + \phi_{13}e^{-\zeta_3\omega_3 t} A_3 \sin(\omega_3 t + \theta_3) \\ u_2 = \phi_{21}e^{-\zeta_1\omega_1 t} A_1 \sin(\omega_1 t + \theta_1) + \phi_{22}e^{-\zeta_2\omega_2 t} A_2 \sin(\omega_2 t + \theta_2) + \phi_{23}e^{-\zeta_3\omega_3 t} A_3 \sin(\omega_3 t + \theta_3) \\ u_3 = \phi_{31}e^{-\zeta_1\omega_1 t} A_1 \sin(\omega_1 t + \theta_1) + \phi_{32}e^{-\zeta_2\omega_2 t} A_2 \sin(\omega_2 t + \theta_2) + \phi_{33}e^{-\zeta_3\omega_3 t} A_3 \sin(\omega_3 t + \theta_3) \end{cases} \tag{7.34}$$

where A_1, A_2, A_3 and θ_1, θ_2, θ_3 can be solved by initial conditions, and the motion is a free vibration one.

The above is illustrated by a numerical problem for the one-bay three-storey shear building as shown in Figure 7.6 as Worked Example 7.2 as follows.

Worked Example 7.2

For Figure 7.6, $m_1 = 44100\text{kg}$; $m_2 = m_3 = 29400\text{kg}$. The columns are 500 mm (W) × 1200 mm (D) so that the I value is 0.072m^4 . The Young's modulus of concrete is $2.5\times10^6\,\text{kN/m}^2$.

$$k_1 = \frac{12EI}{L^3}\times 2 = \frac{12\times 25\times 10^6 \times 0.072\times 2}{5^3} = 345600\text{kN/m} = 345600000\text{kg/sec}^2;$$

$$k_2 = k_3 = \frac{12EI}{L^3}\times 2 = \frac{12\times 25\times 10^6 \times 0.072\times 2}{4^3} = 675000\text{kN/m} = 675000000\text{kg/sec}^2$$

So, the stiffness matrix is

$$\boldsymbol{k} = \begin{bmatrix} k_1 + k_2 & -k_2 & 0 \\ -k_2 & k_2 + k_3 & -k_3 \\ 0 & -k_3 & k_3 \end{bmatrix} = \begin{bmatrix} 1020600000 & -675000000 & 0 \\ -675000000 & 1350000000 & -675000000 \\ 0 & -675000000 & 675000000 \end{bmatrix}$$

$$d = k^{-1}m = \begin{bmatrix} 0.012517969 & 0.008345313 & 0.008345313 \\ 0.012517969 & 0.012618113 & 0.012618113 \\ 0.012517969 & 0.012618113 & 0.016890913 \end{bmatrix}$$

The eigenvalues and eigenvectors are solved numerically as

$$\lambda = \begin{bmatrix} 0.003629 \\ 0.0004252 \\ 0.0001481 \end{bmatrix} \text{ and } \Phi = \begin{bmatrix} -0.443927 & -0.708789 & 0.299317 \\ -0.592824 & -0.003392444 & -0.842934 \\ -0.671929 & 0.705413 & 0.4470701 \end{bmatrix}$$

Each of the three columns of Φ represents each of the three mode shapes shown below in Figure 7.7:

λ is related to the angular frequencies by $\omega = \sqrt{1/\lambda}$. So,

$\omega_1 = \sqrt{1/0.000369968} = 51.98977$ or $f_1 = 51.98977/2\pi = 8.274$ Hz;

$\omega_2 = \sqrt{1/0.000043347} = 151.8868$ or $f_2 = 151.8868/2\pi = 24.1735$ Hz ;

$\omega_3 = \sqrt{1/0.000015095} = 257.38666$ or $f_3 = 257.38666/2\pi = 40.9644$ Hz .

Thus, the structure possesses three natural frequencies with distinct modes of vibrations. The fundamental (first), second, and third natural frequencies are 2.642 Hz, 7.718 Hz, and 13.079 Hz. The motion of the structure is a combination of the three modes.

By Equation 7.24

$$M = \Phi^T m \Phi = \begin{bmatrix} M_1 & 0 & 0 \\ 0 & M_2 & 0 \\ 0 & 0 & M_3 \end{bmatrix} = \begin{bmatrix} 32296.949 & 0 & 0 \\ 0 & 36785.008 & 0 \\ 0 & 0 & 30716.98 \end{bmatrix}$$

By Equation 7.25

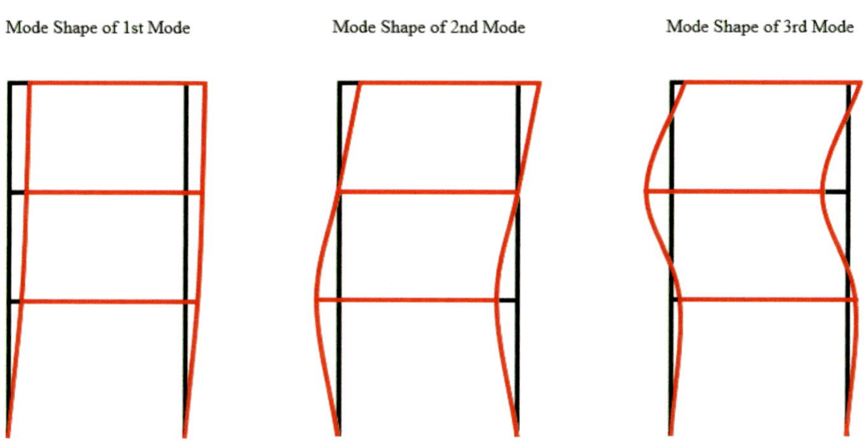

FIGURE 7.7 Mode shapes of the numerical problem – three-storey shear building.

$$\boldsymbol{K} = \Phi^T \boldsymbol{k} \Phi = \begin{bmatrix} K_1 & 0 & 0 \\ 0 & K_2 & 0 \\ 0 & 0 & K_3 \end{bmatrix} = \begin{bmatrix} 87296648 & 0 & 0 \\ 0 & 848615356 & 0 \\ 0 & 0 & 2034934589 \end{bmatrix}$$

At time $t = 0$, if given that the structure is at its greatest displacement (+ve to the right), with $u_1 = 0.001$; $u_2 = 0.001512$; $u_3 = 0.002024$. $\therefore \theta_1 = \theta_2 = \theta_3 = \pi / 2$. By Equation 7.34

$$\begin{cases} u_1 = \phi_{11} e^{-\zeta_1 \omega_1 t} A_1 \cos(\omega_1 t) + \phi_{12} e^{-\zeta_2 \omega_2 t} A_2 \cos(\omega_2 t) + \phi_{13} e^{-\zeta_3 \omega_3 t} A_3 \cos(\omega_3 t) \\ u_2 = \phi_{21} e^{-\zeta_1 \omega_1 t} A_1 \cos(\omega_1 t) + \phi_{22} e^{-\zeta_2 \omega_2 t} A_2 \cos(\omega_2 t) + \phi_{23} e^{-\zeta_3 \omega_3 t} A_3 \cos(\omega_3 t) \\ u_3 = \phi_{31} e^{-\zeta_1 \omega_1 t} A_1 \cos(\omega_1 t) + \phi_{32} e^{-\zeta_2 \omega_2 t} A_2 \cos(\omega_2 t) + \phi_{33} e^{-\zeta_3 \omega_3 t} A_3 \cos(\omega_3 t) \end{cases}$$

$$\Rightarrow \begin{cases} 0.001000 = \phi_{11} A_1 + \phi_{12} A_2 + \phi_{13} A_3 \\ 0.001512 = \phi_{21} A_1 + \phi_{22} A_2 + \phi_{23} A_3 \\ 0.002024 = \phi_{31} A_1 + \phi_{32} A_2 + \phi_{33} A_3 \end{cases}$$

$$\Rightarrow \begin{bmatrix} A_1 \\ A_2 \\ A_3 \end{bmatrix} = \begin{bmatrix} \phi_{11} & \phi_{12} & \phi_{13} \\ \phi_{21} & \phi_{22} & \phi_{23} \\ \phi_{31} & \phi_{32} & \phi_{33} \end{bmatrix}^{-1} \begin{bmatrix} 0.001000 \\ 0.001512 \\ 0.002024 \end{bmatrix} = \begin{bmatrix} -0.00266010896 \\ 0.000287280455 \\ 0.0000759274645 \end{bmatrix}$$

So Equation 7.34 should read numerically as

$$\begin{aligned} u_1 = {} & 0.001180895 e^{-\zeta_1 51.98977t} \cos(51.98977t) \\ & - 0.0002036212 e^{-\zeta_2 151.8868t} \cos(151.8868t) \\ & + 0.00002272636 e^{-\zeta_3 257.38666t} \cos(257.38666t) \end{aligned}$$

$$\begin{aligned} u_2 = {} & 0.001576976 e^{-\zeta_1 51.98977t} \cos(51.98977t) \\ & - 0.0000009745829 e^{-\zeta_2 151.8868t} \cos(151.8868t) \\ & - 0.00006400185 e^{-\zeta_3 257.38666t} \cos(257.38666t) \end{aligned}$$

$$\begin{aligned} u_3 = {} & 0.001787404 e^{-\zeta_1 51.98977t} \cos(51.98977t) \\ & + 0.0002026513 e^{-\zeta_2 151.8868t} \cos(151.8868t) \\ & + 0.0000339449 e^{-\zeta_3 257.38666t} \cos(257.38666t) \end{aligned}$$

From the coefficients of the equations, it can be readily seen that the first mode dominates. Assigning damping ratio = 5% for ζ_1, ζ_2 and ζ_3, the motions of the top storey are plotted for the three modes and the total (sum of displacements of the three modes) in Figure 7.8. It can be seen graphically that the first mode has the lowest frequency but the greatest contribution to the total displacement and vice versa for the second and third modes, i.e., the higher the mode, the greater the frequency

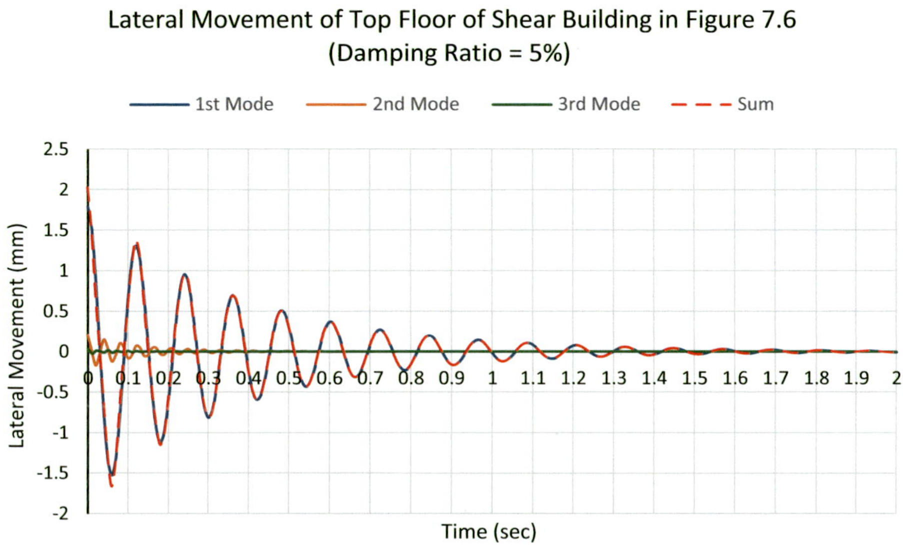

FIGURE 7.8 Lateral movement of the top floor – three-storey shear building.

and the lesser the contribution to the total displacement and hence the velocity and acceleration.

The vibrations are diminishing with time. With a higher damping ratio, the vibrations diminish at a faster rate.

After the determination of the motions, an important parameter named "modal participation mass ratio" is discussed as follows.

First of all, scale Φ is scaled to Φ_u so that $\Phi_u^T \boldsymbol{m} \Phi_u$ becomes a unity matrix. This is done by multiplying the column vectors by factors determined for the modes. Taking the three degrees of freedom structure as an illustration, we list and s values as,

$$\begin{cases} s_1 = \sqrt{\dfrac{1}{m_1\phi_{11}^2 + m_2\phi_{21}^2 + m_3\phi_{31}^2}} \\ s_2 = \sqrt{\dfrac{1}{m_1\phi_{12}^2 + m_2\phi_{22}^2 + m_3\phi_{32}^2}} \\ s_3 = \sqrt{\dfrac{1}{m_1\phi_{13}^2 + m_2\phi_{23}^2 + m_3\phi_{33}^2}} \end{cases} \tag{7.35}$$

So the matrix Φ can be scaled down to

$$\Phi_u = \begin{bmatrix} s_1\phi_{11} & s_2\phi_{12} & s_3\phi_{13} \\ s_1\phi_{21} & s_2\phi_{22} & s_3\phi_{23} \\ s_1\phi_{31} & s_2\phi_{32} & s_3\phi_{33} \end{bmatrix} \tag{7.36}$$

Define
$$\boldsymbol{L} = \boldsymbol{\Phi}_{\boldsymbol{u}}{}^{T}\boldsymbol{m}\begin{bmatrix}1\\1\\1\end{bmatrix} = \begin{bmatrix} s_1\phi_{11}m_1 + s_1\phi_{21}m_2 + s_1\phi_{31}m_3 \\ s_2\phi_{12}m_1 + s_2\phi_{22}m_2 + s_2\phi_{32}m_3 \\ s_3\phi_{13}m_1 + s_3\phi_{23}m_2 + s_3\phi_{33}m_3 \end{bmatrix} \tag{7.37}$$

where $\begin{bmatrix}1\\1\\1\end{bmatrix}$ represents the rigid-body base displacement vector.

The effective masses for the three modes will be

$$\begin{cases} m_{eff1} = L_{1,1}{}^2 \\ m_{eff2} = L_{2,1}{}^2 \\ m_{eff3} = L_{3,1}{}^2 \end{cases} \tag{7.38}$$

It should be noted that $m_{eff1} + m_{eff2} + m_{eff3} =$ the total mass $= m_1 + m_2 + m_3$

The modal participating mass ratios for the three modes are :

$$\begin{cases} r_{eff1} = \dfrac{m_{eff1}}{m_{eff1} + m_{eff2} + m_{eff3}} \\ r_{eff2} = \dfrac{m_{eff2}}{m_{eff1} + m_{eff2} + m_{eff3}} \\ r_{eff3} = \dfrac{m_{eff3}}{m_{eff1} + m_{eff2} + m_{eff3}} \end{cases} \tag{7.39}$$

The modal participating mass ratio carries the physical meaning that it is the ratio of the total mass of the structure being excited in the respective mode. That is, r_{eff1} represents the ratio of the total mass of the structure being excited in the first mode, r_{eff2} the second mode etc.

In the numerical example as discussed in the foregoing, the modal participating mass ratios are calculated as follows:

By Equation 7.35
$$\begin{cases} s_1 = \sqrt{\dfrac{1}{m_1\phi_{11}{}^2 + m_2\phi_{21}{}^2 + m_3\phi_{31}{}^2}} = 5.564412\times 10^{-3} \\ s_2 = \sqrt{\dfrac{1}{m_1\phi_{12}{}^2 + m_2\phi_{22}{}^2 + m_3\phi_{32}{}^2}} = 5.213922\times 10^{-3} \\ s_3 = \sqrt{\dfrac{1}{m_1\phi_{13}{}^2 + m_2\phi_{23}{}^2 + m_3\phi_{33}{}^2}} = 5.705724\times 10^{-3} \end{cases}$$

$$\text{By Equation 7.36 } \Phi_u = \begin{bmatrix} -2.470194\times10^{-3} & -3.695570\times10^{-3} & 1.707819\times10^{-3} \\ -3.298717\times10^{-3} & -1.768794\times10^{-3} & -4.809549\times10^{-3} \\ -3.738888\times10^{-3} & 3.677967\times10^{-3} & 2.550858\times10^{-3} \end{bmatrix}$$

$$\text{By Equation 7.37 } \boldsymbol{L} = \begin{bmatrix} -315.841141 \\ -55.362424 \\ 8.909299 \end{bmatrix}$$

$$\text{By Equation 7.38} \begin{cases} m_{eff1} = 99755.63\text{kg} \\ m_{eff2} = 3064.998\text{kg} \\ m_{eff3} = 79.375606\text{kg} \end{cases}$$

$$m_{eff1} + m_{eff2} + m_{eff3} = 102900 = m_1 + m_2 + m_3$$

$$\text{By Equation 7.39} \begin{cases} r_{eff1} = \dfrac{m_{eff1}}{m_{eff1} + m_{eff2} + m_{eff3}} = 0.969 \\ r_{eff2} = \dfrac{m_{eff2}}{m_{eff1} + m_{eff2} + m_{eff3}} = 0.030 \\ r_{eff3} = \dfrac{m_{eff3}}{m_{eff1} + m_{eff2} + m_{eff3}} = 0.001 \end{cases}$$

Thus the first mode can be readily excited, with 96.9% of the total mass being excited in the mode. On the other hand, the second mode is less easily and the third mode the least.

However, if there are external loads of sinusoidal variation applied at the three storeys of time varying nature, i.e., $P_1 \sin\Omega t$, $P_2 \sin\Omega t$ and $P_3 \sin\Omega t$ acting on the first, second, and third storeys, Equation 7.29 has to be solved with loads not equal to zero. If the frame is subjected to seismic movement at its base with an acceleration of $g = 1\text{m/sec}^2 \approx 0.1g$ at a circular frequency of $\Omega = 51\,/\text{sec}$ which is close to its fundamental circular frequency at 51.9898/sec, in accordance with Section 5.7.3 in Chapter 5, the ground motion is equivalent to applying forces of $m_1 \times 1\sin\Omega t = 44100\sin 51t$ at first storey and $m_2\left(= m_3\right)\times 1\sin\Omega t = 29400\sin 51t$ for second and third storeys. By Equation 7.31, we may list

$$\begin{cases} M_1\ddot{q}_1 + C_1\dot{q}_1 + K_1 q_1 = \phi_{11}P_1 + \phi_{21}P_2 + \phi_{31}P_3 \\ M_2\ddot{q}_2 + C_2\dot{q}_2 + K_2 q_2 = \phi_{12}P_1 + \phi_{22}P_2 + \phi_{32}P_3 \\ M_3\ddot{q}_3 + C_3\dot{q}_3 + K_3 q_3 = \phi_{13}P_1 + \phi_{23}P_2 + \phi_{33}P_3 \end{cases} = \Phi^T \begin{bmatrix} P_1 \\ P_2 \\ P_3 \end{bmatrix}$$

With $\Phi^T \begin{bmatrix} P_1 \\ P_2 \\ P_3 \end{bmatrix} = \begin{bmatrix} -56760.923 \\ -10618.191 \\ 1561.467 \end{bmatrix}$

$$\begin{cases} M_1\ddot{q}_1 + C_1\dot{q}_1 + K_1q_1 = -56760.923\sin 51t \\ M_2\ddot{q}_2 + C_2\dot{q}_2 + K_2q_2 = -10618.191\sin 51t \\ M_3\ddot{q}_3 + C_3\dot{q}_3 + K_3q_3 = 1561.467\sin 51t \end{cases}$$

The solution of Equation 7.31 comprises the complementary solution and particular solution as similar to Equation 7.9. The r and θ are first calculated.

$$r_1 = \frac{\Omega}{\omega_1} = \frac{51}{51.98977} = 0.98096;\ r_2 = \frac{\Omega}{\omega_2} = 0.33578;\ r_3 = \frac{\Omega}{\omega_3} = 0.19815$$

$$\theta_1 = \tan^{-1}\left(\frac{2r_1\zeta}{1-r_1^2}\right) = 1.2038;\ \theta_2 = \tan^{-1}\left(\frac{2r_2\zeta}{1-r_2^2}\right) = 0.03783;$$

$$\theta_3 = \tan^{-1}\left(\frac{2r_3\zeta}{1-r_3^2}\right) = 0.02062$$

The complementary solutions are

$$q_{c1} = e^{-\zeta\omega D_1 t}\left[A_1\cos\left(\omega_{D1}t\right) + B_1\sin\left(\omega_{D1}t\right)\right] = e^{-0.05\times 51.58598t}\left[A_1\cos\left(51.8598t\right)\right.$$
$$\left.+B_1\sin\left(51.8598t\right)\right] = e^{-2.593t}\left[A_1\cos\left(51.8598t\right) + B_1\sin\left(51.8598t\right)\right]$$

$$q_{c2} = e^{-\zeta\omega_{D2}t}\left[A_2\cos\left(\omega_{D2}t\right) + B_2\sin\left(\omega_{D2}t\right)\right] = e^{-0.05\times 151.5071t}\left[A_2\cos\left(151.507t\right)\right.$$
$$\left.+B_2\sin\left(151.5071t\right)\right] = e^{-7.5575t}\left[A_2\cos\left(151.507t\right) + B_2\sin\left(151.507t\right)\right]$$

$$q_{c3} = e^{-\zeta\omega_3 t}\left[A_3\cos\left(\omega_3 t\right) + B_3\sin\left(\omega_3 t\right)\right] = e^{-0.05\times 256.7432t}\left[A_3\cos\left(256.7432t\right)\right.$$
$$\left.+B_3\sin\left(256.7432t\right)\right] = e^{-12.8972t}\left[A_3\cos\left(256.7432t\right) + B_3\sin\left(256.7432t\right)\right]$$

The particular solutions are

$$q_{p1} = \frac{F_{e1}\sin\left(\Omega t - \theta_1\right)}{K_1\sqrt{\left(1-r_1^2\right)^2 + \left(2r_1\zeta\right)^2}}$$
$$= \frac{-56760.923\sin\left(51t - 1.2038\right)}{87296648\sqrt{\left(1-0.98096^2\right)^2 + 4\times 0.05^2\times 0.98096^2}}$$
$$= -0.006187\sin\left(51t - 1.2038\right)$$

$$q_{p2} = \frac{F_{e2}\sin(\Omega t - \theta_2)}{K_2\sqrt{(1-r_1^2)^2 + (2r_1\zeta)^2}}$$

$$= \frac{-10618.191\sin(51t - 0.03783)}{848615356\sqrt{(1-0.3783^2)^2 + 4\times 0.05^2 \times 0.3783^2}}$$

$$= -0.00001409\sin(51t - 0.03783)$$

$$q_{p3} = \frac{F_{e3}\sin(\Omega t - \theta_3)}{K_3\sqrt{(1-r_3^2)^2 + (2r_3\zeta)^2}} = \frac{1561.467\sin(51t - 0.02062)}{2034934589\sqrt{(1-0.02062^2)^2 + 4\times 0.05^2 \times 0.26062^2}}$$

$$= 0.0000007985\sin(51t - 0.02062)$$

The solution $\boldsymbol{q}$ is the sum of the complementary solution and the particular solution as $\boldsymbol{q} = \begin{bmatrix} q_1 \\ q_2 \\ q_3 \end{bmatrix} = \begin{bmatrix} q_{c1} + q_{p1} \\ q_{c2} + q_{p2} \\ q_{c3} + q_{p3} \end{bmatrix}$

The A_1, A_2, A_3; B_1, B_2, B_3 are solved for initial conditions of $u = 0$ and $\dot{u} = 0$ when $t = 0$ by which $A = \dfrac{F_e \sin(-\theta)}{k\sqrt{(1-r^2)^2 + (2r\zeta)^2}}$, $B = \dfrac{1}{\omega_D}\left(\zeta\omega A - \dfrac{F_e\sin(-\theta)\Omega}{k\sqrt{(1-r^2)^2 + (2r\zeta)^2}}\right)$

Solving $A_1 = -0.0429507$, $B_1 = 0.0283658$. Others are effectively zero.

By Equation 7.27 $\boldsymbol{u} = \begin{bmatrix} u_1 \\ u_2 \\ u_3 \end{bmatrix} = \Phi\boldsymbol{q}$, $\begin{bmatrix} u_1 \\ u_2 \\ u_3 \end{bmatrix}$ versus time are obtained and plotted in Figure 7.9 as follows with the ground motion $1/\Omega^2 \sin\Omega t$ (under an acceleration of 1 m/sec^2).

7.3.2 One-Bay Three-Storey Frame with Finite Stiffness in Structural Members

The shear building frame, as shown in Figure 7.6, is now changed to a plane frame with the horizontal members of finite stiffness and analyzed as Worked Example 7.3.

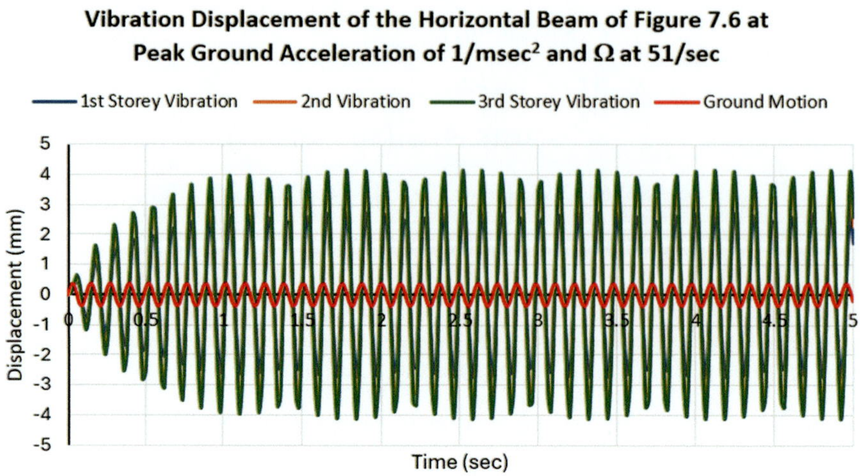

FIGURE 7.9 Horizontal displacement of the frame of Figure 7.6 at ground acceleration of 1 m/sec², $\Omega = 16$/sec, and $\zeta = 0.05$.

Worked Example 7.3

With the same stiffness for the columns and the same geometry as Worked Example 7.2, the analysis is now carried out with the beams of sizes 0.4 m (w) × 0.6 m (D). Basically, the analysis is based on the same principle as for the shear building frame, but with a greater number of degrees of freedom, and the masses are lumped at the

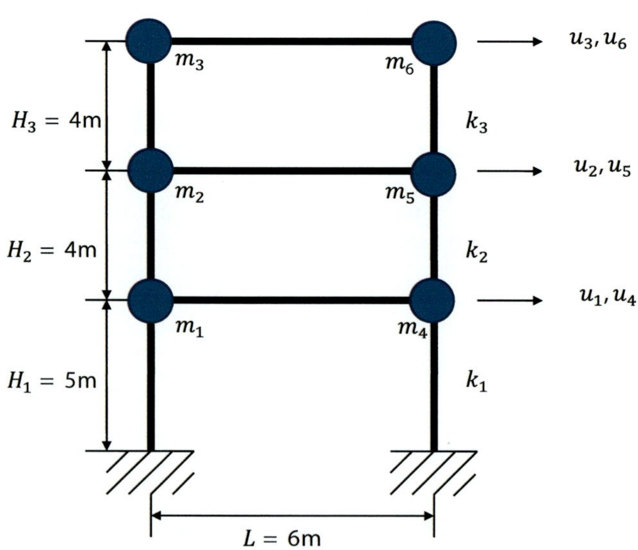

FIGURE 7.10 Plane frame structure for dynamic analysis.

beam column nodes, as shown in Figure 7.10. In the analysis, the masses are only added at horizontal movements while those in the vertical direction are ignored. In the frame, $m_1 = m_4 = 22100\text{kg}$, $m_2 = m_3 = m_5 = m_6 = 14700\text{kg}$.

While the stiffness matrix follows the conventional formulation, the mass matrix will be a diagonal one with array size 18 × 18 and entries only in the X-direction movement of the nodes as

$$m = \begin{bmatrix}
22050 & 0 & 0 & 0 & 0 & 0 & 0 & 0 & 0 & 0 & 0 & 0 & 0 & 0 & 0 & 0 & 0 & 0 \\
0 & 0 & 0 & 0 & 0 & 0 & 0 & 0 & 0 & 0 & 0 & 0 & 0 & 0 & 0 & 0 & 0 & 0 \\
0 & 0 & 0 & 0 & 0 & 0 & 0 & 0 & 0 & 0 & 0 & 0 & 0 & 0 & 0 & 0 & 0 & 0 \\
0 & 0 & 0 & 14700 & 0 & 0 & 0 & 0 & 0 & 0 & 0 & 0 & 0 & 0 & 0 & 0 & 0 & 0 \\
0 & 0 & 0 & 0 & 0 & 0 & 0 & 0 & 0 & 0 & 0 & 0 & 0 & 0 & 0 & 0 & 0 & 0 \\
0 & 0 & 0 & 0 & 0 & 0 & 0 & 0 & 0 & 0 & 0 & 0 & 0 & 0 & 0 & 0 & 0 & 0 \\
0 & 0 & 0 & 0 & 0 & 0 & 14700 & 0 & 0 & 0 & 0 & 0 & 0 & 0 & 0 & 0 & 0 & 0 \\
0 & 0 & 0 & 0 & 0 & 0 & 0 & 0 & 0 & 0 & 0 & 0 & 0 & 0 & 0 & 0 & 0 & 0 \\
0 & 0 & 0 & 0 & 0 & 0 & 0 & 0 & 0 & 0 & 0 & 0 & 0 & 0 & 0 & 0 & 0 & 0 \\
0 & 0 & 0 & 0 & 0 & 0 & 0 & 0 & 0 & 22050 & 0 & 0 & 0 & 0 & 0 & 0 & 0 & 0 \\
0 & 0 & 0 & 0 & 0 & 0 & 0 & 0 & 0 & 0 & 0 & 0 & 0 & 0 & 0 & 0 & 0 & 0 \\
0 & 0 & 0 & 0 & 0 & 0 & 0 & 0 & 0 & 0 & 0 & 0 & 0 & 0 & 0 & 0 & 0 & 0 \\
0 & 0 & 0 & 0 & 0 & 0 & 0 & 0 & 0 & 0 & 0 & 0 & 14700 & 0 & 0 & 0 & 0 & 0 \\
0 & 0 & 0 & 0 & 0 & 0 & 0 & 0 & 0 & 0 & 0 & 0 & 0 & 0 & 0 & 0 & 0 & 0 \\
0 & 0 & 0 & 0 & 0 & 0 & 0 & 0 & 0 & 0 & 0 & 0 & 0 & 0 & 0 & 0 & 0 & 0 \\
0 & 0 & 0 & 0 & 0 & 0 & 0 & 0 & 0 & 0 & 0 & 0 & 0 & 0 & 0 & 14700 & 0 & 0 \\
0 & 0 & 0 & 0 & 0 & 0 & 0 & 0 & 0 & 0 & 0 & 0 & 0 & 0 & 0 & 0 & 0 & 0 \\
0 & 0 & 0 & 0 & 0 & 0 & 0 & 0 & 0 & 0 & 0 & 0 & 0 & 0 & 0 & 0 & 0 & 0
\end{bmatrix}$$

Eigenvalue analysis is carried out on the matrix $\boldsymbol{k}^{-1}\boldsymbol{m}$ which results in 18 degrees of freedom. The $\boldsymbol{\Phi}$ matrix obtained as vectors of $\boldsymbol{k}^{-1}\boldsymbol{m}$ is used to perform "decoupling" for the stiffness and mass matrices by $\boldsymbol{K} = \boldsymbol{\Phi}^T\boldsymbol{k}\boldsymbol{\Phi}$ and $\boldsymbol{M} = \boldsymbol{\Phi}^T\boldsymbol{m}\boldsymbol{\Phi}$. However, unlike the shear building in the previous example where the $\boldsymbol{\Phi}$ matrix is full and contains all the 3 lateral motions which are all required for the decoupling exercise, only 6 out of the 18 DoF of the $\boldsymbol{\Phi}$ matrix are relevant for decoupling. As some software for calculation of eigenvalues analysis arrange the eigenvectors in the order of the ascending or descending values of the eigenvalues instead of matching the $\boldsymbol{m}$ and $\boldsymbol{k}$ matrices, sorting and re-arranging of the column vectors may be required for the matching. In this example, sorting and re-arrangement of the $\boldsymbol{\Phi}$ matrix is done and listed as follows with numerical values in three digits as the original $\boldsymbol{\Phi}$ matrix lists all the relevant column vectors at the 13th to the 18th columns:

$$
\Phi = \begin{bmatrix}
-0.152 & 0 & 0 & -0.447 & 0 & 0 & -0.348 & 0 & 0 & -0.658 & 0 & 0 & 0.135 & 0 & 0 & 0.023 & 0 & 0 \\
-0.003 & 0 & 0 & 0.005 & 0 & 0 & -0.002 & 0 & 0 & 0 & 0 & 0 & 0 & 0 & 0 & 0 & 0 & 0 \\
0.048 & 0 & 0 & 0.063 & 0 & 0 & -0.107 & 0 & 0 & 0.031 & 0 & 0 & 0.119 & 0 & 0 & 0.025 & 0 & 0 \\
-0.368 & 0 & 0 & -0.320 & 0 & 0 & 0.507 & 0 & 0 & -0.216 & 0 & 0 & -0.574 & 0 & 0 & -0.236 & 0 & 0 \\
-0.004 & 0 & 0 & -0.119 & 0 & 0 & -0.005 & 0 & 0 & 0 & 0 & 0 & 0 & 0 & 0 & 0 & 0 & 0 \\
0.053 & 0 & 0 & 0.009 & 0 & 0 & -0.067 & 0 & 0 & 0.129 & 0 & 0 & 0.030 & 0 & 0 & 0.092 & 0 & 0 \\
-0.578 & 0 & 0 & 0.380 & 0 & 0 & -0.186 & 0 & 0 & 0.044 & 0 & 0 & 0.216 & 0 & 0 & -0.651 & 0 & 0 \\
-0.005 & 0 & 0 & 0.012 & 0 & 0 & -0.009 & 0 & 0 & 0 & 0 & 0 & 0 & 0 & 0 & 0 & 0 & 0 \\
0.047 & 0 & 0 & -0.185 & 0 & 0 & 0.267 & 0 & 0 & -0.032 & 0 & 0 & -0.301 & 0 & 0 & -0.651 & 0 & 0 \\
-0.152 & 0 & 0 & -0.447 & 0 & 0 & -0.348 & 0 & 0 & -0.658 & 0 & 0 & 0.135 & 0 & 0 & 0.023 & 0 & 0 \\
-0.003 & 0 & 0 & 0.005 & 0 & 0 & -0.002 & 0 & 0 & 0 & 0 & 0 & 0 & 0 & 0 & 0 & 0 & 0 \\
0.048 & 0 & 0 & 0.063 & 0 & 0 & -0.107 & 0 & 0 & 0.031 & 0 & 0 & 0.119 & 0 & 0 & 0.025 & 0 & 0 \\
-0.368 & 0 & 0 & -0.320 & 0 & 0 & 0.507 & 0 & 0 & -0.216 & 0 & 0 & -0.574 & 0 & 0 & -0.236 & 0 & 0 \\
-0.004 & 0 & 0 & -0.119 & 0 & 0 & -0.005 & 0 & 0 & 0 & 0 & 0 & 0 & 0 & 0 & 0 & 0 & 0 \\
0.053 & 0 & 0 & 0.009 & 0 & 0 & -0.067 & 0 & 0 & 0.129 & 0 & 0 & 0.030 & 0 & 0 & 0.092 & 0 & 0 \\
-0.578 & 0 & 0 & 0.380 & 0 & 0 & -0.186 & 0 & 0 & 0.044 & 0 & 0 & 0.216 & 0 & 0 & -0.651 & 0 & 0 \\
-0.005 & 0 & 0 & 0.012 & 0 & 0 & -0.009 & 0 & 0 & 0 & 0 & 0 & 0 & 0 & 0 & 0 & 0 & 0 \\
0.047 & 0 & 0 & -0.185 & 0 & 0 & 0.267 & 0 & 0 & -0.032 & 0 & 0 & -0.301 & 0 & 0 & 0.106 & 0 & 0
\end{bmatrix}
$$

By matching with the mass and stiffness matrices, decoupling with them is then achieved as similar to the shear building.

As only the lateral movement is interested, the $\boldsymbol{\Phi}$ matrix can be simplified to $\boldsymbol{\Phi}_r$ as a 6 × 6 degrees of freedom representing only the lateral movements of the six beam-column nodes for subsequent analysis after the eigenvalue analysis.

$$
\Phi_r = \begin{bmatrix}
-0.15203816 & -0.44723637 & -0.34760584 & -0.6578999 & 0.135115446 & 0.022927837 \\
-0.36796296 & -0.32035936 & 0.507272372 & -0.21588669 & -0.57402955 & -0.23635017 \\
-0.57796077 & 0.380433904 & -0.18579707 & 0.043647924 & 0.215666277 & -0.65063156 \\
-0.15203816 & -0.44723637 & -0.34760584 & -0.6578999 & 0.135115446 & 0.022927837 \\
-0.36796296 & -0.32035936 & 0.507272372 & -0.21588669 & -0.57402955 & -0.23635017 \\
-0.57796077 & 0.380433904 & -0.18579707 & 0.043647924 & 0.215666277 & -0.65063156
\end{bmatrix}
$$

And the eigenvalues are
$\lambda_1 = 0.003071014$; $\lambda_2 = 0.00017874$; $\lambda_3 = 0.000027629$; $\lambda_4 = 0.0000096778$; $\lambda_5 = 0.0000061582$; $\lambda_6 = 0.0000073148$. By $\omega = 1/\sqrt{\lambda}$,

$\omega_1 = 18.045092$; $\omega_2 = 74.798147$; $\omega_3 = 190.246765$; $\omega_4 = 321.449147$; $\omega_5 = 402.971139$; $\omega_6 = 369.741434$.

By $\boldsymbol{M} = \Phi^T \boldsymbol{m} \Phi$
$M_1 = 14820.798441$; $M_2 = 16093.284365$; $M_3 = 13908.861791$;
$M_4 = 20514.162174$; $M_5 = 11860.130546$; $M_6 = 14111.170238$.
By $\boldsymbol{K} = \Phi^T \boldsymbol{k} \Phi$
$K_1 = 4826026.621$; $K_2 = 90038107.913$; $K_3 = 503415003.566$;
$K_4 = 2119719225.314$; $K_5 = 1925916062.161$; $K_6 = 1929120133.369$
By $C = 2M\omega\zeta$ and taking $\zeta = 0.05$

$C_1 = 26744.267;\ C_2 = 120374.785;\ C_3 = 2646115.597;\ C_4 = 659425.992$

$C_5 = 477929.032;\ C_6 = 521748.432$

The frame is then analyzed to ground motion of magnitude $a_g = 1\text{m/sec}^2$ at circular frequency $\Omega = 18/\text{sec}$. The equivalent forces due to the ground movement are node forces at $P = ma_g \sin\Omega t$ at the nodes with the specified masses m_1 to m_6 as derived in Chapter 5. The six forces have to be converted by $\boldsymbol{F} = \Phi^T \boldsymbol{P}$ after de-coupling of the six motion equations of the nodes. Thus $F_1 = -34515.0405$; $F_2 = -17956.9321$; $F_3 = -5878.0435$; F_4, F_5 and F_6 are effectively zero.

Recalling $r = \dfrac{\Omega}{\omega}$, the phase angle $\theta = \tan^{-1}\left(\dfrac{2r\zeta}{1-r^2}\right)$, as similar to the previous calculations for the shear building, the coefficients A and B in Equation 7.9 are determined by the initial conditions that displacements and velocities are zero at time $t = 0$ by $A = \dfrac{F\sin(-\theta)}{k\sqrt{(1-r^2)^2 + (2r\zeta)^2}}$, $B = \dfrac{1}{\omega_D}\left(\zeta\omega A - \dfrac{F_e \sin(-\theta)\Omega}{k\sqrt{(1-r^2)^2 + (2r\zeta)^2}}\right)$ for each mode.

The results are summarized in Table 7.1.

Upon determination of the coefficients in Table 7.1, the q ordinates of the six modes can be calculated. Then by $\boldsymbol{u} = \Phi\boldsymbol{q}$, the motions of the six nodes can be determined. Due to symmetry, the lateral motions at the same storeys are the same. The motions of the three storeys are plotted at the ground motion of $(1/\Omega^2)\sin\Omega t$ in Figure 7.11. The ground motion is also plotted for reference.

It can be seen that the motion becomes steady at around $t = 4$ with the maximum amplitude of oscillations at ±40 mm.

In the foregoing exercise, all the six modes are used to calculate the motion of the structure. But it can be seen that the last three modes have very little contributions which can be ignored. These modes correspond to small "effective masses" discussed and evaluated under Equation 7.38 and Equation 7.39.

In a real structure, the degrees of freedom will be far greater than six where taking all modes into consideration is unrealistic and unnecessary. It is a practice that the number of modes taken into consideration will be such that the cumulated effective mass ratio reaches a figure of 90% to 95%, depending on the accuracy to be aimed for.

The internal forces of the structure at any time t can be obtained by analyzing the structure to the prescribed lateral displacements determined for the time under consideration.

7.4 NUMERICAL INTEGRATION METHOD

Another popular method for structural dynamics analysis is the numerical integration method. Through the use of the basic equation of the motion of structural

TABLE 7.1
Summarization of Coefficients for Determination of Motions of Plane Frame shown in Figure 7.10

Mode	1	2	3	4	5	6
K (N/m)=	4826027.621	90038107.91	503415003.6	2119719225	1925916062	1929120133
M (kg)=	14820.79844	16093.28437	13908.86179	20514.16217	11860.13055	14111.17024
$\omega(/s)$=	18.045092	74.798147	190.246765	321.449147	402.971139	369.741434
$\omega_D(/s)=$	17.99997927	74.61115163	189.7711481	320.6455241	401.9637112	368.8170804
$\zeta =$	0.05	0.05	0.05	0.05	0.05	0.05
$a_g\left(m/s^2\right) =$	1	1	1	1	1	1
F (N)=	−34515.04047	−17956.93212	−5878.043504	0	0	0
$\Omega(/s) =$	18	18	18	18	18	18
$r =$	0.997501149	0.240647673	0.094613961	0.055996416	0.044668211	0.048682669
$\theta =$	1.520798407	0.025538503	0.009546568	0.005617196	0.004475721	0.004879793
$F/K =$	−0.007151853	−0.000199437	−1.16763E-05	0	0	0
$u_0(\text{m}) =$	0	0	0	0	0	0
$\dot{u}_0(\text{m}) =$	0	0	0	0	0	0
A =	−0.071518613	−5.40407E-06	−1.12469E-07	0	0	0
B =	−6.12431E-06	5.07678E-05	1.11178E-06	0	0	0

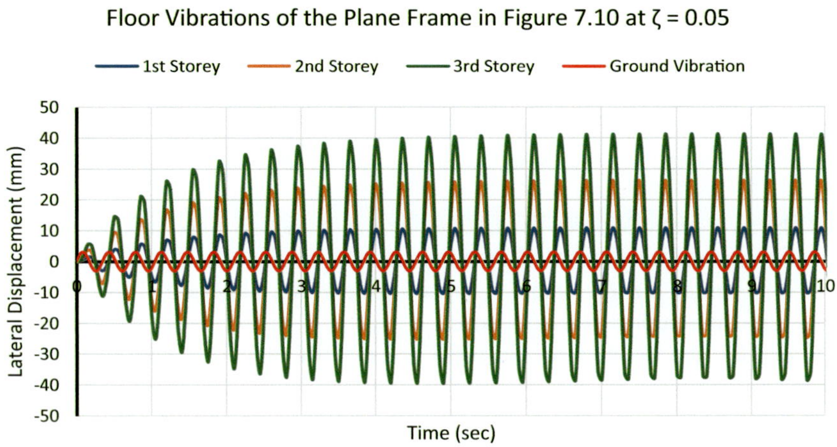

FIGURE 7.11 Floor vibrations of the three-storey plane frame at Figure 7.10 at $\zeta = 0.05$ and ground motion of $(1/\Omega^2)\sin\Omega t$, $\Omega = 18$/sec by the modal method.

dynamics listed in Equation 7.8, the method involves breaking up the motions of a structure into a number of successive time-steps and based on known values of the displacements, velocities, and accelerations of the past time-steps, predict the future ones.

7.4.1 Central Difference Method

The basic "Central Difference Method" is first described. In the method, displacements, velocities, and accelerations of successive time-steps t_{i-1}, t_i, t_{i+1} separated by Δt are related through the finite difference approach as

$$\dot{u}_i = \frac{u_{i+1} - u_{i-1}}{2\Delta t} \tag{7.40}$$

$$\ddot{u}_i = \frac{\dfrac{u_{i+1} - u_i}{\Delta t} - \dfrac{u_i - u_{i-1}}{\Delta t}}{\Delta t} = \frac{u_{i+1} - 2u_i + u_{i-1}}{\Delta t^2} \tag{7.41}$$

Then consider the general equation of motion of a structure expressed in matrix form

$$\boldsymbol{m}\ddot{\boldsymbol{u}} + \boldsymbol{c}\dot{\boldsymbol{u}} + \boldsymbol{k}\boldsymbol{u} = \boldsymbol{F}_e \tag{7.42}$$

Substituting Equations 7.40 and 7.41 into Equation 7.42 and re-arranging

$$\left(\frac{\boldsymbol{m}}{\Delta t^2} + \frac{\boldsymbol{c}}{2\Delta t}\right)\boldsymbol{u}_{i+1} = \left(\frac{2\boldsymbol{m}}{\Delta t^2} - \boldsymbol{k}\right)\boldsymbol{u}_i + \left(\frac{-\boldsymbol{m}}{\Delta t^2} + \frac{\boldsymbol{c}}{2\Delta t}\right)\boldsymbol{u}_{i-1} + \boldsymbol{F}_e \tag{7.43}$$

where u_0, $\dot{u}_0$ and $\ddot{u}_0$ are column matrices. For simplicity, put $A_0 = \left(\frac{m}{\Delta t^2} + \frac{c}{2\Delta t}\right)$, $A_1 = \left(\frac{2m}{\Delta t^2} - k\right)$ and $A_2 = \left(\frac{-m}{\Delta t^2} + \frac{c}{2\Delta t}\right)$, so Equation 7.43 becomes

$$A_0 u_{i+1} = A_1 u_i + A_2 u_{i-1} + F_e \tag{7.44}$$

Thus, the next deflection can be obtained by the previously known deflections.

The numerical accuracy and stability depend on the choice of Δt. To obtain a stable solution, $\Delta t < T_1 / \pi$ where T_1 is the shortest natural period of the structure. If $\Delta t > T_1 / \pi$, the solution may not converge.

A step-by-step procedure for the Central Difference method is outlined as follows:

(1) Form the mass, damping, stiffness matrices and subsequently form the A_0, A_1 and A_0;
(2) By the initial conditions, determinate u_0, $\dot{u}_0$ and $\ddot{u}_0$, then calculate u_1 by Equations 7.40 and 7.41 as $u_1 = u_0 + \dot{u}_0 \Delta t + \frac{\Delta t^2}{2} \ddot{u}_0$ (eliminating u_{-1}). The u_0, $\dot{u}_0$ and $\ddot{u}_0$ are column matrices.
(3) Form the matrices A_0, A_1 and A_2. Solve for u_2 by Equation 7.44 as $A_0 u_2 = A_1 u_1 + A_2 u_0 + F_{e,t_1}$
(4) Calculate $\dot{u}_1$ and $\ddot{u}_1$ by Equations 7.40 and 7.41
(5) Then repeat step (3) to solve u_3 and step (4) for $\dot{u}_2$ and $\ddot{u}_2$.
(6) The process (3) and (4) to be recurrent for the subsequent values of period t_i for u_i, $\dot{u}_i$ and $\ddot{u}_i$ time as desired.

Such a numerical method is therefore a direct analysis on the structure. It does not require the determination of the dynamic properties of the structure which are inherent in the mass and stiffness matrices. In addition, the solutions are solved directly for the loads acting on the structure at any time. The loads need not be sinusoidal as in the modal method. In the modal method, the loads have to be converted to sinusoidal for analysis by such techniques as Fourier Transform. However, lengthy computer time is required by this numerical method as the structure has to be analyzed for each time step.

7.4.2 Newmark Method

The Newmark Method (Newmark 1959) is a more versatile one as compared with the Central Difference Method. It is more commonly used by computer software. The method is outlined as follows:

Consider again the equation $M\ddot{u}+C\dot{u}+Ku=F_e$ as listed in Equation 7.42 and using the same symbols as in the Central Difference Method.

To present $\ddot{u}$ and $\dot{u}$ in terms of u for possible solution of Equation 7.42 by numerical method, Newmark (1959) has made use of Taylor's series listed as

$$u_t = u_{t-\Delta t} + \Delta t\dot{u}_{t-\Delta t} + \frac{\Delta t^2}{2}\ddot{u}_{t-\Delta t} + \frac{\Delta t^3}{6}\dddot{u}_{t-\Delta t} + \ldots \tag{7.45}$$

$$\dot{u}_t = \dot{u}_{t-\Delta t} + \Delta t\ddot{u}_{t-\Delta t} + \frac{\Delta t^2}{2}\dddot{u}_{t-\Delta t} + \ldots \tag{7.46}$$

In Equations 7.45 and 7.46, Δt is a time step between successive "time stations" in which u, $\dot{u}$ and $\ddot{u}$ are to be determined. Newmark truncated these equations by the use of two coefficients β and γ as

$$u_t = u_{t-\Delta t} + \Delta t\dot{u}_{t-\Delta t} + \frac{\Delta t^2}{2}\ddot{u}_{t-\Delta t} + \beta\Delta t^3\dddot{u}_{t-\Delta t} \tag{7.47}$$

$$\dot{u}_t = \dot{u}_{t-\Delta t} + \Delta t\ddot{u}_{t-\Delta t} + \gamma\Delta t^2\dddot{u}_{t-\Delta t} \tag{7.48}$$

Newmark suggests $\gamma = 0.5$ to avoid artificial damping. For the coefficient β, if $\beta = 1/4$, it implies that the acceleration within the time interval Δt is constant. If $\beta = 1/6$, the acceleration within the time interval Δt is linearly varying.

If the acceleration is assumed to be linear within the time step,

$$\dddot{u} = \frac{\ddot{u}_t - \ddot{u}_{t-\Delta t}}{\Delta t} \tag{7.49}$$

Substituting Equation 7.49 into Equations 7.47 and 7.48, Newmark's equations in standard form are listed as:

$$u_t = u_{t-\Delta t} + \Delta t\dot{u}_{t-\Delta t} + \left(\frac{1}{2} - \beta\right)\Delta t^2\ddot{u}_{t-\Delta t} + \beta\Delta t^2\ddot{u}_t \tag{7.50a}$$

$$\dot{u}_t = \dot{u}_{t-\Delta t} + (1-\gamma)\Delta t\ddot{u}_{t-\Delta t} + \gamma\Delta t\ddot{u}_t \tag{7.50b}$$

Wilson (1962) nevertheless formulated Newmark's method in matrix form (as in Equation 7.42). The need for iteration can also be eliminated by introducing the direct solution of equations at each time step. Equations 7.50a and 7.50b can then be re-written as

$$\ddot{u}_t = b_1\left(u_t - u_{t-\Delta t}\right) + b_2\dot{u}_{t-\Delta t} + b_3\ddot{u}_{t-\Delta t} \tag{7.51a}$$

$$\dot{u}_t = b_4\left(u_t - u_{t-\Delta t}\right) + b_5\dot{u}_{t-\Delta t} + b_6\ddot{u}_{t-\Delta t} \tag{7.51b}$$

where the b_1 to b_6 coefficients as related to β and γ are as follows:

$$b_1 = \frac{1}{\beta \Delta t^2}; b_2 = \frac{-1}{\beta \Delta t}; b_3 = 1 - \frac{1}{2\beta}; b_4 = \frac{\gamma}{\beta \Delta t}; b_5 = 1 - \frac{\gamma}{\beta}; b_6 = \left(1 - \frac{\gamma}{2\beta}\right)\Delta t \quad (7.52)$$

Generally, $\gamma = 0.5$ and $\beta = 1/6$ for the assumption that acceleration is linearly varying within the time interval Δt. If $\beta = 0.25$, constant acceleration is assumed within the time interval. The latter possesses the advantage of being stable in computation regardless of the size of time step.

The substitution of Equations 7.51a and 7.51b into Equation 7.42 allows the dynamic equilibrium of the system at time "t" to be written in terms of the unknown node displacement $\boldsymbol{u}_t$. Or

$$(b_1\boldsymbol{m} + b_4\boldsymbol{c} + \boldsymbol{k})\boldsymbol{u}_t = \boldsymbol{F}_t + \boldsymbol{m}\left(b_1\boldsymbol{u}_{t-\Delta t} - b_2\dot{\boldsymbol{u}}_{t-\Delta t} - b_3\ddot{\boldsymbol{u}}_{t-\Delta t}\right) + \boldsymbol{c}\left(b_4\boldsymbol{u}_{t-\Delta t} - b_5\dot{\boldsymbol{u}}_{t-\Delta t} - b_6\ddot{\boldsymbol{u}}_{t-\Delta t}\right) \quad (7.53)$$

In Equation 7.53 $(b_1\boldsymbol{m} + b_4\boldsymbol{c} + \boldsymbol{k})$ is the effective stiffness matrix $\boldsymbol{k}_{eff}$.

So the procedure for solution can be summarized as follows:

(1) Form the mass matrix $\boldsymbol{m}$, damping matrix $\boldsymbol{c}$ and stiffness matrix $\boldsymbol{k}$ in the usual manner.
(2) Choose the values β and γ and calculate the b-coefficients under Equation 7.52.
(3) Referring to Equation 7.53, form the effective matrix $\boldsymbol{k}_{eff} = \boldsymbol{k} + b_1\boldsymbol{m} + b_4\boldsymbol{c}$.
(4) Specify the initial conditions for $\boldsymbol{u}$, $\dot{\boldsymbol{u}}$ and $\ddot{\boldsymbol{u}}$ at time $t = 0 = t_0$ as $\boldsymbol{u}_0$, $\dot{\boldsymbol{u}}_0$ and $\ddot{\boldsymbol{u}}_0$.
(5) Choose a value for time interval Δt, say 0.05 to 0.1 sec.
(6) Referring to Equation 7.53, formulate the effective load vector at the first time increment $t_1 = t_0 + \Delta t$ (then $t_0 = t_1 - \Delta t$) as

$$\boldsymbol{F}_{eff@t_1} = \boldsymbol{F}(t_1) + \boldsymbol{m}\left(b_1\boldsymbol{u}_0 - b_2\dot{\boldsymbol{u}}_0 - b_3\ddot{\boldsymbol{u}}_0\right) + \boldsymbol{c}\left(b_4\boldsymbol{u}_0 - b_5\dot{\boldsymbol{u}}_0 - b_6\ddot{\boldsymbol{u}}_0\right)$$

where $\boldsymbol{F}(t_1)$ is the externally applied force that is time dependent.
(7) Solve the Node displacement $\boldsymbol{u}_1$ by $\boldsymbol{u}_1 = \boldsymbol{K}_{eff}^{-1}\boldsymbol{F}_{eff@t_1}$.
(8) Referring to Equations 7.51a and 7.51b, calculate the node velocities and accelerations at t_1 by

$$\dot{\boldsymbol{u}}_1 = b_4(\boldsymbol{u}_1 - \boldsymbol{u}_0) + b_5\dot{\boldsymbol{u}}_0 + b_6\ddot{\boldsymbol{u}}_0$$

$$\ddot{\boldsymbol{u}}_1 = b_1(\boldsymbol{u}_1 - \boldsymbol{u}_0) + b_2\dot{\boldsymbol{u}}_0 + b_3\ddot{\boldsymbol{u}}_0$$

(9) Repeat step (6) with $\boldsymbol{u}_1$ as the previous $\boldsymbol{u}_0$ to calculate $\boldsymbol{u}_2$ and so on until the time the designer wishes to stop the calculations. For convenience, the equations in steps (6) to (8) are relisted in general terms of t_i as the ith time station as follows:

$$\boldsymbol{F}_{eff(t_i)} = \boldsymbol{F}(t_i) + \boldsymbol{m}\left(b_1\boldsymbol{u}_{t_i-\Delta t} - b_1\dot{\boldsymbol{u}}_{t_i-\Delta t} - b_3\ddot{\boldsymbol{u}}_{t_i-\Delta t}\right) + \boldsymbol{c}\left(b_4\boldsymbol{u}_{t_i-\Delta t} - b_5\dot{\boldsymbol{u}}_{t_i-\Delta t} - b_6\ddot{\boldsymbol{u}}_{t_i-\Delta t}\right)$$

$$\boldsymbol{u}_{t_i} = \boldsymbol{K}_{eff}^{-1}\boldsymbol{F}_{eff(t_i)}$$

$$\dot{\boldsymbol{u}}_{t_i} = b_4\left(\boldsymbol{u}_{t_i} - \boldsymbol{u}_{t_i-\Delta t}\right) + b_5\dot{\boldsymbol{u}}_{t_i-\Delta t} + b_6\ddot{\boldsymbol{u}}_{t_i-\Delta t}$$

$$\ddot{\boldsymbol{u}}_{t_i} = b_1\left(\boldsymbol{u}_{t_i} - \boldsymbol{u}_{t_i-\Delta t}\right) + b_2\dot{\boldsymbol{u}}_{t_i-\Delta t} + b_3\ddot{\boldsymbol{u}}_{t_i-\Delta t}$$

(10) $\boldsymbol{u}$ can still be calculated for certain intervals even when the externally applied force $\boldsymbol{F}(t_i)$ stops applying by simply setting $\boldsymbol{F}(t_i) = 0$.

Upon acquisition of the deflections, the internal forces of the structure can be determined by multiplying the individual members' stiffness by their node displacements.

When $\beta = 0$, $\gamma = 0.5$, Newmark method is reduced to the Central Difference Method.

7.4.3 First Worked Example on Newmark Method – Plane Frame under Lateral Load

Two worked examples based on the Newmark Method are presented as follows for illustration. The first one is the analysis of the plane frame shown in Figure 7.10 as Worked Example 7.4.

Worked Example 7.4

The original stiffness matrix and mass matrix are first formed, which are identical to $\boldsymbol{k}$ and $\boldsymbol{m}$ used in the modal analysis. The damping matrix is also formed as $\boldsymbol{c}$, with the damping constants as $2m_i\omega\zeta$ for the six point masses. ω is taken as $1/\sqrt{\lambda_1}$ where λ_1 is the first eigenvalue arrived by $\boldsymbol{k}^{-1}\boldsymbol{m}$. By the previous modal analysis, $\omega = 18.045095/\text{sec}$. The damping ratio $\zeta = 0.05$ is used as in the modal method. So the damping matrix is as follows, where the entries are in kNsec/m.

$$
\boldsymbol{c} = \begin{bmatrix}
39.789 & 0 & 0 & 0 & 0 & 0 & 0 & 0 & 0 & 0 & 0 & 0 & 0 & 0 & 0 & 0 & 0 & 0 \\
0 & 0 & 0 & 0 & 0 & 0 & 0 & 0 & 0 & 0 & 0 & 0 & 0 & 0 & 0 & 0 & 0 & 0 \\
0 & 0 & 0 & 0 & 0 & 0 & 0 & 0 & 0 & 0 & 0 & 0 & 0 & 0 & 0 & 0 & 0 & 0 \\
0 & 0 & 0 & 26.526 & 0 & 0 & 0 & 0 & 0 & 0 & 0 & 0 & 0 & 0 & 0 & 0 & 0 & 0 \\
0 & 0 & 0 & 0 & 0 & 0 & 0 & 0 & 0 & 0 & 0 & 0 & 0 & 0 & 0 & 0 & 0 & 0 \\
0 & 0 & 0 & 0 & 0 & 0 & 0 & 0 & 0 & 0 & 0 & 0 & 0 & 0 & 0 & 0 & 0 & 0 \\
0 & 0 & 0 & 0 & 0 & 0 & 26.526 & 0 & 0 & 0 & 0 & 0 & 0 & 0 & 0 & 0 & 0 & 0 \\
0 & 0 & 0 & 0 & 0 & 0 & 0 & 0 & 0 & 0 & 0 & 0 & 0 & 0 & 0 & 0 & 0 & 0 \\
0 & 0 & 0 & 0 & 0 & 0 & 0 & 0 & 0 & 0 & 0 & 0 & 0 & 0 & 0 & 0 & 0 & 0 \\
0 & 0 & 0 & 0 & 0 & 0 & 0 & 0 & 0 & 39.789 & 0 & 0 & 0 & 0 & 0 & 0 & 0 & 0 \\
0 & 0 & 0 & 0 & 0 & 0 & 0 & 0 & 0 & 0 & 0 & 0 & 0 & 0 & 0 & 0 & 0 & 0 \\
0 & 0 & 0 & 0 & 0 & 0 & 0 & 0 & 0 & 0 & 0 & 0 & 0 & 0 & 0 & 0 & 0 & 0 \\
0 & 0 & 0 & 0 & 0 & 0 & 0 & 0 & 0 & 0 & 0 & 0 & 26.526 & 0 & 0 & 0 & 0 & 0 \\
0 & 0 & 0 & 0 & 0 & 0 & 0 & 0 & 0 & 0 & 0 & 0 & 0 & 0 & 0 & 0 & 0 & 0 \\
0 & 0 & 0 & 0 & 0 & 0 & 0 & 0 & 0 & 0 & 0 & 0 & 0 & 0 & 0 & 0 & 0 & 0 \\
0 & 0 & 0 & 0 & 0 & 0 & 0 & 0 & 0 & 0 & 0 & 0 & 0 & 0 & 0 & 26.526 & 0 & 0 \\
0 & 0 & 0 & 0 & 0 & 0 & 0 & 0 & 0 & 0 & 0 & 0 & 0 & 0 & 0 & 0 & 0 & 0 \\
0 & 0 & 0 & 0 & 0 & 0 & 0 & 0 & 0 & 0 & 0 & 0 & 0 & 0 & 0 & 0 & 0 & 0
\end{bmatrix}
$$

For consistency in computation, the stiffness matrix $\boldsymbol{k}$ is formed in kN, m, sec and the mass matrix $\boldsymbol{m}$ is formed with the entries in metric T (1000 kg).

In addition, the following parameters are determined:

$\beta = 0.25$ (assume constant acceleration in the extrapolation process, more stable in computation)

$\gamma = 0.5$ (suggestion by Newmark).

$\Delta t = 0.02\,\text{sec}$ (chosen by the authors)

By Equation 7.52, $b_1 = \dfrac{1}{\beta \Delta t^2} = \dfrac{1}{0.25 \times 0.02^2} = 10000;$

$$b_2 = \frac{-1}{\beta \Delta t} = \frac{-1}{0.25 \times 0.02} = -200;$$

$$b_3 = 1 - \frac{1}{2\beta} = 1 - \frac{1}{2 \times 0.25} = -1; \; b_4 = \frac{\gamma}{\beta \Delta t} = \frac{0.5}{0.25 \times 0.02} = 100;$$

$$b_5 = 1 - \frac{\gamma}{2\beta} = 1 - \frac{0.5}{0.25} = -1; \; b_6 = \left(1 - \frac{\gamma}{2\beta}\right)\Delta t = \left(1 - \frac{0.5}{2 \times 0.25}\right) \times 0.02 = 0$$

In the first analysis, a ground movement of 1 m/sec^2 with circular frequency $\Omega = 18/\text{sec}$ as in the previous modal analysis as external excitation is tried. The vibration displacements with time are by the above process and parameters and plotted in Figure 7.12.

Compared with Figure 7.11 under the same ground motion for the same structure, the analysis by the Newmark Method agrees remarkably well with that by the Modal Method (40.4 mm versus 41.4 mm). In fact, for linear systems, if the damping is zero, and there is no numerical inaccuracy resulting from the numerical time integration

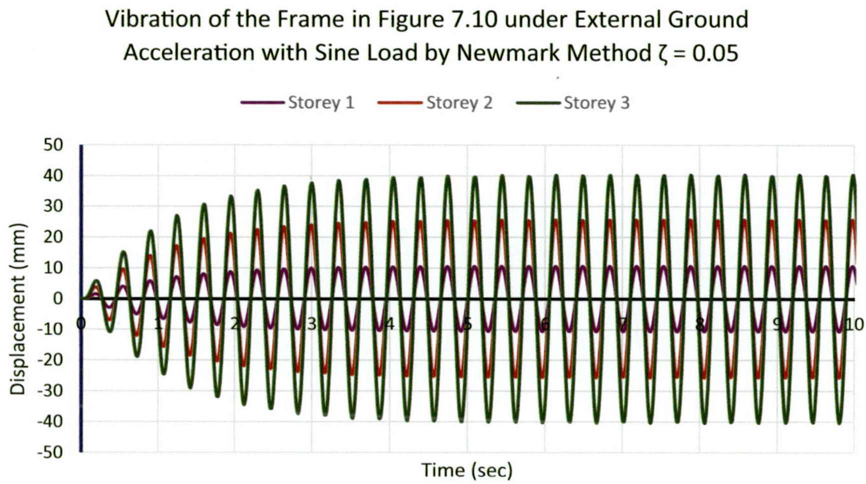

FIGURE 7.12 Floor vibrations of the three-storey plane frame at Figure 7.10 at $\zeta = 0.05$ and ground motion of $\left(1/\Omega^2\right)\sin\Omega t$, $\Omega = 18/\text{sec}$ by the Newmark Method.

process, mathematically, results from modal analysis with all modes being considered should be exactly the same as direct time integration using the full matrix. The modal decomposition process itself is mathematically exact. Nevertheless, the inaccuracy that we observe in daily practice is instead resulted from (i) insufficient number of modes being considered; (ii) the modal combination method (e.g., CQC (Complete Quadratic Combination)) which is mathematically just a rough approximation; and (iii) in the case of seismic analysis, the use of smoothened design spectrum which removes all variations from individual ground motion time history records. In this case of a simple structure where all modes have been used without the approximate modal combination method, it is thus not surprising that very good agreement has been reached.

The Newmark Method can directly allow solution against external forces or ground motion of arbitrary nature as illustrated in Worked Example 7.5.

7.4.4 Second Worked Example to the Newmark's Method

Worked Example 7.5
An analysis of the El Centro S00W acceleration with time history shown in Figure 7.13 for the first 25 sec is carried out by which the displacements of the three storeys are shown in Figure 7.14.

The modal method can also be used for analysis of the frame to seismic action. However, the seismic time history function on the time domain needs to be converted to a series of sine functions by Fourier transform which, comparatively, is rather indirect. And as discussed in the foregoing, some accuracy may be lost in the process of modal combinations.

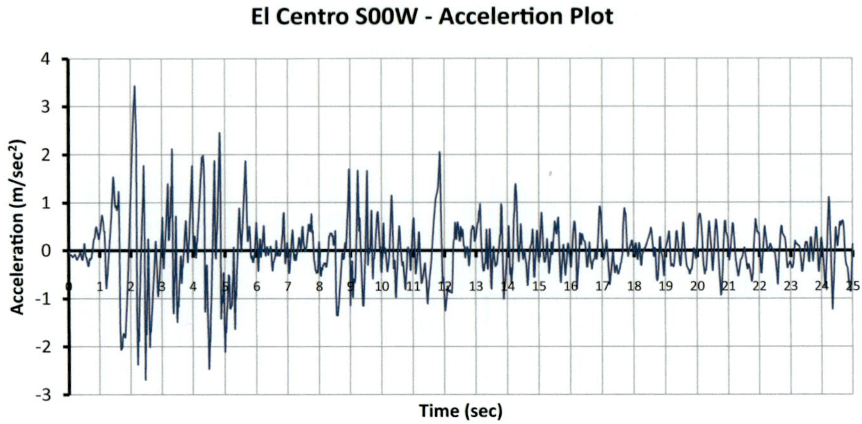

FIGURE 7.13 El Centro S00W time history record.

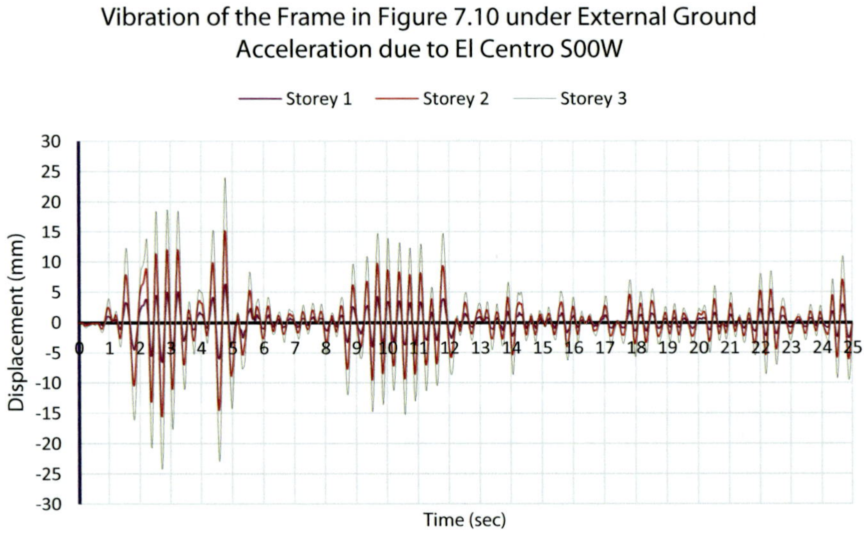

FIGURE 7.14 Vibration of the frame in Figure 7.10 under external ground acceleration of El Centro S00W.

7.4.5 Third Worked Example – Beam Vibration under Pedestrian Walking

The next worked example for the Integration Method is an analysis of a beam under footfall analysis as Worked Example 7.6.

Worked Example 7.6

FIGURE 7.15 Model of the beam for dynamic analysis.

A beam is to undergo footfall analysis by occupants. The beam is discretized into a number of segments with point masses representing the mass of the beam. Loads due to occupants are added onto the beam for analysis of the dynamic motion.

The beam is of structural size 400 mm (w) × 750 mm (d). The span is 15 m and its two ends are restrained from movement and rotation. It is divided into 20 segments with each segment of length 0.75 m as shown in Figure 7.15. The mass of the beam is 18T/m. So each of the dots carries $18 \times 0.75 = 13.5\text{T}$.

The natural frequency of the fixed-end beam is, by formula

$$f = 3.56\sqrt{\frac{EI}{qL^4}} = 3.56\sqrt{\frac{25000000 \times 0.4 \times 0.75^3 / 12}{18 \times 15^4}} = 2.211\text{Hz}$$

where EI is the flexural rigidity of the beam, q is the mass per unit length and L is the length of the beam.

Newmark method is used to analyze the beam against the following two load cases in which the loads are acting normal to the beam:

(i) A load of sinusoidal variation with time at a frequency of 2.2 Hz acting uniformly onto the beam with a maximum magnitude 18 kN/m;
(ii) An occupant walking with a speed of 1.5 m/sec along the beam with a footfall function discussed in Chapter 5.

The damping ratio is taken as 0.01 as per recommendation by Murray et al. (2016) for floors with no non-structural components or furnishings and few occupants.

As the analyses are with loads perpendicular to the beam, the stiffness matrix of the beam is formulated with the axial deformation ignored. So, the beam is a structure of $19 \times 2 = 38$ degrees of freedom.

Using the Newmark's method, using the time interval $\Delta t = 0.02\text{sec}$, the parameters β, γ, b_1 to b_6 are determined with values which are identical to the previous portal frame.

The variation of displacement with time at mid-span of the beam with time is as attached in Figure 7.16.

For floor vibration which is directly related to human comfort of the occupants, acceleration is more important. The variation of the mid-span acceleration with time is also plotted in Figure 7.17.

As the natural frequency of the beam is very close to that of the exciting load, the acceleration of 1.2 m/sec^2 is high, as discussed in Chapter 5. So revision of the structural layout or installation of "dampers" is required.

The next exercise involves a single occupant walking along the beam at a speed of 1.5 m/sec with a footfall function discussed in Chapter 5 which is reproduced in

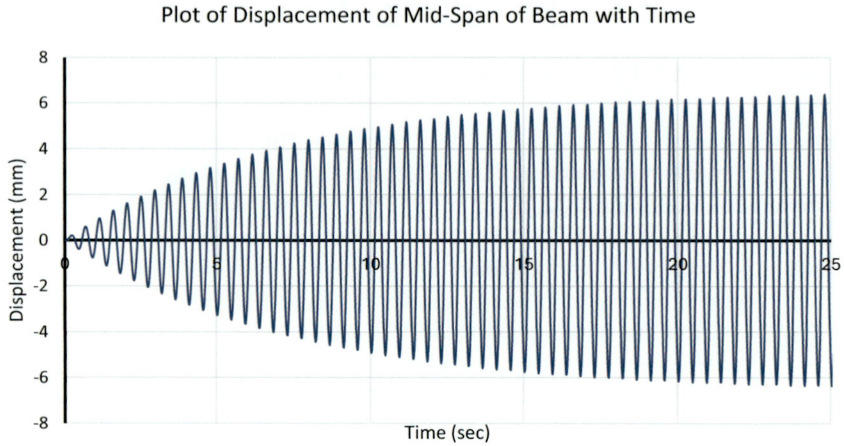

FIGURE 7.16 Variation of mid-span displacement with time.

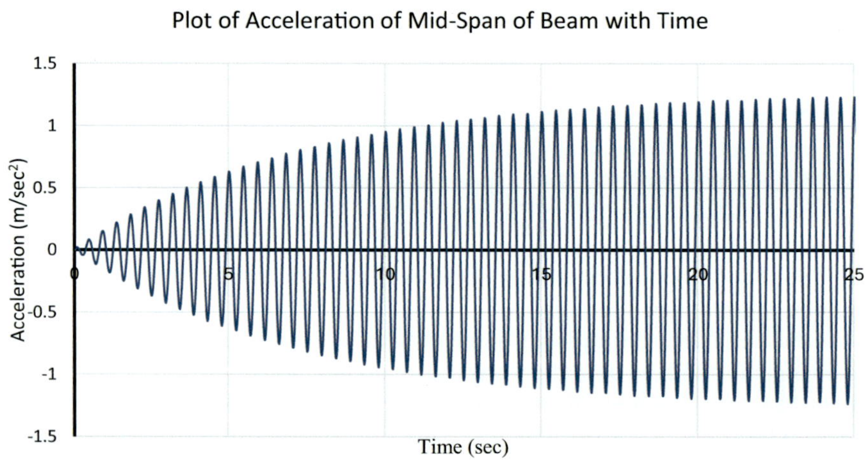

FIGURE 7.17 Variation of mid-span acceleration with time.

Figure 7.18. The maximum "dynamic weight" of the occupant is taken as 1.25 kN as compared with his static weight in the order of 0.7 kN.

During the analysis, each footfall rests on a node of the beam model for a duration of 0.4 sec with variation as shown in Figure 7.19. The successive footsteps are separated from each other by 0.75 m / 1.5 m/sec = 0.5 sec.

The displacement and acceleration of the mid-span due to the footfall of the single occupant are shown in Figures 7.20 and 7.21.

As discussed in Chapter 5, the crowd effect may have to be accounted for which may be a scaling up of the effects due to a single walking person or a detailed analysis of a crowd of walking persons.

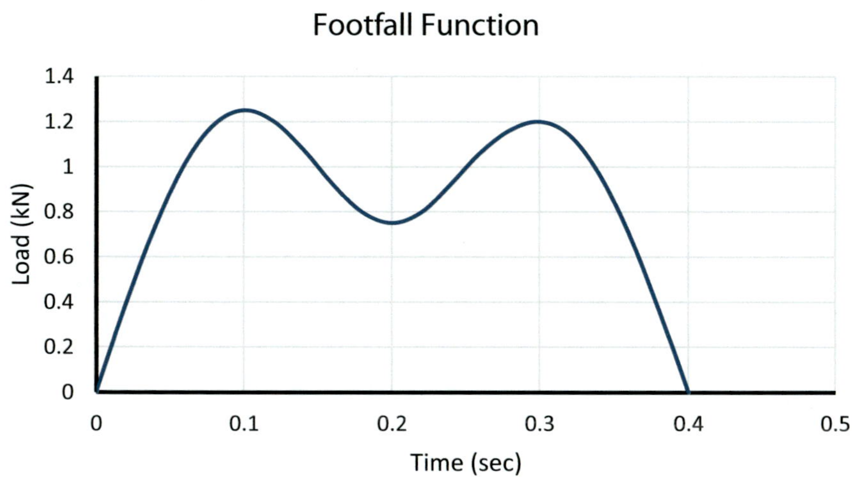

FIGURE 7.18 Footfall function.

FIGURE 7.19 Demonstration of load applications simulating footfall on the beam structure.

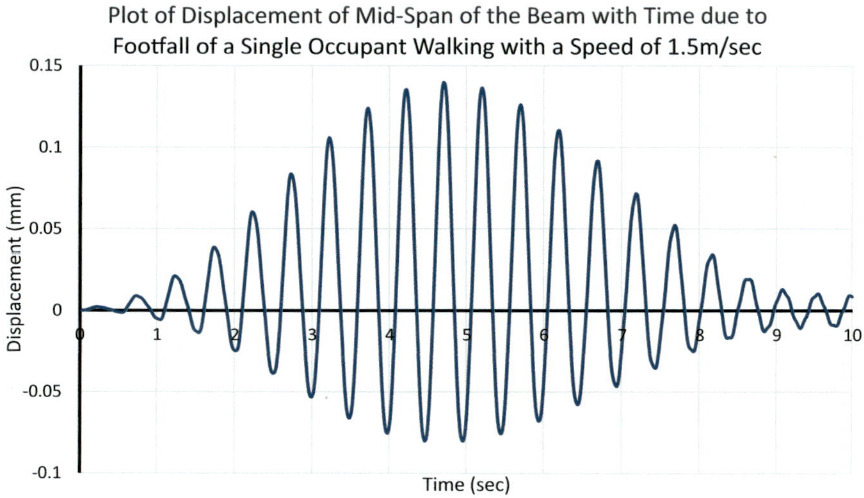

FIGURE 7.20 Variation of mid-span displacement with time due to footfall of a single occupant walking with a speed of 1.5 m/sec.

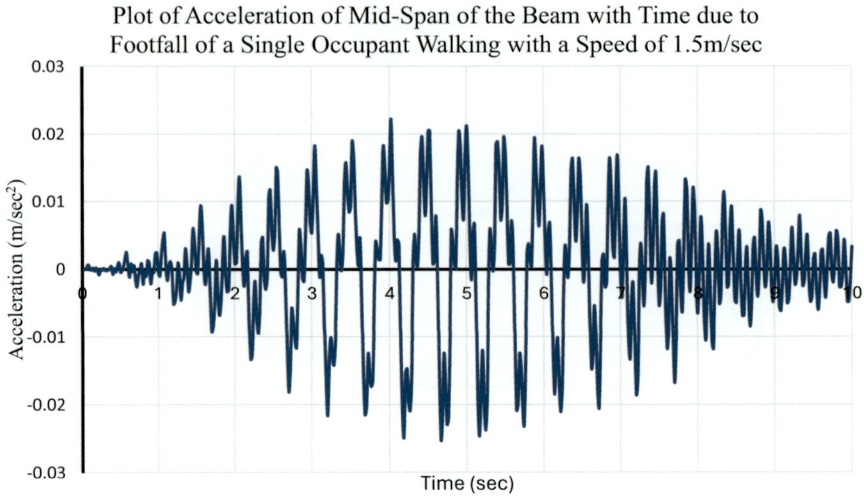

FIGURE 7.21 Variation of mid-span scceleration with time due to footfall of a single occupant walking with a speed of 1.5 m/sec.

7.5 ANALYSIS OF BUILDINGS BY COMPUTER SOFTWARE

The foregoing sections describe the underlying principles and approaches in the analysis of structures against dynamic effects. The demonstrations were done by hand calculations and the use of spreadsheets, which can only be confined to simple structures. For 3-dimensional real structures involving thousands or more degrees of freedom, computer software is required. Yet the analysis generally follows the aforementioned principles.

Analysis by computer method is discussed in this section, with highlights on the procedures and acquisition of results.

A 40-storey tube-in-tube building structure is taken for illustration, which is shown in Figure 7.22 as Worked Example 7.7.

Worked Example 7.7

The following steps should be adopted in employing computer software for dynamic analysis of structure:

(1) The dynamic properties of the structure should first be acquired. To obtain the dynamic properties, the user has to specify the "mass" used for the analysis. Normally, the mass should include the permanent dead load comprising the structure's own weight and the permanent dead load. A portion of live load should be used which normally lies between 0.3 to 0.4 of the full live load.

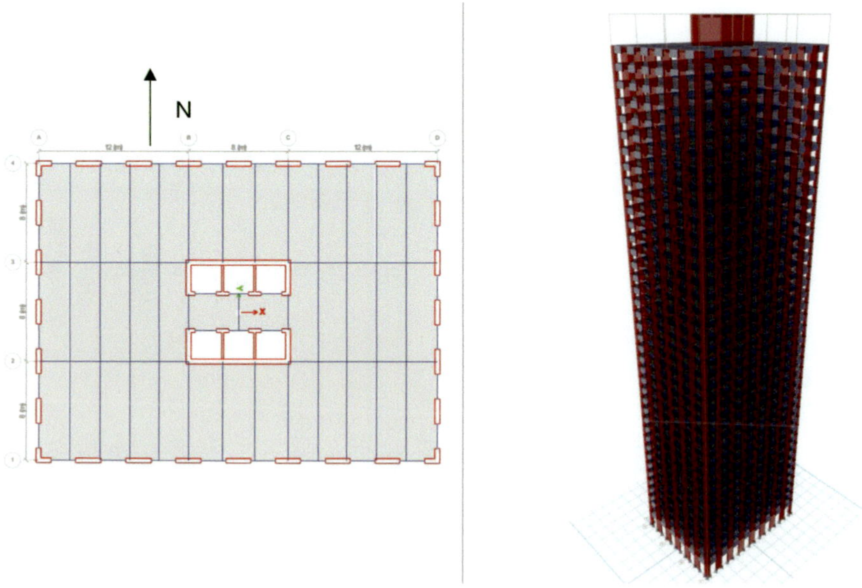

FIGURE 7.22 A tube-in-tube building structure for illustration of dynamic analysis by the computer method.

(2) The basic dynamic properties of the structure comprise at least (1) natural frequencies (or natural periods) generally in ascending order with the lowest one having the lowest natural frequency also named the fundamental frequency; (2) modes of each of the natural frequencies; and (3) "modal participating mass ratios" as associated with the natural frequencies. (1) and (2) have been explained in the foregoing. (3) is useful information explained in Equation 7.39. The ratio of a mode for a movement (translation or rotation) indicates the percentage of the total mass being mobilized in the mode. For example, in Table 7.2 which lists the modal participating mass ratios of the tube-in-tube building shown in Figure 7.22, the first mode of natural period 2.252 sec (or natural frequency 1/2.252 = 0.444 Hz) is 0.6678 indicating that 0.6678 of the total mass (assigned in step 1) vibrates with the frequency in the Y-direction. This lowest mode is also known as the fundamental frequency of the structure. The designer may also check that the building should possess the least stiffness in the Y-direction as under the concept $f = \dfrac{1}{2\pi}\sqrt{\dfrac{K}{M}}$ for the lowest frequency. In addition, as both the modal method analysis (dynamic) and the response spectrum analysis utilize frequency and/or mode shapes of the modes in analysis, an adequate number of modes should be calculated for accurate results. Generally, it is required that the cumulated modal participating mass for the movements should not be less than 90% to 95%. As such, 25 modes as listed in Table 7.2

TABLE 7.2
Modal Participating Mass Ratios of the Tube-in-Tube Building

TABLE: Modal Participating Mass Ratios								
Case	Mode	Period sec	UX	UY	UZ	RX	RY	RZ
Modal	1	2.252	0	0.6678	0	0.3416	0	0.00001166
Modal	2	1.828	0.6654	0	0	0	0.3438	0
Modal	3	0.548	0	0.0000124	0	0.00002215	0	0.8253
Modal	4	0.515	0	0.2079	0	0.3351	0	0.00002144
Modal	5	0.413	0.2082	0	0	0	0.3292	0
Modal	6	0.24	0	0.0546	0	0.1235	0	0.000001518
Modal	7	0.192	0.0554	0	0	0	0.1246	0
Modal	8	0.183	0	0	0	0	0	0.0916
Modal	9	0.156	0	0.0243	0	0.0655	0	0
Modal	10	0.124	0.0251	0	0	0	0.0671	0
Modal	11	0.115	0	0.0131	0	0.0366	0	0.000004011
Modal	12	0.109	0	0.000001096	0	0.000002635	0	0.0327
Modal	13	0.091	0	0.0081	0	0.0239	0	0
Modal	14	0.091	0.0138	0	0	0	0.0384	0
Modal	15	0.078	0	0	0	0.00000101	0	0.0163
Modal	16	0.075	0	0.0054	0	0.016	0	0.000001549
Modal	17	0.071	0.0086	0	0	0	0.0251	0
Modal	18	0.064	0	0.0038	0	0.0115	0	6.441E-07
Modal	19	0.061	0	0	0	0.000000877	0	0.0096
Modal	20	0.059	0.0057	0	0	0	0.0167	0
Modal	21	0.056	0	0.0028	0	0.0083	0	0
Modal	22	0.05	0.0039	0	0	0	0.0118	0.000001644
Modal	23	0.05	0.000001044	0.000001315	0	0.000004321	0.000003159	0.0061
Modal	24	0.049	0	0.002	0	0.0063	0	0.000006493
Modal	25	0.044	0	0.0015	0	0.0046	0	0
		Sum	0.986101044	0.991314811		0.972930993	0.956703159	0.981648959

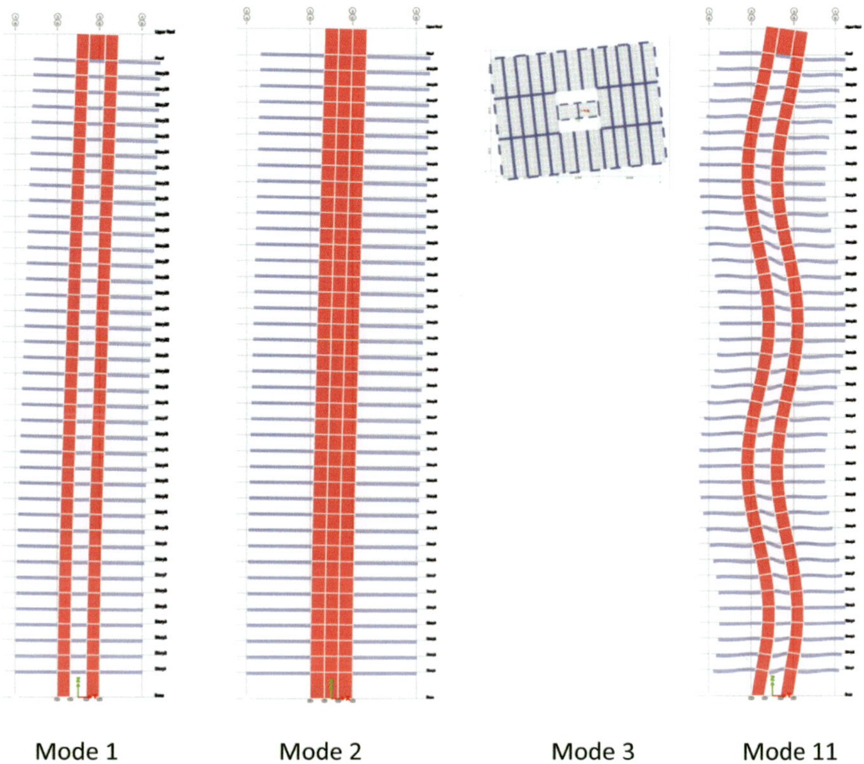

FIGURE 7.23 Mode shapes of the tube-in-tube building.

are listed. The vertical motion UZ has been ignored in the determination of mass which is a normal practice in seismic design.

An interesting point to note is the mode shapes as shown in Figure 7.23. Modes 1 and 2 refer to the shapes of vibration of the lowest modes in the X and Y directions, which are simple or can be approximated as linear. So does Mode 3, which is torsional as RZ in Table 7.2. In higher modes, the shapes are more complicated, such as mode 11 shown, which is the vibration in the Y-direction. Nevertheless, the participating mass ratios of these complicated modes are small, being 0.0131 for model 11.

(3) Time histories for the dynamic analysis in terms of accelerations with time should then be defined and inputted. They may be some periodic functions or the earthquake time histories. Upon input, the software will assign periodic force to the structure in accordance with the approaches described in the foregoing, which were demonstrated by the simple structures.

(4) Generally, displacement, velocity, and acceleration of any points can be listed, together with the member forces. As these results vary with time,

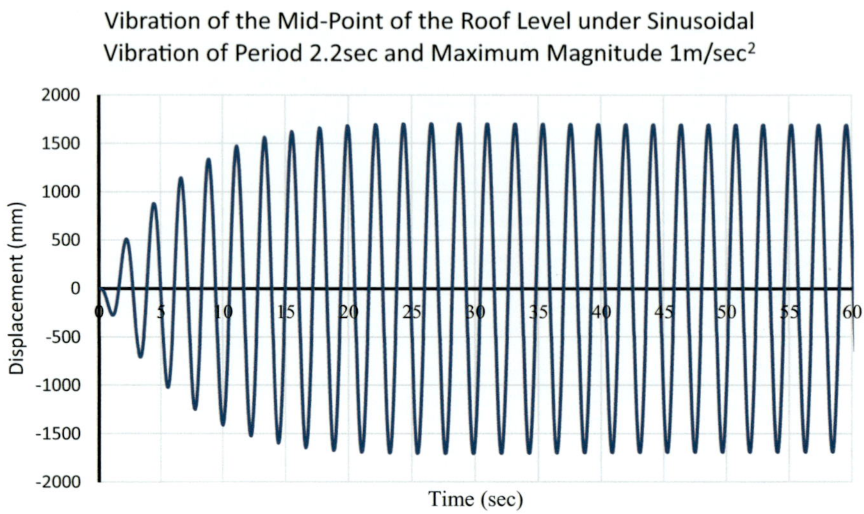

FIGURE 7.24 Vibration of the mid-point of the roof under sinusoidal vibration of period of 2.2 sec and maximum magnitude 1 m/sec^2.

the results can therefore be given in the time domain. For ease of design, maximum and minimum (maximum negative) values can generally be outputted.

Two load cases acting in the Y-direction are analyzed: (i) a sinusoidal ground vibration of period 2.2 sec (close to the building's fundamental period of 2.25 sec in the Y-direction) with maximum magnitude 1 m/sec^2; (ii) the El Centro NS vibration. The vibrations of the mid-point at storey 40 are plotted in Figures 7.24 and 7.25 for the 2 cases.

By Figure 7.24, the building has been brought to steady state at around 20 sec after the start of the external excitation. The high magnitude of vibration is due to the close proximity of the excitation vibration to the natural frequency of the building, creating resonance. The large magnitudes of vibration would likely bring the structure to the plastic stage. That is, the elastic analysis assumed in the analysis may not be very valid.

The variation of the maximum and minimum bending moments (maximum negative) of the northern part of the core wall (Figure 7.22) with building heights due to the seismic effect are extracted and plotted in Figure 7.26 which may be adopted for the structural design of the core wall.

7.6 USE OF TUNED MASS DAMPER

When the vibration of a building structure is very excessive and the variation of its stiffness or mass to reduce vibration becomes expensive or impractical, a "tuned mass damper" can help reduce vibration effectively at a reasonable cost.

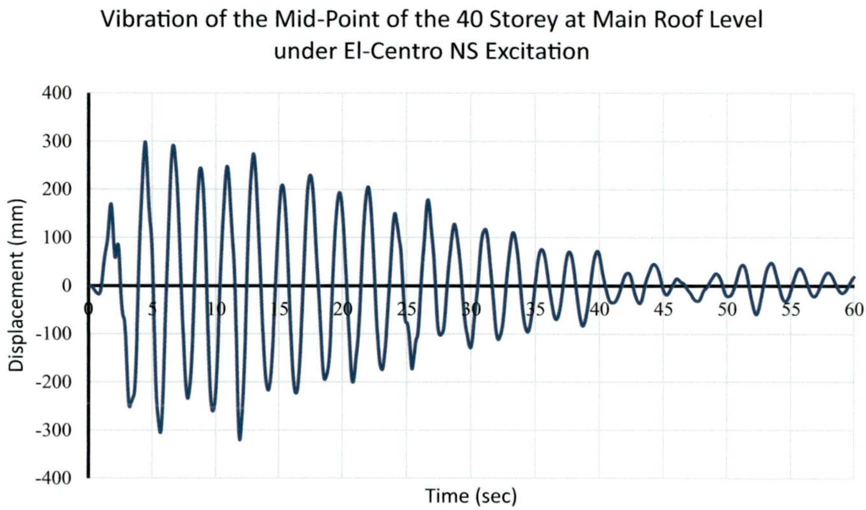

FIGURE 7.25 Vibration of the mid-point of the main roof under El Centro NS ground vibration.

A tuned mass damper (TMD) is a passive device that mitigates vibrations in structures like buildings. It consists of a mass-spring-damper system tuned to the structure's fundamental natural frequency. When the building oscillates from effects like wind or earthquakes, the TMD moves out of phase with the structure, absorbing and dissipating energy. This action reduces the building's vibrational amplitude, thereby enhancing its stability, safety, and occupant comfort.

7.6.1 Fundamentals

Some fundamentals of the tuned mass damper are given as follows. Consider a single degree freedom (SDOF) of mass M under stiffness K and damping constant C being connected to a tuned mass damper of mass m_d with stiffness k_d and damping constant c_d as shown in Figure 7.27.

If the mass M is subjected to an external force F_e, the following equations of motion on the two masses can be formulated separately as:

$$M\ddot{x}+C\dot{x}+Kx-k_d\left(x_d-x\right)-c_d\left(\dot{x}_d-\dot{x}\right)=F_e$$

$$\Delta M\ddot{x}+\left(C+c_d\right)\dot{x}+\left(K+k_d\right)x-c_d\,\dot{x}_d-k_d x_d=F_e \tag{7.54}$$

$$m_d\ddot{x}_d+c_d\left(\dot{x}_d-\dot{x}\right)+k_d\left(x_d-x\right)=0$$

$$\Delta m_d\ddot{x}_d+c_d\dot{x}_d+k_d x_d-c_d\dot{x}-k_d x_d=0 \tag{7.55}$$

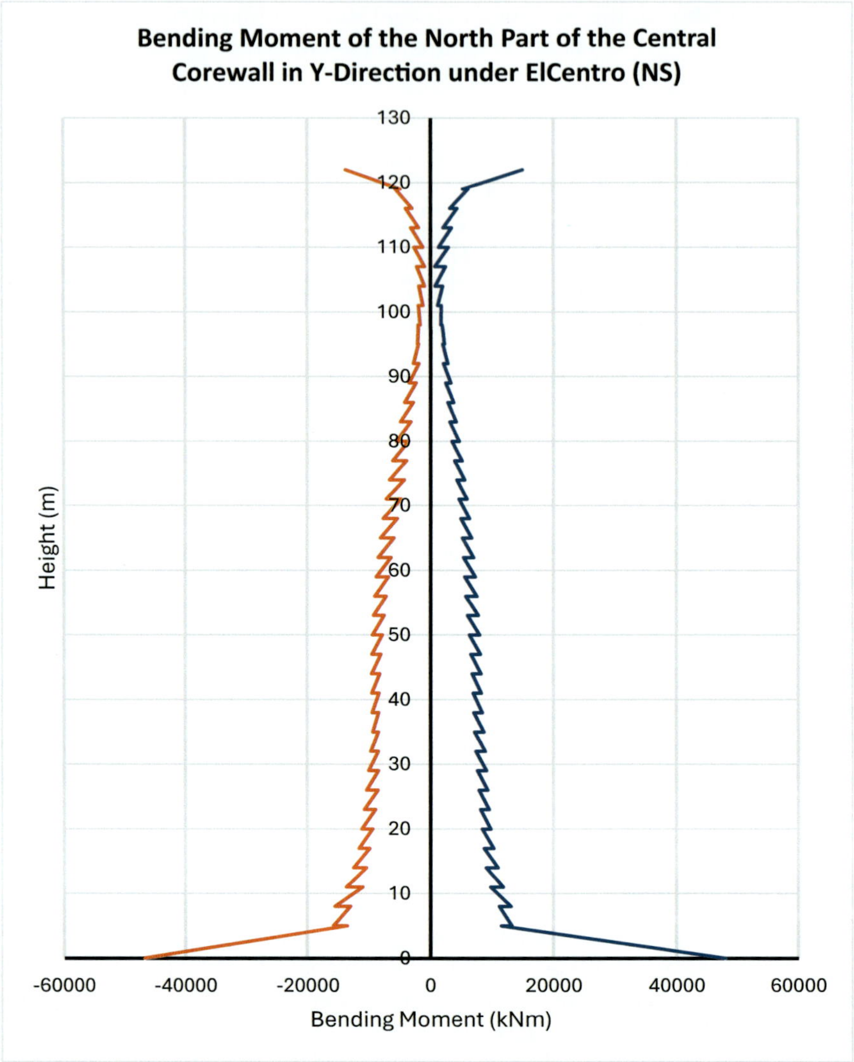

FIGURE 7.26 Variation of the bending moment of the northern part of the central core wall in the Y-direction with building height under the El Centro (NS) excitation.

With F_e taken as a periodic force expressed in a complex form as $F_e = F_0 e^{i\Omega t}$, the two equations can be combined in matrix form as

$$\begin{bmatrix} M & 0 \\ 0 & m_d \end{bmatrix} \begin{bmatrix} \ddot{x} \\ \ddot{x}_d \end{bmatrix} + \begin{bmatrix} C + c_d & -c_d \\ -c_d & c_d \end{bmatrix} \begin{bmatrix} \dot{x} \\ \dot{x}_d \end{bmatrix} + \begin{bmatrix} K + k_d & -k_d \\ -k_d & k_d \end{bmatrix} \begin{bmatrix} x \\ x_d \end{bmatrix} = \begin{bmatrix} F_e \\ 0 \end{bmatrix} \tag{7.56}$$

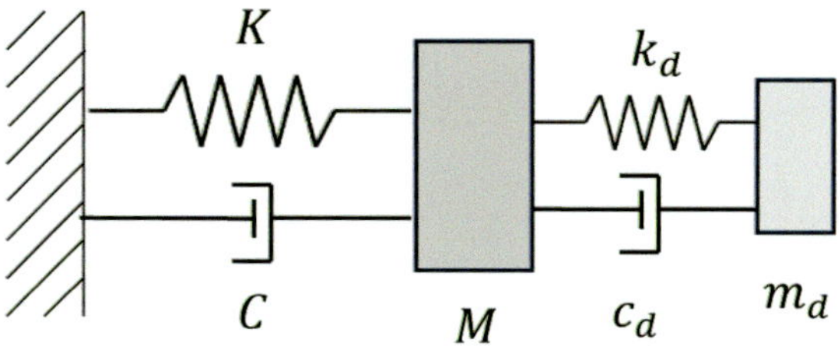

FIGURE 7.27 SDOF mass with TMD.

$$\text{So}\begin{bmatrix} x \\ x_d \end{bmatrix} = \begin{bmatrix} xe^{i\Omega t} \\ x_d e^{i\Omega t} \end{bmatrix} x; \begin{bmatrix} \dot{x} \\ \dot{x}_d \end{bmatrix} = -i\omega \begin{bmatrix} xe^{i\Omega t} \\ x_d e^{i\Omega t} \end{bmatrix}; \begin{bmatrix} \ddot{x} \\ \ddot{x}_d \end{bmatrix} = -\omega^2 \begin{bmatrix} xe^{i\Omega t} \\ x_d e^{i\Omega t} \end{bmatrix} \tag{7.57}$$

Substituting Equation 7.57 into Equation 7.56 and cancelling out the $e^{i\Omega t}$ terms

$$-\Omega^2 \begin{bmatrix} M & 0 \\ 0 & m_d \end{bmatrix} \begin{bmatrix} x \\ x_d \end{bmatrix} + i\Omega \begin{bmatrix} C + c_d & -c_d \\ -c_d & c_d \end{bmatrix} \begin{bmatrix} x \\ x_d \end{bmatrix} + \begin{bmatrix} K + k_d & -k_d \\ -k_d & k_d \end{bmatrix} \begin{bmatrix} x \\ x_d \end{bmatrix} = \begin{bmatrix} F_e \\ 0 \end{bmatrix} \tag{7.58}$$

$$\text{Solving for } x = F_e \sqrt{\frac{a^2 + b^2}{c^2 + d^2}} \tag{7.59}$$

where $a = k_d - \Omega^2 m_d$; $b = \Omega c_d$;

$$c = \Omega^4 M m_d - \Omega^2 \left[m_d \left(K + k_d \right) + M k_d + C c_d \right] + K k_d$$

$$d = \Omega \left(c_d K + C k_d \right) - \Omega^3 \left[m_d \left(C + c_d \right) + M c_d \right]$$

Defining $\omega = \sqrt{\frac{K}{M}}$, circular frequency of the mass M

$\xi - \frac{C}{2\omega M}$, damping ratio of the mass M

$\omega_d = \sqrt{\frac{k_d}{m_d}}$, circular frequency of the TMD alone

$\xi_d = \frac{C_d}{2\omega_d m_d}$, damping ratio of the TMD, and the following ratios

$q = \frac{\omega_d}{\omega}$, $\mu = \frac{m_d}{M}$, $r = \frac{\Omega}{\omega}$, Equation 7.59 can be written as

$$\bar{x} = \sqrt{\frac{\bar{a}^2 + \bar{b}^2}{\bar{c}^2 + \bar{d}^2}} \tag{7.60}$$

where $\bar{a} = q^2 - r^2$, $\bar{b} = 2\xi_d{}^2 rq$,

$$\bar{c} = r^4 - r^2\left(q^2\left(1+\mu\right) + 1 + 4\xi\xi_d q\right) + q^2$$

$$\bar{d} = 2rq\xi_d + 2rq^2\xi - r^3\left(2\xi + 2\xi_d q\mu + 2\xi_d q\right)$$

Optimum parameters of the TMD system ξ_d and q are to be searched for a specific mass ratio μ to minimize the maximum displacement of the mass M. Den Hartog (1956) suggests the optimal values for the frequency ratio and the damping ratio of the TMD are as follows:

$$q_{opt} = \frac{1}{1+\mu} \tag{7.61}$$

$$\xi_{d,opt} = \sqrt{\frac{3\mu}{8\left(1+\mu\right)}} \tag{7.62}$$

The stiffness of the damper is $k_d = \omega_d{}^2 m_d = \left(2\pi f_d\right)^2 m_d$ (7.63)

The damping constant is $c_d = 2\sqrt{k_d m_d}\,\xi_{d,opt}$ (7.64)

The two equations can also be applied to structures of multi-degree of freedom with q_{opt} referring to the fundamental frequency of the system.

7.6.2 Types of TMD

7.6.2.1 Passive TMD

The commonest type of tuned mass damper is in the form of a device as shown in Figure 7.28 extracted from Purdue University (2002) which comprises a "spring"

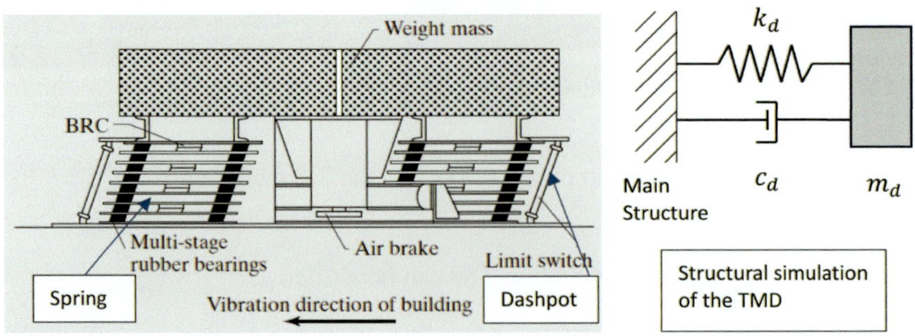

FIGURE 7.28 Illustration and structural simulation of TMD.

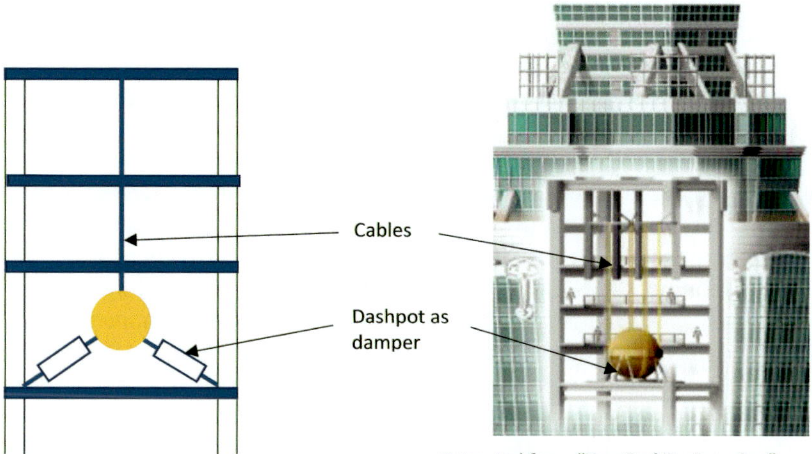

Extracted from "Practical Engineering" February 2016 "Tuned Mass Damper in Skyscrapers" for Taipei 101 Tower

FIGURE 7.29 Pendulum tuned mass damper.

and a "damper" connecting a weight mass to the main structure as shown in Figure 7.28. The TMD performs damping to the horizontal motion of the building by its out-of-phase motion. To be effective, the weight mass should be at least 1% to 2% of the building. For simulation of the TMD in structural analysis, a spring of stiffness k_d and a dashpot of damping constant c_d can be connected to a point with a mass m_d . In some computer software, there are "link" elements which can replace the spring and the dashpot for convenient simulation of TMD.

7.6.2.2 Pendulum Tuned Mass Damper

Another type of tuned mass damper is the pendulum tuned mass damper, which comprises a heavy mass hung by steel wire to the roof of the building. The heavy mass is underdamped by dashpots. A schematic illustration of the system and its practical use in the famous Taipei 101 tower is contained in Figure 7.29. It performs its damping functions by the swinging motion of the mass, which is out of phase with the building's horizontal motion. Energy from the building is dissipated to the damper.

In a pendulum TMD, the restoring force to the motion of the mass is the horizontal component of the weight of the mass m_d is $m_d gx / L$ where x is the displacement of the mass from the equilibrium position and L is the length of the pendulum. So, the stiffness of the damper is $k_d = m_d g / L$. The circular frequency is $\omega = \sqrt{k_d / m_d}$. With $\omega = 2\pi / T_d$, where T_d is the structural period of the damper,

$$T_d = 2\pi\sqrt{L / g} \tag{7.65}$$

The structural simulation of the PTMD may be performed as shown in the left part of Figure 7.29 by linking a point mass by a line member of very strong stiffness while being hinged at the building structure. Alternatively, it may be simulated by an assembly of a spring, a dashpot and a point mass, as shown on the right side of Figure 7.28.

7.6.2.3 Tuned Liquid Damper by Roof Water Tank

A tuned liquid damper (TLD), also known as a tuned sloshing damper, is a type of vibration control device that uses the motion of a liquid inside a tank to reduce the sway or vibrations of a structure. It is the liquid-based equivalent of a conventional tuned mass damper (TMD) which uses a solid mass. So, for buildings, a roof water tank can then serve dual purposes in water storage/supply and dampers.

The fundamental principle is the same as for any tuned mass damper by which energy is transferred through resonance. The TLD is "tuned" so that the natural frequency of the sloshing liquid matches the natural frequency of the structure. When the structure sways, it causes the liquid in the tank to slosh. The goal is to make the liquid slosh out-of-phase with the structure's motion, creating forces that oppose and dampen the building's movement.

Energy is dissipated through the followings:

(i) Wave Breaking: The chaotic action of waves crashing against the tank walls and themselves.
(ii) Friction: The fluid friction within the liquid itself and against the tank walls.
(iii) Baffles: Many TLDs have obstructions (baffles) placed inside the tank to intentionally increase turbulence and energy dissipation.

Figure 7.30 illustrates the actions by a tuned liquid damper:

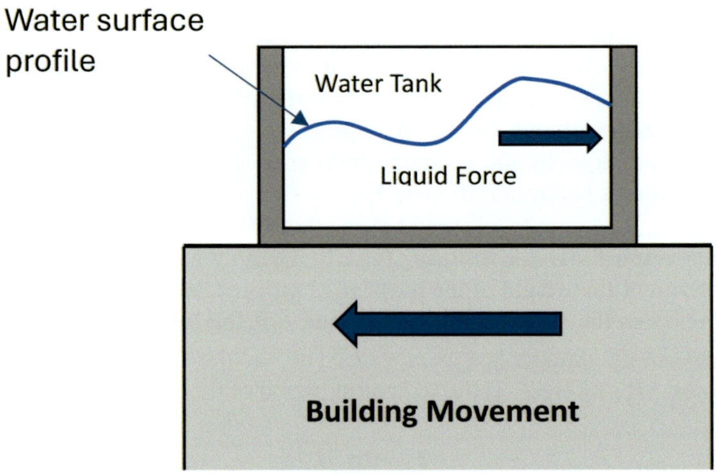

FIGURE 7.30 Working principle of a tuned liquid damper.

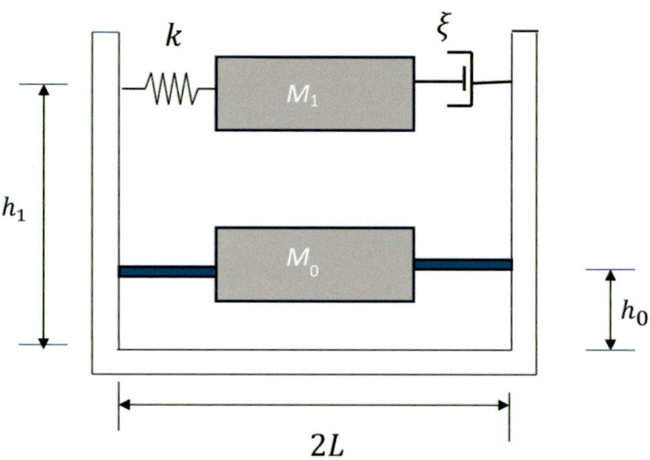

FIGURE 7.31 Equivalent dynamic system of a TLD.

The fundamental mode of the natural frequency of a rectangular tuned liquid damper can be found by the following formula as quoted by Fujino (1992) which is based on small amplitude sloshing and a rigid tank.

$$f = \frac{1}{2\pi}\sqrt{\frac{\pi g}{L}\tanh\frac{\pi h}{L}} \tag{7.66}$$

For analysis, a simplified method which may be known as the "lumped mass method" can be used to simulate the behaviour of TLD. With the lumped mass method, Housner (1963) assumes the TLD walls to be rigid and, during excitation, a mass M_0 of water at a height of h_0 above the bottom of the tank is rigidly attached to it, which will move synchronically with the tank. The motion of the tank will, however, excite the upper part of the water into oscillations which in turn exert an oscillating force on the tank. This oscillating force is the equivalent to that of mass M_1 at height $h_1$oscillating under a restraining spring, together with a dashpot as shown in Figure 7.31.

For more accurate representation, Ibrahim (2005) splits M_1 into a series of masses, each with its own spring and dashpot.

Approximating by a single mass M_1 and referring to Figure 7.31, these parameters can be estimated by the following equations for rectangular water tanks as listed in Housner (1963).

$$M_0 = \frac{\tanh\left(\sqrt{3}L/h\right)}{\sqrt{3}L/h}M \tag{7.67}$$

$$M_1 = \frac{0.83\tanh\left(1.6h/L\right)}{1.6h_f/L}M \tag{7.68}$$

$$h_0 = \frac{3}{8}h\left[1+\alpha\left(\frac{m_f}{M_0}\left(\frac{L}{h}\right)^2 - 1\right)\right] \tag{7.69}$$

$$h_1 = h_f\left[1 - \frac{1}{3}\frac{M}{M_1}\left(\frac{L}{H_1}\right)^2 + 0.63\beta\frac{L}{h}\sqrt{0.28\left(\frac{ML}{M_1 h}\right)^2 - 1}\right] \tag{7.70}$$

where M is the total mass of the contained fluid, h is the liquid height, $\alpha = 1.33$ and $\beta = 2.0$.

The stiffness can be calculated by the relation $\omega^2 = (2\pi f)^2 = k / M_1$

$$k = (2\pi f)^2 M_1 \tag{7.71}$$

The damping ratio can be estimated by the following:

$$\xi_f = \sqrt{\frac{v_f \omega}{2}}\left[1 + \frac{2h}{B} + S\right]\frac{L}{h\sqrt{gh}} \tag{7.72}$$

where $v_f \approx 10^{-6}$m²/sec is the viscosity of water. S is a surface contamination factor that can vary between 0 and 2.

With the determination of all the dimensions and water levels of the tank, damping can be achieved.

7.6.3 Worked Example

So, the action by the tuned mass damper can finally be described by its mass, a spring and a dashpot providing resistances proportional to displacement and velocity respectively. A design with analysis for the building in Figure 7.22 is demonstrated in the following Worked Example 7.8.

Worked Example 7.8

The mass of the building, contributed by dead load comprising own weight and superimposed dead load plus 40% of design live load, is $M = 44672$T. The fundamental frequencies in the Y (lowest) and X directions are 0.444 Hz and 0.567 Hz respectively. Two options are tried with mass ratios (the ratio of the mass of the TMD to the building structure) $\mu = 0.02$ and $\mu = 0.01$. The frequencies of the dampers and subsequently the stiffnesses and the damping constants are worked out as follows:

	Mass Ratio $\mu = 0.02$		**Mass Ratio $\mu = 0.01$**	
	Y-Direction	**X-Direction**	**Y-Direction**	**X-Direction**
Frequency (Hz) (Equation 7.61)	0.4353	0.5363	0.4397	0.5416
Mass of Damper (T) μM	893.449	893.449	446.724	446.724
k_d kN/m (Equation 7.63)	6684.86	10145.58	3408.94	5173.74
c_d kNsec/m (Equation 7.64)	410.90	506.21	207.49	255.61

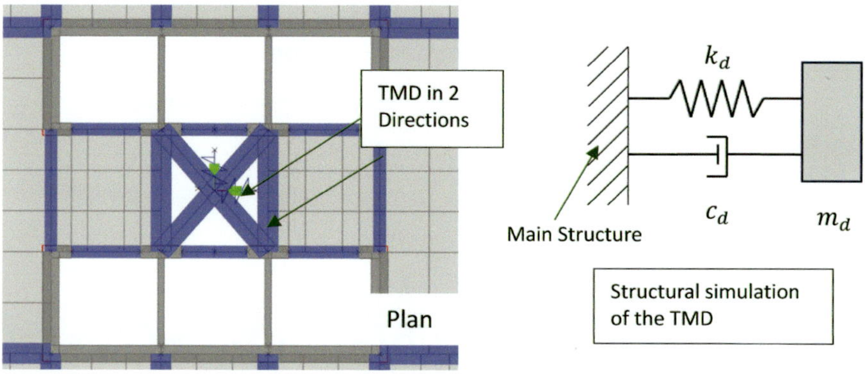

FIGURE 7.32 Locations of the tuned mass damper.

The tuned mass dampers are proposed to be installed at the centre of the roof level over the lift core as shown in Figure 7.31. They can either be simulated by masses, springs and dashpots as shown on the right part of Figure 7.32. The authors have however, used the "link" elements provided in the software for the simulation.

The structure is re-analyzed with the dampers. The deflections in the Y-direction at the centre of the roof level are plotted with time for the ground excitations by (i) periodic excitation at period 2.2 sec (0.4545 Hz) with maximum magnitude 1 m/sec^2; and (ii) the El Centro time history. The following (Figures 7.33 and 7.34) displacements versus time graphs are plotted.

The analysis demonstrates that a significant reduction in vibration amplitude can be achieved, with higher damper mass yielding greater attenuation. However,

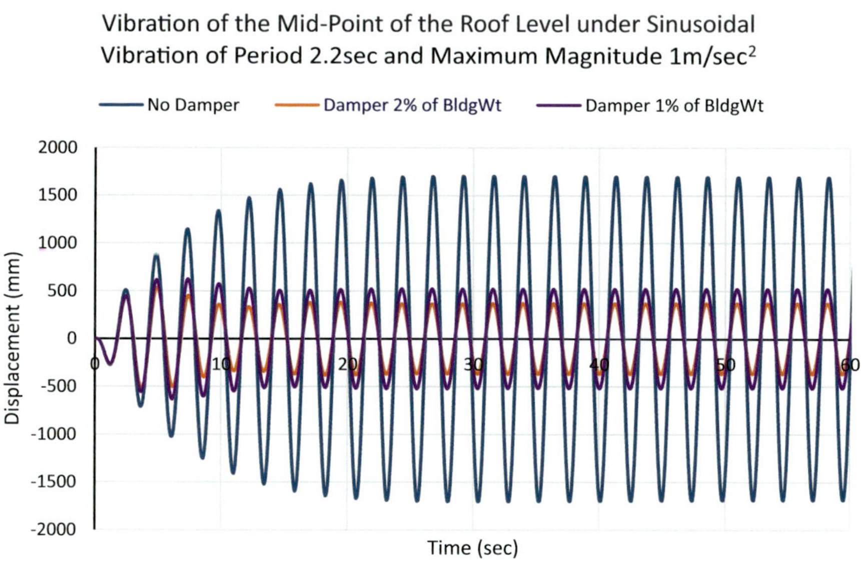

FIGURE 7.33 Vibration of the mid-point of the roof level under sinusoidal vibration of period 2.2 sec and maximum magnitude 1 m/sec^2.

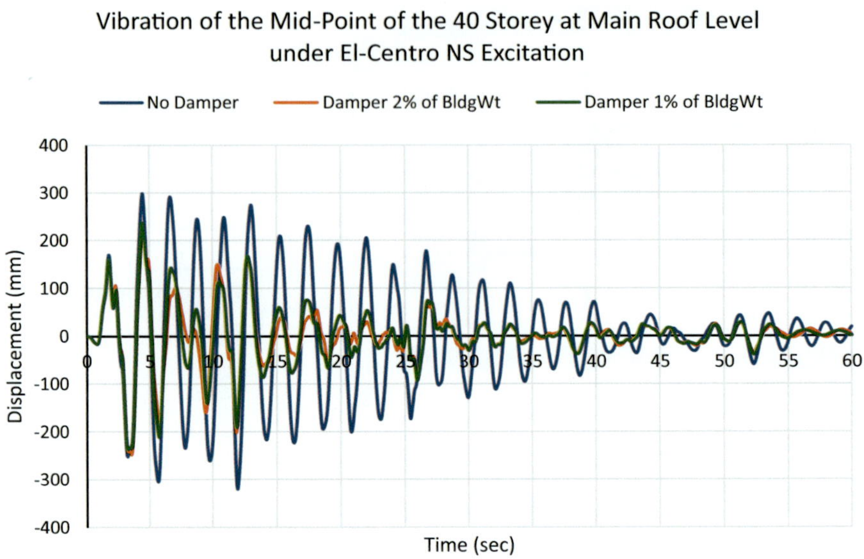

FIGURE 7.34 Vibration of the mid-point of the 40th storey at main roof level under El Centro NS excitation.

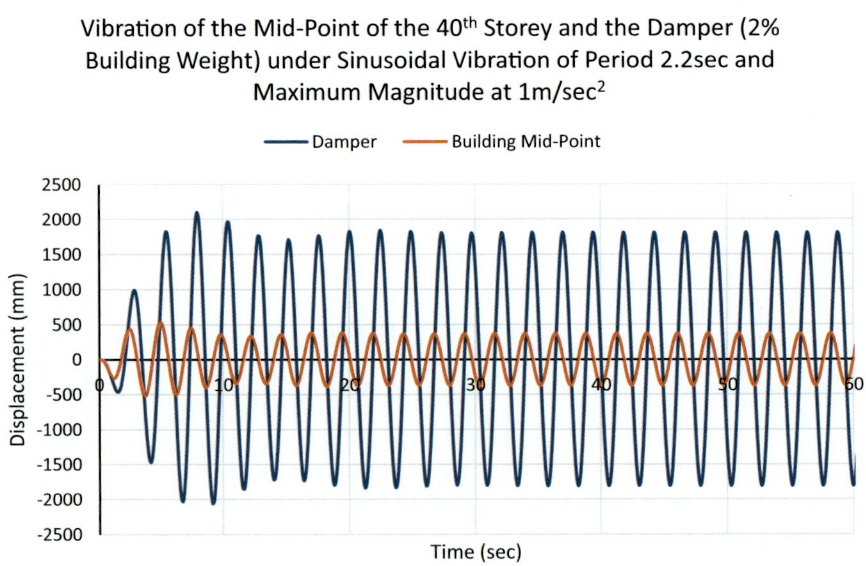

FIGURE 7.35 Vibration of the mid-point of the 40th storey and the damper (2% building weight) under sinusoidal vibration of period 2.2 sec and maximum magnitude at 1 m/sec^2.

the improvement observed when increasing the mass from 0.01 to 0.02 is less than double, indicating diminishing returns in damping efficiency with additional mass.

The vibration of the damper mass and that of the building centre point at roof level are also plotted in the same graph in Figure 7.35, showing the amplitude of vibration of the damper is much larger than the parent structure and the vibrations are out-of-phase.

7.7 CONCLUSION

Starting from the fundamental theory of structural dynamics, basic approaches for dynamic analysis of structures have been discussed and applied in solving various problems. The demonstrations include analysis of simple structures with procedures more or less identical to that of computer software. Readers can follow and perform their analyses by hand calculations and simple mathematical software which would help build a deep and intuitive understanding of the topic. Application of computer methods on large structures are also included.

REFERENCES

Den Hartog, J.P. (1956) Mechanical Vibration. New York: McGraw-Hill.

Fujino, Y., Sun, L., Pacheco, B.M. and Chaiseri, P. (1992) Tuned Liquid Damper (TLD) for Suppressing Horizontal Motion of Structures. Journal of Engineering Mechanics, ASCE, 2017–2030.

Housner, G.W. (1963) The Dynamic Behaviour of Water Tanks. Bulletin of the Seismological Society of America, 53(2), 381–387.

Ibrahim, R.A. (2005) Liquid Sloshing Dynamics: Theory and Application. Cambridge University Press.

Mario, M. and Willam, L. (2004) Structural Dynamics: Theory and Computation, 5th Edition. Kluwer Academic Publishers.

Murray, T.M., Allen, D.E. and Ungar, E.E. (2016) Steel Design Guide 11: Vibrations of Steel Framed Structural Systems Due to Human Activity. American Institute of Steel Construction.

Newmark, N.M. (1959) A Method of Computation for Structural Dynamics. ASCE Journal of the Engineering Mechanics Division, 85(EM3), 67–94.

Purdue University (2002) *Chapter 4 Tuned Mass Damper Systems*. Purdue University.

Wilson, E.L. (1962) *Dynamic Response by Step-By-Step Matrix Analysis*. Proceedings of the Symposium on the Use of Computers in Civil Engineering. Laboratorio Nacional de Engenharia Civil, Lisbon, Portugal, 1–5 October.

8 Ductility, Robustness, Durability, and Fire Resistance of Reinforced Concrete Buildings

8.1 INTRODUCTION

Structural design of building structures is not confined to strength and deformation only though they are the primary concern. Other criteria including ductility, robustness, durability, and fire resistance also affect the design of the structural layout and sizes of the building. Meeting specific design requirements for these criteria would ensure satisfactory performance of the structure under these criteria. Consequently, these criteria should be integrated into the holistic design of the building's layout and structural members. Knowledge of them in some depth is therefore required for design engineers.

In this chapter, design considerations relating to ductility, robustness, durability, and fire resistance of reinforced concrete building structures are discussed in some details.

8.2 DUCTILITY

Ductility is a critical property in structural engineering that refers to a material's or structure's ability to undergo large inelastic deformations without significant loss of strength before failure. In building structure design, ductility is essential for ensuring resilience, especially under extreme loads such as earthquakes, wind forces, or accidental impacts. It is therefore a preferable property of building structures, as in addition to those satisfying the ultimate and serviceability limit states.

A structure exhibits its ductility by effectively absorbing and dissipating energy generated from external loads through controlled yielding and deformation. Sudden collapse can then be avoided so that occupants are provided with vital time to evacuate. Strictly speaking, a structure possesses certain inherent ductility. But for the achievement of certain standards, design codes impose design requirements at various levels. These design requirements, however, are often geared to the seismicity of the region as the main concern for ductility design, with more stringent requirements in localities of higher seismicity.

DOI: 10.1201/9781003636380-8

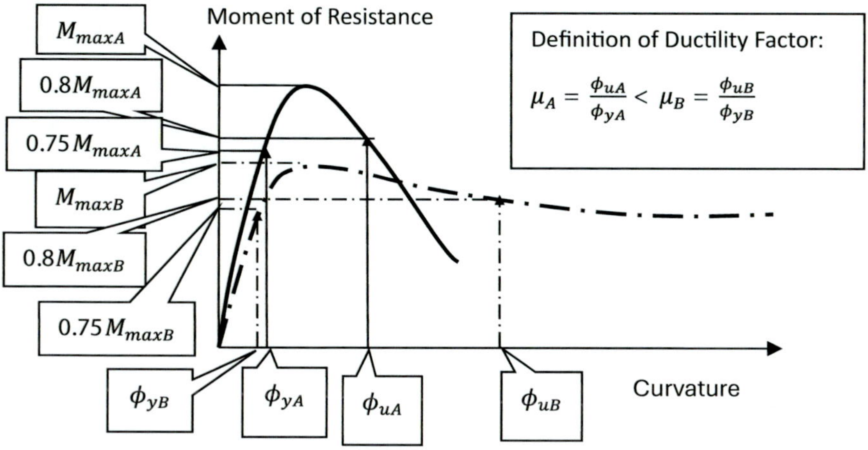

FIGURE 8.1 Illustration of ductility by bending of beams.

Figure 8.1 demonstrates behaviour of ductile and non-ductile (or brittle) structural elements. The figure shows a plot of the moments of resistance against the curvatures of two beams, Beam A and Beam B. Their ductilities are demonstrating different degrees of ductile behaviours.

In Figure 8.1, although Beam A can achieve a higher maximum moment of resistance. But its bending strength falls quickly with an increase in curvature in the "post-peak" portion. On the contrary, Beam B can sustain a considerable moment of resistance under the post-peak curvature, though it has a smaller maximum moment of resistance. Beam B is regarded as one with higher "ductility" than Beam A.

Ductility can be quantified by a "ductility factor" defined as $\mu = \phi_u / \phi_y$ where ϕ_u is the ultimate curvature taken as that corresponding to 80% of the maximum moment of resistance in the post-maximum region and ϕ_y is the yield curvature as that corresponding to 75% of the maximum moment of resistance in the pre-maximum region, as discussed by J.C.M. Ho et al. (2003) which is demonstrated in Figure 8.1. Nevertheless, it seems more popular that ϕ_u is taken as 85% of the maximum moment of resistance, as indicated in BSEN 1998-1:2004 (2004). It can also be seen that by the slopes of moment of resistance versus curvature for both beams in the post-peak regions, that of Beam B is much flatter than Beam A. The difference can then be quantified by their ductility factors μ with $\mu_B > \mu_A$. In more common words, Beam B can tolerate a larger angular rotation (deformation in terms of curvature) than Beam A prior to failure, even though A has a higher maximum moment of resistance.

Considering ductility as an important design parameter, it is thus desirable for a structure to attain certain levels of ductility so that its failure will not be all of a sudden, which can then give warning to the users to repair or to evacuate. The requirement for ductility is more important in seismic structural design, where extensive structural damage or even collapse can result. The levels of ductility required also

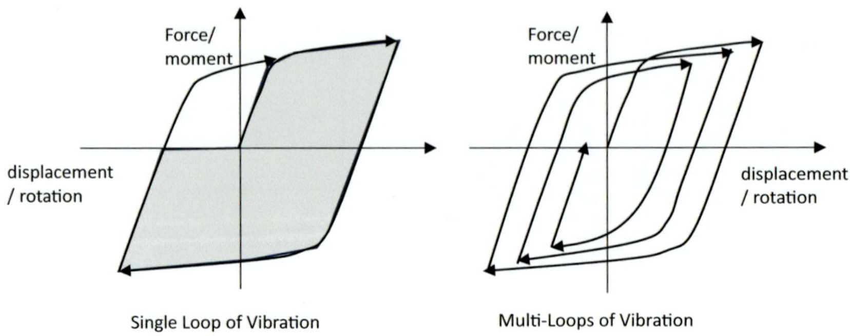

FIGURE 8.2 The phenomenon of "Hysteresis."

depend on the severity of the seismic actions (which vary from location to location) and the importance of the structure.

8.2.1 General Ductility Requirements

Some general discussions of ductility behaviour and its requirements in building structures are given as follows:

(i) A structure's ductility is its ability to undergo large deformations without catastrophic failure, which is crucial for withstanding seismic vibrations. Energy is dissipated most effectively through plastic deformation at the joints via a process called hysteresis.
As shown in the force-displacement diagram (Figure 8.2), the shaded area of the hysteresis loop represents the energy dissipated in a single movement cycle. Repeated cycles dissipate energy multiple times until the motion ceases. Therefore, joint designs must be detailed to undergo stable, repetitive hysteresis, as this provides the necessary ductility for seismic resistance.

(ii) Ductility is directly related to the structure's ability to undergo large deformation as discussed in the foregoing.

(iii) Besides offering significant deformation capacity before failure, ductility design is also associated with preferred failure modes, particularly under seismic action. A preferred mode of failure refers to the preferred sequence of failure of structural members, e.g., beam failure takes precedence over column which is the well-known "strong column-weak beam" design concept. This sequence of failure is preferred as the reverse, in which the column fails first, will be more drastic and can lead to large areas of collapse. Some design codes, such as the New Zealand Code NSZ3101:2006 (2006) and the China Seismic Code GB50011 (2010), require the sum of moments of resistance of the columns to exceed 1.2 times the sum of moments of resistance of the beams meeting at a column-beam

$$\sum M_c > 1.2 \sum M_b \qquad (8.1)$$

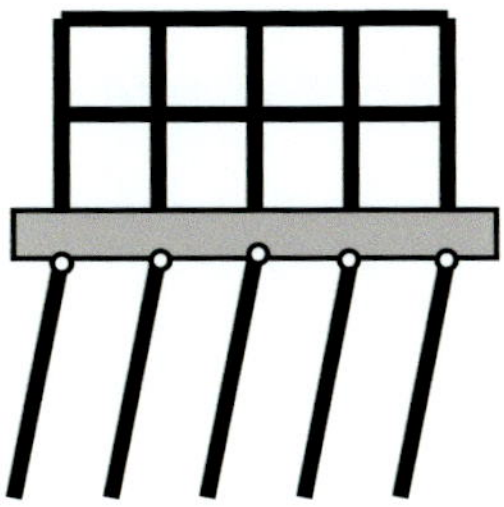

FIGURE 8.3 Formation of soft storey.

where M_c and M_b are the moments of resistance of the columns and beams respectively. This condition ensures beam failure prior to column failure as a "capacity design" concept, by which the design of an element should take into account the design strength (capacity) of the adjoining elements instead of only the load it has to carry.

(iv) Beam-column joints are susceptible to shear failure. To prevent a brittle joint failure from occurring before the connecting beams yield, the joint must be designed to be stronger than the beams. This capacity design principle requires designing the joint for an unbalanced moment greater than the beams' probable flexural strength. For example, the New Zealand Code NZS 3101:2006 (2006) specifies using a factor of 1.25 times the yield strength of the beam reinforcement to calculate this design moment.

(v) A "soft storey," which is a floor level significantly more flexible and weaker than those above it, should be avoided in building design. This condition often occurs when flexible columns are connected to a very stiff floor diaphragm. Under high lateral load, as shown in Figure 8.3, these columns are forced to absorb most of the deformation, leading to flexural failure at their connections to the rigid floor.

8.2.2 Ductility Design Considerations of Reinforced Concrete Structure

Particular ductility design requirements of reinforced concrete structure are listed as follows:

(i) Ductility behaviour of concrete

Concrete itself is not a ductile material. It possesses very low tensile strengths and tensile failure is very brittle. Its compressive stress-strain curve generally takes the shape as show in Figure 8.4 which is based on Attard and Setunge (1996) for concrete grades of characteristic cube strengths from 35 to 60 MPa. The peak strengths of the concretes are taken as $0.72f_{cu}$. The listing of the formulae of Attard and Setunge's

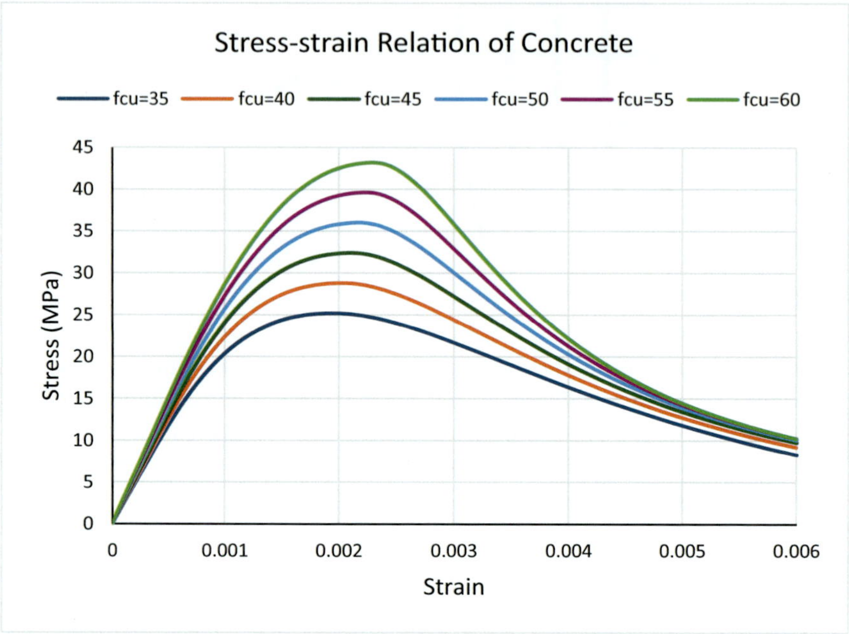

FIGURE 8.4 Stress-strain relation of concrete by Attard and Setunge (1996).

concrete stress-strain relation which can take into account the effect of confinement are contained in Appendix F.

By the definition in Figure 8.1, the ductility factors of the concrete grades vary from 3.43 to 2.51, which decrease with an increase in concrete strength. So high-grade concretes tend to be more brittle.

The ductility of reinforced concrete structures therefore has to rely on the embedded reinforcement, which is a much more ductile material.

(ii) Ductility provided by reinforcement in reinforced concrete structure

The stress strain curve at steel is typically shown in Figure 8.5 as follows which is often idealized as a linear relationship for design purposes until the yield stress is reached. In the idealized relationship, the strength stays constant at the yield stress beyond the yield strain.

The stress-strain curves of steel are similar in tension and compression.

The stress-strain curves of Concrete C60 and Grade 500 steel reinforcement with design strength at $0.87f_y$ $\left(f_y = 500\text{MPa}\right)$ are plotted in Figure 8.6 with the left vertical scale for concrete stress and the right for steel.

From Figure 8.6, the decrease in strength of concrete beyond strain around 0.0022 can be compensated by steel reinforcement with design strength of $0.87f_y$,

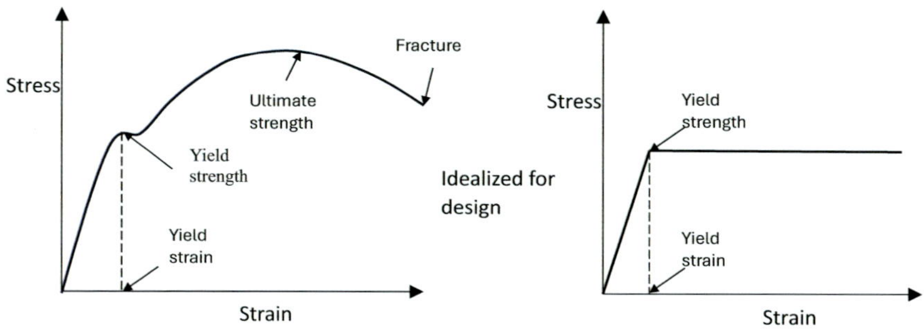

FIGURE 8.5 Typical stress-strain curve of steel.

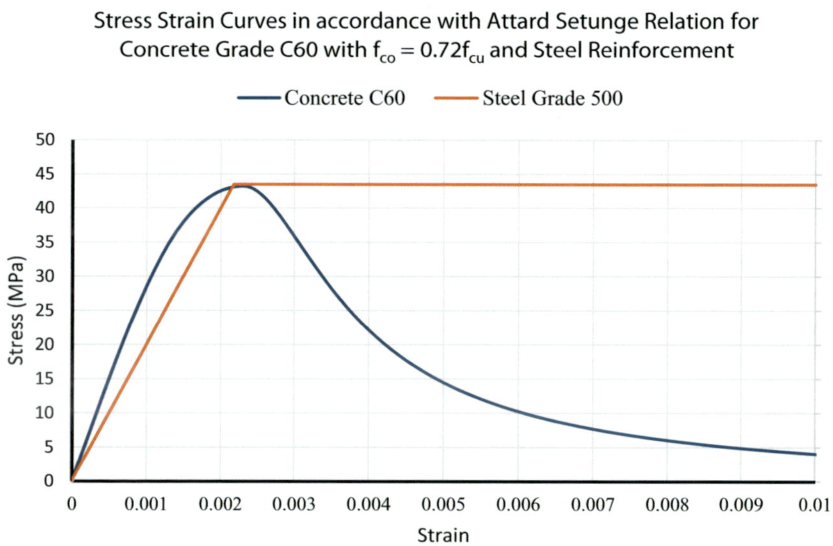

FIGURE 8.6 Stress-strain curve of concrete and steel reinforcement.

demonstrating the role of reinforcement in improving ductility of the reinforced concrete members, particularly beyond the post-peak strain.

(iii) Dense transverse reinforcements to achieve better concrete confinement

Close and strong transverse reinforcements help confine concrete which would increase both concrete strength and ductility in compression in beams, columns and walls. Details of the origin and the quantification are discussed in Chapter 11. Figure 8.7 shows a typical stress-strain relation of concrete under confinements (mostly by lateral reinforcement ties) based on the mathematical equations by Attard and Setunge (1996) and Mander et al. (1988) for concrete grade C60 $\left(f_{cu} = 60\text{MPa}\right)$,

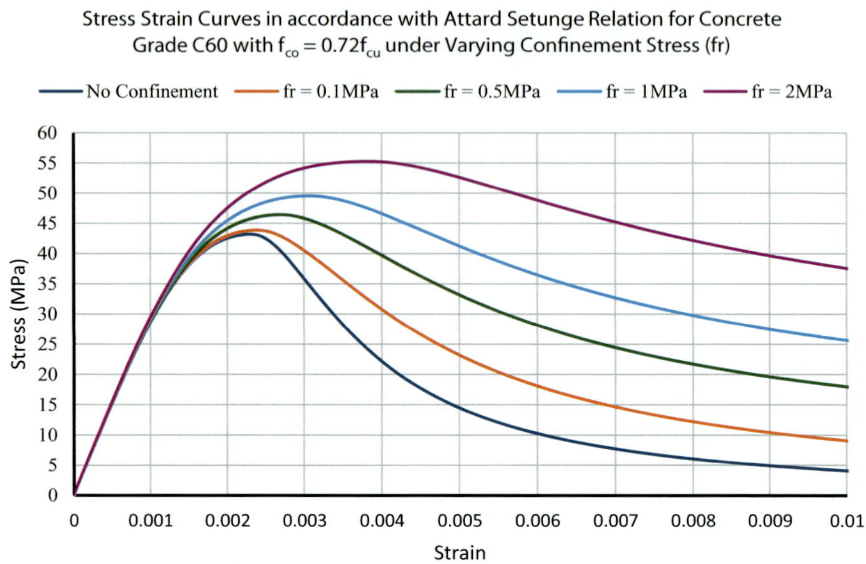

FIGURE 8.7 Stress-strain curves of concrete C60 under confinement.

details of which can be found in Chapter 11. By Attard and Setunge (1996), the concrete strength under confinement can be increased significantly, as shown in the figure under confinement stresses even in small magnitudes starting from 0.1 MPa. From the figure, not only the peak strengths are increased with confinement, the post-peak strength increases at large strains (deformation), showing a significant increase in ductility, which can be quantified by the ductility factors.

Another demonstration in the increase of strength and ductility by confinement is the flexural strength of a column under different confinements by transverse reinforcements as shown in Figure 8.8. Heavier transverse reinforcement can provide higher lateral confinement stress as higher load in the transverse reinforcement can be mobilized.

So, it follows that design codes generally require minimum lateral reinforcements at locations of high bending moment as a ductility requirement, so as to avoid failure of the structure by bending in these locations.

(iv) Axial stress decreases ductility of column and wall

While it is a well-known phenomenon that high axial load in a column and wall will decrease the moment of resistance as that generally provided in reinforcement design charts of wall and column, the decrease is also exhibited in ductility as shown by a numerical example in Figure 8.9. By the figure, ductility starts to decrease drastically as shown by drops in the bending strengths of the column at post-peak curvatures when the axial stress exceeds $0.4f_{co}$ where $f_{co} = 0.72f_{cu}$.

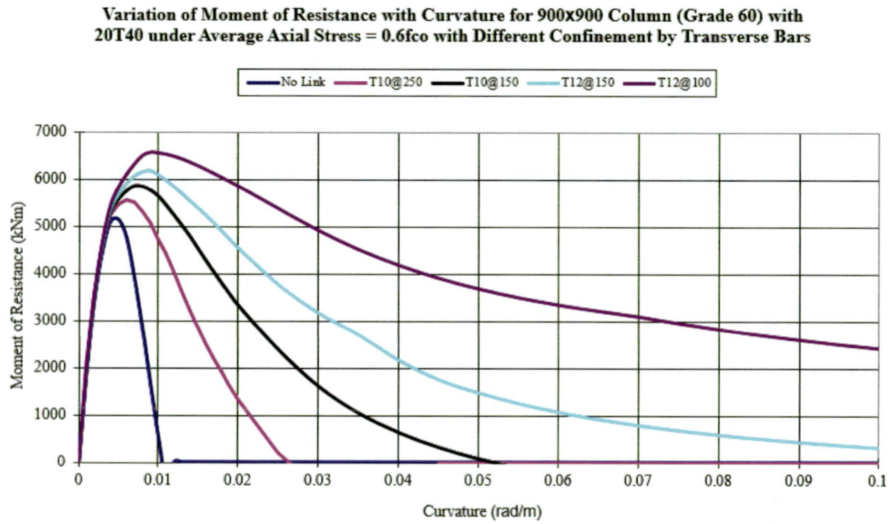

FIGURE 8.8 Variation of moment of resistance with curvature of a 900 mm square column under different confinement by transverse reinforcement.

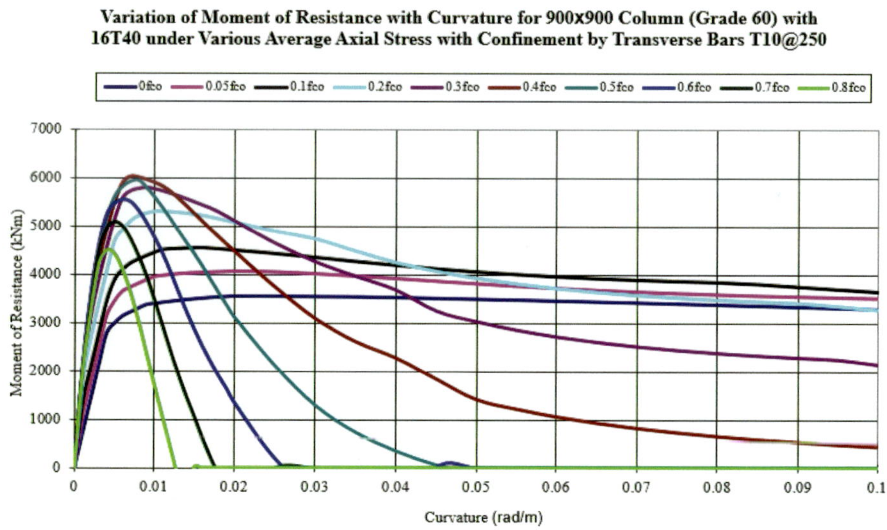

FIGURE 8.9 Increase of ductility with increase of axial stress in a reinforced concrete column.

Thus, it is often a design requirement to limit the axial stress of columns and walls in seismic design. The Eurocode BSEN1998-1:2004 (2004) limits the "normalized axial force" (the ratio of the design axial force to the product of the column's cross-sectional area and the design compressive strength of concrete) of columns to

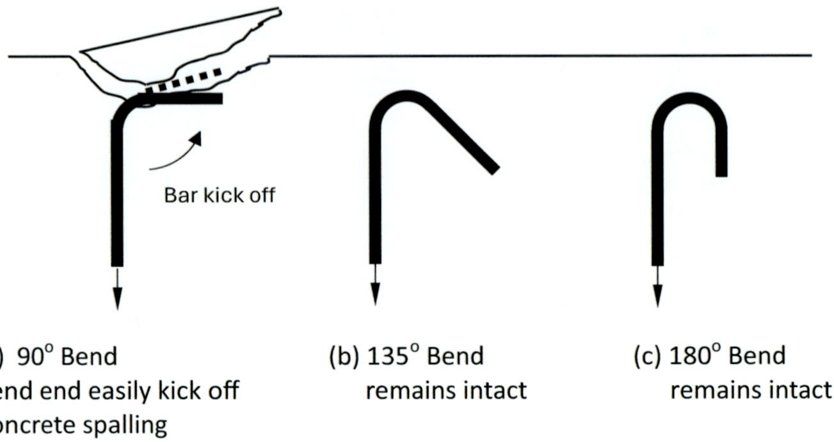

FIGURE 8.10 Performance of links at different bend angles.

0.65 for ductility class DCM (medium ductility) and 0.55 for ductility class DCH (high ductility). For primary seismic walls in DCH, the Eurocode limits the normalized axial force for walls to 0.35. The same requirements are also found in the China Seismic Code GB50011-2010 (2010) by limiting the "axial compression ratio" similarly defined as the design axial load (ultimate) divided by the concrete sectional strength (concrete sectional area multiplied by the concrete design strength) from 0.6 to 1.05 for columns and 0.4 to 0.6 for walls, depending on the structural type and seismic design category.

(v) Stronger anchorage by transverse reinforcements in concrete by using end hooks with bent angles not less than 135°

Transverse reinforcements with end hooks of bent angles not less than 135° can ensure better performance in reinforcing bar anchorage. This is illustrated in Figure 8.10 by which a 135° bend shows better performance than a 90°. The 180° performs even better.

BSEN1998-1:2004 (2004) generally specifies close stirrups to be with 135° hooks for hoops used as transverse reinforcements in beams, columns and walls. The New Zealand Code NZS3101:2006 (2006) and some other codes have similar requirements.

Nevertheless, if there is confinement, say by concrete mass such as a beam or a thick slab restraining open-up action of the link, a 90° bend should be acceptable as far as prevention of concrete spalling is concerned.

(vi) More stringent requirements in restraining and containing longitudinal reinforcing bars in compression against buckling by closer and stronger transverse reinforcements with hooks of included angles not less than 135°.

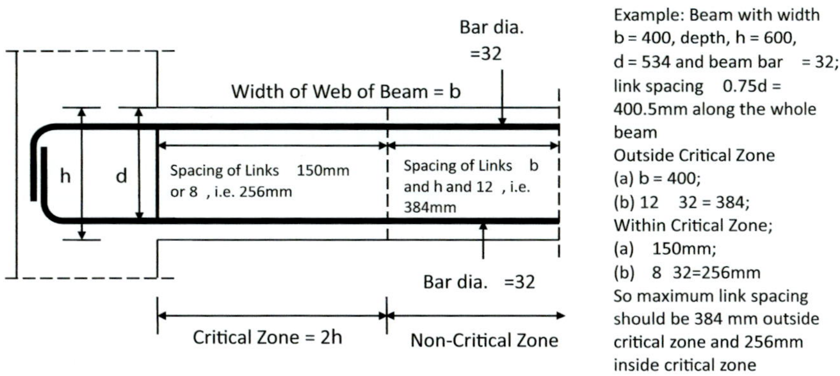

FIGURE 8.11 Maximum spacing of shear links in the span direction of beam as a ductility requirement by the New Zealand Code NZS3101:2006.

Such requirements are especially applicable in ends of beams and columns resisting high moments which also help increase strength of concrete by confining it as explained in (ii). An example of a codified requirement by the New Zealand Code NZS3101 for densified transverse reinforcement within a zone of twice structural beam depth at beam ends is shown in Figure 8.11. The zone is termed "critical zone."

(vii) Longer anchorage length of beam reinforcing bar into column to ensure that failure by yielding will take precedence over bond slippage

Such a failure sequence is preferred because bond failure is more brittle. Design codes for seismic actions often use lower bond strength of concrete, which is identical to requiring over-strength of reinforcing bar by 25% to 30% to effect the preferred failure sequence, as illustrated in Figure 8.12.

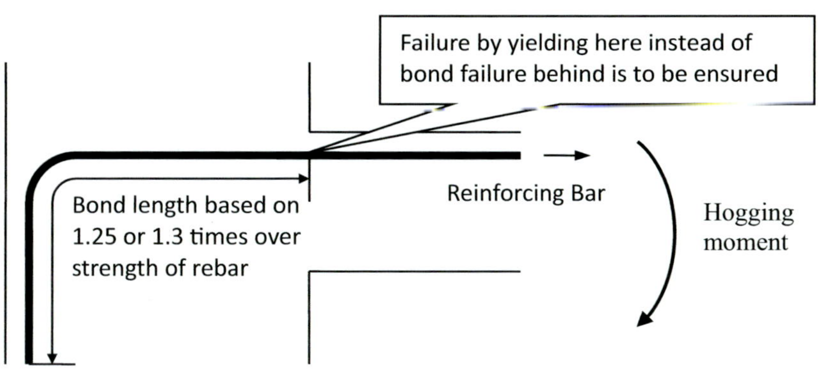

FIGURE 8.12 Longer anchorage length of reinforcing bars to avoid bond failure prior to bar yielding.

(viii) Restricting location of reinforcing bars lapping

Lapping of reinforcement bars is potentially weak in strength design. Therefore, lapping should be avoided in locations where high bending moments would take place, which would create high bond stresses. So, there is a practice of not permitting lapping of bars close to the beam ends and column ends where there are high moments generally. For example, lapping of the reinforcement bars in columns shall only be carried out in the middle third part of the column height according to the New Zealand Code NZS3101:2006 (2006). Nevertheless, the requirement can be relaxed when couplers that can resist cyclic loads are used.

(ix) Tension reinforcements in flexural members to be limited and certain quantities of compressive reinforcements in beams to be provided

The requirement is to avoid over-provision of tension reinforcement in a beam, which will increase the beam's neutral axis depth, leading to greater concrete strain and thus easier concrete failure. As a result, ductility is decreased, as illustrated in Figure 8.13(a). This phenomenon has been discussed by Kwan et al. (2002) and Law (2010). On the other hand, the increase of compression reinforcement will increase ductility, as discussed by Law (2010). The plots of moment of resistance of a 700(D) × 400(B) beam of concrete cube strength $f_{cu} = 45$ N/mm^2 in Figure 8.13(a) and 8.13(b) with different tension and compression steel ratios demonstrate how the ductility is affected. In Figure 8.13(b), the compressive reinforcement contents are set at the constant ratio of 0.5 of the tensile reinforcement contents. It can be seen that at low tensile steel percentage as low as 0.5% for the singly reinforced beam, the ductility factor is almost at infinity (theoretically), likewise for doubly reinforced beam. Nevertheless, the curvature cannot be at infinity. The final control will be the maximum tolerable concrete strain.

(x) Denser transverse reinforcements at ends of Walls

Unlike beams and columns, which require denser transverse reinforcement at their ends to resist high moments and provide ductility, walls typically do not develop significant in-plane bending moments at each floor level. This is due to their high in-plane stiffness relative to the floor diaphragms. Consequently, the bending moment diagram for a wall under lateral load differs from that of a column, as illustrated in Figure 8.14.

Consequently, high stresses become concentrated at the planar ends of the wall. To address this, ductility requirements often mandate the provision of "confined boundary elements" at these locations. These are zones of a specified length with a high reinforcement ratio, including both longitudinal and transverse steel, which may sometimes be formed as enlarged heads.

The extents of these boundary elements, as illustrated in Figure 8.15 (extracted from the Hong Kong Code of Practice for the Structural Use of Concrete 2013 (2013), which make references to the China Seismic Code GB50011-2010 (2010), are

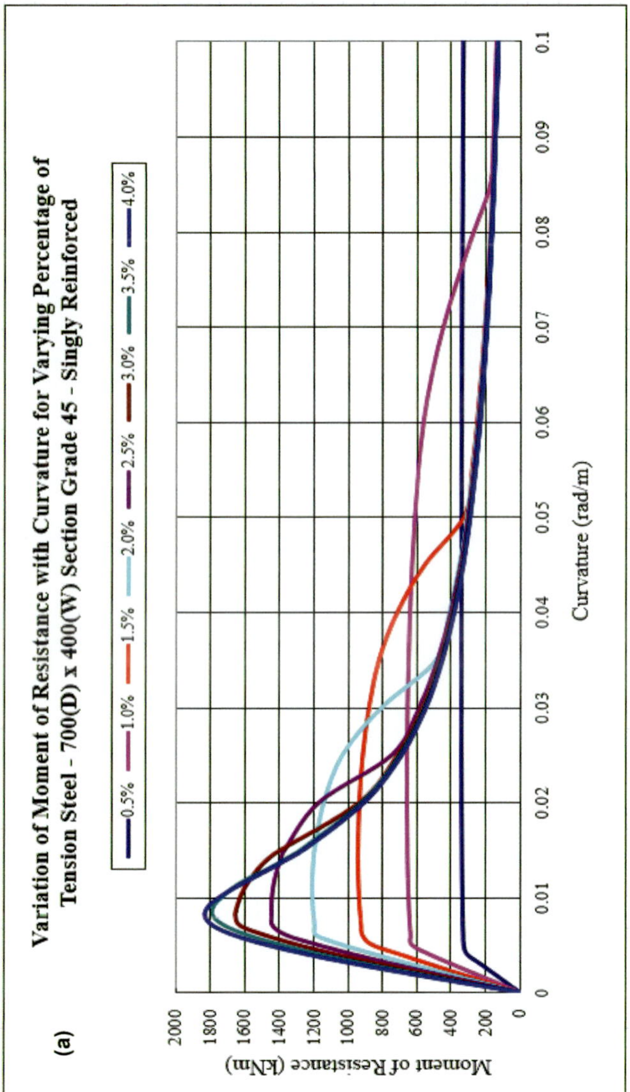

FIGURE 8.13(a) Ductility of beam at different tension reinforcement ratios.

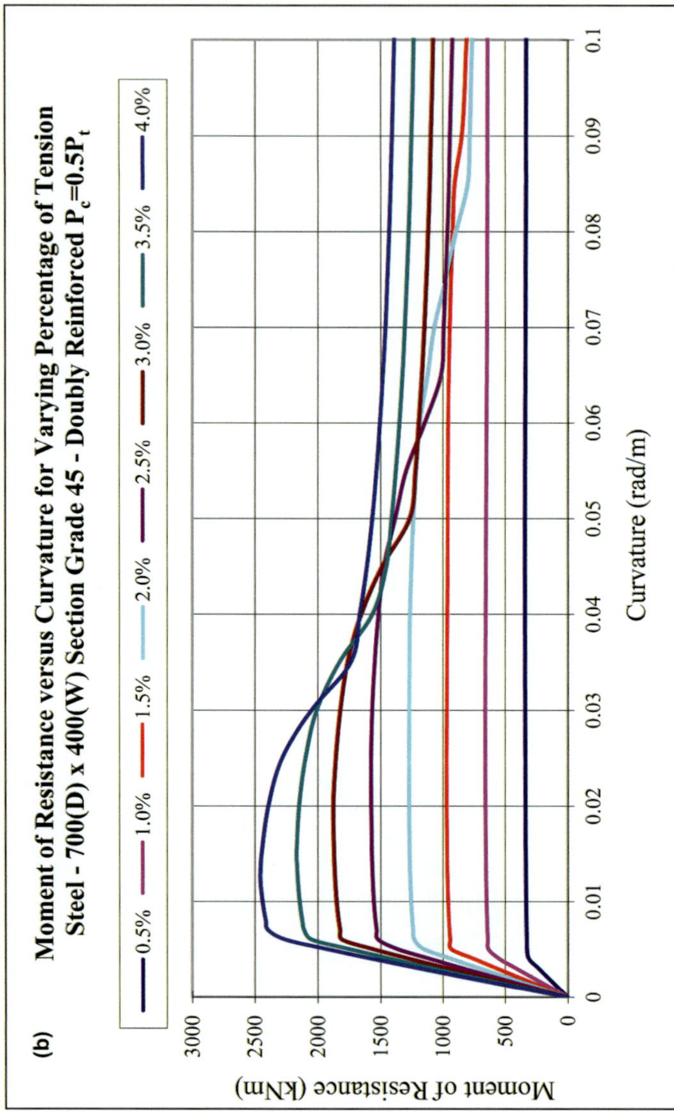

FIGURE 8.13(b) Ductility of beam under compressive reinforcing bars.

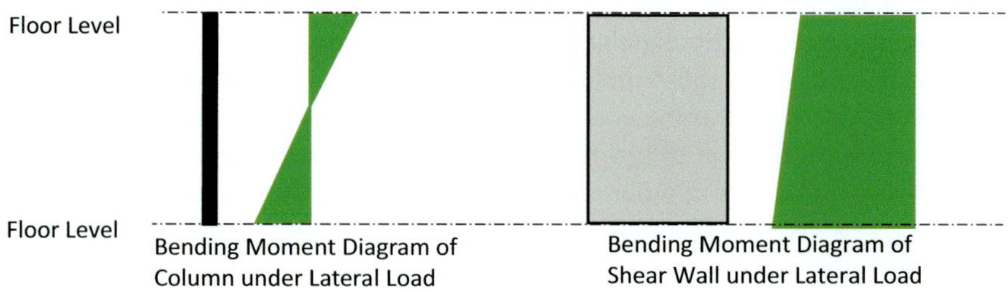

FIGURE 8.14 Typical bending moment diagram of column and shear wall under lateral load.

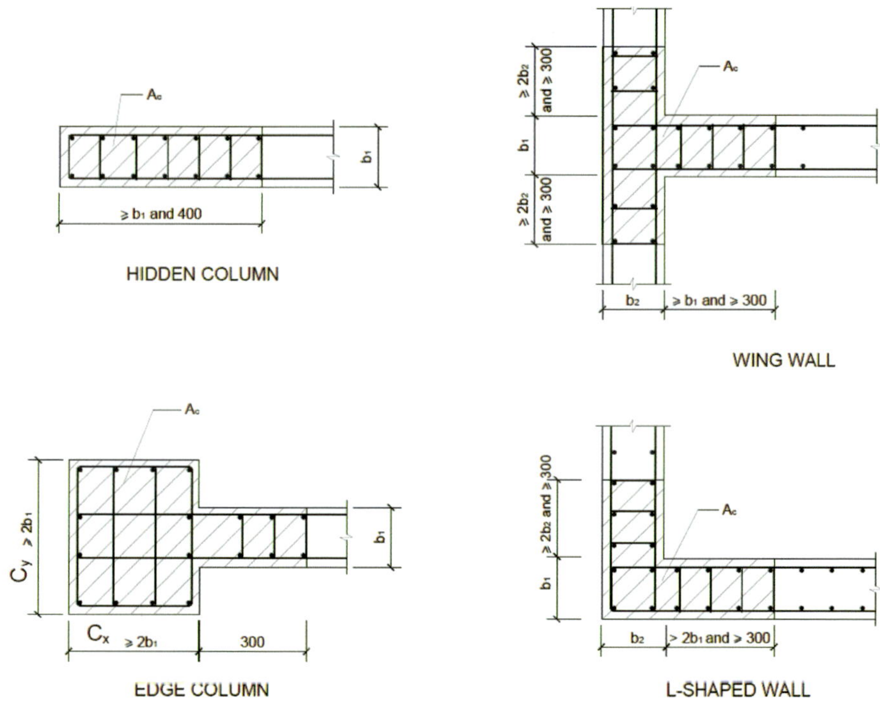

FIGURE 8.15 Confined boundary elements in shear walls according to the Hong Kong Concrete Code 2013.

typically determined based on the wall's axial compression ratio (ACR). A higher ACR necessitates more stringent detailing requirements for the boundary elements to ensure adequate confinement.

8.2.3 Further Consideration of Ductility

Ductility requirements may impose restrictions on structural strengths in certain elements, for example, limiting the bending capacity of a reinforced concrete beam by capping the allowable tensile steel percentage. Even if excess reinforcement is not factored into strength calculations, exceeding these limits can still lead to non-compliance with code requirements, as the ductility factor may be reduced.

However, ductility is influenced by multiple factors. While non-compliance with a specific code provision might reduce ductility, this reduction can sometimes be offset by over-provision in other areas, thereby maintaining the code's implied ductility level. For instance, Law (2010) demonstrated that in beams where tensile steel exceeds the 2.5% limit specified by the *Hong Kong Concrete Code 2013* (2013), the resulting decrease in ductility factor can be compensated by additional compressive reinforcement, restoring the required ductility. Similarly, in columns with insufficient transverse reinforcement (reducing confinement), ductility can be preserved by increasing longitudinal reinforcement, as illustrated in Figure 8.16.

Figure 8.16 illustrates how ductility is affected by changes in longitudinal and transverse reinforcement. The section with 8T20 bars and T12@250 links (as required by the *Hong Kong Concrete Code* (2013)) achieves a code implied ductility, represented by the dark line. When transverse reinforcement (links) is entirely removed, ductility deteriorates significantly—even with an increase in longitudinal reinforcement to 8T25 bars. This is evident from the brown line, where the moment resistance drops rapidly in the post-peak region, performing far worse than the code-compliant section.

Increasing the longitudinal reinforcement to 8T32 bars (green line) improves behaviour, though ductility still remains inferior beyond a curvature of 0.026 rad/m. However, with a further increase to 8T40 bars (blue line), the ductility performance

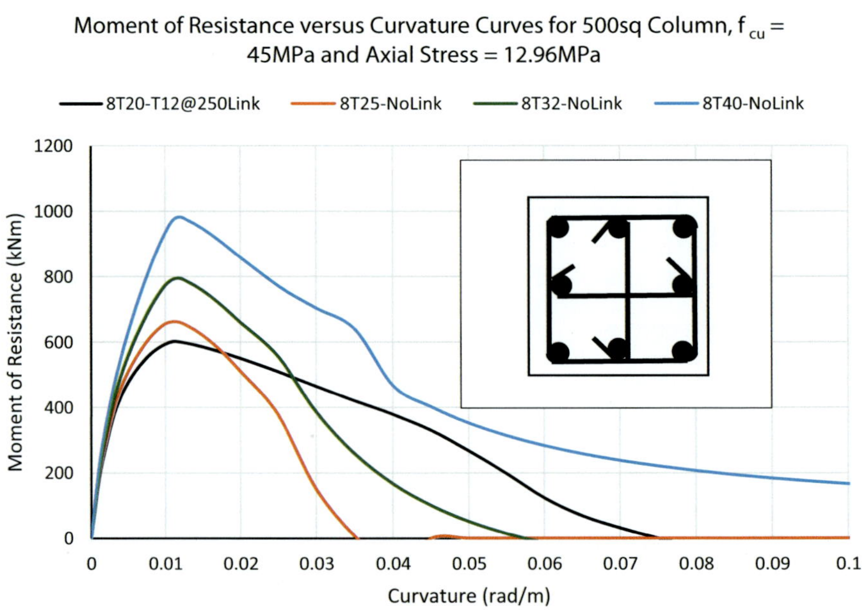

FIGURE 8.16 Moment of resistance versus curvature curves for 500 sq column.

fully surpasses that of the original section, demonstrating that higher longitudinal reinforcement can compensate for the absence of transverse reinforcement, albeit inefficiently.

However, it should be borne in mind that the transverse reinforcement does not only play the function of confinement in increasing strength and ductility, it also serves to anchor the longitudinal reinforcement by preventing them from buckling. The foregoing examples with no transverse reinforcement only illustrate alternative provisions to compensate for the loss of ductility when some code requirements are not fulfilled.

In addition, in the comparison of structural performance of the sections of 8T20 with links and 8T32 without links in Figure 8.14, it can be argued that the latter, though with lesser ductility, may not perform poorer than the former as by the higher strength it can withstand higher load for the structure compared with the former without moving to the post-peak region. This can also be considered as achieving a higher safety margin.

Nevertheless, in terms of intended sequencing of failure such as beam failure prior to column as a ductility requirement, limiting strength of some elements relative to others is still a preferred practice.

8.3 ROBUSTNESS

Robustness of a structure is defined by the Eurocode EN1991-1-7:2006 (2006) as the ability of a structure to withstand events like fire, explosion, impact of the consequences of human error, without being damaged to an extent disproportionate to the original cause. This principle is fundamentally linked to the prevention of progressive or disproportionate collapse.

A seminal example of disproportionate collapse is the 1968 Ronan Point gas explosion in the UK. The incident involved a 22-storey residential tower where an explosion in a corner apartment triggered the collapse of a large portion of the building, as illustrated in Figure 8.17 extracted from the New Civil Engineer of the UK.

The Eurocode BSEN1991-1-7:2006 (2006) states two strategies for robustness design. The first strategy is to identify the extreme events including the followings:

(i) Design the structure to have sufficient robustness.
(ii) Prevent or reduce the occurrence of the action, e.g., protective barrier against vehicular collision to a column.
(iii) Design the structure to sustain the action.

The second strategy based on limiting the extent of local failures includes the following:

(i) Enhance redundancy (alternative load paths).
(ii) Key element designed to sustain the notional accidental action.
(iii) Prescriptive rules, e.g., integrity and ductility.

FIGURE 8.17 The Ronan Point explosion accident demonstrating disproportionate collapse.

The design loads to achieve robustness have been discussed in Section 5.4 of Chapter 5, together with structural provisions in terms of ties and disproportionate collapse.

8.4 DURABILITY OF CONCRETE STRUCTURES

8.4.1 General

A building should have a long service life, performing its intended functions with minimal maintenance or repair. The Eurocode BSEN1990:2002 (2002) defines the service life, or often termed "design working life" as an "assumed period for which a structure or part of it is to be used for its intended purpose with anticipated maintenance but without major repair being necessary." For buildings, this is typically 50 years.

However, it is a misconception that a building becomes unusable after this period. The design working life only represents a set of minimum design criteria and conservative environmental factors specified by codes. As most buildings are designed to exceed these minimum standards, they often perform well beyond their nominal design life with little maintenance.

To ensure durability, the selection of durable construction materials is essential. Concrete, generally more durable than steel or timber, is often the preferred choice for long-life, low-maintenance structures. Additionally, the structural form itself must be considered, as it affects stress distributions and the exposure of materials to corrosive environments.

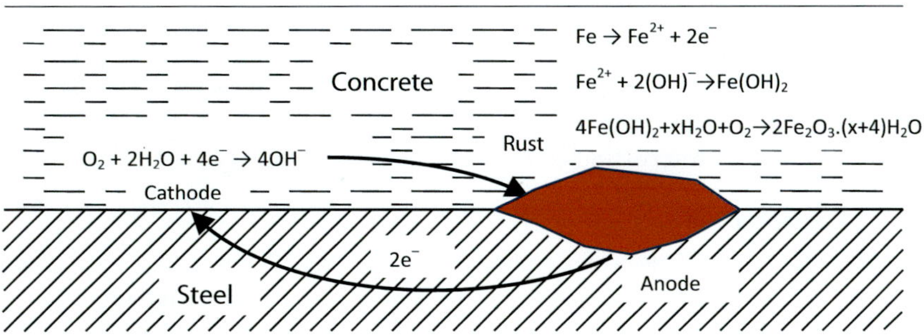

FIGURE 8.18 Chemical process of corrosion of reinforcing bar in concrete.

8.4.2 Corrosion of Reinforced Concrete Structures

Concrete is a highly durable material, largely inert to corrosive agents like water, carbon dioxide, oxygen, and chlorides. While it can slowly react chemically with some of these agents, many, like carbon dioxide and chlorides, do not significantly damage the concrete itself.

The primary deterioration mechanism for reinforced concrete is instead the corrosion of the embedded steel reinforcement. The most common form of corrosion is the electrochemical reaction between steel, water, and oxygen, which produces rust, as illustrated in Figure 8.18.

8.4.2.1 General Corrosion Process

The corrosion of reinforced concrete structures starts with the corrosion of the steel reinforcing bars as illustrated in Figure 8.18.

In the steel rusting process, the difference in electric potential is first set up due to the irregularity of the reinforcing bar. Then the oxygen and water that may be present in the atmosphere or within the concrete carry out the chemical reactions by the equations shown in Figure 8.16 by first forming ferrous hydroxide ($Fe(OH)_2$) and ferric hydroxide ($Fe(OH)_3$). These hydroxides further undergo hydration to yield hydrated iron oxides which are commonly known as iron rust in accordance with the following chemical reactions. The rusting process of steel reinforcement is an electrochemical reaction. It begins with a difference in electrical potential along the bar, often due to surface irregularities or impurities. In the presence of water and oxygen, this creates anodic and cathodic sites.

At the anode, iron dissolves into ferrous ions (Fe^{2+}). These ions react with hydroxyl ions (OH^-) from the cathode to form ferrous hydroxide ($Fe(OH)_2$), which is further oxidized by oxygen to form ferric hydroxide ($Fe(OH)_3$). These compounds then undergo hydration to form various hydrated iron oxides, collectively known as rust, as detailed in the reactions in Figure 8.16.

$$O_2 + 2H_2O + 4e^- \rightarrow 4(OH)^- \quad (\text{at cathode}) \tag{8.2}$$

$$Fe(OH)_2 \leftrightarrow FeO + H_2O \tag{8.3}$$

$$4Fe(OH)_2 + O_2 + xH_2O \rightarrow 2Fe_2O_3(x+4)H_2O \quad (x \text{ is a whole number}) \tag{8.4}$$

The hydrated ferric oxide ($Fe_2O_3xH_2O$) is the "rust" that is expansive in nature. Accordingly, the concrete may crack, spall, or even delaminate, leading to higher porosity so that the reinforcing bars are more prone to further rusting. The cycle goes on, which may bring significant harm to the structure.

So, it follows that iron rusting cannot take place without oxygen and water.

8.4.2.2 Carbonation

As Neville (1995) explains, steel embedded in concrete is protected by a state of "passivation." The hydrating cement paste forms a thin, strongly adherent layer of iron oxide on the steel surface that protects it from corrosion.

To maintain this passivation, a highly alkaline environment (pH around 12.6–13.5) is required. This is provided by the presence of lime ($Ca(OH)_2$) in the pore water of the hardened cement paste, which offers excellent protection to the steel.

Unfortunately, this alkalinity can be reduced through carbonation. Atmospheric carbon dioxide dissolves in pore water to form carbonic acid, which reacts with lime ($Ca(OH)_2$) in the concrete to form neutral calcium carbonate. When the resulting pH drops to approximately 10, the passive layer breaks down and corrosion protection is lost.

8.4.2.3 Action by Chloride

As similar to carbon dioxide, chloride ions may also de-passivate and destroy the protective passivity layer of reinforcement by the following chemical reactions:

$$Fe^{++} + 2Cl^{-} \rightarrow FeCl_2\ (\text{ferrous chloride}) \tag{8.5}$$

$$FeCl_2 + 2H_2O \rightarrow Fe(OH)_2 + 2HCl \tag{8.6}$$

$$2HCl + O^{2-} \rightarrow H_2O + 2Cl^{-} \tag{8.7}$$

The resulting hydroxides undergo hydration to yield hydrated iron oxides as described in the foregoing.

As shown in Equation 8.7, the chloride ion (Cl^-) is not consumed in the rusting process and is regenerated to continue catalyzing the reaction. Chloride thereby acts as a catalyst, significantly accelerating the corrosion of steel.

Furthermore, chloride-induced corrosion typically results in "pitting," a highly localized form of attack, as opposed to the more uniform surface rusting caused by carbonation. This concentrated pitting creates severe stress concentrations and leads to a rapid loss of cross-sectional area at discrete points. Consequently, corrosion in the presence of chloride is far more detrimental, as it can cause structural failure much faster.

8.4.2.4 Action by Sulphate

Sulphate ions which may be present within the concrete or externally in soil or underground water can react with the hydration products of cement, primarily calcium hydroxide and aluminates (including tricalcium aluminates C_3A). A key reaction involves the formation of ettringite (calcium sulfoaluminate hydrate), which is a crystalline compound that expands significantly during its formation. In addition, gypsum (calcium sulphate dihydrate) can also be formed, which is also expansive. Magnesium sulphate is in particular aggressive by reacting with calcium hydroxide to form magnesium hydroxide (brucite) and further attacks the cement paste. Water is required in the reaction as sulphates need to be dissolved in water to react.

The consequence of the expansive reactions will increase the internal stresses in concrete, with subsequent cracking, reduction in concrete strength and increase in permeability (affecting long-term durability).

To reduce the risk of sulphate attack in high sulphate environments, sulphate resistant cement, which has low C_3A content, should be used in the casting of the concrete. The use of C_3A content may necessitate replacement by pulverized fuel ash (PFA) or ground granulated blast furnace slag (GGBS). Production of concrete with low porosity is also helpful in reducing the risk of ingress of sulphate ions.

8.4.2.5 Alkali-aggregate Reaction AAR

This reaction is initiated by attack by alkalis (Na_2O and K_2O) from cement or other sources on the silica minerals or argillaceous dolomitic limestone contained in aggregates that may lead to cracking of concrete. Effective means to reduce AAR comprise the use of low alkali cement, limiting the cement contents.

8.4.3 Rate of Corrosion

The aforementioned corrosion agents require a porous medium to penetrate into the interior of the concrete mass.

The diffusion of gases/vapours/ions through a porous medium generally follows Fick's Law, which states the rate of diffusion is proportional to the concentration gradient as

$$\frac{dq}{dt}\frac{1}{A} = D\frac{dc}{dL} \tag{8.8}$$

where $\frac{dq}{dt}$ is the rate of diffusion in terms of mass per time, A is the sectional area, D is the coefficient of diffusivity of the medium and $\frac{dc}{dL}$ is the concentration gradient.

According to Kwan (2005), D is affected by the permeability or porosity of the concrete and also the relative humidity or the degree of saturation of the pores. The outermost surface of concrete in contact with the atmosphere will first be carbonated and the carbonation depth progresses into the concrete generally at a reduced rate as CO_2 has to diffuse through the pore system and the already carbonated surface zone. Under steady conditions, the carbonation depth may be depicted by the following equation:

$$d_c = C\sqrt{t} \tag{8.9}$$

where d_c is the carbonation depth, C is the carbonation coefficient dependent on both the external environment and the quality, particularly the permeability of the concrete and t is the time of exposure.

8.4.4 Measures to Enhance Durability of Reinforced Concrete Structure

Carbonation itself, in fact, does no harm to the concrete. Instead, the carbonated concrete has an even greater hardness than uncarbonated concrete. However, it would be critical if d_c in Equation 8.10 exceeds the concrete cover and thus depassivates the embedding reinforcements. According to Kwan (2005), broadly speaking, in concrete with a W/C ratio of 0.6, a carbonation depth of 15 mm would be reached after 15 years, but in concrete with a W/C ratio of 0.45, the same carbonation depth would not be reached until after 100 years.

Similarly, chloride does no harm to concrete. But it corrodes reinforced concrete structures by destroying the protective passivity layer on the surface of the embedded reinforcing bar and thereby causes depassivation as described in the foregoing. There are two sources of chloride: internally from the ingredients of the concrete mix and externally from the environment. The chloride content for use of concrete is therefore limited to some 0.2% to 0.4% by mass of cement by the Eurocode BSEN 206-1 (2000) and similar figures by other codes. The external environment will be rich in chloride if in close proximity to the sea where not only direct contact with seawater like the marine structure but also the airborne seawater droplets can bring chloride into the concrete structure. According to Neville (1995), airborne chlorides travelling for substantial distances of up to 2 km have been reported. The chloride-rich environment may also include lavatories using seawater for flushing. Yet it should be noted that without oxygen, corrosion will not start. So, a reinforced concrete structure fully immersed in seawater will not be corroded although the environment is very rich in chloride ions. However, concrete structures in the "splash and tidal" zone subject to alternate wetting and drying by sea waves are very prone to corrosion under the free supply of oxygen.

Thus, the main aspects in design for reinforced concrete structures against corrosion due to carbonation and chlorination and also sulphate attack are (a) better concrete quality; (b) greater cover to reinforcing bars; and (c) protective coating. Better concrete quality can be achieved as follows :

(i) Use of concrete mixes with low W/C ratio and incorporation of additives like silica fume – lesser water content (as a corroding agent) and denser cement paste can be achieved such that not only concrete strength is increased but also the rate of entry of corrosion agents and subsequently the rate of corrosion are reduced.

(ii) Adequate concrete cover is essential to protect reinforcing bars from corrosion. Greater cover thickness enhances protection by shielding the reinforcement from environmental factors. Ideally, the concrete cover should exceed the carbonation depth throughout the building's design life.

According to ACI 365.1R-00 (2000), the building's service life ends when the carbonation front reaches the reinforcement, marking the end of the "initiation period." Codified requirements for concrete cover vary based on concrete quality and exposure conditions, with thicker cover required for lower-quality concrete or harsher environments. Additionally, the cover must exceed the bar diameter to ensure both corrosion protection and sufficient bond strength with the concrete.

(iii) Better workmanship in concreting – High-quality workmanship in concreting significantly enhances concrete durability. Proper compaction without segregation, smooth surface finishing to limit the ingress of corrosive agents, and thorough curing to promote cement hydration, especially in the surface zone, reduce concrete permeability, resulting in a more durable concrete material.

(iv) Protective coating – Protective coatings on concrete surfaces act as barriers to block corrosive agents, such as carbon dioxide, thereby slowing or preventing carbonation. These coatings are typically applied as paint. Reinforcing bars can also be coated with protective layers, such as zinc (galvanizing) or epoxy. Zinc coatings function as sacrificial anodes, preferentially corroding over the steel reinforcement by reacting with oxygen and water, as zinc is a stronger reducing agent than iron in the electrochemical series. Epoxy coatings, in contrast, provide a non-sacrificial barrier to protect the reinforcement. However, epoxy-coated bars may exhibit reduced bond strength with concrete, necessitating a longer bond length.

(v) Use of sulphate resistance Portland cement (SRPC). SRPC is a cement with reduced tricalcium aluminate (C_3A) which is a component most susceptible to attack by sulphate to be used if concrete is exposed to environments with considerable quantities of sulphates.

(vi) Avoid harmful environments and misuse of the structure – do not expose the structure to harmful environments such as seawater rich in chloride contents or soil rich in sulphates unless the concrete has been specifically designed to withstand such environments. In addition, misuse of the structure such as overloading, application of abrasive loads such as that by heavy lorries which will induce cracking or reduce cover of the concrete shall also be avoided unless the structure has been designed for such loads.

8.4.5 Codified Approach

For simplification in design control, codes generally require minimum concrete cover to reinforcements in accordance with the different classes of exposure they define for durability design. BSEN1992-1-1:2023 (2023) has comprehensive lists of classes of exposure to carbonation, chloride attack etc. to which minimum covers are specified. Service lives of buildings are also taken into account. Other design codes have similar provisions.

8.5 FIRE RESISTANCE

During its design life, a building may be exposed to fire, which can damage structural components and reduce their strengths, potentially compromising the building's performance. To mitigate this, buildings must be designed and constructed with adequate fire resistance to limit damage under specific "fire load" conditions for a defined duration. Fire load represents the potential severity of a fire within a given space, typically quantified as heat output per unit floor area (kJ/m²), calculated from the calorific value of materials present. For practical analysis and design, standardized temperature-time curves, known as "fire curves," are used to simulate real fire conditions. These are detailed in Eurocode BS EN 13501-2:2023 (2023), which outlines various fire scenarios, corresponding fire curves, testing procedures, and compliance criteria, as discussed in section 8.5.1.

8.5.1 Fire Curves

To standardize fire-resistant design, "fire curves," time-temperature relationships simulating temperature changes in real fires, are used for analysis, design, and establishing prescriptive structural provisions in codes and standards. These curves typically assume uniform temperature distribution within a compartment which is an assumption generally valid for fully developed fires. However, many fire curves only account for the rising temperature phase, omitting the cooling phase, which can lead to conservative designs in certain scenarios. The Eurocode BS EN 1991-1-2:2024 (2024) and its predecessor BS EN 1991-1-2:2002 (2002) define "nominal fire curves," comprising three curves consistent with ISO 834-8 (2002), as outlined in the equations below.

(i) Standard time-temperature curve : $\theta_g = 20 + 345\log_{10}(8t+1)$;
(ii) External fire curve: $\theta_g = 660\left(1 - 0.687e^{-0.32t} - 0.313e^{-3.8t}\right) + 20$;
(iii) Hydrocarbon curve : $\theta_g = 1080\left(1 - 0.325e^{-0.167t} - 0.675e^{-2.5t}\right) + 20$

where θ_g is the gas temperature of the compartments for (i) and (iii), the gas temperature near the member for (ii); t is the time. The three curves take 20°C as the original temperature.

The Eurocode defines these three fire curves, with the standard time-temperature curve being the most widely used, serving as the basis for most experimental studies on fire protection materials. The hydrocarbon curve applies to scenarios involving small petroleum fires, such as those from car fuel tanks, petrol or oil tankers, or certain chemical tankers. The external fire curve is used for testing materials exposed to fire through a building's façade, reflecting lower temperatures due to less severe fire conditions outside compared to within a compartment.

Similar fire curves exist in other standards for various building compartments, such as the RWS and HCM tunnel curves for tunnel design. Key curves, including the three Eurocode curves and the American Standard ASTM E119 (2000), are summarized in Figure 8.19, with their equations or tabulated data provided in Appendix G.1.

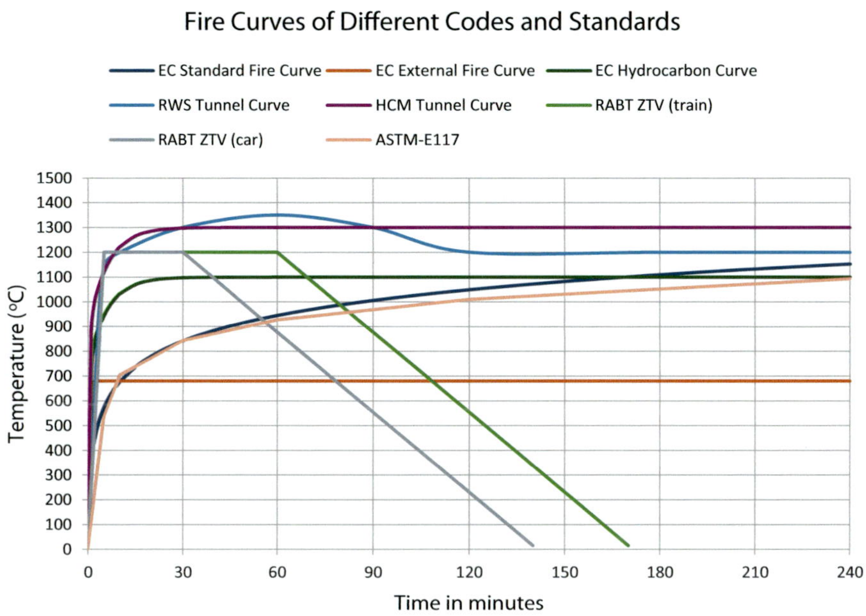

FIGURE 8.19 Nominal fire curves.

8.5.2 Strength Reduction of Concrete and Reinforcement at Elevated Temperatures

Fire poses a significant threat to structural safety by reducing the strength of concrete and reinforcement at elevated temperatures. According to the Concrete Society of the UK's Technical Report Note 68 (2008), concrete cracking or spalling during fire exposure primarily results from vapour pressure generated by the release of physically and chemically bound water through concrete pores early in the heating process. Hager (2013) explains that water evaporates in sequence based on its binding energy: (1) free water, (2) capillary water, (3) chemically bound water, and (4) water produced by chemical reactions. This evaporation increases pore pressure, potentially causing cracking and spalling. Additionally, differential thermal expansion between cement paste and aggregates, as well as between steel reinforcement and concrete (due to steel's higher thermal expansion coefficient), contributes to cracking. Poon et al. (2001) note that calcium hydrate dissolution begins at 400°C and progresses until the complete breakdown of calcium-silicate-hydrate at approximately 900°C. These processes, combined with cracking and chemical decomposition, lead to gradual or rapid reductions in concrete strength and elastic modulus. Eurocode BS EN 1992-1-2:2023 (2023), table 5.1, quantifies the percentage reduction in concrete strength at various temperatures for design purposes, as extracted in Table 8.1.

Generally, strength reduction of concrete due to cracking is not reversible. However, the contribution by chemical decomposition may be partly recovered by

TABLE 8.1
Main Parameters of the Stress-Strain Relations for Normal Weight Concrete with Siliceous or Calcareous Aggregates at Alevated Temperature from BSEN1991-1-2:2023 ($K_{c,\theta} = f_{c,\theta} / f_{ck}$)

Concrete Temp (°c)	$K_{c,\theta}$ ($f_{c,k}\theta < 70$ MPa) Siliceous Aggregate	$K_{c,\theta}$ ($f_{c,k} < 70$ MPa) Calcareous Aggregate	$K_{c,\theta}$ ($f_{c,k} > 70$ MPa) Any Type of Aggregate	$\varepsilon_{c1,\theta}$	$\varepsilon_{c1u,\theta}$
20	1.00	1.00	1.0	0.0025	ε.0θ
ε00	θ.00	1.00	0.75	0.0040	0.0225
200	0.95	0.97	0.75	0.0055	0.025
300	0.85	0.91	0.75	0.0070	0.0275
400	0.75	0.85	0.75	0.01	0.03
500	0.6	0.74	0.6	0.015	0.0325
600	0.45	0.6	0.45	0.25	0.035
700	0.3	0.43	0.3	0.25	0.0375
800	0.15	0.27	0.15	0.25	0.04
900	0.08	0.15	0.08	0.25	0.0425
1000	0.04	0.06	0.04	0.25	0.045
1100	0.01	0.02	0.01	0.25	0.0475
1200	0.00	0.00	0.00	–	–

"post-fire curing" in accordance with Poon et al. (2001). Some discussions are outlined in Section 8.5.5.

Similar reduction in reinforcing steel is also found which is accounted for by the increased mobility of the atoms and "dislocations" within the steel's crystal structure. A dislocation within the molecular lattice structure of iron is simply a "defect" in the lattice structure in which a few ions in a layer are missing. This causes the neighbouring layers to be displaced slightly to minimize the strain from defect. Table 5.3 of BSEN1992-1-2:2023 (2023) in relation to strength reduction of steel due to elevated temperature is extracted as Table 8.2 for reference.

In contrast with concrete, steel regains most of its properties, including strength upon cooling, provided that the elevated temperatures it has suffered do not exceed certain limits which are in the order of 500°C to 600°C.

The previous version BSEN 1992-1-2:2004 (2004) contains charts showing similar trends of strength reductions for concrete and reinforcing steel.

8.5.3 Fire Resistance Rating and Fire Resistance Requirements

In this chapter, the provisions and requirements of the Eurocode BSEN1992-1-2:2023 (2023) are mainly chosen for illustration of general structural fire design.

TABLE 8.2

Values for Parameters of the Stress-Strain Relations for Hot Tolled and Cold Worked Reinforcing Steel at Elevated Temperature from – Table 5.3 of BSEN1991-1-2:2023

Steel Temp (°c)	$K_{sy,\theta} = f_{sy,\theta} / f_{yk}$ Hot Rolled	$K_{sy°\theta} = f_{s\theta,\theta} / f_{y\theta}$ Cold Worked	$K_{sp,\theta} = f_{\theta p,\theta} / f_{\theta k}$ Hot Rolled	$K_{sp,\theta} = f_{s\theta,\theta} / f_{y\theta}$ Cold Worked	$K_{se,\theta} = f_{\theta e,\theta} / f_{\theta k}$ Hot Rolled or Cold Worked	$K_{Es\theta\theta} = E_{s,\theta} / E_s$ Hot Rolled	$K_{Es,\theta} = E_{s,\theta}\theta/ E_s$ cold Worked
20	1.00	1.θ0	1.00θ	1.00	1.00	1.00	1.00
100	1.00	1.00	1.00	0.96	1.00	1.00	1.00
200	1.00	1.00	0.81	0.92	0.95	0.90	0.87
300	1.00	1.00	0.61	0.81	0.90	0.80	0.72
400	1.00	0.94	0.42	0.63	0.85	0.70	0.56
500	0.78	0.67	0.36	0.44	0.60	0.60	0.40
600	0.47	0.40	0.18	0.26	0.35	0.31	0.24
700	0.23	0.12	0.07	0.08	0.10	0.13	0.08
800	0.11	0.11	0.05	0.06	0.08	0.09	0.06
900	0.06	0.08	0.04	0.05	0.06	0.07	0.05
1000	0.04	0.05	0.02	0.03	0.04	0.04	0.03
1100	0.02	0.03	0.01	0.02	0.02	0.02	0.02
1200	0.00	0.00	0.00	0.00	0.00	0.00	0.00

BSEN1992-1-2:2023 (2023) is the second generation of the Eurocode, with its previous version at BSEN1992-1-2:2004 (2004) as the first generation.

Fire resistance rating (FRR) in Eurocode for concrete structures (BSEN1991-1-2:2002 (2002)) refers to the required functions and levels of performance assigned to an assembly or a structural element against fire. It is expressed in minutes by which the compartment or the structural element can comply with criteria including load bearing (R), integrity (E), and insulation (I) within the specified duration. These criteria are explained as follows.

Criterion R (loadbearing) is the ability of a structure or a member to sustain specified loads during fire, according to defined criteria. The defined criteria in accordance with other codes such as the former British Code BS476-20:1987 (1987) are more specific including (1) for vertical members: support the design load without failure; and (2) for horizontal members: deflection not to be greater than span/20 or rate of deflection not to exceed clear span (in mm)^2/(9000 times structural depth (in mm)), the latter does not apply when span/30 is exceeded.

Criterion E (Integrity) is the ability to have a fire-separating function, to withstand fire exposure on one side only, without the transmission of fire to the unexposed side.

Criterion I (insulation) refers to the insulation fire resisting rating. BS476-20:1987 (1987) gives more concrete yardsticks that the Criterion may be assumed to be satisfied where the average temperature rise over the whole of the non-exposed surface is limited to 140°C, and the maximum temperature rise at any point of that surface does not exceed 180°C.

The number following R, E, or I refers to the time in minutes the structure can sustain the criteria. So, the notations of R120 and E90 mean loadbearing for 120 minutes and integrity for 90 minutes respectively. These criteria may be required individually or in combination with an assembly to a structural member.

In BSEN1992-1-2:2023 (2023) states in its Introduction, "Required functions and levels of performance can be specified either in terms of nominal (standard) fire resistance rating, generally given in national fire regulations or by referring to fire safety engineering for assessing passive and active measures." So, the term fire resistance rating (FRR) is still being used in the new version. The criteria of "loadbearing" (R), integrity (E), and insulation (I) are also similarly defined. The fire resistance of a structural member may be expressed as REI 60 or R 60 which require compliance with R, E and I for 60 minutes and compliance with R for 60 minutes respectively.

8.5.4 Design Approaches

Using a standard fire curve defined by a code or standard, along with specified compliance criteria, "prescriptive approaches" establish requirements such as minimum structural dimensions and concrete cover for reinforcement to meet fire resistance ratings, such as R240 for beams or columns. These requirements, detailed in codes or standards for user convenience, often result in conservative designs. In BSEN1992-1-2:2023 (2023), these are presented as "tabulated values," specifying minimum structural sizes and concrete cover (expressed as "axis distances" in the Eurocode) for concrete structures.

Alternatively, check of structure under fire can be done by design calculations in which the design load can be reduced as compared with the standard design load for ultimate design. The factor η_{fi} listed in Equation 8.11 accounts for reduced load for accidental actions, as outlined in the previous version, BS EN 1992-1-2:2004 (2004).

$$\eta_{fi} = \frac{E_{d,fi}}{E_d} = \frac{G_k + \psi_{fi} Q_{k,1}}{\gamma_G G_k + \gamma_{Q1} Q_{k,1}} \tag{8.10}$$

In Equation 8.11,G_k is the characteristic value of the permanent action; $Q_{k,1}$ is the characteristic value of the leading variable action, γ_G is the partial factor for the permanent action; γ_{Q1} is the partial factor for the leading variable action 1; ψ_{fi} is the combination factor taken as $\psi_{1,1}$ or $\psi_{2,1}$ as the frequent or quasi-permanent value. So if $Q_{k,1}/G_k$ is 0.5, and γ_G, γ_{Q1} and $\psi_{1,1}$ take their usual values at 1.35, 1.5, and 0.7 respectively according to BSEN1990:2002 (2002) (see Chapter 5), $\eta_{fi} = \dfrac{G_k + 0.7\times(0.5G_k)}{1.35G_k + 1.5\times(0.5G_k)} = 0.64$, i.e., the design load for fire is only 64% of that of ultimate design. By Equation 8.10, the designer can assess the factor for individual cases. However, in BSEN1992-1-2:2023 (2023), η_{fi} is taken as 0.7 for general design rules in using "tabulated design data." There is an illustration by Example G.1 in Appendix G for modification of the axis distance to the tabulated values in accordance with the actual provision of reinforcements.

BSEN1992-1-2:2023 (2023) offers advanced design calculation methods for assessing structural elements under fire conditions. In section 7, "Simplified Design Methods," equations are provided to calculate elevated temperatures in structural elements exposed to the standard fire curve for durations exceeding 30 minutes (1800 seconds). These equations account for exposure conditions, such as whether one, two, three, or four sides of the element are exposed to fire, which are reproduced in Appendix G with a demonstration of application. The derivation of these equations is based on simplified thermal models that apply fundamental principles of heat transfer with the conservative parameters as listed in section 5.2 of BSEN1992-1-2:2023 (2023). These parameters are thermal properties of concrete including emissivity, conductivity, specific heat, and density, which are all temperature dependent. Environmental factors such as convection under fire conditions have also been taken into account. The equations can finally determine temperatures at any point within a concrete structure based on the standard fire curve. The authors have used them to generate temperature profiles and contours as illustrated in Figures 8.20 to 8.22, which depict: (i) the temperature profile for a slab with one face exposed to fire; (ii) the temperature contour for a 500 mm (width) by 600 mm (height) beam after 240 minutes of fire exposure; and (iii) the temperature contour for a 500 mm (width) by 600 mm (height) column after 240 minutes of fire exposure.

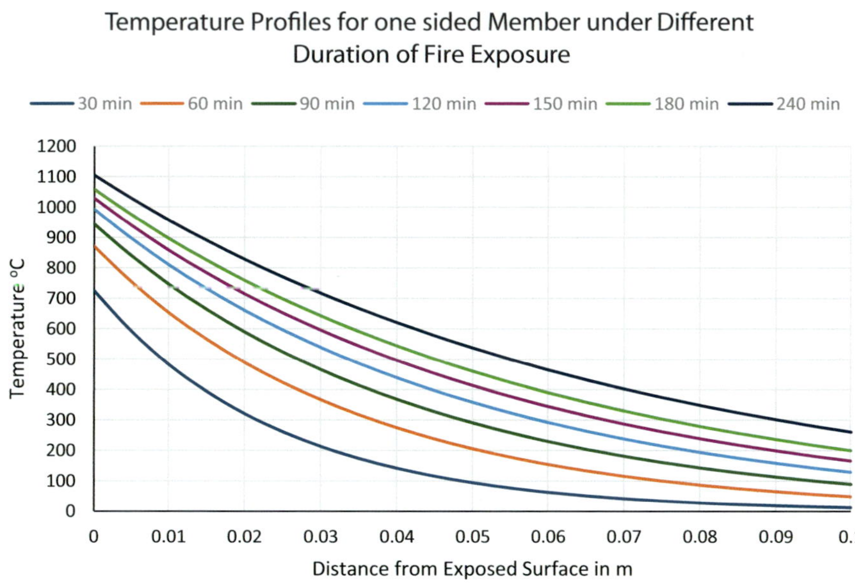

FIGURE 8.20 Temperature profiles of slab (one face exposure) versus time duration.

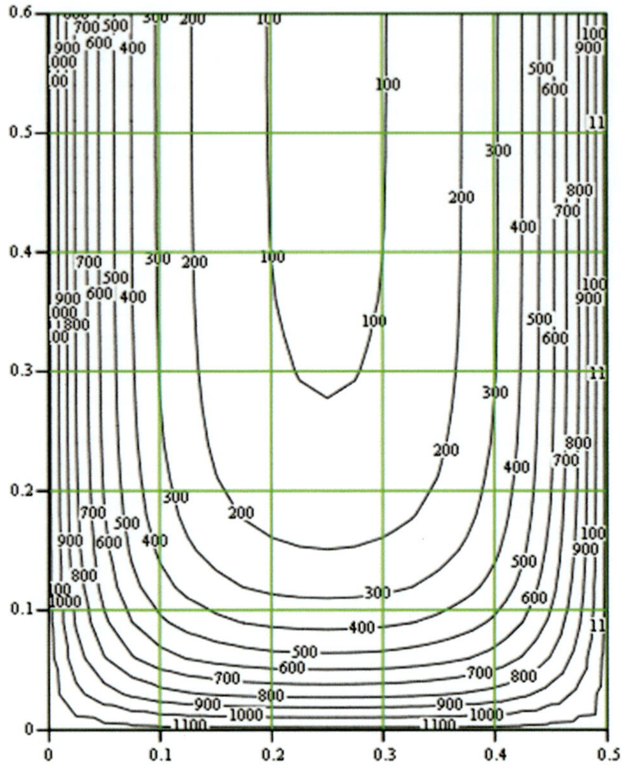

FIGURE 8.21 Temperature contours for 500 mm by 600 mm Beam with three Faces exposure for 240 minutes duration.

These charts show that it takes hours for the interior of the structure in terms of depths of 100 mm or more to reach a few hundred degrees Celsius, demonstrating that concrete is a very poor conductor of heat. So, a concrete structure can comply with R, E, and I criteria for hours after fire.

Upon determination of the temperatures at various locations of the structure by which the reduced strengths can be determined in accordance with the code (tables 5.1 and 5.3 of BSEN1991-1-2:2023 (2023) extracted in Tables 8.1 and 8.2). The strength of the structural member can then be estimated and compliance with the established criteria R can be checked subsequently. Worked examples based on the "Simplified design methods" for checking of reinforced structures to elevated temperatures are included in Appendix G.

In addition, BSEN1992-1-2:2023 (2023) also allows the use of "Advanced design methods" which are based on fundamental physical behaviour, employing local equilibrium equations which are satisfied at every point in the structure. Numerical models based on finite element analysis or others. The Eurocode lists criteria for the employment of the "Advanced design methods."

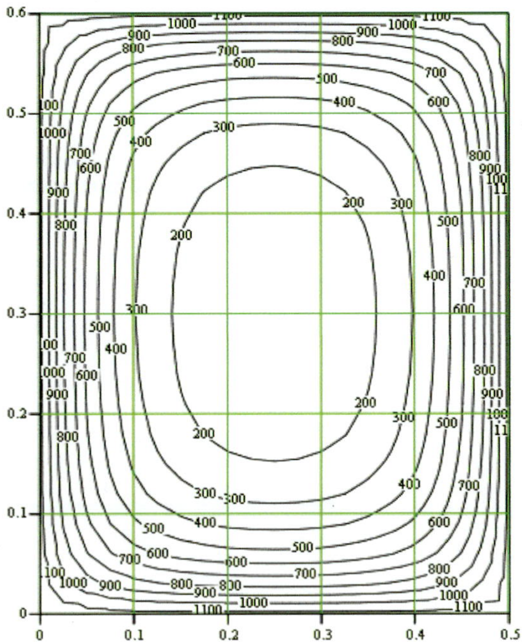

FIGURE 8.22 Temperature contours for 500 mm by 600 mm Column with four faces exposure for 240 minutes duration.

The re-analysis of the structure after fire damage can be as simple as one done to a simply supported beam whose material strengths have been reduced at certain depths from exposed surfaces due to fire (standard fire or the actual fire estimated). It can also be an analysis on a sub-frame or a global analysis of the entire structure with change of design parameters, e.g., reduced elastic moduli due to fire leading to buckling problem.

8.5.5 Reinforced Concrete Structures in the Post-fire State

As explained in 8.4.2, concrete and reinforcing steel lose strength in fire. As cracks are formed during elevated temperatures in concrete which are not reversible, strengths in concrete cannot be totally recovered. In addition, further cracking will be incurred as the concrete cools down which will lead to further loss of strength. The Technical Report Note 68 (2008) therefore comments that the calculation methodologies presented in EN 1991-1-2:2002 (2002) may not be applicable to specific situations since effects such as cooling of the structure or restraint and residual stresses need consideration after a fire. The Note therefore proposes a more conservative factor to assess the strength of the concrete in its Figure 8.3. Based on the data in the figure, the residual strength factor of concrete versus maximum temperature it has been subjected to is plotted in Figure 8.23. By the figure, residual concrete strength will be zero if the concrete has been subjected to temperature in excess of 300°C.

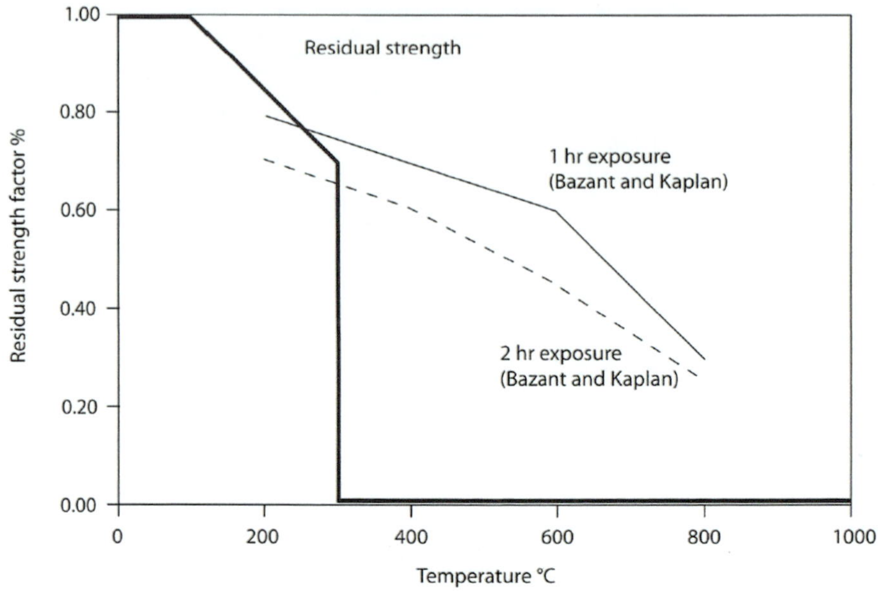

FIGURE 8.23 Typical effect of heat upon the compressive strength of dense aggregate concrete after cooling (figure 3 of TRN 68).

The work by Abrams (1971) on concrete compressive strengths on cylinder specimens, as reported by Bazant and Kaplan (1996) comprised the following tests:

(i) "Unstressed," the unloaded specimen was heated to the required temperature and then loaded to failure while hot.
(ii) "Stressed," the specimen was loaded and then heated to the required temperature. The load was then increased in increments until failure.
(iii) "Unstressed residual," the unloaded specimen was heated to the required temperature and then allowed to cool down slowly to room temperature. After seven days the specimen was loaded to failure.

The test results are contained in Bazant (1996). As some of the factors of residual concrete strength are greater than unity, there are likely experimental errors or deviations among the samples. Based on the data presented in Bazant (1996), the residual factors versus temperature are plotted in Figure 8.24. In the figure, the "Unstressed residual Strength" curve is significantly lower than the other two, which demonstrates the substantial strength reduction under cooling.

In addition to the physical change related to water vaporization and cracking, chemical changes in concrete also occur, significantly impacting its properties. One notable chemical reaction is the dehydration of calcium hydroxide ($Ca(OH)_2 \rightarrow CaO + H_2O\uparrow$), which releases water vapour. This process contributes to concrete cracking due to evaporation and results in a brighter concrete colour. Additionally, prolonged exposure to temperatures above 300°C oxidizes iron compounds in aggregates,

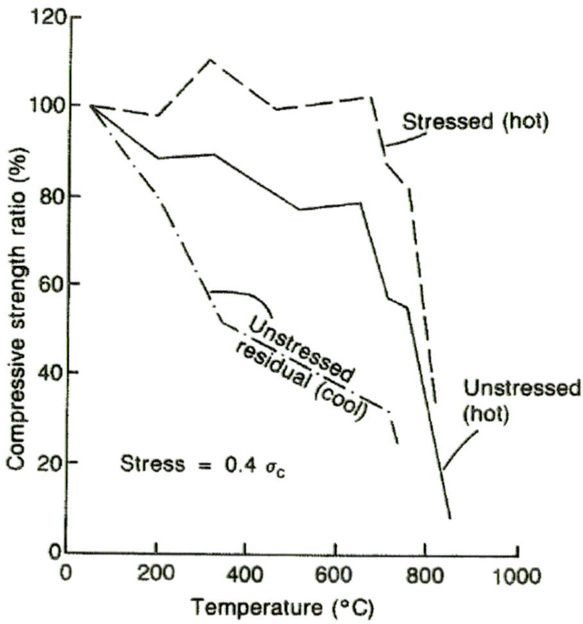

FIGURE 8.24 Figure 5.2 of Bazant (1996) showing strength ratio of calcareous aggregate concrete under various loading conditions.

imparting a reddish hue. According to Hager (2013), the intensity of this colour change varies with temperature, serving as an indicator of the maximum temperature the concrete has endured.

Other chemical changes include the decomposition of limestone aggregate, which weakens the concrete structure. Accelerated carbonation tests by Oliveira and Ribeiro (2019) demonstrate that high temperatures and rapid cooling exacerbate concrete durability issues by advancing the carbonation front. Rapid cooling, such as quenching, promotes pore rehydration and cracking, facilitating CO_2 penetration and accelerating carbonation. In contrast, slow cooling hinders CO_2 diffusion due to the loss of free water in concrete pores, limiting carbonic acid formation, pH reduction, and subsequent carbonation.

Fire also affects reinforcing steel, reducing its strength and Young's modulus, as discussed in Section 8.4.2. Unlike concrete, these reductions are typically temporary for steel if temperatures remain below specific thresholds. After cooling, steel largely regains its pre-fire properties. The UK Concrete Society's TRN 68 (2008) provides a "fire-damaged factor" chart based on which Figure 8.25 is prepared for temperatures up to 700°C, indicating strength reductions starting at 500°C for cold-worked bars and 600°C for hot-rolled bars. For temperatures exceeding 700°C, TRN 68 notes that phase changes in steel introduce additional property variations, and consultation with the manufacturer is recommended.

Poon et al. (2001) investigated the recovery of strength and durability in fire-damaged concrete through post-fire curing by testing specimens. Their findings

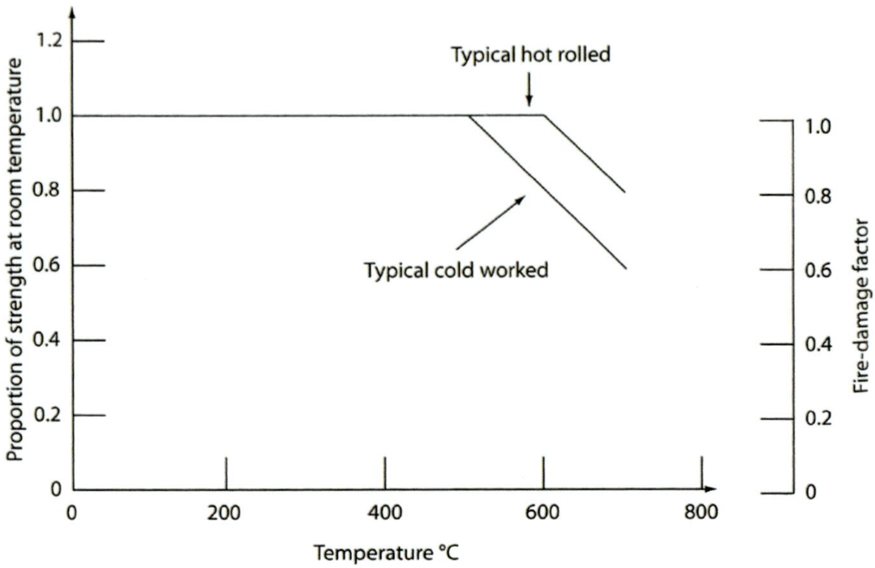

FIGURE 8.25 Yield strength of reinforcing steels at room temperature after heating to an elevated temperature extracted from figure 9 of the TRN 68

revealed significant recovery, with some specimens regaining up to 93% of their original strength and durability. The authors attribute this recovery to rehydration processes that regenerate calcium-silicate-hydrate (C-S-H), which is broken down during fire-induced dehydration. These new rehydration products, smaller than the original hydration products, fill internal cracks, honeycombs, and capillaries formed during the fire, thereby restoring a portion of the concrete's strength. However, Poon et al.'s experiments were conducted under controlled curing conditions. For in situ concrete structures, effective recovery would require thorough and prolonged post-fire curing.

8.5.6 Impacts on Bond Strength between Concrete and Reinforcement Due to Fire

Fire compromises the bond strength between concrete and reinforcing bars, primarily due to the thermal degradation of concrete. This deterioration reduces the concrete's capacity to anchor the reinforcement, even in the absence of spalling. Bazant and Kaplan (1996) examined the bond behaviour at elevated temperatures, presenting their findings in a bond strength ratio curve (shown in Figure 8.26). The bond strength ratio is defined as the bond strength at elevated temperatures divided by the bond strength at 20°C.

Building on Bazant and Kaplan's research (1996), Huang (2010) developed equations for ribbed bars, represented by a conservative dotted line that aligns with the "Model used for Ribbed bars" curve in Figure 8.26.

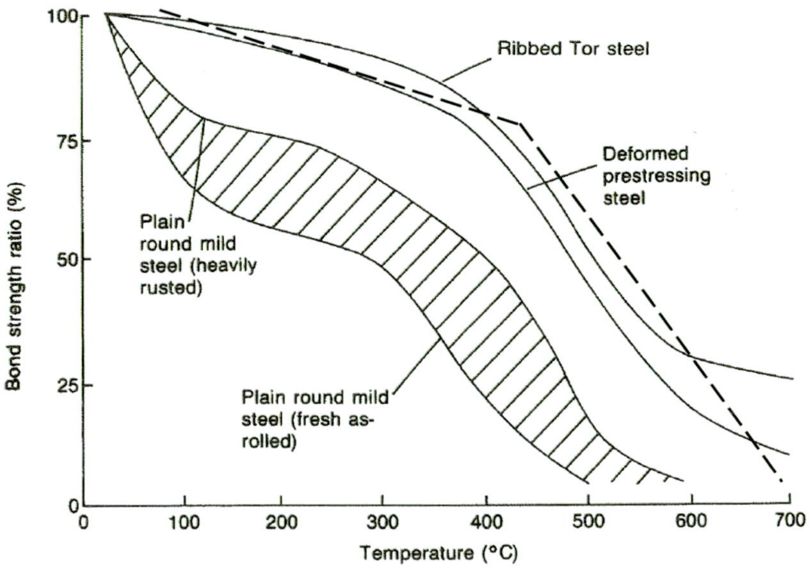

Figure 5.11 Strength of bond of steel bars with concrete at various temperatures

FIGURE 8.26 Degradation of bond strength between concrete and reinforcing steel by Bazant and Kaplan (1996).

$$\tau_{max,T} = \tau_{max,20}\left(1.0 - \frac{0.22}{360}(T-20)\right) \text{ for } 20^{\circ}C \le T \le 380^{\circ}C$$

$$\tau_{max,T} = \tau_{max,20}\left(0.78 - \frac{0.75}{270}(T-380)\right) \quad \text{for} \quad 380^{\circ}C \le T \le 650^{\circ}C \qquad (8.11)$$

Bazant and Kaplan (1996) provide the following key findings regarding bond strength at elevated temperatures:

(i) Bond strength decreases with rising temperatures, with a more significant reduction rate compared to concrete compressive strength.
(ii) Ribbed bars exhibit a greater percentage reduction in bond strength than plain round bars.
(iii) Variations in bar diameter (for both plain and deformed bars) have minimal impact on bond strength reduction.
(iv) Experimental procedures influence bond test outcomes.
(v) The type of aggregate used affects bond test results.
(vi) Smaller concrete cover leads to greater bond strength reduction.

Additionally, studies by Aslani and Samali (2013) and Arunita et al. (2019) indicate further bond strength reduction after cooling. Aslani and Samali (2013) analyzed historical test data on "residual bond strength" and correlated it with residual

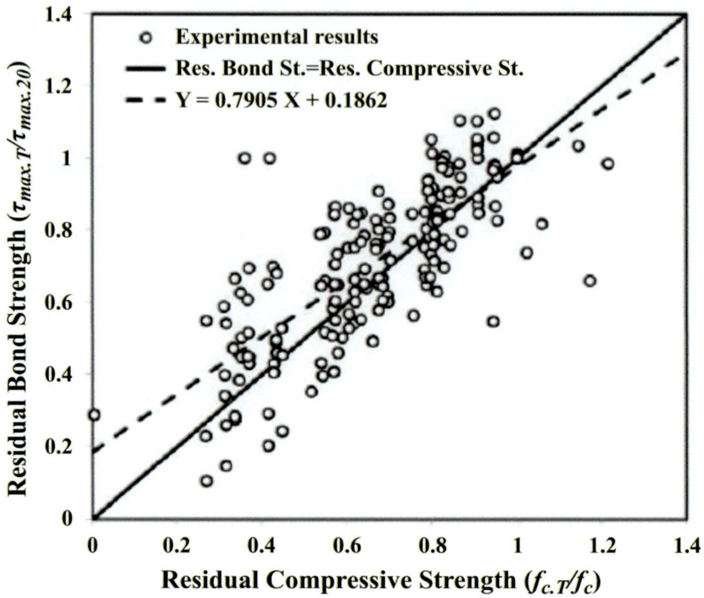

FIGURE 8.27 Residual bond strength versus residual compressive strength as extracted from figure 8 of Aslani and Samali (2013).

compressive strength. Their findings, presented in Figure 8.27, plot the ratio of post-fire bond strength $\tau_{max.T}$ to bond strength at 20°C $\tau_{max.20}$ against the ratio of residual compressive strength at elevated temperature $f_{c.T}$ to compressive strength at 20°C $f_{c.20}$. Through regression analysis, they derived a linear relationship: Y = 0.7905X + 0.1862, where Y represents the residual bond strength ratio $\tau_{max.T}$.T / $\tau_{max.20}$ and X represents the residual compressive strength ratio $f_{c.T} / f_{c.20}$.

However, according to Technical Report Note 68 (2008), restoring the full concrete cover and shear reinforcement during repairs allows the assumption of complete bond restoration between the existing reinforcement and the repair material. As a result, lap and anchorage lengths remain uncompromised. Nevertheless, if the existing bars have buckled due to fire damage, their bond with the surrounding concrete is inadequate. In such cases, sufficient concrete must be removed from behind the bars to allow the repair material to fully encase the reinforcement, ensuring complete composite action.

8.5.7 Summary for Fire Resistance

Summing up, according to Eurocode BS EN 1992-1-2:2023 (2023), the design of reinforced concrete structures to resist fire can utilize two approaches:

(i) "tabulated data" specifying minimum structural dimensions and "axis distances" (concrete cover to the centre of reinforcing bars), and

(ii) analytical design calculations based on estimated temperatures and reduced material properties (strength and Young's modulus) using the "Simplified Design Method" or "Advanced Design Method." Appendix G (Sections G2 and G3) provides worked examples for analyzing a slab and a simply supported reinforced concrete beam under fire, employing tabulated data, the "500°C Isotherm Method," and the "Zone Method" (the latter not applicable to slabs).

However, these Eurocode-based checks, as detailed in Appendix G, apply specifically to structures exposed to a standard fire, which is considered conservative for most real fire scenarios. During fire exposure, a reduced load factor of $\eta_{fi} = 0.7$ is permitted. Post-fire concrete may experience further strength reduction due to cracking by cooling, while reinforcement bars typically recover most or all of their strength if the maximum temperature remains below 500°C to 600°C, as discussed previously. Consequently, a separate structural assessment is required after a fire, using standard load factors. For conservative estimation of residual concrete strength, refer to the UK Concrete Society's Technical Report No. 68 (2008), as illustrated in Figure 8.22.

During the cooling phase after a fire, concrete experiences further strength reduction due to additional cracking. However, studies, including Poon et al. (2001), demonstrate that post-fire curing can facilitate strength recovery through the reformation of calcium-silicate-hydrate (C-S-H) via rehydration processes. Nevertheless, for effective strength regain in in situ concrete structures, prolonged and thorough curing is essential.

8.6 CONCLUDING REMARKS

The four attributes namely ductility, robustness, durability and fire resistance discussed in this chapter form essential pillars of holistic design, apart from that due to imposed loads on the structure. By integrating them from concept through detailing, material selection, and construction, engineers can deliver reinforced concrete structures that perform reliably under normal conditions and remain resilient against earthquakes, accidents, aggressive environments, and fire, often well beyond their nominal design life.

REFERENCES

Abrams, M.S. (1971) Compressive Strength of Concrete at Temperatures to 1600°F. In: Temperature and Concrete, ACI Special Publication No. 25, Paper SP25-2. American Concrete Institute, Detroit, pp. 33–58.

American Society for Testing and Materials (2000) ASTM E119: Standard Test Methods for Fire Tests of Building Construction and Materials. ASTM International.

Aslani, F. and Samali, B. (2013) Predicting the Bond Between Concrete and Reinforcing Steel at Elevated Temperatures. Structural Engineering and Mechanics, 48(5), 643–660.

Attard, M.M. and Setunge, S. (1996) The Stress Strain Relationship of Confined and Unconfined Concrete. ACI Materials Journal, 93(5), 432–442.

Bazant, Z. and Kaplan, M. (1996) Concrete at High Temperatures: Material Properties and Mechanical Models. Longman Group Limited.

British Standard Institution (1987) BS 476: Part 20:1987 Fire Tests on Building Materials and Structures – Part 20: Method for Determination of the Fire Resistance of Elements of Construction (General Principles). British Standard Institution.

British Standard Institution (2000) BS EN 206-1:2000 Concrete – Part 1: Specification, Performance and Conformity. British Standard Institution.

British Standard Institution (2002) BS EN 1990:2002 Eurocode – Basis of Design. British Standard Institution.

British Standard Institution (2004) Eurocode BS EN 1992-1-2:2024 Eurocode 2 – Design of Concrete Structures Part 1-2: Structural Fire Design. British Standard Institution.

British Standard Institution (2004) BS EN 1998-1:2004 Eurocode 8: Design of Structures for Earthquake Resistance – Part 1: General Rules, Seismic Actions and Rules for Buildings. British Standard Institution.

British Standard Institution (2006) BS EN 1991-1-7:2006 Eurocode 1 – Actions on Structures – Part 1-7: General Actions – Accidental Actions. British Standard Institution.

British Standard Institution (2023) BS EN 13501-2:2023 Concrete Fire Classification of Construction Products and Building Elements. British Standard Institution.

British Standard Institution (2024) Eurocode BS EN 1991-1-2:2024 Eurocode 1 – Actions on Structures Part 1-2: General Actions – Actions on Structures Exposed to Fire. British Standard Institution.

British Standard Institution (2023) Eurocode BS EN 1992-1-2:2023 Eurocode 2 – Design of Concrete Structures Part 1-2: Structural Fire Design. British Standard Institution.

Buildings Department of Hong Kong (2013) Code of Practice for the Structural Use of Concrete 2013. Buildings Department, Hong Kong SAR Government.

China Academy of Building Research (2010) GB50011: Code for Seismic Design of Buildings (2010). China Academy of Building Research.

Concrete Design Committee P3101 (New Zealand) (2006) NZS 3101: Part 1 & 2:2006 Concrete Structures Standard. Concrete Design Committee, New Zealand.

Concrete Society (2008) Technical Report No. 68: Assessment, Design and Repair of Fire-Damaged Concrete Structures. Concrete Society.

Das, A., Bosnjak, J. and Sharma, A. (2019) Numerical Investigation on Post-Fire Bond Behaviour of Reinforcement in Concrete. RILEM Technical Letters, 4, 1–8.

Hager Izabela (2013) Colour Change in Heated Concrete. Fire Technology.

Ho, J.C.M., Kwan, A.K.H. and Pam, H.J. (2003) Theoretical Analysis of Post-Peak Flexural Behaviour of Normal- and High-Strength Concrete Beams. The Structural Design of Tall and Special Buildings, 12, 109–125.

Huang, Z.H. (2010) Modelling the Bond Between Concrete and Reinforcing Bar in a Fire. Engineering Structures, 32(11), 3660–3669.

Kwan, A.K.H. (2005) *Durability of Reinforced Concrete Structure: Theory vs Practice*. Standing Committee of Concrete Technology Paper.

Kwan, A.K.H., Ho, C.M. and Pam, H.J. (2002) Flexural Strength and Ductility of Reinforced Concrete Beams. Proceedings of the Institution of Civil Engineers, Structures and Buildings, 152(4), 361–369.

Law, C.W. (2010) Discussion on the Maximum Longitudinal Steel Ratio as a Ductility Requirement for Reinforced Concrete Beams by the Code of Practice for the Structural Use of Concrete 2004. HKIE Transactions, 18(3), 14–21.

Mander, J.B., Priestley, M.J.B. and Park, R. (1988) Theoretical Stress-Strain Model for Confined Concrete. Journal of Structural Engineering, ASCE, 114(8), 1804–1825.

Neville, A.M. (1995) Properties of Concrete, 4th Edition. Prentice Hall.

Oliveira, J. and Ribeiro, J.L.R. (2019) Durability of Concrete After Fire Through Accelerated Carbonation Tests. Material Research, 22, 1–7.

Poon, C.S., Azhar, S., Anson, M. and Wong, Y.L. (2001) Strength and Durability Recovery of Fire-Damaged Concrete After Post-Fire-Curing. *Cement and Concrete Research*, 31(9), 1307–1318.

Technical Committee ISO/TC 92, Fire Safety, Subcommittee SC2 Fire Containment (2002) ISO 834-8 (2002). International Organization for Standardization.

9 Reinforced Concrete Sectional Design of Common Structural Elements

9.1 INTRODUCTION

The strength design of a reinforced concrete member is based on a sectional analysis at its critical sections. The capacity of each section is verified against the total internal forces, which include axial loads, bending moments, and shear forces determined from structural analysis.

Modern sectional design is predominantly governed by the ultimate limit state (ULS) method. This approach calculates the ultimate capacity of a section and compares it to the factored ultimate loads it must support. The ultimate limit state capacity is defined as the maximum load a section can carry before failure.

To ensure adequate safety margins, design codes apply partial factors. The material's ultimate strength is reduced by a partial material factor to obtain a "design strength," while the applied characteristic loads are increased by partial load factors to obtain "factored loads." At the ultimate limit state, the concrete and reinforcement are assumed to be fully or partially plastic, meaning stresses remain constant once the yield strain is exceeded.

This chapter focuses on the application of the ultimate limit state method for the sectional design of reinforced concrete members, such as slabs, beams, columns, and walls.

A structural section is typically subjected to a combination of axial load, out-of-plane bending moments, shear, and torsion. For design purposes, loads producing the same type of stress will be combined together in design. Therefore, axial load and bending moment will be combined as they both produce axial stresses. Similarly, shear and torsion producing shear stresses are combined in the design check.

The design approaches and parameters discussed in this chapter are aligned with several reputable international codes of practice, which are referenced throughout. However, these references are intended mainly to illustrate underlying principles and should not be considered to be a complete design guide. Essential parameters,

DOI: 10.1201/9781003636380-9

such as minimum reinforcement requirements and critical detailing rules, are omitted. For complete design implementation, readers must consult the original code documents.

9.2 STRESS-STRAIN RELATIONSHIP OF CONCRETE

Design codes provide compressive stress-strain relationships which are similar among themselves for the ultimate reinforced concrete design, which are described in detail in this section. Tensile strength is generally ignored for design purpose.

9.2.1 Eurocode BSEN1992-1-1:2023 (2023)

The Eurocode provides an equation (equation 5.6) to describe the stress-strain relationship as reproduced in Equation 9.1, which may be used to model the response of concrete to short-term uniaxial compression.

$$\frac{\sigma_c}{f_{cm}} = \frac{k\eta - \eta^2}{1+(k-2)\eta} \tag{9.1}$$

$$k = 1.05E_{cm}\varepsilon_{c1} / f_{cm}\ ;\ \eta = \varepsilon_c / \varepsilon_{c1};\ \varepsilon_{c1} = 0.7f_{cm}^{\ \ 1/3} / 1000 \le 0.0028;$$

$$\varepsilon_c < \varepsilon_{cu1} = \left(2.8 + 1.4\left(1 - f_{cm} / 108\right)^4\right) / 1000 \le 0.0035$$

In Equation 9.1, σ_c is the concrete stress at strain ε_c. f_{cm} is mean concrete cylinder strength and other parameters are defined as above. The stress-strain curve for the concrete C35/45 (cylinder strength/cube strength) with $f_{cm} = 35 + 8 = 43\text{N/mm}^2$ is plotted in Figure 9.1 as the short-term compression curve.

Nevertheless, for general design purposes, the Eurocode uses a design value of concrete compressive strength f_{cd} which is a discounted value from f_{cm}. f_{cd} is given by equation 5.3 of BSEN1992-1-1:2003 as

$$f_{cd} = \eta_{cc}k_{tc}\frac{f_{ck}}{\gamma_c} \tag{9.2}$$

where $\eta_{cc} = \left(\frac{f_{ck,ref}}{f_{ck}}\right)^{\frac{1}{3}} \le 1$, with $f_{ck,ref} = 40$, $k_{tc} = 0.85$, $\gamma_c = 1.5$

So, for the C35/45 concrete, $f_{cd} = 19.83\ \text{N/mm}^2$. The design stress-strain profile is defined by equation 8.4 of the Eurocode which is reproduced as Equation 9.3 as follows:

$$\sigma_{cd} = \begin{cases} f_{cd}\left[1 - \left(1 - \dfrac{\varepsilon_{cd}}{0.002}\right)^2\right] & \text{where } 0 \le \varepsilon_{cd} \le 0.002 \\ f_{cd} & \text{where } 0.002 < \varepsilon_{cd} \le 0.0035 \end{cases} \tag{9.3}$$

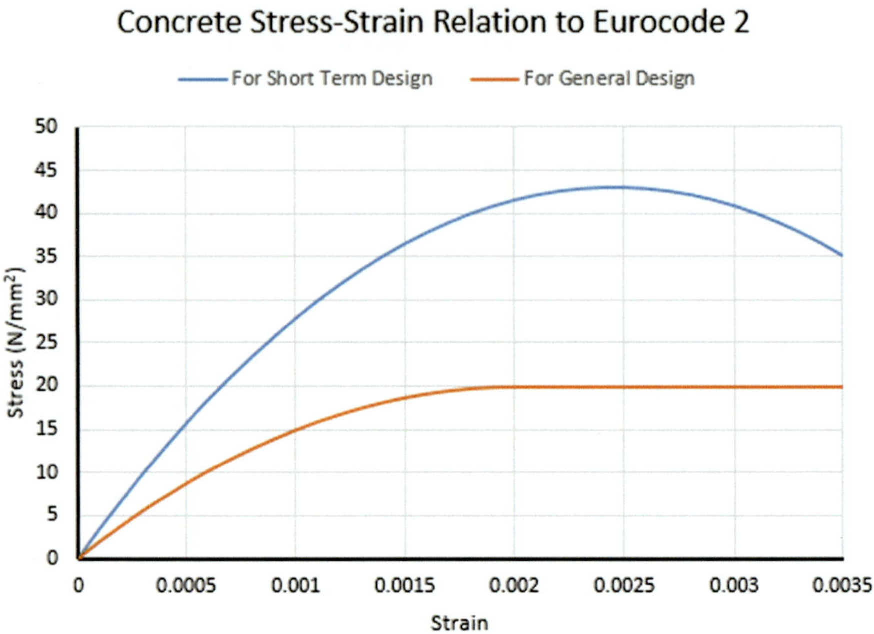

FIGURE 9.1 Design concrete stress-strain curves in BSEN 1992-1-1.

The curve for σ_{cd} is also plotted in light brown in Figure 9.1 for comparison purpose.

The equation 9.1 takes the same form as the former British Code BS8110-2:1985 (1985) with the maximum value at $0.8f_{cu}$ where f_{cu} is the cube strength and other parameters slightly differ with different numerical values. It is stipulated in the BS that the curve is used for "rigorous analysis for non-critical sections" while there is another formula in the BS similar to Equation 9.3 for general design with the maximum concrete strength at $0.45f_{cu}$. In this respect, the Eurocode requirements are very similar to BS8110.

In addition, to simplify the design calculation, Equation 9.3 comprising a portion with parabolic relation and another portion with constant relation (rectangular stress distribution) is replaced by one with constant relation of stress range which extends only for 0.8 of the neutral axis depth. The transition is as shown presented in Figure 9.2.

9.2.2 China Code GB50010-2010 (2010)

The China Code gives a concrete stress-strain relationship similar to the Eurocode given in Equation 9.4 as listed as below:

$$\sigma_c = \begin{cases} f_c\left[1-\left(1-\dfrac{\varepsilon_c}{\varepsilon_o}\right)^n\right] & \text{where } 0 \le \varepsilon_{cd} \le \varepsilon_o \\ f_c & \text{where } \varepsilon_o < \varepsilon_{cd} \le \varepsilon_{cu} \end{cases} \tag{9.4}$$

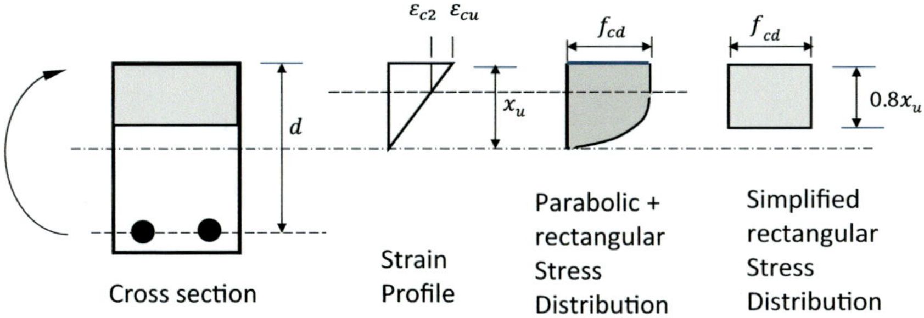

FIGURE 9.2 Extract from Figure 8.2 of the Eurocode BSEN1992-1-1:2023.

where $n = 2 + \frac{1}{60}\left(f_{cu,k} - 50\right)$;

$$\varepsilon_o = 0.002 + 0.5\left(f_{cu,k} - 50\right) \times 10^{-5} \geq 0.002;$$

$$\varepsilon_{cu} = 0.0033 - \left(f_{cu,k} - 50\right) \times 10^{-5}.$$

In the China Code, f_c is the design strength of concrete, which is around 45–50% of the cube strength $f_{cu,k}$ as listed in the Code. So, although the formulae are more complicated, they are similar to the Eurocode.

The China Code GB50010:2010 (2010) also allows a rectangular stress block to be used as shown in the figure (figure 6.2.10 of the Code) extracted below :

In addition to α_1 defined in Figure 9.3 for determination of the design concrete strength for the rectangular stress block as $\alpha_1 f_c$, another coefficient β_1 by which the rectangular concrete stress block depth x is related to the neutral axis depth x_n by $x = \beta_1 x_n$. In accordance with 6.2.6 of the China Code, when the concrete grade does not exceed C50, $\alpha_1 = 1$ and $\beta_1 = 0.8$. When the concrete grade reaches C80,

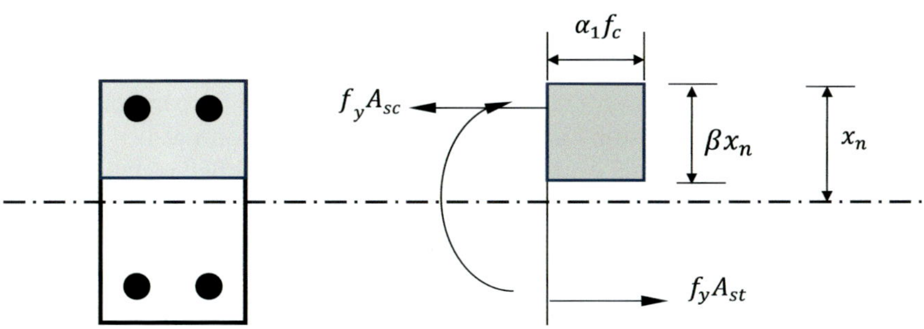

FIGURE 9.3 Rectangular stress block by the China Code GB50010-2010.

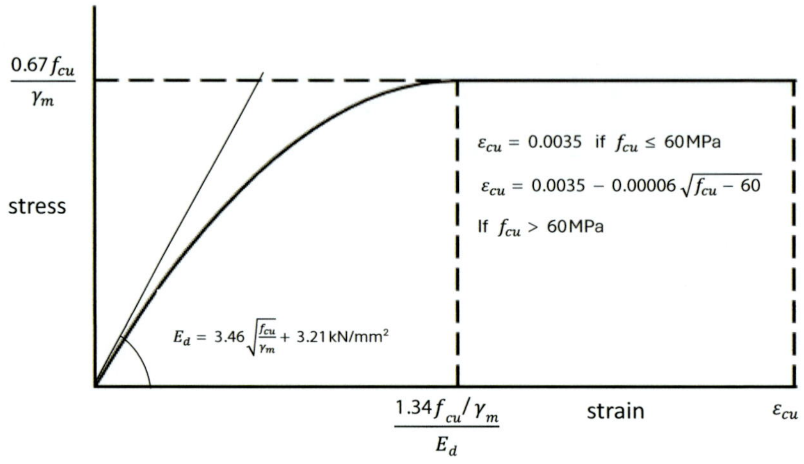

FIGURE 9.4 Stress-strain relationship of Concrete in the Hong Kong Code.

$\alpha_1 = 0.94$ and $\beta_1 = 0.74$. Intermediate values between C50 and C80 can be linearly interpolated.

9.2.3 The American Code ACI318-M19 (2019)

The American Code gives a similar rectangular stress block in its 22.2.2.4.1 where the constant stress is $0.85 f_c$' with $f_c^{'}$ as the specified compressive strength of concrete. For design purposes, a strength reduction factor ϕ which ranges from 0.6 to 0.9, depending on the nature of the loads (moments or axial load or shear) and the strain level is applied. The rectangular stress block stretch for $\beta_1$times the neutral axis depth which falls from 0.85 to 0.65 between concrete strengths $f_c^{'} = 4000$psi (27.57 MPa) and $f_c^{'} = 8000$psi (55.16 MPa) under linear interpolation. β_1 stays constant at 0.85 when $f_c^{'}$ below 4000 psi and 0.65 when $f_c^{'}$ exceeds 8000 psi.

9.2.4 The Code of Practice for the Structural Use of Concrete 2013 of Hong Kong

The Hong Kong Code has a similar stress-strain relationship with parabolic + rectangular stress-strain relationship for concrete as extracted in Figure 9.4.

The Hong Kong Concrete Code comprises a similar simplified rectangular stress block as extracted in Figure 9.5.

Other national or international codes have similar concrete stress-strain relationships for reinforced concrete section design.

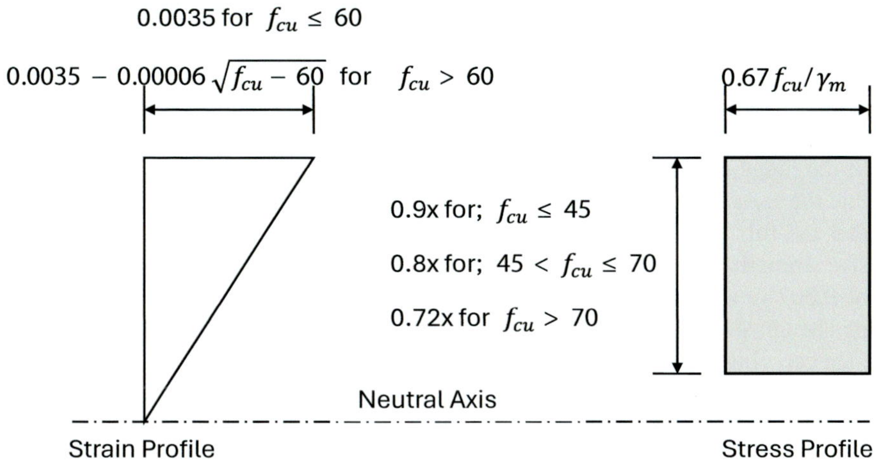

FIGURE 9.5 Simplified rectangular concrete stress block of the Hong Kong Code.

9.3 STRESS-STRAIN RELATIONSHIP OF REINFORCEMENT

The stress-strain relationship of reinforcement for design purposes is relatively simple, which generally takes the form as illustrated in Figure 9.10 which is extracted from the Hong Kong Concrete Code (2013). Unlike concrete, reinforcement can resist both compression and tension (Figure 9.6).

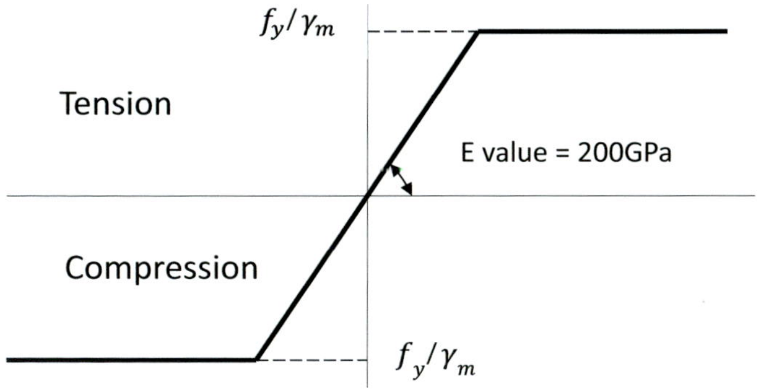

FIGURE 9.6 Stress-strain relationship for reinforcement of the Hong Kong Code.

9.4 DESIGN FOR AXIAL LOAD AND BENDING MOMENT

9.4.1 General

As discussed in Section 9.2, in the ultimate design of a reinforced concrete section, the strain of the extreme fibre is assumed to have reached an ultimate value which is in the range of 0.003 to 0.0035 as specified by the respective design codes. In case that the externally applied load comprises only an axial load, it may then be assumed that the full section has reached the ultimate strain by which the design compressive strength of the concrete can be attained. As the yield strain of steel in the order of 0.002 is well below the ultimate strain of concrete, the reinforcement has also reached its design compressive capacity in accordance with Section 9.3. So the axial load carrying capacity of the section is simply $F_{Rd} = f_c A_c + f_y A_s$ where f_c and f_y are the design strengths of concrete and steel (after the application of material factors as required by the Code being adopted) and A_c and A_s are the concrete and steel cross-sectional areas. F_{Rd} is then checked against the applied load F_{Ed} and the section is considered adequate when $F_{Rd} \geq F_{Ed}$.

9.4.2 Rectangular Section under Axial Load and Uniaxial Bending

In case that an axial load and a uniaxial bending moment act on a section, the design for a rectangular section is first discussed as illustrated in Figure 9.7 where tension occurs. In the figure, d is the effective depth of the section measured from the compressive edge to the tensile steel, x is the neutral axis depth, f_{cd} and f_{st} are the design strengths of concrete and steel, A_{st} is the steel area, ε_{cu} is the ultimate concrete strain, β_1 is the ratio of the depth of the rectangular stress block to x. Linear strain

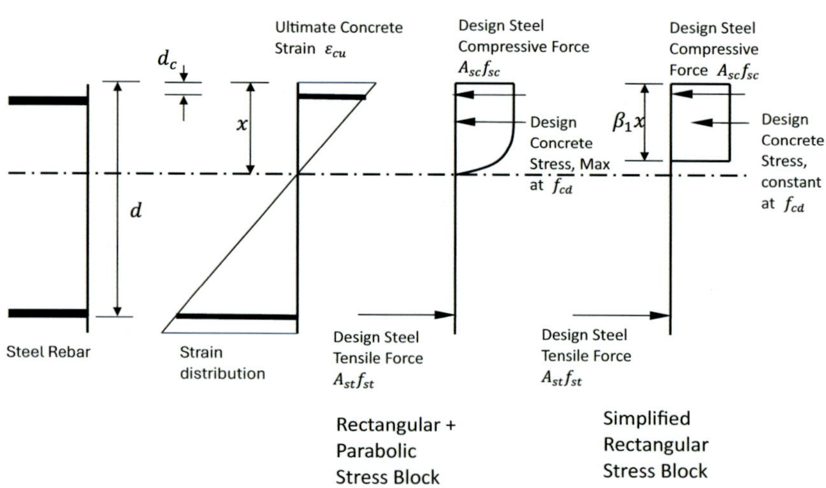

FIGURE 9.7 A typical stress-strain diagram for sectional design of a rectangular section.

distribution along the section is assumed as shown in the figure. It should, however, be noted that when high compressive axial load is applied, tension may not exist, and the neutral axis depth extends beyond the section depth.

Figure 9.7 can first be used to demonstrate the reinforcement design of a singly reinforced beam where the top bar is absent and the net force on the section is zero. The designer can take the moment about the level of the reinforcing steel bar. x can then be solved by $M = f_{cd} b\beta_1 x\left(d - 0.5\beta_1 x\right)$ where M is the design moment and b is the beam width under the use of rectangular stress block. Under the same principle, x can also be solved under the use of rectangular plus parabolic stress block. By equating the force by reinforcement to that of concrete, the reinforcement area can then be worked out. In addition, codes often limit the neutral axis depth to be not greater than $0.5d$. If the applied moment exceeds the moment provided by concrete under the limitation of $x = 0.5d$, compressive reinforcement added in the concrete compression zone together with added tensile reinforcement can provide the extra moment. This is a typical design method adopted by most of the codes.

However, if the net force is not zero, the section design becomes a column design. Under the known value of ε_{cu} as limited by the code being followed, the strains of the reinforcement bars can be expressed in terms of x as $\varepsilon_{st} = \left(d - x\right)\varepsilon_{cu} / x$ for the tensile bar and $\varepsilon_{sc} = \left(x - d_c\right)\varepsilon_{cu} / x$ for the compressive bar. Subsequently, the reinforcement stresses and forces F_{st} and F_{sc} are determined, together with the concrete force F_c, all in terms of x. Equating $F_c + F_{sc} = F_{st}$, x can then be solved by trial and error and subsequently the moment of resistance M_R of the section. The section will be considered satisfactory if $M_R \geq M_A$ as the applied moment.

If there are reinforcements between the top and bottom bars, they can be taken into account in the determination of the strength of the section. The stresses in these bars shall follow the linear strain profile which may not reach the ultimate values.

Worked Example 9.1

The worked example is based on a column section of 1500 mm (W) by 2000 mm (D) of concrete grade C35/45 with 4 no. of T40 corner bars; 58 nos. of T40 bars as shown in Figure 9.8 is to be analyzed. $f_{ck} = 35\,\text{MPa}$.

In accordance with BSEN1992-1-1:2023 (2023), the reinforcement bars are with strength at $f_{yk} = 500\text{MPa}$ which has to be divided by the material factor $\gamma_s = 1.15$ as per table 4.3 (NDP) for design. The concrete design strength is $f_{cd} = \eta_{cc} k_{tc} f_{ck} / \gamma_c = 19.83\,\text{MPa}$ as per equation 5.3 after the application of the strength factor at $\eta_{cc} = \left(40 / f_{ck}\right)^{(1/3)} \leq 1$, $\gamma_c = 1.5$ and $k_{tc} = 0.85$. The "N–M" curves are constructed as shown in Figure 9.9 which relate the axial load capacity and the uniaxial moment of resistance about the X and Y axes. The curves are constructed by calculating the axial load capacities and moments of resistance for varying values of neutral axis depths in the X and Y directions respectively with the extreme fibre strains constant at 0.0035. By Figure 9.9, any "load case" comprising an axial load and a uniaxial moment about the X or Y direction is within the capacity of the column if the point representing the load pair is within the respective N_{Rd}–M_{Rd} curves. But in the general design for a load case comprising an axial load N_{Ed} and

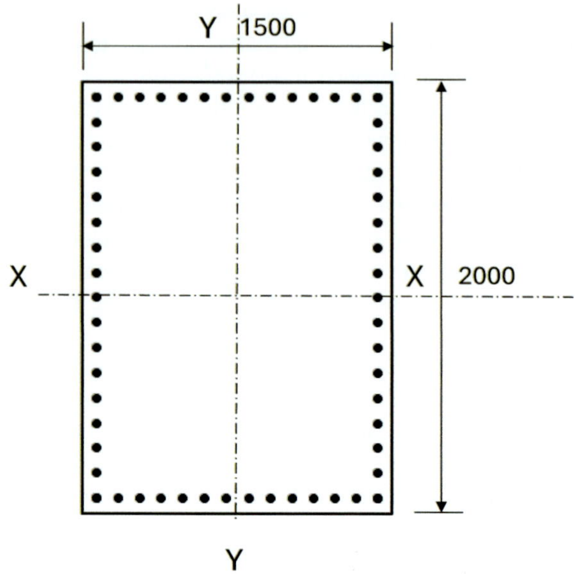

FIGURE 9.8 Column section of Worked Example 9.1.

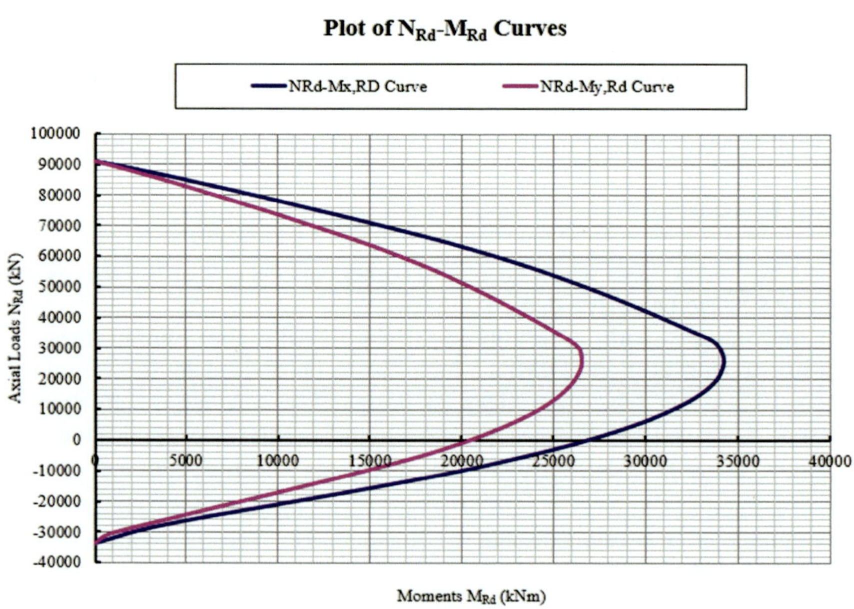

FIGURE 9.9 N_{Rd}–M_{Rd} curves in accordance with BSEN1992-1-1:2023 for a C35/45 rectangular column of 1500 mm by 2000 mm with 58 nos. of T40 bars.

moment M_{Ed}, it is not necessary to form these curves. The designer needs only to try the neutral axis depth until the sum of the forces provided by the concrete and reinforcement match N_{Ed}. Then the moment of resistance M_{Rd} is calculated in accordance with the balanced neutral axis depth and the section is adequate if M_{Rd} exceeds M_{Ed}. The approach is outlined in greater detail in Appendix H together with Worked Example H1.

Detailed description of the design procedure and the formulae relating strain, stress to forces and moments of the concrete and reinforcements under axial load and moments are contained in Appendices E and F of the "Manual for Reinforced Concrete Analysis and Design to the Code of Practice for the Structural Use of Concrete 2013 (of Hong Kong)." The manual can be downloaded from the internet. Although the manual aims at providing design in accordance with the design parameters of the Hong Kong Code, the basic theory and procedure are applicable to design by other codes with a change of design parameters. Due to similarities of the concrete and steel stress-strain relationships to other codes, the formulae contained in the manual can be readily modified for use in other codes.

9.4.3 Rectangular Section under Axial Load and Biaxial Bending

The foregoing discussion restricts itself to that of uniaxial bending. However, if under biaxial moment, the orientation of the neutral axis will not be parallel to the X and Y axes as shown in Figure 9.10(c) while Figures 9.10(a) and (b) show the neutral axis orientation when the moment is uniaxial about Y and X.

In Figure 9.10(c), it should also be noted that the orientation of the neutral axis may not necessarily be perpendicular to the plane where the resultant moment acts. So, two unknowns including the neutral axis depth and the orientation are required to be solved under biaxial moments and axial load. The solution process shall often be by trial and error on these two unknowns until the sum of the forces of concrete

(a) Moment about Y, neutral axis parallel X

(b) Moment about X, neutral axis parallel Y

(c) Moment and neutral axis not about X and Y

FIGURE 9.10 Orientation of neutral axis and direction of moment.

and reinforcement is equal to the applied load and the resultant moment of resistance aligns with the external moment. The section will then be adequate if the aligned resultant moment provided by the section exceeds the applied moment.

As the trial and error process described in the previous paragraph is often tedious, there are codified methods such that the solution is reduced to a uniaxial bending problem for conservative estimation. The Eurocode BSEN1992-1-1:2023 (2023) recommends the fulfilment of its equation 8.2 extracted as follows:

$$\left(\frac{|M_{Edz}|}{M_{Rdz,N}}\right)^{\alpha_N} + \left(\frac{|M_{Edy}|}{M_{Rdy,N}}\right)^{\alpha_N} \leq 1.0 \tag{9.5}$$

where M_{Edz} and M_{Edy} are the applied moments about the z and y axes (which are Y and X axes as given in Figure 9.10). $M_{Rdz,N}$ and $M_{Rdy,N}$ are the moments of resistance of the section about the z and y axes determined as uniaxial moment problems under the applied load N.

α_N is given by the following table for rectangular section, allowing linear interpolation for intermediate values. N_{Ed} is the applied axial load and $N_{Rd,0}$ is the full axial load capacity of the section without moment which is $f_{cd}A_c + f_{yd}A_s$

$\lvert N_{Ed}\rvert / N_{Rd,0}$	0.1	0.7	1.0
α_N	1.0	1.5	2.0

Worked Example 9.2
Consider the following loads and moments acting on the column section shown in Figure 9.8: $N_{Ed} = 70676\text{kN}$; $M_{Edy} = 13172\text{kNm}$ (about X) and $M_{Edz} = 3824\text{kNm}$ (about Y), check if the section is adequate to resist the design loads.

As by the process described in the foregoing $M_{Rdy,N} = 15316\text{kNm}$ and $M_{Rdz,N} = 11650\text{kNm}$ after trials on the neutral axis depths. The reader may also estimate these values from the chart in Figure 9.9. With $N_{Rd,0} = (19.83\times1500\times2000 + 58\times12569.64\times500/1.15)\div1000 = 91179$ kN, $|N_{Ed}|/N_{Rd,0} = 0.775$, $\alpha_N = 1.625$ can be found by interpolation. Plugging these values into Equation 9.5,

$$\left(\frac{|M_{Edy}|}{M_{Rdy,N}}\right)^{\alpha_N} + \left(\frac{|M_{Edz}|}{M_{Rdz,N}}\right)^{\alpha_N} = \left(\frac{13172}{15316}\right)^{1.625} + \left(\frac{3824}{11650}\right)^{1.625} = 0.946 < 1.0.$$

So, the section is satisfactory.

The ACI318-08 (R10.3.6 and R10.3.7) (2008) and the China Code GB50010:2010 (2015 Edition) (2015) (eqution 6.2.21-3), however, adopt another approach by checking

$$1/P_{ni} = 1/P_{nx} + 1/P_{ny} - 1/P_o \tag{9.6}$$

where P_{ni} = nominal strength at given moments about X and Y axes;
P_{nx} = nominal strength at given moments about the X-axis;
P_{ny} = nominal strength at given moments about the Y-axis;
P_o = nominal strength with no moments.

Based on the N–M relations obtained by the Eurocode, for the load case $P = 70676$ kN; $M_x = 13172\text{kNm}$ and $M_y = 3824\text{kNm}$, the neutral axis depths are determined such that M_x they M_y are matched. The neutral axis depths determined are 2196.10 mm and 2251.26 mm are matching M_x and M_y. P_{nx} and P_{ny} calculated which are respectively 73823 kN and 85093 kN. $P_o = 91179$ kN as determined previously. So,

$$P_{ni} = 1 \div \left(\frac{1}{P_{nx}} + \frac{1}{P_{ny}} - \frac{1}{P_o} \right) = 1 \div \left(\frac{1}{73823} + \frac{1}{85093} - \frac{1}{91179} \right) = 69782\text{kN}$$

As $P_{ni} = 69782$ kN $< P = 70676$kN, the section is marginally unsatisfactory.

It may be noted that Equation 9.6 is not found in the later versions of ACI318. Exact interactions between axial load and the biaxial moments may be required by the ACI318.

The former British Code BS8110-1:1997 (1997) and the Hong Kong Concrete Code have adopted a simpler approach by comparing M_x / h' and M_y / b' where M_x and M_y are the moments about X-axis and Y-axis respectively; h' and b' are the effective depths of the column section along which M_x and M_y act. If $M_x / h' > M_y / b'$, the column will be designed as a uniaxial moment problem about the X-axis under a greater moment of $M_x + \beta M_y h'/b'$ where β is a coefficient given in the codes depending on axial stress due to the axial load. Conversely, the design moment is $M_y + \beta M_x b'/h'$.

Despite the foregoing provisions for designing columns under biaxial bending with simplified procedures, the accurate method of searching for the neutral axis depth and orientation is always acceptable, which will generally have to be handled by computer methods.

9.4.4 Circular Section under Axial Load and Moment

The design for circular columns follows the same principle. Although the resultant moment acting on the section is now always uniaxial after vector summation of that in the X and Y directions, the design calculation is still more complicated due to the geometry of the section with varying widths. Nevertheless, the "Manual for Reinforced Concrete Analysis and Design to the Code of Practice for the Structural Use of Concrete 2013" outlines an approach and contains formulae for the design

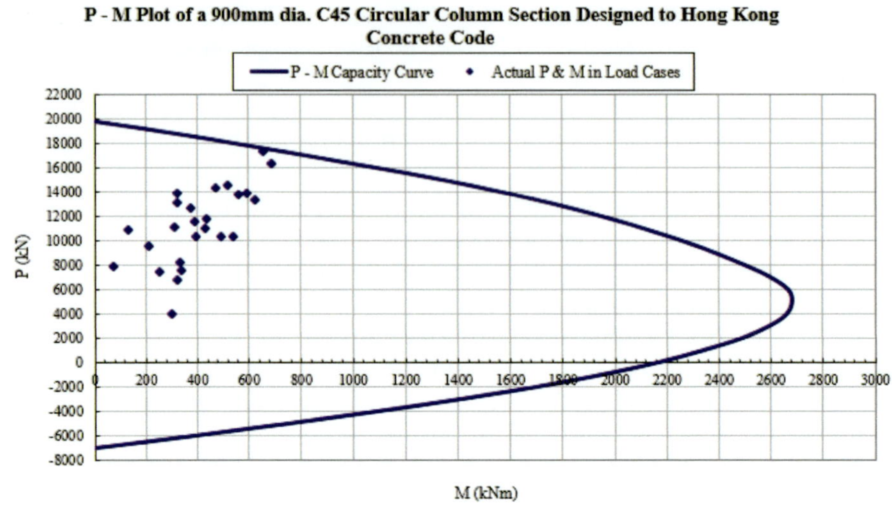

FIGURE 9.11 PM diagram of circular column designed to Hong Kong Code.

calculations. A plot of the design of a 900 mm diameter circular column with 20 nos. of T32 bars designed to the Hong Kong Concrete Code 2013 (2013) is contained in Figure 9.11. The dots represent the actual load cases which are enveloped by the P-M (axial load versus moment) curve.

9.4.5 Irregular Section under Axial Load and Moment

For irregular sections, the same principle of linear strain distribution applies. The outline of the section and subsequently its area and first moment of area is, however, most conveniently defined by coordinate geometry as demonstrated in Figure 9.12.

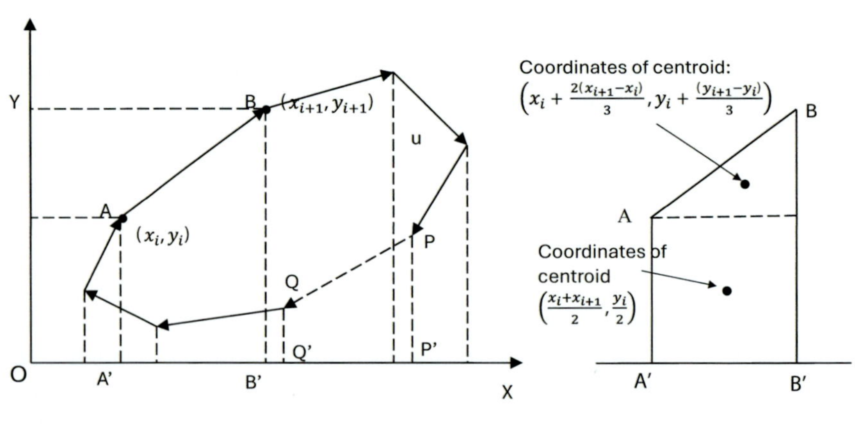

FIGURE 9.12 Defining section by coordinate geometry.

In the figure, the section as a polygon is defined by specifying coordinates of the corners of the figure in consecutive order. For example, the inputs of the coordinates A to B, a trapezium area ABB'A' is defined by which its area and centroid can be calculated. A similar trapezium is also formed by P to Q as PQQ'P'. However, as the trapezium from A to B is positive and from P to Q is negative, the net area of the section can be defined upon a closed loop around the corner coordinates for the section.

The concrete stress profile has to be defined in accordance with the Code to be designed for, generally with a parabolic + rectangular concrete stress profile or simplified rectangular stress profile and zero tensile strength. Upon pre-determined values for neutral axis depth and orientation, together with extreme concrete fibre strain at the ultimate value, the strains and subsequently the stresses and forces in the concrete and reinforcements are all defined. For convenience in computation, it is most convenient to transform all the coordinates from XY to uv by which u is parallel to the neutral axis and the origin of the uv system is the centroid of the section as shown in Figure 9.13. Then the force and biaxial moments of the cross-section can be computed by

$$F_c = \Sigma \int_{v_1}^{v_2} \int_{u_1}^{u_2} \sigma_c(v)\, dudv \tag{9.7}$$

$$M_{cu} = \Sigma \int_{v_1}^{v_2} \int_{u_1}^{u_2} \sigma_c(v)\, vdudv \tag{9.8}$$

$$M_{cv} = -\Sigma \int_{v_1}^{v_2} \int_{u_1}^{u_2} \sigma_c(v)\, ududv \tag{9.9}$$

where u_1, v_1, u_2, v_2 represent coordinates of two consecutive nodes with reference to rotated uv axes. In the figure $V_{min}(\theta)$ is the outermost v-coordinate of the section. From Equations 9.7 to 9.9, the integrals $\int_{v_1}^{v_2} \int_{u_1}^{u_2} \sigma_c(v)\, dudv$ are the combined values of the geometry of the section and the stress read from the stress-strain relationship of concrete and steel.

The moments of the concrete section can then be converted back to the XY axes as

$$M_{cX} = M_{cu} \cos\theta - M_{cv} \sin\theta \tag{9.10}$$

$$M_{cY} = M_{cu} \sin\theta + M_{cv} \cos\theta \tag{9.11}$$

Reinforcements in terms of bar diameters and locations are pre-determined in the section. The forces and moments contributed by the reinforcements can be similarly calculated which can conveniently be done about the X and Y axes. The forces and moments are then added to that of concrete as

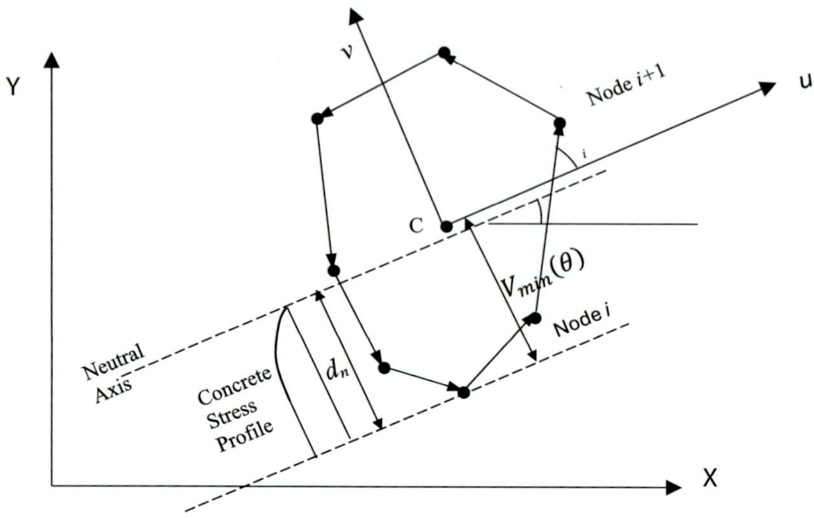

FIGURE 9.13 Calculation of the concrete forces and moments on the irregular section.

$$F = F_c + F_s \tag{9.12}$$

$$M_X = M_{cX} + M_{sX} \tag{9.13}$$

$$M_Y = M_{cY} + M_{sY} \tag{9.14}$$

So, functions for F, M_X and M_Y are defined for pre-determined neutral axis depth d_n and the neutral axis orientation θ. Trial for the values is then carried out such that F matches the externally applied load F_a and the resultant of M_X and M_Y aligns with that of the externally applied moment. The trials are in two dimensions. The trial θ should be within a range of some $\pm 30°$ of the orientation of the externally applied moment ϕ_a. At each trial of θ, the d_n is determined so that $F = F_a$. The orientation of M_X and M_Y is calculated and its orientation ϕ checked with ϕ_a and difference noted. The trial for θ should be systematic under increments from an extreme value of the trial range until there is a sign change in $\phi - \phi_a$. Interpolation for accurate value can be carried out with subsequent determination of d_n.

A spreadsheet on Excel has been prepared for irregular column (or wall) section with design in accordance with the Hong Kong Concrete Code 2013 (2013). Figures 9.14(a) and 9.14(b) show the M_X and M_Y contours for a core wall section and a column section under different axial load levels, both are of concrete grade C45 and steel at 500B.

The spreadsheet aims to check the capacity of the section for individual load cases, each comprising an axial load and biaxial moments.

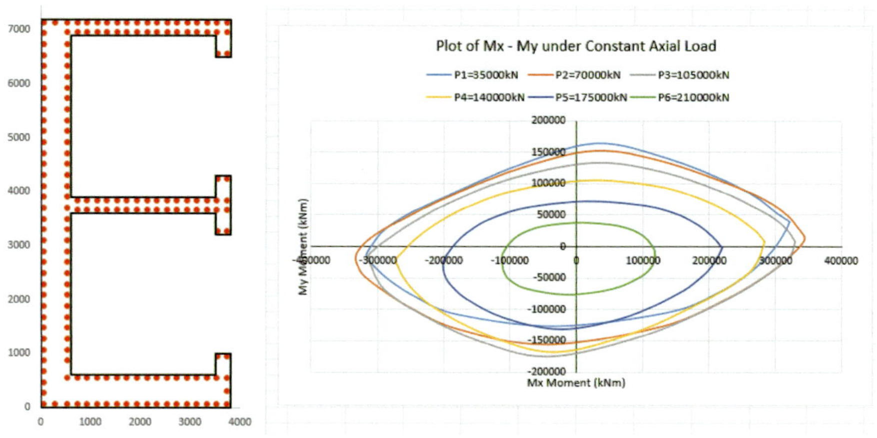

(a)

FIGURE 9.14(A) Plot and moment envelopes of a core wall.

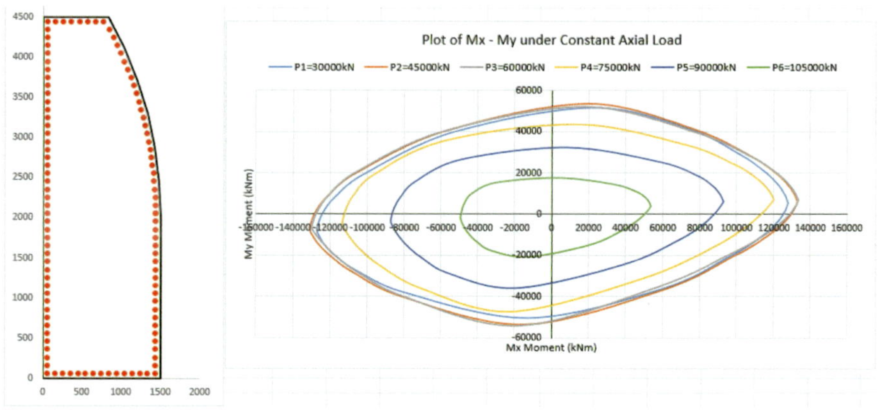

(b)

FIGURE 9.14(B) Plot and moment envelopes of an irregular column.

9.5 DESIGN FOR SHEAR

Traditionally, shear in reinforced concrete strut (beam or column) is considered to be creating diagonal compression and tension in the beam due to the principal compression and tension resulted from the shear stress, as shown in Figure 9.15. Approaches based on this assumption by designing shear reinforcement crossing the diagonal tension cracks have been formulated, which are widely used in various codes of practice, as will be discussed in coming sections.

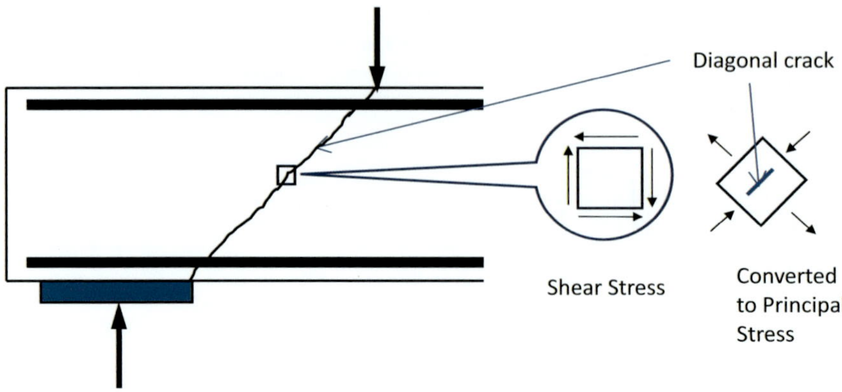

FIGURE 9.15 Diagonal stresses due to shear in reinforced concrete beam.

9.5.1 Shear Strength by Concrete – General

Shear in a reinforced concrete strut is a complex phenomenon. Theories have been developed to account for its behaviour for structural design. For convenience in design, the shear stress v by which the shear force V on a section is averaged over an effective area is often taken as a design parameter. v is either expressed by $V/(b_w d)$ (e.g., ACI318) or $V/(b_w z)$ (Eurocode) where b_w is the width of the section, d is the effective depth and z is the lever arm. Codes often directly or indirectly restrict v (i) to a maximum value beyond which the section has to be increased and (ii) to a certain threshold value beyond which shear reinforcement has to be added.

The total resistance to shear is contributed by the concrete and the shear reinforcement (if any). The general approach by the codes is to require shear reinforcement when the concrete strength in shear is exhausted as listed in ACI318-M19 (22.5.1.1) reproduced as Equation 9.15 as below

$$V_n = V_c + V_s \tag{9.15}$$

where V_n is the nominal shear strength of the section, V_c and V_s are those contributed by concrete and steel.

Generally, shear resistance contributed by concrete arises from the shear strength of concrete in the form of diagonal tensile strength, aggregate interlock, and dowel action of the longitudinal bars. The following factors are discussed with reference to provisions in some design codes:

(i) Shear strength of concrete – Generally expressed in terms of the concrete strength as $kf_{ck}{}^n$ where f_{ck} is the compressive strength of concrete k and n are constants and both are less than one. The term generally refers to the tensile strength of concrete.
(ii) Size effect – Shear strength per unit area of the cross-section in beam a strut decreases when the sectional size increases. Some theories attribute

this phenomenon to the greater chance of crack formation and propagation in the concrete section when the section is large.

(iii) Reinforcement content – dowel action by the longitudinal bars increases shear resistance.

(iv) Axial load – axial load increases shear resistance by delaying crack formation.

9.5.2 Codified Provisions and Requirements

To meet the aforementioned factors, the following codes provide approaches and formulae for determining shear strength of concrete as follows :

(i) The Eurocode BSEN1992-1-1:2023 (2023) and BSEN1992-1-1:2004 (2004)

BSEN1992-1-1:2023 (2023) (8.20) states the minimum shear stress resistance as

$$\tau_{Rd,min} = \frac{11}{\gamma_V}\sqrt{\frac{f_{ck}}{f_{yd}}\frac{d_{dg}}{d}} \tag{9.16}$$

where γ_V is the partial factor for shear design according to table 4.3 of the Eurocode which is 1.4 for persistent and transient design situations;

f_{ck} is the characteristic strength of concrete;
f_{yd} is the design value of yield strength of longitudinal reinforcement;
d is the effective depth of the section.
d_{dg} is a size parameter describing the failure zone roughness, which depends on the concrete type and its aggregate properties. d_{dg} may be taken as:

$-16\text{mm} + D_{lower} \le 40\text{mm}$ for $f_{ck} \le 60\text{MPa}$;

$-16\text{mm} + D_{lower}\left(60/f_{ck}\right)^2 \le 40\text{mm}$ for $f_{ck} > 60\text{MPa}$.

And D_{lower} is the smallest value of the upper sieve size D in an aggregate for the coarsest fraction of aggregates in the concrete permitted by the specification of concrete according to the Eurocode BSEN206-1 (2000):2000. In case D_{max} which is the "maximum nominal upper aggregate size" known, D_{lower} can be replaced by D_{max}.

The design value of the shear stress resistance by equation 8.27 of BSEN1992-1-1:2023 (2023) is

$$\tau_{Rd,c} = \frac{0.66}{\gamma_V}\left(100\rho_l.f_{ck}.\frac{d_{dg}}{d}\right) \ge \tau_{Rd,min} \tag{9.17}$$

where $\rho_l = \dfrac{A_{sl}}{b_w d}$ with A_{sl} as the effective area of tensile reinforcement at the distance d beyond the section considered.

The Eurocode BSEN1992-1-1:2004 adopts a more conventional approach that is described as follows:

equation 6.2a of the Eurocode states

$$V_{Rd,c} = \left[C_{Rd,c} k \left(100 \rho_l f_{ck}\right)^{1/3} + k_1 \sigma_{cp}\right] b_w d \text{ (in N)} \tag{9.18}$$

where $C_{Rd,c} = 0.18 / \gamma_c$;

γ_c is the partial factor for concrete taken as 1.5 according to table 2.1N of the Eurocode;

$k = 1 + \sqrt{200 / d}$ (size effect);

$\rho_l = A_{sl} / \left(b_w d\right) \le 2.0$ as the longitudinal bar percentage;

$k_1 = 0.15$; $\sigma_{cp} = N_{Ed} / A_c$ which is the axial stress induced by external axial load; b_w : width of strut;

d : depth of strut.

(ii) The ACI318-M19 (2019) table 22.5.5.1 gives the following:

Criteria	V_c (nominal shear strength provided by concrete)	
$A_v \ge A_{v,min}$	Either of:	$\left[0.17\lambda\sqrt{f_c'} + \dfrac{N_u}{6A_g}\right] b_w d$
		$\left[0.66\lambda\left(\rho_w\right)^{1/3}\sqrt{f_c'} + \dfrac{N_u}{6A_g}\right] b_w d$
$A_v < A_{v,min}$		$\left[0.66\lambda_s\lambda\left(\rho_w\right)^{1/3}\sqrt{f_c'} + \dfrac{N_u}{6A_g}\right] b_w d$ (9.19)

Note: In ACI318-M19 (2019), nominal strength is defined as the strength of a member or cross-section calculated in accordance with the provisions and assumptions of the strength design method of the Code before the application of any strength reduction factors.

In the above table, A_v is the shear reinforcement area within spacing s,

$A_{v,min}$ is the minimum value of A_v which is given by its table 9.6.3.4 where $A_{v,min} / s$ to be the greater of $0.062\sqrt{f_c'} b_w / f_{yt}$ and $0.035 b_w / f_{yt}$ with b_w as the beam width;

f_{yt} the specified yield strength of the transverse reinforcement;

λ is the modification factor for lightweight concrete. $\lambda = 1$ for normal weight concrete;

f_c' is the specified compressive strength of concrete in N/mm^2;

N_u is the factored axial load normal to the section in N;

A_g is the gross section area in mm^2; d is the distance from the extreme compression fibre to the centroid of the reinforcement, i.e., effective depth in mm;

ρ_w is the longitudinal steel ratio $A_s / \left(b_w d\right)$;

$\lambda_s = \sqrt{\dfrac{2}{1+0.004d}} \le 1$ as size effect factor

(iii) The Code of Practice for the Structural Use of Concrete Hong Kong 2013 (2013) gives the following for v_c as the ultimate resistance shear stress in concrete in its table 6.3, which is identical to that in the former British Code BS8110-1:1997 (1997).

$$v_c = 0.79\left(\frac{100A_s}{b_v d}\right)^{1/3}\left(\frac{400}{d}\right)^{1/4}\frac{1}{\gamma_m}\left(\frac{f_{cu}}{25}\right)^{1/3} \quad (9.20)$$

where A_s is the longitudinal reinforcement;
b_v is the strut width;
d is the effective depth;
$\gamma_m = 1.25$ as the partial strength factor; f_{cu} is the cube strength.
In addition, v_c is enhanced to v_c' under axial compressive load in accordance with its 6.1.2.5(k) as

$$v_c' = v_c\sqrt{1+\frac{N}{Av_c}} \quad (9.21)$$

where N is the axial load on the section and A is the cross-sectional area.

(iv) The China Code GB50010:2010 gives the following by its equation 6.3.3-1 for shear strength by concrete under no shear reinforcement:

$$V_c = 0.7\beta_h f_t b h_0 \quad (9.22)$$

where $\beta_h = \left(\dfrac{800}{h_0}\right)^{1/4}$ with $h_0 \le 2000\,\text{mm}$

h_0 is the distance from the extreme compression fibre to the centroid of the reinforcement, i.e., effective depth;
f_t is the design tensile strength read from table 4.1.4-1,
b is the width of the beam.
By equation 6.3.4-1, for a beam with shear links, GB50010:2010 (2010) provides an allowance for axial compression and shear span ratio as

$$V \le \alpha_{cv} f_t b h_0 \quad (9.23)$$

where $\alpha_{cv} = 0.7$ under general loading but $\alpha_{cv} = \dfrac{1.75}{\lambda+1}$ when a concentrated load contributes more than 75% of the total shear of a support and λ is the shear span ratio defined as a / h_0 where a is the distance between the concentrated load and the support face or the section under consideration. In addition, $1.5 \le \lambda \le 3$.

Enhancement of V by axial load N is by $0.07N$ but limited to $N \le 0.3 f_c A$ by equation 6.3.12 listed as

$$V \le \frac{1.75}{\lambda + 1} f_t b h_0 + f_{yv} \frac{A_{sv}}{s} h_0 + 0.07N \tag{9.24}$$

The second term of Equation 9.24 is the shear capacity by shear reinforcement.

Codes generally require the use of shear reinforcement when the concrete shear strength is exceeded, which will be discussed separately in detail.

There is no explicit allowance for longitudinal steel ratio for enhancing shear capacity. Instead, GB50010-2010 (2010) implicitly considers longitudinal reinforcement through minimum shear reinforcement and shear-span ratio adjustments.

Summing up, these aforementioned codes and others generally provide concrete strengths in resisting shear in accordance with the criteria (i) to (iv) stated under the Section 9.5.1 "Shear Strength by Concrete – General."

9.5.3 Ultimate Concrete Strength in Resisting Shear

A design code often imposes a limit on shear force on a section or shear stress based on concrete strength. Exceedance beyond such a limit will require revision of the structural size. The limit is to avoid heavy stress on the concrete, which may lead to brittle failure. Such failures are in the form of local crushing of the compressive strut of the concrete member or extensive cracking leading to brittle failure.

(i) The Eurocode BSEN1992-1-1:2023 (2023) gives the following for maximum shear stress resistance for a planar member with shear reinforcement in its equation 8.35

$$\tau_{Rdc,max} = 2.15.\tau_{Rdc,0} \left(\frac{\alpha_{cs,0}}{d} \right)^{1/6} \le 2.7.\tau_{Rdc,0} \tag{9.25}$$

where $\alpha_{cs,0} = \left| \frac{M_{Ed}}{V_{Ed}} \right| \ge d$ as given by equation 8.30 of the Eurocode with M_{Ed} and V_{Ed} as the design moment and shear force in the section;

$\tau_{Rdc,0}$ is calculated by Equation 9.17.

The former version of Eurocode BSEN 1992-1-1:2004 (2004) utilizes the popular concept based on the "truss analogy" developed by Ritter (1899) and Morsch (1909) for the determination of ultimate shear capacity of a reinforced concrete beam as shown in Figure 9.16.

The truss analogy has been adopted by many codes of practice. In the Eurocode BSEN1991-1-1:2004 (2004), the maximum shear force that can be sustained by a

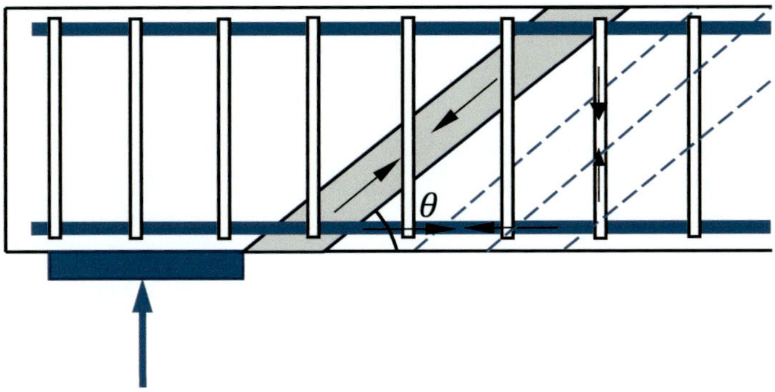

FIGURE 9.16 Truss analogy in shear design of reinforced concrete beam.

beam or column member, limited by crushing of the compression struts is given by its equation 6.9 reproduced as Equation 9.26 as below

$$V_{Rd,max} = \alpha_{cw} b_w z v_1 f_{cd} / \left(\cot\theta + \tan\theta\right) \qquad (9.26)$$

where α_{cw} is a coefficient taking account of the state of the stress in the compression chord which can be taken as one for non-prestressed member;

b_w is the width of the structural member;

z is the lever arm which can generally be taken $0.9d$ for design purpose;

v_1 is a strength reduction factor for concrete cracked in shear which is taken as 0.6 for $f_{ck} \le 60\text{MPa}$ and $0.9 - f_{ck} / 200 > 0.5$ for $f_{ck} \ge 60\text{MPa}$;

f_{cd} is the design value of concrete compressive strength of concrete;

θ is the inclination of the inclined concrete strut limited between $\tan\theta = 0.40$ and 1, i.e., 21.8° and 45°.

The expression of $1/\left(\cot\theta + \tan\theta\right)$ is actually equal to $\cos\theta\sin\theta$. $\cos\theta$ accounts for the width of the compressive strut as related to the length of the member along its longitudinal axis and $\sin\theta$ accounts for the vertical component of the inclined strut force in resisting the transverse shear to the member. So Equation 9.26 shows the maximum force that can be provided by the concrete inclined strut which is taken as the maximum shear that the concrete section can provide.

The Eurocode BSEN1991-1-1:2004 (2004) allows freedom for the designer to choose the value of θ between 21.8° and 45° (i.e., $\tan\theta = 0.4$ to 1). The greatest value of $V_{Rd,max}$ can be achieved when $\theta = 45^o$. However, the designer may choose other values to achieve economical shear link design as will be explained in the section for shear reinforcement design.

(ii) The ACI318-M19 (2019) provides an expression in its equation 22.5.1.2 as follows for requiring the least cross-sectional dimension to resist factored shear force V_u (in N) so as to minimize the likelihood of diagonal compression failure and cracking:

$$V_u \le \varnothing\left(V_c + 0.66\sqrt{f_c}\text{'}b_w d\right) \tag{9.27}$$

where$\varnothing$ is the material factor,

V_c is that determined by Equation 9.19,

f_c' is the specified compressive strength of concrete in N/mm^2,

b_w and d are respectively the beam width and effective depth in mm.

(iii) The China Code GB50010-2010 (2010) states in its equation 6.3.16 the maximum concrete shear as

$$V \le 0.25\beta_c f_c b h_0 \tag{9.28}$$

where f_c is in the concrete design compressive strength obtained from table 4.1.4-1 of the Code;

β_c is the concrete strength index which is 1 for concrete grade not greater than C50 and $\beta_c = 0.8$ upon reaching C80. Interpolation can be carried out for concrete between C50 and C80.

(iv) The Hong Kong Concrete Code 2013 (2013), following the former British Code BS8110-1:1997 (1997) gives, in its equation 6.1.2.5(a), a value of maximum shear strength v_{tu} as related to the cube strength of concrete f_{cu} in N/mm^2.

$$v_{tu} = 0.8\sqrt{f_{cu}} \le 7\text{N/mm}^2 \tag{9.29}$$

9.5.4 Shear Reinforcement Design

When the concrete design strength is exceeded, shear reinforcement has to be added. The following contains brief descriptions of provisions and requirements of four codes:

(i) The Eurocode BSEN1992-1-1:2023 (2023) and BSEN1992-1-1:2004 (2004) are essentially identical. BSEN1992-1-1:2004 (2004) in its equation 6.13 states,

$$V_{Rd,s} = \frac{A_{sw}}{s} z f_{ywd}\left(\cot\theta + \cot\alpha\right)\sin\alpha \tag{9.30}$$

where $V_{Rd,s}$ is the design shear force provided by shear reinforcement;

A_{sw} is the reinforcement area across a section;

s is the link spacing;

z is the lever arm;

f_{ywd} is the strength of the shear reinforcement;

θ is defined in Figure 9.17;

α is the angle of inclination of the link to the longitudinal axis of the beam.

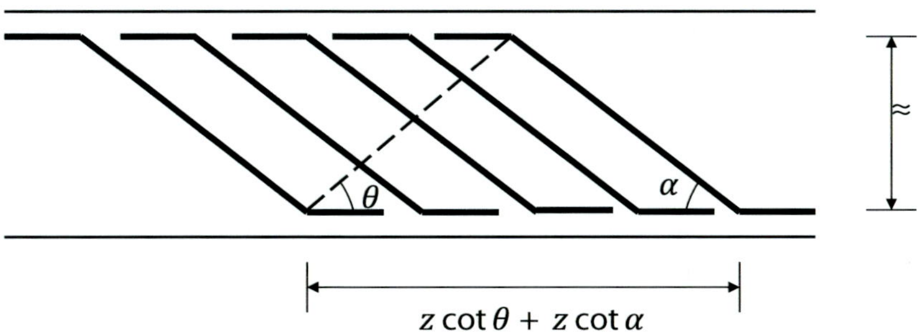

FIGURE 9.17 Proof of Equation 9.30.

The formula can be proved as follows with reference to Figure 9.17.

In Figure 9.17, the dotted line represents the failure surface at an angle θ.

So, the total number of links across the failure surface is $\left(z\cot\theta + z\cot\alpha\right)/s = z\left(\cot\theta + \cot\alpha\right)/s$.

The total link area is therefore $A_{sw}z\left(\cot\theta + \cot\alpha\right)/s$.

The total force in the links at angle α is $A_{sw}f_{ywd}z\left(\cot\theta + \cot\alpha\right)/s$.

As only the vertical component is effective in resisting the vertical shear, the force is therefore $\frac{A_{sw}}{s}f_{ywd}z\left(\cot\theta + \cot\alpha\right)\sin\alpha$ as given in Equation 9.30

In the particular case when $\alpha = 90^o$, Equation 9.30 is reduced to $\frac{A_{sw}}{s}f_{ywd}z\cot\theta$.

So, when $\theta = 21.8^o$ as the minimum value allowed by the Eurocode, $\frac{A_{sw}}{s}$ becomes the minimum required shear reinforcement for a constant value of $V_{Rd,s}$.

(ii) The ACI318-M19 gives in its equation 22.5.8.5.4 as

$$V_s = A_v f_{yt}\left(\cos\alpha + \sin\alpha\right)d/s \tag{9.31}$$

under the assumption that the failure inclined plane is at 45° to the longitudinal axis of the member.

α is the angle of the stirrups made with the longitudinal axis of the member;
d is the effective depth;
A_v is area of all bars legs of each stirrup;
s is the spacing of stirrups;
f_{yt} is the tensile strength of the stirrup.

(iii) The China Code GB50010 [3] gives the following for vertical links in its equation 6.3.12

$$V_s = \frac{A_v}{s}f_{yv}h_0 \tag{9.32}$$

where h_0 is the effective depth

A_v is area of all bars legs of each stirrup;
s is the spacing of stirrups;
f_{yv} is the tensile strength of the stirrup

(iv) The Hong Kong Concrete Code 2013 (2013) in its equation 6.20 gives practically the same expression as the Eurocode for shear resistance provided by reinforcement

$$V_b = A_{sb}\left(0.87 f_{yv}\right)\left(\cos\alpha + \sin\alpha \cot\beta\right)\left(d - d'\right)/ s_b; \tag{9.33}$$

where A_{sb} is the cross-sectional area of the shear reinforcement;

f_{yv} is the strength of the shear reinforcement with 0.87 as the partial material strength factor;
α is the inclination of the shear reinforcement;
β is the inclination of the shear failure surface.
$(d-d)'$ is the effective depth minus the concrete cover to the reinforcement at the opposite surface.

9.5.5 Shear Capacity of Beam Close to Support

Shear capacity of beam close to support is generally stronger because shear failure if occurs will take place at a greater angle with the longitudinal axis. In addition, some of the loads close to the support will be directly transferred to the support without incurring shears on the beam. Some codes of practice often allow shear strength enhancements or shear force reductions described as follows:

(i) The Eurocode BSEN1992-1-1:2004 (2004), 6.2.1(8) states that for members subject to predominantly uniformly distributed loading, the design shear force need not be checked at a distance less than d from the face of the support. However, any shear reinforcement required should continue to the support. In addition, it should be verified that the shear at the support does not exceed $V_{Rd,max}$. Furthermore, 6.2.2(6) for beams without shear reinforcement states, "For members with loads applied on the upper side within a distance $0.5d \le a_v \le 2d$ from the edge of a support (or centre of bearing where flexible bearings are used), the contribution of this load to the shear force V_{Ed} may be multiplied by $\beta = a_v / 2d$. This is only valid provided that the longitudinal reinforcement is fully anchored at the support. For $a_v \le 0.5d$ the value $a_v = 0.5d$ should be used."

In addition, for beams with shear reinforcement, 6.2.3(8) of the Eurocode states similarly as, "For members with loads applied on the upper side within a distance

$0.5d \leq a_v \leq 2d$ the contribution of this load to the shear force V_{Ed} may be reduced by $\beta = a_v / 2d$." The Eurocode BSEN1992-1-1:2004 (2004) considers that the reduced portion of loads are directly transferred to the support without contributing shear to the beam. The newer version BSEN1992-1-1:2023 (2023) adopts similar principles.

(ii) The ACI318-M19 (2019) 9.4.3.2 permits the critical section for shear to be taken at a distance *d* (effective depth) from the face of the support if (1) the support reaction introduces compression into the beam; (2) the loads are near the top surface of the beam; and (3) no concentrated load occurs *d* from the support face. This allows a higher shear strength calculation close to the support due to arching action.

(iii) The China Code GB50010-2010 (2010) allows shear capacity of a rectangular, T or I section to be enhanced under shear span ratio, which has been discussed in Equations 9.22 and 9.23.

(iv) The Hong Kong Concrete Code 2013 (2013) 6.1.2.5(g) similarly considers that shear failure at sections of beams and cantilevers without shear reinforcement will normally occur on a plane inclined at an angle of about 30° to the horizontal. If the angle of the failure plane is forced to be inclined more steeply than this, the shear force required to produce failure is increased by increasing the design concrete shear stress v_c to $2dv_c / a_v$ where *d* is the effective depth, a_v is the length of that part of a member traversed by a shear failure plane to the support face, provided that the shear stress at the face of the support remains less than the lesser of $0.8\sqrt{f_{cu}}$ or 7.0 N/mm² (this limit includes a $\gamma_m = 1.25$.) 6.1.2.5(g) further adds that the enhancement can be taken in any situation where the section considered is closer to the face of a support or concentrated load than twice the effective depth, *d*.

9.5.6 Biaxial Shear

When there are shears in the two orthogonal directions of a beam, generally the principle of "reciprocal formula" may apply such that

$$\left(\frac{V_x}{V_{Rx}}\right)^{\alpha} + \left(\frac{V_y}{V_{Ry}}\right)^{\alpha} \leq 1 \tag{9.34}$$

where V_x, V_y are the applied shear in the two orthogonal directions X and Y; V_{Rx}, V_{Ry} are the resistance of the shear of the section to the respective directions; α is a power coefficient generally greater than one. In the absence of further justification, it may be conservatively taken as one.

The ACI318M-19 [4] 22.5.1.10 gives more detailed provisions by its equation 22.5.1.10 a and b

$$\frac{v_{u,x}}{\phi v_{n,x}} \leq 0.5 \quad \text{and} \quad \frac{v_{u,y}}{\phi v_{n,y}} \leq 0.5 \tag{9.35}$$

where $v_{u,x}$, $v_{u,y}$ are the maximum factored shear stresses;

$v_{n,x}$, $v_{n,y}$ are the nominal shear stresses;

ϕ is the partial factor for material strength.

If Equation 9.35 is satisfied, the interaction of shear forces can be neglected, i.e., the design can be separately done for each direction with consideration of the other.

If Equation 9.35 is not satisfied, the following equation has to be satisfied.

$$\frac{v_{u,x}}{\phi v_{n,x}} + \frac{v_{u,y}}{\phi v_{n,y}} \leq 1.5 \tag{9.36}$$

The China Code GB50010-2010 (2015) 6.3.17 gives

$$V_x \leq \frac{V_{ux}}{\sqrt{1+\left(\frac{V_{ux}\tan\theta}{V_{uy}}\right)^2}} \quad \text{and} \quad V_y \leq \frac{V_{uy}}{\sqrt{1+\left(\frac{V_{uy}\tan\theta}{V_{ux}}\right)^2}} \tag{9.37}$$

V_x, V_y are the factored applied shears;

$$\theta = \tan^{-1}\left(V_y / V_x\right);$$

V_{ux}, V_{uy} are the shear strengths in the X and Y directions calculated by

$$V_{ux} = \frac{1.75}{\lambda+1} f_t b h_0 + f_{yv}\frac{A_{svx}}{s} h_0 + 0.07N \text{ and}$$

$$V_{uy} = \frac{1.75}{\lambda+1} f_t b b_0 + f_{yv}\frac{A_{svy}}{s} b_0 + 0.07N \tag{9.38}$$

where f_t is the concrete design tensile strength;

b_0 and h_0 are the width and depth of the beam respectively;

A_{svx} and A_{svy} are the effective link cross-sectional areas for resisting shears in the X and Y directions;

N is the axial load in the beam.

9.6 DESIGN FOR TORSION

In accordance with the Eurocode BSEN 1992-1-1:2004 (2004), design for torsion in a structural member of a building is required when the static equilibrium of a structure depends on the torsional resistance of the structural member. A similar statement is also found in the Hong Kong Concrete Code 2013 (2013) and the former BS8110-1:1997 (1997) which add that specific calculations for torsion are not usually necessary in normal slab-and-beam or framed construction. Nevertheless, these codes require certain minimum reinforcements in the form of stirrups and longitudinal bars to prevent excessive cracking.

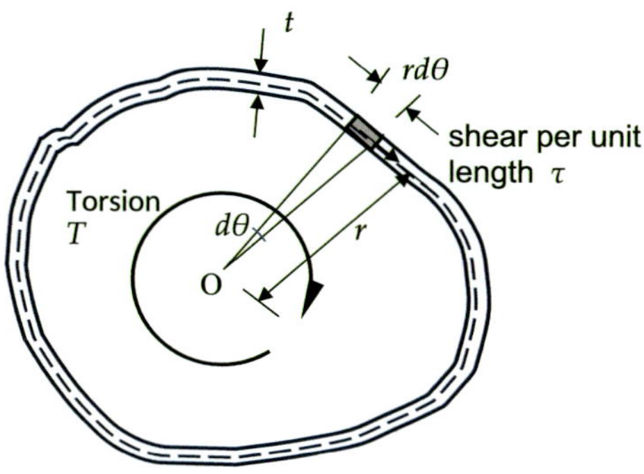

The shear force by the shaded element is $\tau\,(rd\theta)$.

The contribution of the element in resisting Torsion is $dT = \tau\,(rd\theta)\,r = \tau . r^2 d\theta \Rightarrow T = \tau \int r^2 d\theta$

As $r^2 d\theta$ is twice of the area of the triangle formed by O and the mid-points of the ends of the element, $\int r^2 d\theta$ is twice the area enclosed by the mid-line of the cell as shown by the dotted line. $\int r^2 d\theta = 2A$

$\tau = 2T/A$. So the shear stress is $\tau_s = \frac{\tau}{t} = \frac{2T}{At}$

FIGURE 9.18 Derivation of shear stress on thin-celled structure.

9.6.1 General Theory

Where the design of torsion in a beam is necessary, the design of a reinforced concrete solid beam against torsion may be based on the derivation of torsional shear stress on a thin wall section as shown in Figure 9.18. The derivation of the formula for the torsional shear stress in a thin-wall structure is shown in the figure. For equilibrium, the shear flow per metre run along the thin wall is constant at τ so that the shear stress is $\tau_s = \tau / t$ where t is the thickness of the thin wall.

While a reinforced concrete beam resists shear in forms of strut-and-tie truss as shown in Figure 9.16, the behaviour of a reinforced concrete beam under torsion is similar in 3-dimensions to a "space truss" as shown in Figure 9.19. The space-truss analogy as the first theory for determining torsional strength of reinforced concrete members was proposed by Rausch (1929) and further developed by fellow researchers. This theory assumes that the behaviour of a solid concrete beam under a torsion

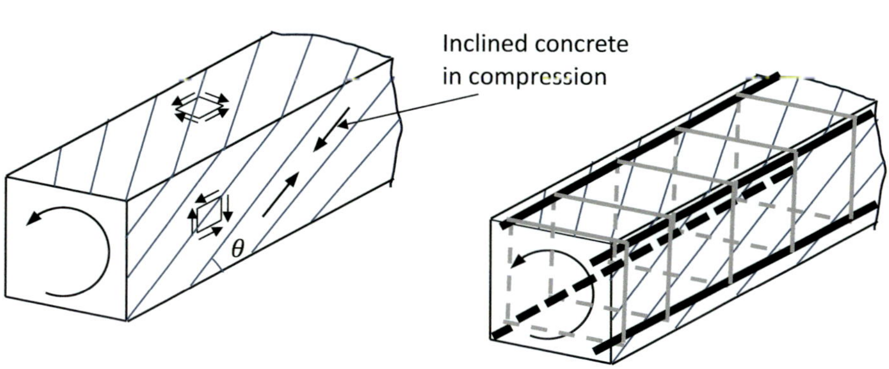

FIGURE 9.19 Torsion on a reinforced concrete beam.

was similar to a thin-walled box beam with a constant shear flow in the wall cross-section as derived in the foregoing. This assumption was proved to be correct by later experiments which confirm that, once cracking has occurred, the central core of the concrete members has little contribution in resisting torsional shears while the outer part as a "tube" structure produces a closed shear flow to resist the torsion. It will be shown in the following sections that some design codes have utilized this assumption in deriving torsional strength of reinforced concrete beams. In addition, after cracking, the tube is idealized as a space truss consisting of closed links, longitudinal steel bars in the corners, and concrete compression diagonals inclined at an angle θ and approximately centred on the links as shown in Figure 9.19.

9.6.2 Code Design Requirements

(i) Design requirements by the Eurocode BSEN 1992-1-1:2004 (2004) (BSEN 1992-1-1:2023 (2023) similar)

The strength of a reinforced concrete beam against torsion without reinforcement can be determined by

$$T_{Rd,c} = f_{ctd} \cdot 2A_k t_{ef} \tag{9.39}$$

where f_{ctd} is the design tensile strength of concrete; A_k is the area enclosed by the centrelines of the "walls"; t_{ef} is the effective wall thickness which may be taken as A/u with A as the area of the cross-section and u as the perimeter of the section.

But t_{ef} should not be less than twice the distance between edge and centre of longitudinal reinforcement. Figure 9.20 illustrates the determination of A_k and t_{ef} for a rectangular beam. So BSEN 1992-1-1:2004 (2004) follows the principle of simulating thin wall assumption in torsion discussed in the previous paragraph.

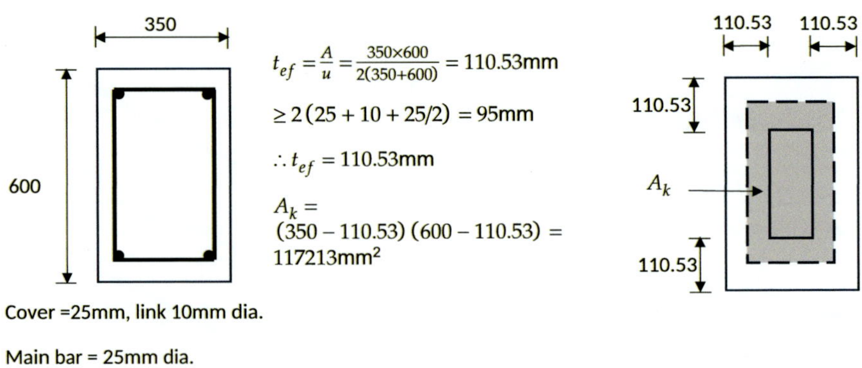

FIGURE 9.20 Determination of A_k and t_{ef} for torsion in rectangular beam.

Combination with action due to vertical shear should be carried out and checked as by equation 6.29 of the Eurocode reproduced as Equation 9.40 below

$$T_{Ed} / T_{Rd} + V_{Ed} / V_{Rd} \leq 1 \tag{9.40}$$

If Equation 9.40 is not satisfied, shear reinforcement has to be added; otherwise, minimum reinforcement can be used.

The maximum torsional resistance of the beam adopts similar principles by its equation 6.30 reproduced as in Equation 9.41. While checking if shear reinforcement is required is based on the concrete's tensile strength, check on the maximum torsional resistance, where exceedance would lead to a revision of the section, is based on the concrete's compressive crushing strength as:

$$T_{Rd,max} = 2v\alpha_{cw} f_{cd} A_k t_{ef} / \left(\cot\theta + \tan\theta\right) \tag{9.41}$$

where the definition of α_{cw}, f_{cd} can be referred to Equation 9.26, v is the concrete strength reduction factor due to cracking taken as $0.6\left(1 - f_{ck} / 250\right)$with f_{ck} in N/mm^2, A_k and t_{ef} are defined as for Equation 9.39).

BSEN1992-1:2004 (2004) requires longitudinal bar $\sum A_{sl}$ of strength f_{yd} once the applied torsion $T_{Ed} > T_{Rd,c}$ by Equation 9.42

$$\frac{\sum A_{sl} f_{yd}}{u_k} = \frac{T_{Ed}}{2A_k} \cot\theta \tag{9.42}$$

where u_k is the perimeter of the area A_k.

Since BSEN1992-1-1:2004 (2004) hasn't provided formula for torsional links, the formula from Eurocode 2 is used as follows, which is identical to equation 8.82 of BSEN1992-1-1:2023 (2023) after conversion from torsional stress to torsion.

$$\frac{A_{sw}}{s} = \frac{T_{Ed}}{2A_k f_{yd} \cot\theta} \tag{9.43}$$

(ii) Design Requirements by the ACI318M-19 (2019)

ACI318M-19 (2019) provides similar formulae based on the "thin wall" theory in its table 22.7.4.1(a) as threshold torsion for reinforcement by solid sections as follows:

$$T_{th} = 0.083\sqrt{f_c'}\left(\frac{A_{cp}^{\ 2}}{P_{cp}}\right) \tag{9.44}$$

where A_{cp} is the area enclosed by the perimeter and P_{cp} is the perimeter of the section. A_{cp} includes the area of any voids within the perimeter.

And table 22.7.4.1(b) is for hollow sections:

$$T_{th} = 0.083\sqrt{f_c'}\left(\frac{A_g^{\ 2}}{P_{cp}}\right) \tag{9.45}$$

where A_g is the gross area of concrete section (the hollow part is not included).

By (22.7.6.1(a) and 22.7.6.1(b)), T_n as the nominal torsional moment strength should be the lesser of

$$T_n = \frac{2A_o A_t f_{yt}}{s} \cot\theta \tag{9.46}$$

$$T_n = \frac{2A_o A_l f_{yt}}{p_h} \tan\theta \tag{9.47}$$

where A_o is the gross area enclosed by the torsional shear flow path to be determined by analysis, A_t is the area of a single leg of the torsional link, s is the link spacing, θ should not be taken less than 30° or greater than 60°, A_l is the area of the longitudinal bar, p_h is the perimeter of the centreline of the outermost link.

The ACI318M-19 (2019) also imposes in its equations 22.7.7.1a and 22.7.7.1b when shear and torsion are in coexistence to require adequate structural sizes:

(1) For solid sections:

$$\sqrt{\left(\frac{V_u}{b_w d}\right)^2 + \left(\frac{T_u}{1.7A_{oh}^{\ 2}}\right)^2} \le \varnothing\left(\frac{V_c}{b_w d} + 0.66\sqrt{f_c'}\right) \tag{9.48}$$

(2) For hollow section:

$$\left(\frac{V_u}{b_w d}\right)^2 + \left(\frac{T_u}{1.7A_{oh}^{\ 2}}\right)^2 \le \varnothing\left(\frac{V_c}{b_w d} + 0.66\sqrt{f_c'}\right) \tag{9.49}$$

(iii) Design Requirements by the China Code GB50010-2010 (2010).

The China Code GB50010-2010 (2010) has indicated torsional capacity of a beam in its equation 6.4.6-1 reproduced as Equation 9.50.

$$T \le \left(0.35 f_t + 0.07\frac{N}{A}\right) W_t + 1.2\sqrt{\zeta}\, f_{yv} \frac{A_{st1} A_{cor}}{s} \tag{9.50}$$

where $\zeta = \dfrac{f_y A_{stl} s}{A_{st1} u_{cor}}$, f_t is the concrete tensile strength;

$W_t = b^2(3h - b)/6$ for rectangular section with b and h as width and depth of beam;

ζ is the ratio of the strength of longitudinal reinforcement to link reinforcement with $0.6 \le \zeta \le 1.7$;

A_{stl} is the area of the longitudinal reinforcement;

A_{st1} is the single leg area of the stirrup reinforcement;
A_{cor} is the perimeter of the core of the beam enclosed by the links;
u_{cor} is the perimeter of the inner core which is $2(b_{cor}+h_{cor})$ for rectangular section;
s is the link spacing,
N is the design axial load not to be taken as greater than $0.3f_cA$ and A is the cross-sectional area.

Combining shear V and torsion T, structural size for rectangular, T- and I-section and hollow section should be adequate by its equation 6.4.1-1 as

$$\frac{V}{bh_0}+\frac{T}{0.8W_t}\leq 0.25\beta_c f_c \quad \text{for} \quad h_w/b\leq 4\left(\text{or}\, h_w/t_w\leq 4\right) \tag{9.51}$$

$$\frac{V}{bh_0}+\frac{T}{0.8W_t}\leq 0.2\beta_c f_c \quad \text{for} \quad h_w/b\leq 4\left(\text{or}\, h_w/t_w\leq 4\right) \tag{9.52}$$

where h_w, b and t_w are defined in Figure 9.21 extracted from GB50010 section 6.4.1 with $b=2t_w$ for hollow section.

where f_c and β_c carry the same definitions as those after Equation 9.28.

GB50010-2010 (2015) allows torsion to be ignored under combined actions with shear and bending as in its equation 6.4.2-2.

$$\frac{V}{bh_0}+\frac{T}{W_t}\leq 0.7f_t+0.07\frac{N}{bh_0} \tag{9.53}$$

where N is the applied axial force acting in coexistence with V and T. N is limited to $0.3f_cA$ where A is the cross-sectional area.

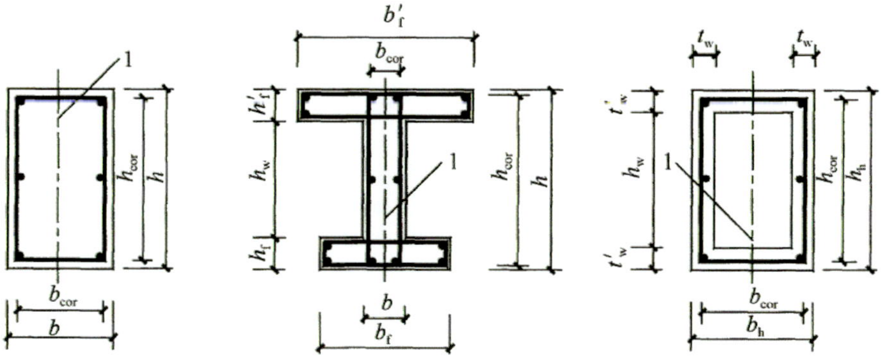

FIGURE 9.21 Dimension definition for torsion.

(iv) Hong Kong Concrete Code 2013 (2013)

The Hong Kong Concrete Code 2013 (2013) again follows the former British Code BS8110-1:1997 (1997) in calculating shear stress v_t due to torsion T by its equation 6.65 on a rectangular section of dimensions h_{max} and h_{min} which are the greater and lesser dimensions of the section.

$$v_t = \frac{2T}{h^2\left(h_{max} - \frac{h_{min}}{3}\right)} \tag{9.54}$$

If v_t exceeds $0.067\sqrt{f_{cu}}$, torsional shear reinforcement is required.

v_t obtained will be added to v as the shear stress due to vertical shear and checked. If $v + v_t$ exceeds $0.8\sqrt{f_{cu}}$, section has to be revised.

Torsional shear reinforcement required as given in its equations 6.67 (for stirrups) and (6.68) (for longitudinal bars) of the Hong Kong Concrete Code reproduced as Equations 9.55 and 9.56 as follows :

$$\frac{A_{sv}}{s_v} > \frac{T}{0.8x_1y_1\left(0.87f_{yv}\right)} \tag{9.55}$$

$$A_{sl} > \frac{A_{sv}f_{yv}\left(x_1 + y_1\right)}{s_v f_y} \tag{9.56}$$

where T is the applied torsion

A_{sv} is the cross-sectional of the outermost stirrups (two legs);
s_v is the spacing of the stirrups;
x_1 and y_1 are the lateral dimensions of the stirrup;
f_{yv} is the strength of the stirrup, with 0.87 as the partial material strength factor;
f_y is the strength of the longitudinal bar;
A_{sl} is the total cross-sectional area of the longitudinal bars.

The proof of Equation 9.55 is extracted from figure 3.33 of Reinforced Concrete Design Manual of the Housing Department Hong Kong (2013) as in Figure 9.22.

In Figure 9.22, the failure surface extends for a distance $\left(x_1 + y_1\right)$ from the location where the torsion acts. $x_1$and y_1 are the lateral dimensions of the closed. The stirrups in both $x_1$and y_1 provide a torsional resistance of $0.5A_{sv}0.87f_{yv}x_1y_1 / s_v$ and therefore give a total torsion of $A_{sv}0.87f_{yv}x_1y_1 / s_v$.

For the longitudinal reinforcement A_{sl}, Equation 9.56

$\Rightarrow A_{sl}f_y\left(1\right) = 0.5A_{sv}f_{yv}2\left(x_1 + y_1\right)\frac{1}{s_v}$, so

$$A_{sl}f_y = 0.5A_{sv}\frac{f_{yv}}{f_y}2\left(x_1 + y_1\right)\frac{1}{s_v} \tag{9.57}$$

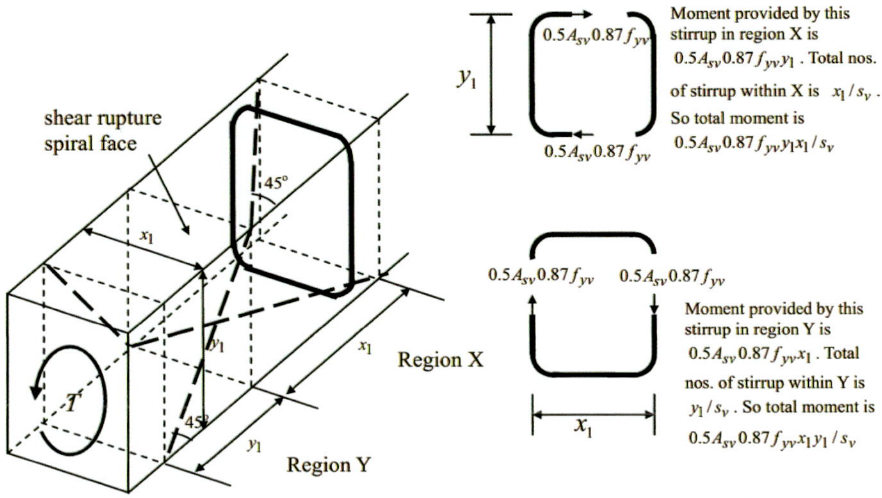

FIGURE 9.22 Proof of design formula of the Hong Kong Concrete Code [5] for stirrups against torsion.

which means that a unit length of $A_{sl}f_y$,i.e., volume of longitudinal bar times its reinforcement strength is equal to the total volume of stirrups times the stirrup reinforcement strength within a unit length.

9.7 CONCLUDING REMARKS

The principles of sectional reinforcement design for out-of-plane axial load and moments and in-plane shears and torsional moments have been discussed in the foregoing, with reference to requirements and provisions by a number of reputable national and international codes of practice. Similarities were found among these codes of practice as they are based on the same or similar engineering principles.

REFERENCES

American Concrete Institute ACI Committee 318 (2019) Building Code Requirements for Structural Concrete ACI 318-08. American Concrete Institute.

American Concrete Institute ACI Committee 318 (2019) Building Code Requirements for Structural Concrete ACI 318M-19. American Concrete Institute.

British Standard Institution BSI (1985) BS 8110-2:1985 Structural Use of Concrete. British Standard Institution.

British Standard Institution (1997) BS 8110-1:1997 Structural Use of Concrete. British Standard Institution.

British Standard Institution BSI (2000) *Eurocode 2: Concrete – Concrete — Part 1: Specification, Performance, Production and Conformity BSEN206-1:2000*. British Standard Institution.

British Standard Institution BSI (2004) *Eurocode 2: Design of Concrete Structures Part 1-1: General Rules and Rules for Buildings BSEN1992-1-1:2004*. British Standard Institution.

British Standard Institution BSI (2023) *Eurocode 2: Concrete – Specifications, Performance, Production and Conformity BSEN1992-1-1:2023+A2:2021*. British Standard Institution.

Hong Kong Special Administrative Region Buildings Department (2013) Code of Practice for the Structural Use of Concrete 2013. Buildings Department, HKSAR Government.

Hong Kong Special Administrative Region Hong Kong Housing Department (2013) Manual for Reinforced Concrete Analysis and Design to the Code of Practice for the Structural Use of Concrete 2013. Hong Kong Housing Department.

The Ministry of Housing and Urban-rural Development (MOHUD) and the General Administration of Quality Supervision, Inspection and Guarantine (AQSIQ) of the Peoples' Republic of China (2015) Code for Design of Concrete Structures GB 50010-2010 (2015 Edition). MOHUD/AQSIQ.

Mörsch, E. (1909) Concrete-Steel Construction (Der Eisenbetonbau), English Translation of the 3rd German Edition. New York: McGraw-Hill Book Co., 368 pp.

Rausch, E. (1929) Design of Reinforced Concrete in Torsion. Berlin: Technische Hochschule, p. 53. (in German).

Ritter, W. (1899) Die Bauweise Hennebique. Schweizerische Bauzeitung (Zürich), 2 February.

10 Reinforced Concrete Design Based on "Stresses" at Nodes

10.1 INTRODUCTION

Sectional reinforcement design principles and methods have been addressed in Chapter 9. However, such methods often rely on the assumption of linearly varying deformation across the section such as "plane section remains plane" which may not be valid when the section is large. On the contrary, numerical approaches like finite element analysis (FEA) enable the computation of stress distributions at nodal points under "fine meshes." These nodal stresses reflect localized loads acting over finite structural regions, allowing designers to conduct more precise, node-specific reinforcement designs. By adopting this node-based approach, localized stress concentrations can then be more accurately captured, facilitating both practical and optimized designs, and the approach is particularly useful in configurations with complex geometries though greater computational rigour and generally a more in-depth understanding of stress-strain fundamentals from the designer are required.

The design for planar structures including (i) 2-dimensional membrane structure can be based on 2-dimensional stress field comprising in-plane direct stresses and shear stresses; (ii) 2-dimensional plate structure for out-of-plane bending comprising bending moment and twisting moment and (iii) combination of (i) and (ii) which becomes 3-dimensional. For more advanced design, 3-dimensional stresses field comprising three direct stresses and three shear stresses can be adopted for a 3-dimensional structure. This chapter will systematically explore these methodologies, beginning with 2D membrane and plate analysis and progressing to 3D structural systems.

10.2 DESIGN FOR IN-PLANE LOAD

An in-plane structure generally comprises a "membrane" element as normally termed subject to in-plane loads, the analysis of which gives in-plane stresses comprising direct stresses and shear stresses as illustrated in Figure 10.1.

Design for the direct stress is more familiar to a structural designer by which he can directly check if the plain concrete can resist the load and add reinforcements as required by the codes being used. However, design for the shear stress is not so straightforward in a membrane structure. Theoretically, design for shear stress can

DOI: 10.1201/9781003636380-10

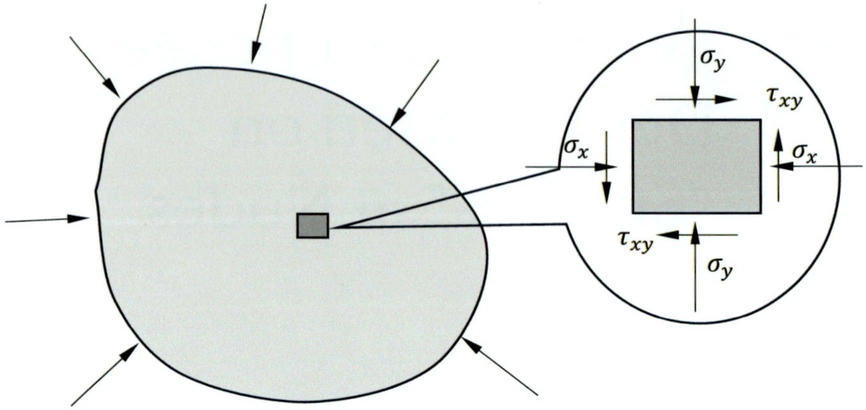

FIGURE 10.1 Illustration of an in-plane structure and the in-plane stresses.

be avoided by computing the "principal stresses" instead and designing the reinforcement to the principal stresses which are direct stresses in the principal directions. But as the principal directions change with locations and load cases, such a design method is generally not practical.

10.3 CLARK'S APPROACH

Clark (1976) has based on force equilibrium, derived equations to include direct stresses and shear stress as σ_x, σ_y and τ_{xy} simultaneously in reinforced concrete design with reinforcements in two pre-determined directions which may not be orthogonal. He has employed the assumptions that concrete does not possess tensile strength and shear is resisted by concrete alone. The equations are simpler in the case of orthogonal reinforcement which is commonly adopted in practical design. The design equations for orthogonal reinforcements by Clark are extracted in this chapter under 9 different cases as listed in Table 10.1.

Notes on the use of Table 10.1:

(i) σ_x and σ_y are the direct stresses taken as positive when tensile and negative when compressive, the direction of τ_{xy} is immaterial as it appears either as τ_{xy}^2 or $|\tau_{xy}|$ in the design equations;

(ii) ρ_x and ρ_y are the reinforcement ratios as reinforcement areas per unit area of concrete. Positive value means tensile reinforcement and negative means compressive reinforcement;

(iii) σ_1 and σ_2 are the principal stresses on concrete also taken as positive when tensile and negative when compressive;

(iv) f_s and f_s' are respectively the tensile and compressive strengths of reinforcement always with positive values which may carry the same or different numerical values;

TABLE 10.1

Reinforcement Design Requirements for In-plane Stresses in accordance with Clark (1976) in Orthogonal Directions

Case	ρ_x	ρ_y	σ_1	σ_2	$\tan\theta$
1	$\frac{1}{f_s}\left(\sigma_x+\lvert\tau_{xy}\rvert\right)$	$\frac{1}{f_s}\left(\sigma_y+\lvert\tau_{xy}\rvert\right)$	0	$-2\lvert\tau_{xy}\rvert$	$\frac{-\tau_{xy}}{\lvert\tau_{xy}\rvert}$
2	0	$\frac{1}{f_s}\left(\sigma_y-\frac{\tau_{xy}^{2}}{\sigma_x}\right)$	0	$\sigma_x+\frac{\tau_{xy}^{2}}{\sigma_x}$	$\frac{\sigma_x}{\tau_{xy}}$
3	$\frac{1}{f_s}\left(\sigma_x-\frac{\tau_{xy}^{2}}{\sigma_y}\right)$	0	0	$\sigma_y+\frac{\tau_{xy}^{2}}{\sigma_y}$	$\frac{\tau_{xy}}{\sigma_y}$
4	$\frac{1}{f_s'}\left(\sigma_{xf}-\lvert\tau_{xy}\rvert\right)$	$\frac{1}{f_s'}\left(\sigma_{yf}-\lvert\tau_{xy}\rvert\right)$	$f_c+2\lvert\tau_{xy}\rvert$	f_c	$\frac{-\tau_{xy}}{\lvert\tau_{xy}\rvert}$
5	0	$\frac{1}{f_s'}\left(\sigma_{yf}-\frac{\tau_{xy}^{2}}{\sigma_{xf}}\right)$	$\sigma_x+\frac{\tau_{xy}^{2}}{\sigma_{xf}}$	f_c	$\frac{-\tau_{xy}}{\sigma_{xf}}$
6	$\frac{1}{f_s'}\left(\sigma_{xf}-\frac{\tau_{xy}^{2}}{\sigma_{yf}}\right)$	0	$\sigma_y+\frac{\tau_{xy}^{2}}{\sigma_{yf}}$	f_c	$\frac{-\sigma_{yf}}{\tau_{xy}}$
7	$\frac{1}{f_s}\left[\sigma_x-\frac{f_c}{2}(1-\beta)\right]$	$\frac{1}{f_s'}\left[\sigma_y-\frac{f_c}{2}(1+\beta)\right]$	0	f_c	$\frac{2\tau_{xy}}{f_c(1+\beta)}$
8	$\frac{1}{f_s'}\left[\sigma_x-\frac{f_c}{2}(1+\beta)\right]$	$\frac{1}{f_s}\left[\sigma_y-\frac{f_c}{2}(1-\beta)\right]$	0	f_c	$\frac{2\tau_{xy}}{f_c(1-\beta)}$
9	0	0	$\sigma_{1(9)}$	$\sigma_{2(9)}$	$(\tan\theta)_9$

(v) f_c is the design compressive strength of concrete carrying a negative value;

(vi) $\sigma_{xf}=\sigma_x-f_c$; $\sigma_{yf}=\sigma_y-f_c$ which are the extra direct stresses over concrete compressive strength

(vii) $\beta=\sqrt{1-4\tau_{xy}^{2}/f_c^{2}}$ in cases 7 and 8

(viii) $\sigma_{1(9)}=\dfrac{\sigma_x+\sigma_y+\sqrt{(\sigma_x-\sigma_y)^2+4\tau_{xy}^{2}}}{2}$; $\sigma_{2(9)}=\dfrac{\sigma_x+\sigma_y-\sqrt{(\sigma_x-\sigma_y)^2+4\tau_{xy}^{2}}}{2}$ in case 9.

(ix) θ is the orientation of the major principal concrete stress to x axis,

$$(\tan\theta)_9=\frac{\sigma_x-\sigma_y-\sqrt{(\sigma_x-\sigma_y)^2+4\tau_{xy}^{2}}}{2\tau_{xy}}$$ in case 9.

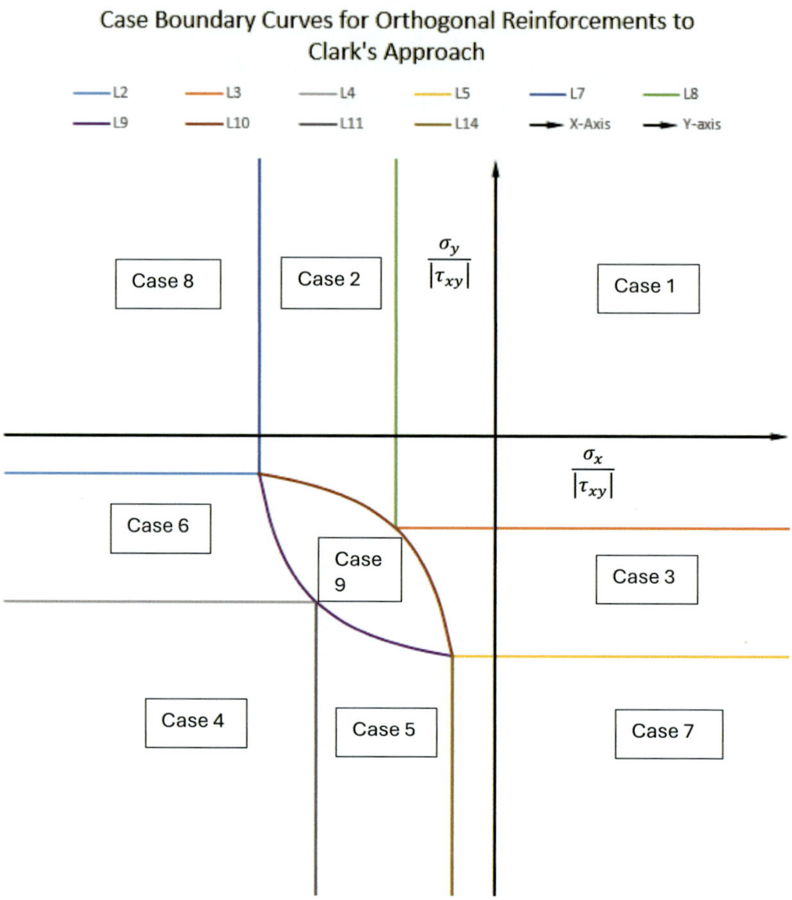

FIGURE 10.2 Case boundary graph for orthogonal reinforcement design to Clark's approach.

The different cases as referred to in Table 10.1 are under different values of $\sigma_x / |\tau_{xy}|$ and $\sigma_y / |\tau_{xy}|$ delineated by different curves on the chart as shown in Figure 10.2. The equations of the curves separating the cases are given in Table 10.2.

The derivations of Clark's equations can be shown as follows. Consider the triangular elements in Figure 10.3 which are expansion of a point in a membrane structure with stress σ_x, σ_y, τ_{xy} in the X and Y directions and principal stresses σ_1 and σ_2.

Consider (a), equilibrium in X-direction and Y-direction gives

$$\sigma_x = \rho_x f_s + \sigma_1 \cos^2\theta + \sigma_2 \sin^2\theta \tag{10.1}$$

$$\tau_{xy} = (\sigma_2 - \sigma_1)\cos\theta \sin\theta \tag{10.2}$$

TABLE 10.2
Separating Curves of the Respective Cases in Design according to Clark's Approach

Curve	Equation	Curve	Equation
L1	$\frac{\sigma_y}{\lvert\tau_{xy}\rvert} = +\infty$	L8	$\frac{\sigma_x}{\lvert\tau_{xy}\rvert} = -1$
L2	$\frac{\sigma_x}{\lvert\tau_{xy}\rvert} = \frac{1}{2}\left(\frac{f_c}{\lvert\tau_{xy}\rvert} + \sqrt{\left(\frac{f_c}{\lvert\tau_{xy}\rvert}\right)^2 - 4}\right)$	L9	$\frac{\sigma_{xf}}{\lvert\tau_{xy}\rvert}\frac{\sigma_{yf}}{\lvert\tau_{xy}\rvert} = 1$
L3	$\frac{\sigma_y}{\lvert\tau_{xy}\rvert} = -1$	L10	$\frac{\sigma_x}{\lvert\tau_{xy}\rvert}\frac{\sigma_y}{\lvert\tau_{xy}\rvert} = 1$
L4	$\frac{\sigma_y}{\lvert\tau_{xy}\rvert} = \frac{f_c}{\lvert\tau_{xy}\rvert} + 1$	L11	$\frac{\sigma_x}{\lvert\tau_{xy}\rvert} = \frac{f_c}{\lvert\tau_{xy}\rvert} + 1$
L5	$\frac{\sigma_y}{\lvert\tau_{xy}\rvert} = \frac{1}{2}\left(\frac{f_c}{\lvert\tau_{xy}\rvert} - \sqrt{\left(\frac{f_c}{\lvert\tau_{xy}\rvert}\right)^2 - 4}\right)$	L12	Inapplicable
L6	$\frac{\sigma_y}{\lvert\tau_{xy}\rvert} = -\infty$	L13	Inapplicable
L7	$\frac{\sigma_x}{\lvert\tau_{xy}\rvert} = \frac{1}{2}\left(\frac{f_c}{\lvert\tau_{xy}\rvert} - \sqrt{\left(\frac{f_c}{\lvert\tau_{xy}\rvert}\right)^2 - 4}\right)$	L14	$\frac{\sigma_x}{\lvert\tau_{xy}\rvert} = \frac{1}{2}\left(\frac{f_c}{\lvert\tau_{xy}\rvert} + \sqrt{\left(\frac{f_c}{\lvert\tau_{xy}\rvert}\right)^2 - 4}\right)$

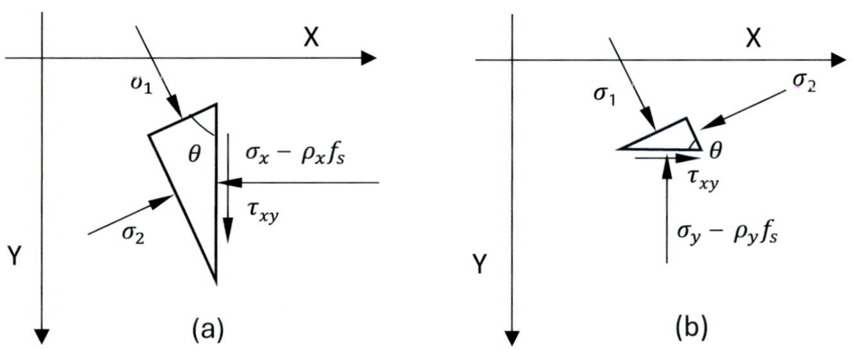

FIGURE 10.3 Proof of Clark's equations.

Similarly, for (b), equilibrium in Y-direction

$$\sigma_y = \rho_y f_s + \sigma_1 \sin^2\theta + \sigma_2 \cos^2\theta \tag{10.3}$$

The three equations are basic ones for deriving the equations in Table 10.1 under various conditions.

Consider case 1 where σ_x and σ_y are in tension, the reinforcements have to be so provided that $\sigma_1 = 0$ and σ_2 is compressive. Substituting $\sigma_1 = 0$ into Equations 10.1 to 10.3 and eliminating σ_2 by expressing it in terms of τ_{xy}, we may list

$$\begin{cases} \sigma_x = \rho_x f_s + \tau_{xy} \tan\theta \\ \sigma_y = \rho_y f_s + \tau_{xy} / \tan\theta \end{cases} \tag{10.4}$$

θ should be chosen to minimize $\rho_x + \rho_y$ which can be expressed as

$$\rho_x + \rho_y = \frac{1}{f_s}\left[\sigma_x + \sigma_y - \tau_{xy}\left(\tan\theta + \frac{1}{\tan\theta}\right)\right] \tag{10.5}$$

By setting $\frac{d}{d\theta}\left(\rho_x + \rho_y\right) = 0$ for minimization of $\rho_x + \rho_y$, $\theta = 45^o$, so

$$\begin{cases} \sigma_x = \rho_x f_s + \tau_{xy} \\ \sigma_y = \rho_y f_s + \tau_{xy} \end{cases} \Rightarrow \begin{cases} \rho_x = \frac{1}{f_S}\left(\sigma_x - \tau_{xy}\right) \\ \rho_y = \frac{1}{f_S}\left(\sigma_y - \tau_{xy}\right) \end{cases} \tag{10.6}$$

which agrees with Table 10.1.

As another example, consider case 3 where $\rho_y = 0$ and $\sigma_1 = 0$ according to Table 10.1. Substituting into Equations 10.1 to 10.3, we may list

$$\begin{cases} \sigma_x = \rho_x f_s + \sigma_2 \sin^2\theta \\ \tau_{xy} = \sigma_2 \cos\theta \sin\theta \\ \sigma_y = \sigma_2 \cos^2\theta \end{cases} \tag{10.7}$$

The task is to eliminate σ_2 and θ for solving ρ_x. This can be done by dividing the second equation by the third equation, $\frac{\tau_{xy}}{\sigma_y} = \tan\theta$ so that $\sin^2\theta = \frac{\tan^2\theta}{1 + \tan^2\theta} = \frac{\tau_{xy}^{\ 2}}{\sigma_y^{\ 2} + \tau_{xy}^{\ 2}}$ and $\cos^2\theta = \frac{1}{1 + \tan^2\theta} = \frac{\sigma_y^{\ 2}}{\sigma_y^{\ 2} + \tau_{xy}^{\ 2}}$ and $\sigma_2 = \frac{\sigma_y^{\ 2} + \tau_{xy}^{\ 2}}{\sigma_y}$. Back-substituting into the first equation, we obtain

$$\rho_x = \frac{1}{f_s}\left(\sigma_x - \left(\frac{\sigma_y^{\ 2} + \tau_{xy}^{\ 2}}{\sigma_y}\right)\frac{\tau_{xy}^{\ 2}}{\sigma_y^{\ 2} + \tau_{xy}^{\ 2}}\right) = \frac{1}{f_s}\left(\sigma_x - \frac{\tau_{xy}^{\ 2}}{\sigma_y}\right) \tag{10.8}$$

Again, the equations in Table 10.1 for case 3 are proved.

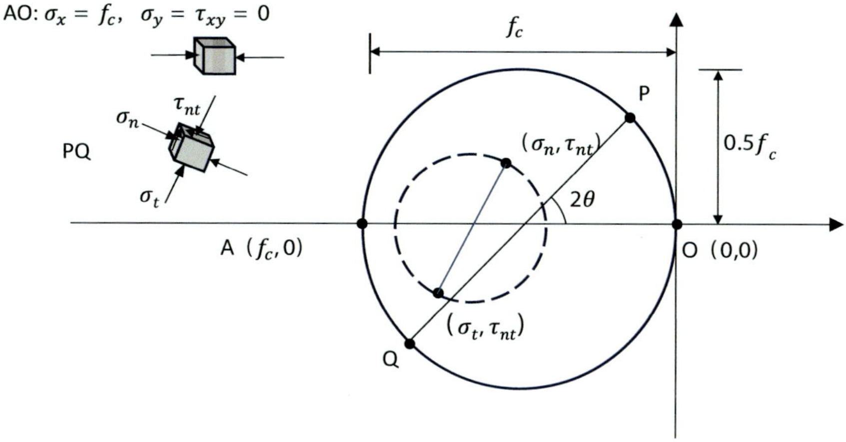

FIGURE 10.4 Construction of the strength circle of concrete.

10.4 PROOF OF CLARK'S EQUATIONS BY GRAPHICAL METHOD

Clark's equations can in fact be proven by graphical method under the consideration of Mohr's Circle as for a stress state under σ_x, σ_y, τ_{xy}, a Mohr Circle can be constructed such that the direct stresses and shear stress vary with orientation as demonstrated by the circle on the right in Figure 10.5. Subedi (1975) has some demonstrations. As illustrations, proofs of the design formulae in cases 1 and 3 are outlined as follows. Nevertheless, prior to the proofs for the two cases, a "strength circle" of concrete of compressive strength f_c is first constructed based on the concept of Mohr Circle as demonstrated in Figure 10.4.

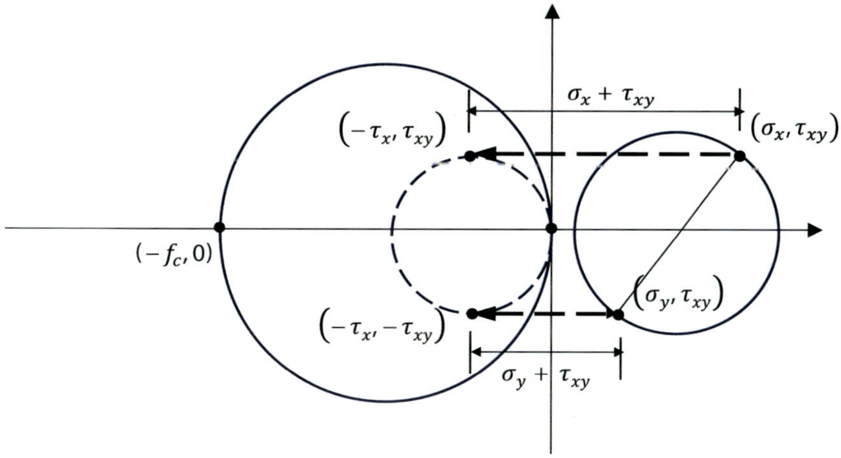

FIGURE 10.5 Diagrammatic proof of case 1.

In Figure 10.4, a stress state at $\sigma_x = f_c$, $\sigma_y = \tau_{xy} = 0$ can be denoted the nodes A and O as two diametrically opposite nodes of a Mohr Circle of stress. The stress state is at marginal failure. So, any pair of diametrically opposite nodes on this Mohr Circle represents the stress state at a different orientation in accordance with the Mohr Circle of 2-dimensional stress theory. For example, the nodes P and Q denote the stress state at an orientation of θ from the X-axis. As AO and PQ are the same stress, the Mohr Circle so constructed represents an envelope of states of marginal failure which can be termed the "strength circle" in this book or "Limiting Circle of Stress" by Subedi (1975). It then follows that any Mohr Circle within the strength circle represents a stress state which would not create failure even without reinforcement such as the dotted circle shown in Figure 10.4. Looking at another angle, the principal stresses of the dotted circle are all less than f_c. So, they are within the capacity of the concrete. Thus, in this 2-dimensional stress state design, the approach is to add reinforcements in X or Y or both or none to take up some of the stresses so that the Mohr Circle of the remaining stresses on the concrete is entirely within the strength circle of concrete. After defining the strength circle of concrete, the followings are graphical proofs of case 1 and case 3 of Clark's equations.

Case 1 – σ_x and σ_y are in tension, $\tau_{xy} \neq 0$.

Consider a stress state comprising σ_x, σ_y and τ_{xy} where σ_x and σ_y are tension. So, the Mohr Circle representing this stress state is drawn in Figure 10.5 with (σ_x, τ_{xy}) and (σ_y, τ_{xy}) located diametrically opposite. For an optimal design, if reinforcements in the X and Y directions are so added to take up some stresses that the concrete stress state becomes $(-\tau_{xy}, \tau_{xy})$ and $(-\tau_{xy}, -\tau_{xy})$ as shown, then the Mohr Circle of the concrete stress becomes the dotted circle which lies entirely within the strength circle of concrete. This ensures that the concrete can sustain the applied stresses in all directions. The stresses to be taken up by reinforcements are respectively $\sigma_x + \tau_{xy}$ and $\sigma_y + \tau_{xy}$ and the reinforcement percentages are calculated accordingly. This is case 1 in Figure 10.2.

Case 3 – σ_x in tension σ_y in compression, $\tau_{xy} \neq 0$.

Consider a stress state comprising σ_x, σ_y and τ_{xy} where σ_x is in tension and σ_y is in compression and within the concrete strength circle as shown in Figure 10.5. So, clearly, the Mohr Circle representing this stress state with (σ_x, τ_{xy}) and (σ_y, τ_{xy}) located diametrically opposite cannot be enveloped by the strength circle. So, reinforcement in the X-direction has to be added to bring the concrete stress into the concrete strength circle as shown in Figure 10.6.

To evaluate the stress required to be taken up by reinforcement, consider the radius of the dotted circle (which is the stress circle of concrete after reinforcement is added) taken as r, the following equation can be set up (bearing in mind that σ_y carries a negative value).

$$(-\sigma_y - r)^2 + \tau_{xy}^{\ 2} = r^2 \Rightarrow \sigma_y^{\ 2} + 2\sigma_y r + \tau_{xy}^{\ 2} = 0 \Rightarrow r = \frac{-1}{2}\left(\sigma_y + \frac{\tau_{xy}^{\ 2}}{\sigma_y}\right)$$

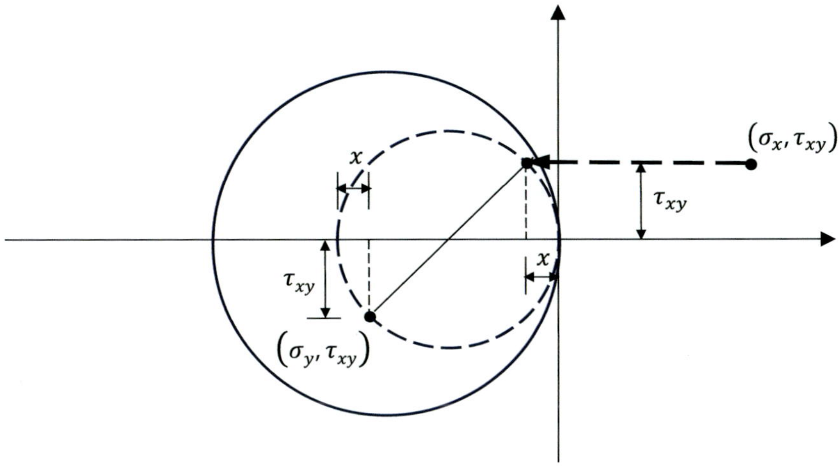

FIGURE 10.6 Diagrammatic proof of case 3.

From Figure 10.5, $x = 2r - (-\sigma_y) = 2r + \sigma_y$.

So, stress required to be taken up by reinforcement for σ_x is $\sigma_x + x$

$$\sigma_x + x = \sigma_x + 2r + \sigma_y = \sigma_x + \left(-\sigma_y - \frac{\tau_{xy}{}^2}{\sigma_y}\right) + \sigma_y = \sigma_x - \frac{\tau_{xy}{}^2}{\sigma_y}.$$

From the foregoing, it is evident that the applied shear stress cannot exceed $f_c / 2$, as otherwise it is not possible to bring the stress state into the concrete strength circle.

10.5 WORKED EXAMPLES FOR IN-PLANE STRESS

The following Worked Example 10.1 demonstrates the use of Clark's Equations as listed in Table 10.3 which comprises design examples based on $f_c = 14\text{N/mm}^2$ and reinforcement strength $f_y = 435\text{N/mm}^2$

Clark's equations can be used in the design of reinforced concrete wall structures or deep beam structures or the like which are mainly undergoing in-plane loads incurring direct and shear stresses, as revealed by analysis such as the finite element method. Nevertheless, these structures may also undergo out-of-plane loads, the design of which will be discussed in the coming sections.

10.6 DESIGN FOR OUT-OF-PLANE BENDING

10.6.1 Review of Moment Field

It is a very common practice for designing a slab with "two-way" bending moments which are out-of-plane actions. Apart from the bending moments in two mutually perpendicular directions as well-known by design engineers, a twisting moment

TABLE 10.3
Worked Example 10.1 for Clark's Equations

Example	σ_x	σ_y	τ_{xy}	σ_x/τ_{xy}	σ_y/τ_{xy}	Case	ρ_x	ρ_y
1	10	8	5	2	1.6	1	3.448	2.989
2	−6	4	5	−1.2	0.8	2	0	1.877
3	4	−6	5	0.8	−1.2	3	1.877	0
4	−10	−13	5	−2	−2.6	4	0.230	0.920
5	−5	−13	5	−1	−2.6	5	0	0.409
6	−20	−5	5	−4	−1	6	2.018	0
7	10	−15	5	2	−3	7	2.782	0.713
8	−20	10	5	−4	2	8	1.862	2.782
9	−7.5	−9	5	−1.5	−1.8	9	0	0

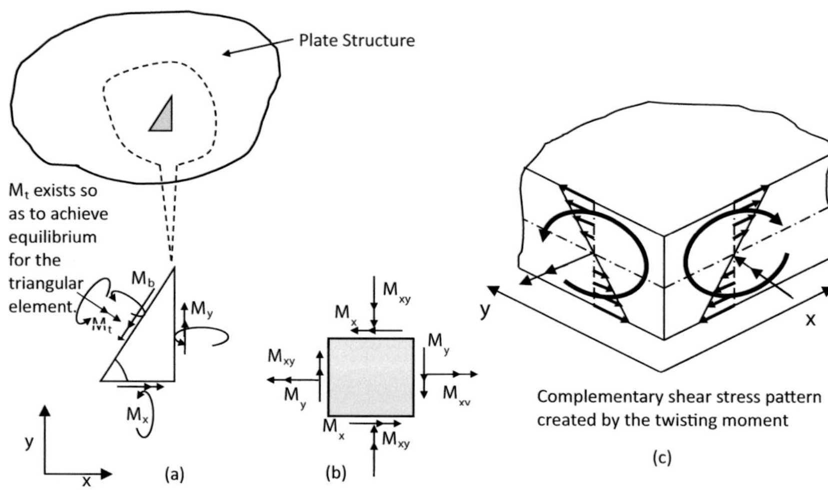

FIGURE 10.7 Derivation and nature of the "twisting moment."

which may be familiar to some engineers, can be proved to be in existence by the plate bending theory as follows.

Consider a small triangular element in a plate bending structure with two of its sides aligning with the global X and Y directions, as shown in (a) of Figure 10.7 where moments M_x and M_y (both in moment per unit width) are bending respectively about X and Y. To achieve equilibrium, a moment M_b is required to be bending about the hypotenuse. However, treating all moments as vectors, as shown in the diagram, as the resultant of M_x and M_y does not necessarily align with M_b, there will generally be a moment acting in the direction perpendicular to M_b to achieve equilibrium, which is denoted as M_t. By the normal direction of M_t to the face of the hypotenuse, M_t is producing torsion or "twisting action" on the hypotenuse. As the hypotenuse can be in any direction of the plate structure, it follows that at any point in the plate bending structure, there will generally be two bending moments, say M_x and M_y together with a twisting moment M_{xy} as shown in (b) of Figure 10.7. The phenomenon is in exact analogy to the in-plane stress problem where generally two direct stresses σ_x and σ_y coupled with a shear stress τ_{xy} exist and these components vary with directions. As the shear stress produced has to be "complementary," the shear flow on the structure has to follow (c) of Figure 10.7. Unlike shear stress in the in-plane stress problem, the shear flow has to be in opposite directions in the lower and upper parts of the plate so as to achieve equilibrium with the twisting moment. The distribution of shear stress is often taken to be linearly varying in accordance with the elastic theory. As such, the twisting moments on the two perpendicular faces are identical, which is again analogous to the shear stress in the in-plane stress problem.

10.6.2 Mathematical Presentation of "Moment Field"

A "moment field" can be described for a plate under out-of-plane bending which comprises M_x, M_y and M_{xy}. When the rectangle in Figure 10.7 shrinks to a point, its state of bending stress can be defined by the M_x, M_y and M_{xy} acting on it at a specific orientation. These values will change with orientation as shown in Figure 10.8, as similar to the in-plane stress problem.

Mathematically, the state of bending stress changes by the following equations by considering equilibrium of the triangle in Figure 10.8:

$$M_b = M_x \cos^2\theta + M_y \sin^2\theta + 2M_{xy} \cos\theta \sin\theta \tag{10.9}$$

$$M_t = \left(M_x - M_y\right)\cos\theta \sin\theta + 2M_{xy}\left(\cos^2\theta - \sin^2\theta\right) \tag{10.10}$$

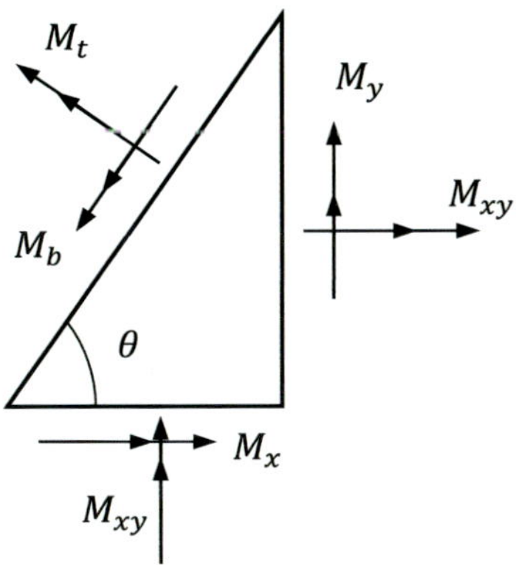

FIGURE 10.8 Resolution of bending and twisting moment.

If θ is so varied that M_t vanishes when $\theta = \phi$, then the element will be having pure bending in the direction. At the same time, at the orthogonal direction to ϕ, the twisting moment is also zero with pure bending. The two moments are termed the "principal moments" and denoted as M_1 and M_2, which again are in exact analogy with the in-plane stress problem having principal stresses at orientations where shear stresses are zero. The angle ϕ can be worked out by

$$\phi = \frac{1}{2}\tan^{-1}\left(\frac{2M_{xy}}{M_x - M_y}\right) \quad (10.11)$$

10.6.3 Wood-Armer Equations

The design approach would involve formulae for providing reinforcing steels in two directions (mostly in orthogonal directions) adequate to resist the "moment field" comprising the bending moments and twisting moments. The most popular one is that developed by Woods (1968) which are commonly referred to as the "Wood-Armer" Equations, the derivation of which is based on the "normal yield criterion." The criterion requires the provided reinforcing steel bars at any point to be adequate to resist the normal moment which is the bending moment M_b in any direction as calculated from Equation 10.9. If the reinforcement bars are adequate to provide flexural strengths along the X and Y to resist M_x^* and M_y^* respectively, then by the Johansen's Yield Criterion (1962), it can resist a bending moment of $M_x^*\cos^2\theta + M_y^*\sin^2\theta$ along an orientation θ. The Criterion can be proved as follows. Consider two sets of orthogonal reinforcements adequate to provide bending strength of M_x^* and M_y^* by unit widths as shown in Figure 10.9, the effective width in the orientation θ is $\cos\theta$. In addition, the effective strength is the component of the reinforcement in the orientation which is also $\cos\theta$. So, the strength that can be provided by the X-reinforcement is $M_x^*\cos\theta\cos\theta = M_x^*\cos^2\theta$. Similarly, the strength provided by Y-direction is $M_y^*\sin^2\theta$, giving a total strength of $M_x^*\cos^2\theta + M_y^*\sin^2\theta$.

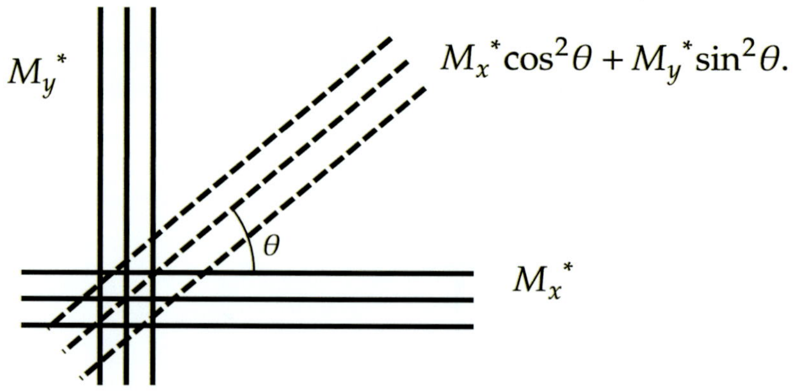

FIGURE 10.9 Diagrammatic proof of Johansen's yield criterion.

So, the design aims at calculating M_x^* and M_y^* such that the following inequality is satisfied:

$$M_x^* \cos^2\theta + M_y^* \sin^2\theta \geq M_x \cos^2\theta + M_y \sin^2\theta + 2M_{xy} \cos\theta \sin\theta \qquad (10.12)$$

Based on Equation 10.12, Wood (1968) derived the following equations for M_x^* and M_y^* listed as follows with sagging moment carrying positive sign and hogging moment carrying negative sign.

For sagging steel reinforcement provisions:

$M_x^* = M_x + |M_{xy}|$ if $M_x^* \geq 0$; $M_y^* = M_y + |M_{xy}|$ if $M_y^* \geq 0$

If $M_x^* < 0$, then $M_x^* = 0$; and $M_y^* = M_y + \left|\dfrac{M_{xy}^2}{M_x}\right|$

If $M_y^* < 0$, then $M_y^* = 0$; and $M_x^* = M_x + \left|\dfrac{M_{xy}^2}{M_y}\right|$

For hogging steel reinforcement provisions:

$M_x^* = M_x - |M_{xy}|$ if $M_x^* \leq 0$; $M_y^* = M_y - |M_{xy}|$ if $M_y^* \leq 0$

If $M_x^* > 0$, then $M_x^* = 0$; and $M_y^* = M_y - \left|\dfrac{M_{xy}^2}{M_x}\right|$

$$\text{If } M_y^* > 0, \quad \text{then } M_y^* = 0; \quad \text{and} \quad M_x^* = M_x - \left|\frac{M_{xy}^2}{M_y}\right| \qquad (10.13)$$

The following worked examples are presented for illustration:

Worked Example 10.2

$M_x = 60$; $M_y = 200$; $M_{xy} = 90$

For sagging:

$M_x^* = M_x + |M_{xy}| = 60 + 90 = 150 \geq 0$;

$$M_y^* = M_y + |M_{xy}| = 200 + 90 = 290 \geq 0$$

The results which are calculated by Equation 10.12 can be verified by the plot in Figure 10.10. In the plots, the dotted lines represent the "normal moments" given by the Wood-Armer Equations which envelop the normal moments of the "moment field."

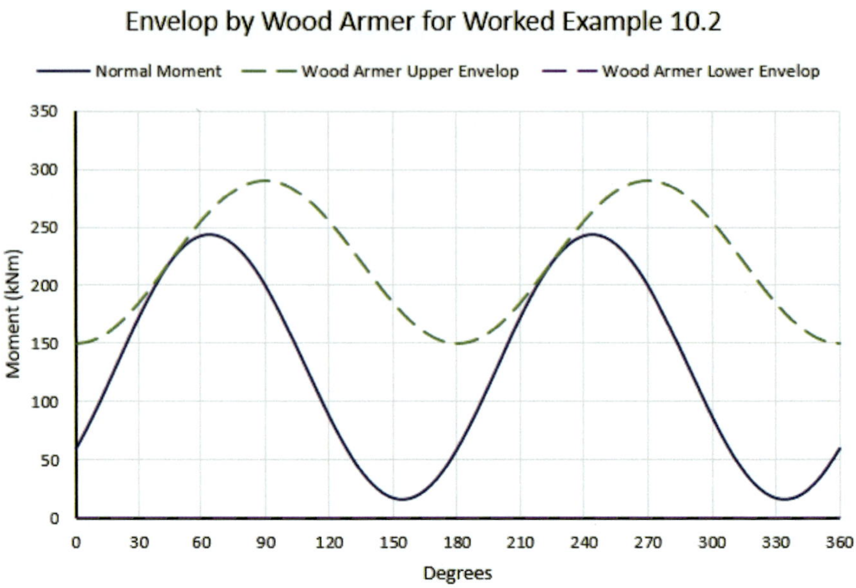

FIGURE 10.10 Plots of Worked Examples 10.2.

Worked Example 10.3

$$M_x = 80;\ M_y = -240;\ M_{xy} = 100$$

For sagging:

$$M_x^* = M_x + |M_{xy}| = 80 + 100 = 180 \geq 0;$$

$$M_y^* = M_y + |M_{xy}| = -240 + 90 = -150 \leq 0$$

So $M_y^* = 0$ and $M_x^* = M_x + \left|\frac{M_{xy}{}^2}{M_y}\right| = 80 + \left|\frac{100^2}{-240}\right| = 121.67$

For hogging:

$$M_x^* = M_x - |M_{xy}| = 80 - 100 = -20 \leq 0;$$

$$M_y^* = M_y - |M_{xy}| = -240 - 100 = -340 \leq 0$$

Summing up: for sagging $M_x^* = 121.67$, $M_y^* = 0$;

for hogging $M_x^* = -20$, $M_y^* = -340$

Plots similar to Worked Example 10.2 are presented in Figure 10.11

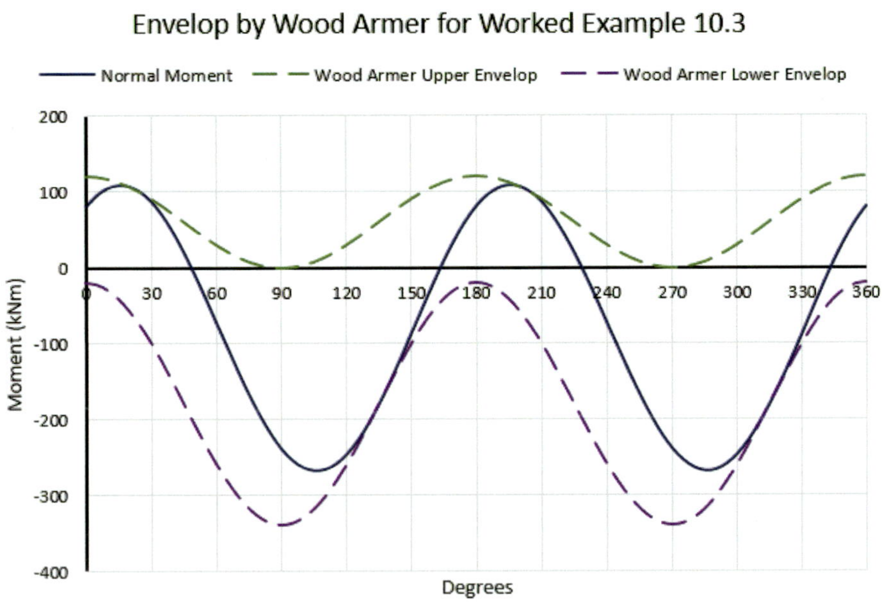

FIGURE 10.11 Plot of Worked Examples 10.3.

10.7 OUT-OF-PLANE SHEAR

In co-existence of the bending and twisting moments, there is generally an out-of-plane shear. As similar to the line element simulating a beam where moment M and shear S can be related by $S = \frac{dM}{dx}$ with x is the distance along the beam, out-of-plane shears (in terms of shear per unit width) in two orthogonal directions X and Y in a plate structure as S_{xz} and S_{yz} can be derived by

$$\begin{cases} S_{xz} = \dfrac{\partial M_x}{\partial x} + \dfrac{\partial M_{xy}}{\partial y} \\ S_{yz} = \dfrac{\partial M_y}{\partial y} + \dfrac{\partial M_{xy}}{\partial x} \end{cases} \tag{10.14}$$

where M_x, M_y are the bending moments along the X and Y axes and M_{xy} is the twisting moments. The differential equations are derived from the equilibrium of forces in accordance with the infinitesimal rectangular element shown in Figure 10.12.

It can readily be seen that the determination of S_{xz} and S_{yz} are based on the orientations of the coordinate axes X and Y. Their values obviously would change with orientations. Consider again a triangular element cut from a plate structure as shown in Figure 10.13 where shear (force per unit width) of S_{xz} at face parallel to Y-axis and S_{yz} at face parallel to X-axis are acting. To achieve equilibrium, a shear of $S_{\theta z}$ acts along the hypotenuse at an angle θ as shown.

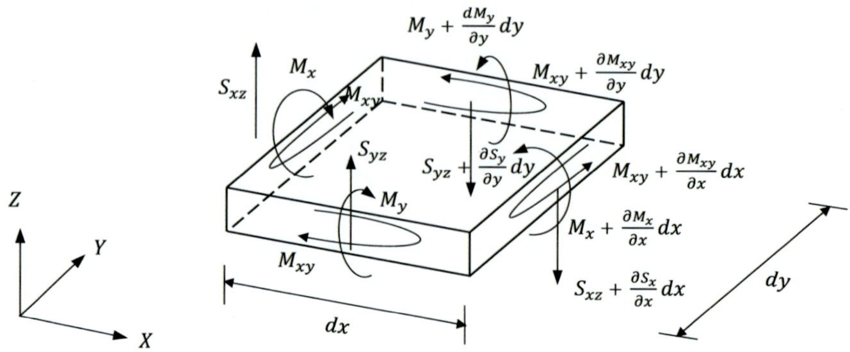

FIGURE 10.12 Diagrammatic illustration of the derivation of the out-of-plane shear.

If the length of the hypotenuse is unity, the equation for balance of force can be written as

$$S_{\theta z} = S_{xz} \sin\theta + S_{yz} \cos\theta \tag{10.15}$$

To find the θ such that $S_{\theta z}$ is maximum at S_{max}, set $\frac{d}{d\theta} S_{\theta z} = 0$,

$$\theta = \tan^{-1}\left(\frac{S_{xz}}{S_{yz}}\right) \tag{10.16}$$

$$S_{max} = \sqrt{S_{xz}^{\ 2} + S_{yz}^{\ 2}} \tag{10.17}$$

So, by the above, the maximum out-of-plane shear S_{max} can be found as the square root of sum of squares of the shears at any two orthogonal faces, and this S_{max} is the shear force to be designed for accordingly. In addition, the orientation identified by Equation 10.16 also indicates the orientation of the failure surface on the plate.

As similar to the shear in strut members, transverse reinforcements which may be in the form of links should be used as reinforcement in case the shear capacity of concrete is exceeded. The design, however, should be based on the S_{max} values as discussed in the foregoing.

10.8 DESIGN FOR COMBINED IN-PLANE LOADS AND OUT-OF-PLANE BENDING

In 3-dimensional analysis with six degrees of freedom, in-plane and out-of-plane stresses are analyzed as outputs generally. A holistic design which can take into consideration all the stresses is therefore required for structural adequacy. A planar element resisting both in-plane and out-of-plane load is often termed “shell” element.

The in-plane shear can be catered for by the in-plane reinforcement as discussed in Section 10.2. So, it does not conflict with the out-of-plane shear. They can therefore be considered separately.

However, the flexural stresses incurred by the out-of-plane bending moment are basically of the same nature of that incurred by the in-plane direct stresses. They have to be combined together to check if the concrete strength is exceeded and to add longitudinal reinforcements as required. The design concept is similar to the design of column or wall structures under axial loads and bending moments. Two approaches are discussed as follows:

10.8.1 Separate Sets of Reinforcement for In-Plane and Out-of-Plane Loads

The proposed approach comprises the following:

(1) The point is checked for the in-plane stresses and provides reinforcement as necessary as if the out-of-plane bending does not exist.
(2) Two sets of additional flexural reinforcements are to be provided to resist the design moments determined by the Wood-Armer Equations M_x^* and M_y^*. The reinforcements for M_x^* comprise two layers, each being close to a surface of the shell separated by a distance d_x and of cross-sectional area A_{stx} which is calculated by $M_x^* / (f_y d_x)$ where f_y is the strength of the reinforcement. Similarly for the Y-direction where $A_{sty} = M_y^* / (f_y d_y)$. As the design moments M_x^* and M_y^* are resisted solely by the added reinforcement A_{stx} and A_{sty} without any exhaustion of that provided for the in-plane stresses, structural adequacy is achieved.

Nevertheless, the approach is generally conservative even if the out-of-plane bending dominates the design, as the concrete compressive strength may not be completely exhausted by the in-plane stresses. A worked example adopting this approach is given as follows:

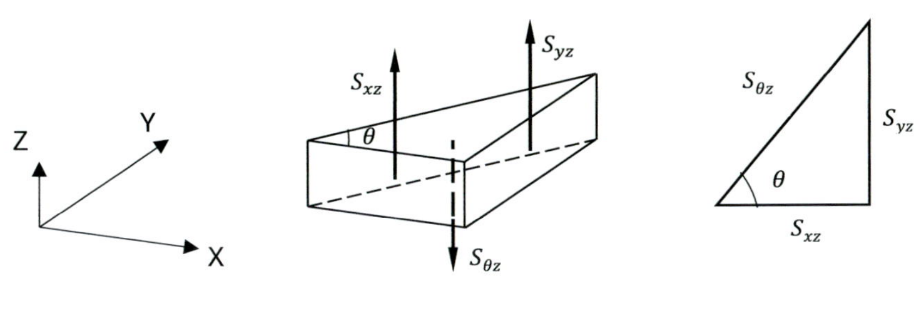

FIGURE 10.13 Diagrammatic illustration of out-of-plane shear.

Worked Example 10.4

Element thickness = 500mm, $f_c = 14\text{N/mm}^2$; $f_y = 435\text{N/mm}^2$; cover = 40 mm;

In-plane load: $F_x = 1200\text{kN/m}$; $F_y = -1800\text{kN/m}$; $S_{xy} = 900$ kN/m;

Out-of-plane bending: $M_x = 80\text{kNm/m}$; $M_y = -240\text{kNm/m}$; $M_{xy} = 100\text{kNm/m}$

Solution:

$\sigma_x = 1200/0.5 = 2400\text{kN/m}^2 = 2.4\text{N/mm}^2$;

$\sigma_y = -1800/0.5 = -3600\text{kN/m}^2 = -3.6\text{N/mm}^2$

$\tau_{xy} = 900/0.5 = 1800\text{kN/m}^2 = 1.8\text{N/mm}^2$

$\sigma_x/|\tau_{xy}| = 1.333$; $\sigma_y/|\tau_{xy}| = -2$.

So, the in-plane stress state is case 3 according to Figure 10.2 and Table 10.2.

By Table 10.1, $\rho_x = \dfrac{1}{f_y}\left(\sigma_x - \dfrac{\tau_{xy}^2}{\sigma_y}\right) = 0.007586$; $\rho_y = 0$

$A_{stx,m} = 0.007586 \times 500 \times 1000 = 3793\text{mm}^2\text{/m}$ or 1897 mm²/m each face.

$M_x = 80$; $M_y = -240$; $M_{xy} = 100$

For sagging:

$$M_x^* = M_x + |M_{xy}| = 80 + 100 = 180 \geq 0;$$

$$M_y^* = M_y + |M_{xy}| = -240 + 90 = -150 \leq 0$$

So $M_y^* = 0$ and $M_x^* = M_x + \left|\dfrac{M_{xy}^2}{M_y}\right| = 80 + \left|\dfrac{100^2}{-240}\right| = 121.67$

For hogging:

$$M_x^* = M_x - |M_{xy}| = 80 - 100 = -20 \leq 0;$$

$$M_y^* = M_y - |M_{xy}| = -240 - 100 = -340 \leq 0$$

Summing up: for sagging $M_x^* = 121.67$, $M_y^* = 0$;

for hogging $M_x^* = -20$, $M_y^* = -340$

Concrete cover taken as 40 mm; X-layer of reinforcement chosen to be the outermost. So, assuming 20 mm dia. bar $d_x = 500 - 40 \times 2 - 20 = 400$

$$A_{stx,b} = \frac{121.67 \times 10^6}{400 \times 435} = 699\text{mm}^2\text{/m}.$$

Adding $A_{stx,m} + A_{stx,b} = 1897 + 699 = 2596\,\text{mm}^2$, use T20-120 (EF).

Assuming 20 mm dia. inner bar, $d_y = 400 - 20 \times 2 - 20 = 340\,\text{mm}$

$A_{sty,b} = \dfrac{340 \times 10^6}{340 \times 435} = 2299\,\text{mm}^2/\text{m}$. As $A_{sty,m} = 0$, use T20-135 (EF).

10.9 SANDWICH MODEL

The sandwich model was developed by Brondum (1974) and incorporated into the Eurocode BSEN1992-1-14:2023 (2023) for design of shell structures resisting both in-plane and out-of-plane loads simultaneously. The model divides a shell element into two outer layers centred on the mid-planes of the outer reinforcement layers and an inner core layer as shown in Figure 10.14.

The working principle of the "Sandwich Model" is to transform the effects of the bending and twisting actions applied on the full depth of the shell element into membrane actions (direct and shear stresses) in the outer layers for design. Such effects are added to those due to the direct stresses and shear stresses originally created by the applied in-plane loads. The distribution of the forces and moments onto the three layers are determined as follows :

(1) The thicknesses of the outer layers are determined by treating their mid-depths at the mid-planes of the reinforcements. So, it would be twice the concrete cover to the reinforcements plus the reinforcement diameter. As the reinforcements in the two orthogonal directions are different, the thicknesses of the outer layers are also different in the two directions.

(2) The core layer is assumed to be not affected by the bending and twisting moments. Its stresses can be directly designed in accordance with Clark's approach. Nevertheless, the reinforcements required will be moved and added to those of the top and bottom outer layers, with halves to each.

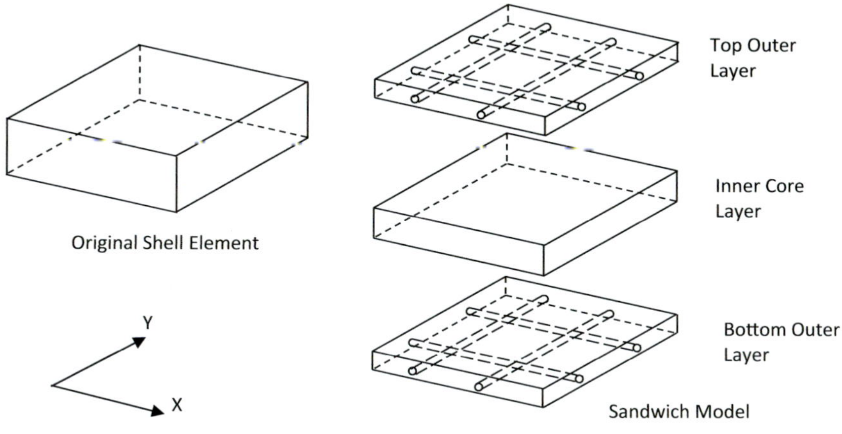

FIGURE 10.14 Sandwich Model.

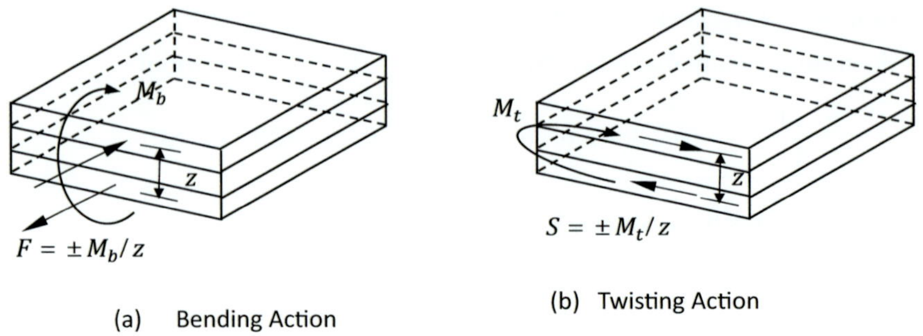

FIGURE 10.15 Transforming bending/twisting actions to membrane actions.

(3) Each of the bending moments will be converted to axial force in the outer layers as shown in Figure 10.15(a) where z is the height between the mid-planes of the two outer layers.
(4) The twisting moment will be converted to shears in the outer layers as shown in Figure 10.15(b). However, as the thicknesses of the outer layers are different in the two directions, average values will be adopted.
(5) The axial forces determined in (3) and shears determined in (4) will be added to the direct and shear forces in the outer layers for design to Clark's approach.
(6) The final reinforcement in the two layers will be the sum of reinforcements obtained in (2) and (5).

The design is demonstrated by the following Worked Example 10.5 with the applied forces and moments identical to Worked Example 10.4.

Worked Example 10.5
Element thickness = 500mm, $f_c = 14\text{N/mm}^2$; $f_y = 435\text{N/mm}^2$; cover = 40 mm;

In-plane load: $F_x = 1200\text{kN/m}$; $F_y = -1800\text{kN/m}$; $\tau_{xy} = 900\text{kN/m}$;

Dividing by 500 mm, $\sigma_x = 2.4\text{N/mm}^2$; $\sigma_y = -3.6\text{N/mm}^2$; $S_{xy} = 1.8\text{ N/mm}^2$

Out-of-plane bending: $M_x = 80\text{kNm/m}$; $M_y = -240\text{kNm/m}$; $M_{xy} = 100\text{kNm/m}$
Reinforcement bars diameter assumed are 20 mm diameter.

(1) The thicknesses of the outer and inner core layers are as follows :

	X-Direction	Y-direction
Outer Layers	$t_{ox} = 40\times2+20 = 100$ mm	$t_{oy} = (40+20)\times2+20 = 140$ mm
Core Layer	$t_{ix} = 500-100\times2 = 300$ mm	$t_{iy} = 500-140\times2 = 220$ mm.
Lever Arm	$z_x = 300+100 = 400$ mm	$z_y = 220+140 = 360$ mm

(2) The axial and shear stresses due to in-plane load are identical to that in Worked Example 10.4, requiring 0.7586% reinforcement in the X-direction only.

(3) The axial forces created by M_x = 80kNm/m are M_x/z_x = 80/0.4 = 200kN/m which is compression on the top outer layer giving a stress of –200/0.1= –2000kN/m² = –2N/mm². The same stress in tension is created on the lower outer layer as 2N/mm².

Similarly, for M_y = –240kNm/m, the axial stress on the outer layers are 240/0.36/0.14/1000 = 4.762N/mm² on the top outer layer and –4.762N/mm² in the lower layer.

(4) Using the averaged value of z_x and z_y as 380mm, the shear forces created by M_{xy} = 100kNm/m on the outer layers are 100/0.38 = 263.158kN/m or 263.158/0.12 = 2193kN/m² = 2.193N/mm².

(5) Adding the direct stresses and shear stresses, the design stresses of the upper and lower outer layers are as follows:

		Stress (N/mm²)	By Clark's Equations
Upper Outer Layer	σ_x	$2.4-2=1.4$	Case 3, $\rho_x=0.772\%$; Steel Area = 1010 mm²/m
	σ_y	$-3.6-4.762=1.162$	$\rho_y=1.185\%$
	τ_{xy}	$1.8+2.193=3.993$	Steel Area = 1659 mm²/m
Lower Outer Layer	σ_x	$2.4+2=4.4$	Case 3, $\rho_x=1.450\%$; Steel Area = 1450 mm²/m
	σ_y	$-3.6-4.762=-8.362$	$\rho_y=0$
	τ_{xy}	$1.8+2.193=3.993$	

(6) Finally, adding the share from the inner core obtained in (2) which is 0.007586 × 1000 × 300 ÷ 2 = 1138mm² on the outer layers, the steel reinforcement in X-direction is 1010 + 1138 = 2148mm² at the upper layer and 1450 + 1138 = 2588mm² at the lower layer. In the Y-direction, the reinforcement is 1659mm² in the top layer and 0 in the bottom layer.

10.10 DESIGN TO 3-DIMENSIONAL STRESSES

In prescribed directions, generally a 3-dimensional stress field comprises three direct stresses in mutually perpendicular directions together with three shear stresses as shown in Figure 10.16 which can be taken as an extension of the 2-dimensional stress field. As similar to the 2-dimensional problem, the "cube" as shown in Figure 10.16

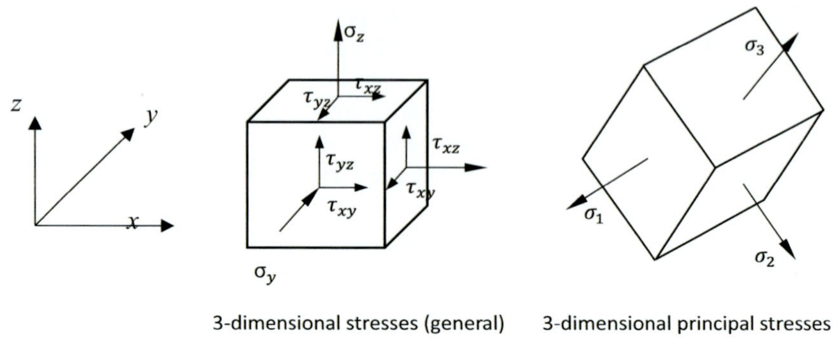

FIGURE 10.16 Three-dimensional stress.

can also be so orientated that the shear stresses vanish, leaving the axial stresses as principal stresses as indicated in the figure.

The principal stresses can be computed by finding the eigenvalues of the following matrix

$$\begin{bmatrix} \sigma_1 \\ \sigma_2 \\ \sigma_3 \end{bmatrix} = \text{eigenvalue} \begin{bmatrix} \sigma_x & \tau_{xy} & \tau_{xz} \\ \tau_{xy} & \sigma_y & \tau_{yz} \\ \tau_{xz} & \tau_{yz} & \sigma_z \end{bmatrix} \tag{10.18}$$

While the 2-dimensional stress analysis can capture local stresses in planar elements, 3-dimensional stress can do the same in a 3-dimensional structural element. In comparison, generally more realistic results can be acquired as the assumption of linear stress/strain distributions across the depth of the planar element can be avoided. But much greater efforts in analysis and design are generally required.

To acquire 3-dimensional stresses for more accurate design, the structure is simulated as an assembly of "bricks" in finite element analysis as in an example shown in Figure D13 of Appendix D and the results involve the six components of stresses as shown in Figure 10.16.

10.11 APPROACH BY LAW ET AL. (2007)

The design for 3-dimensional stresses can follow design principles similar to that of Clark for the 2-dimensional stress but much more complicated. Law et al. (2007) has derived an approach based on the extension of Clark's principle which involves seeking optimal solutions in solving 15 unknowns from the following 12 equations.

$$\sigma_x - \rho_x f_y = \sigma_1 l_x^2 + \sigma_2 m_x^2 + \sigma_3 n_x^2 \tag{10.19}$$

$$\sigma_y - \rho_y f_y = \sigma_1 l_y^2 + \sigma_2 m_y^2 + \sigma_3 n_y^2 \tag{10.20}$$

$$\sigma_z - \rho_z f_y = \sigma_1 l_z^2 + \sigma_2 m_z^2 + \sigma_3 n_z^2 \tag{10.21}$$

$$\tau_{xy} = \sigma_1 l_x l_y + \sigma_2 m_x m_y + \sigma_3 n_x n_y \tag{10.22}$$

$$\tau_{xz} = \sigma_1 l_x l_z + \sigma_2 m_x m_z + \sigma_3 n_x n_z \tag{10.23}$$

$$\tau_{yz} = \sigma_1 l_y l_z + \sigma_2 m_y m_z + \sigma_3 n_y n_z \tag{10.24}$$

$$l_x^2 + m_x^2 + n_x^2 = 1 \tag{10.25}$$

$$l_y^2 + m_y^2 + n_y^2 = 1 \tag{10.26}$$

$$l_z^2 + m_z^2 + n_z^2 = 1 \tag{10.27}$$

$$l_x l_y + m_x m_y + n_x n_y = 0 \tag{10.28}$$

$$l_x l_z + m_x m_z + n_x n_z = 0 \tag{10.29}$$

$$l_y l_z + m_y m_z + n_y n_z = 0 \tag{10.30}$$

From the above equations, ρ_x, ρ_y and ρ_y are the required reinforcements and f_y is the design strength of reinforcement. σ_x, σ_y, σ_z, τ_{xy}, τ_{xz}, τ_{yz} are the applied direct stresses and shear stresses. σ_1, σ_2 and σ_3 are "principal stresses" under the action of the reinforcements but ignoring the tensile strength of concrete. They will be equal to the "principal stresses" calculated based on σ_x, σ_y, σ_z, τ_{xy}, τ_{xz}, τ_{yz} only in the absence of reinforcements and no tension. l_x, m_x, n_x etc. are the direction cosines of the plane where σ_x, σ_y, σ_z act with respect to the directions of σ_1, σ_2, σ_3. Figure 10.17 defines the symbols. In the figure, the UVW axes are the "principal planes" with respect to XYZ axes.

Of the 12 equations, the first 6 are force equilibrium equations and the remaining ones are geometry equations. Figure 10.17 extracted from Law et al. (2007) is that under which Equation 10.21 is derived.

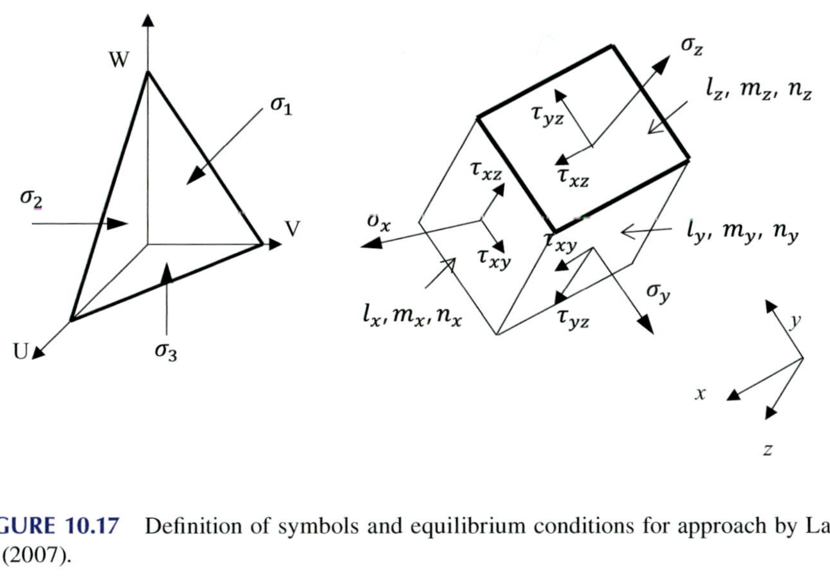

FIGURE 10.17 Definition of symbols and equilibrium conditions for approach by Law et al. (2007).

As there are more unknowns than equations, many solutions do exist. A systematic search process has been established by Law et al. (2007) to search for the optimal solution. Readers may refer to the paper for the detailed derivations. The design procedures are, however, extracted in this chapter for reference:

(1) Determine the "theoretical" principal stresses by Equation 10.18. If all of them are within the range f_c of the design compressive strength of concrete, no reinforcement is required. Otherwise, follow steps (2) to (13);

(2) Set values of l_x and m_x for trial. Extreme values of l_x, say –0.95 may first be used. At $l_x = -0.95$, extreme value of m_x will be around –0.312 as $m_x \geq -\sqrt{1-l_x^2}$. At constant value of $l_x = -0.95$, m_x may also be set at +0.05 increments up to 0. After that, the next $l_x = -0.9$ may be tried (by increment of $+0.05$) again with varying values of m_x.

(3) With the pre-determination of l_x and m_x in step (2) (m_x in the range of $-\sqrt{1-l_x^2}$ to 0), set $n_x \geq -\sqrt{1-l_x^2-m_x^2}$.

(4) l_y is to be obtained from trial and error within the range $-\sqrt{m_x^2+n_x^2}$ and $\sqrt{m_x^2+n_x^2}$. With the trial value of $l_y = -\sqrt{m_x^2+n_x^2}$, solve m_y by

$$m_y = \frac{-l_x l_y \pm \sqrt{\left(l_x l_y m_x\right)^2 - \left(m_x^2+n_x^2\right)\left(l_y^2 n_x^2 + l_x^2 l_y^2 - n_x^2\right)}}{\left(m_x^2+n_x^2\right)}.$$

So, there are two "branches" for trial under m_y. The designer may complete all trials under

$$m_y = \frac{-l_x l_y + \sqrt{\left(l_x l_y m_x\right)^2 - \left(m_x^2+n_x^2\right)\left(l_y^2 n_x^2 + l_x^2 l_y^2 - n_x^2\right)}}{\left(m_x^2+n_x^2\right)}$$

and start another series of trials by

$$m_y = \frac{-l_x l_y - \sqrt{\left(l_x l_y m_x\right)^2 - \left(m_x^2+n_x^2\right)\left(l_y^2 n_x^2 + l_x^2 l_y^2 - n_x^2\right)}}{\left(m_x^2+n_x^2\right)}.$$

(5) Solve n_y by $n_y = \dfrac{-l_x l_y - m_x m_y}{n_x}$

(6) Solve $l_z = \dfrac{\pm 1}{\sqrt{1+\left(\dfrac{l_y n_{x-} l_x n_y}{m_x n_{y-m_y n_x}}\right)^2+\left(\dfrac{l_x m_{y-} l_y m_x}{m_x n_{y-m_y n_x}}\right)^2}}$

(7) Solve m_z by $m_z = \dfrac{l_y n_{x-} l_x n_y}{m_x n_{y-m_y n_x}} l_z$ and $n_z = \dfrac{l_x m_{y-} l_y m_x}{m_x n_{y-m_y n_x}} l_z$

(8) Back substitute the calculated direction cosines from steps (2) to (7) into

$$T = l_z m_z\left(l_x m_y - l_y m_x\right)\tau_{xy} + l_y m_y\left(l_z m_x - l_x m_z\right)\tau_{xz} + l_x m_x\left(l_y m_z - l_z m_y\right)\tau_{yz}$$

to test if it is zero. If not, try the next value of $l_y = -\sqrt{m_x^2 + n_x^2} + \Delta$ where $\Delta$is a small increment, say 0.001 until the expression T changes sign. Then interpolate a more accurate value of l_y. As such, we can very accurately obtain l_y satisfying $T = 0$ which is a derived condition from Equations 10.19 to 10.30.

(9) With the determination of l_y in step (8), refine the values of other direction cosines by steps (4) to (7);

(10) If the "theoretical" principal stresses are all compressive, calculate

$$\sigma_{1f} = \frac{m_z\tau_{xy} - m_y\tau_{xz}}{(l_y m_z - l_z m_y) l_x}; \quad \sigma_{2f} = \frac{l_z\tau_{xy} - l_y\tau_{xz}}{(l_z m_x - l_x m_z) m_x} \quad \text{with } \sigma_3 = f_c \quad \text{where } \sigma_1 = \sigma_{1f} + f_c$$

and $\sigma_2 = \sigma_{2f} + f_c$. If at least one of them is tensile, calculate $\sigma_1 = \dfrac{m_z\tau_{xy} - m_y\tau_{xz}}{(l_y m_z - l_z m_y) l_x}$;

$\sigma_2 = \dfrac{l_z\tau_{xy} - l_y\tau_{xz}}{(l_z m_x - l_x m_z) m_x}$ and $\sigma_3 = 0$. Reject the whole set if any of σ_1 or σ_2 falls outside the range of 0 and f_c.

(11) If both σ_1 and σ_2 are within the range of 0 and f_c, the reinforcement ratios required in the three directions can be worked out as

$$\rho_x = \frac{1}{f_s}\left(\sigma_{xf} - \sigma_{1f} l_x^2 - \sigma_{2f} m_x^2\right) \text{ or } \rho_x = \frac{1}{f_s}\left(\sigma_x - \sigma_1 l_x^2 - \sigma_2 m_x^2\right)$$

$$\rho_y = \frac{1}{f_s}\left(\sigma_{yf} - \sigma_{1f} l_y^2 - \sigma_{2f} m_y^2\right) \text{ or } \rho_y = \frac{1}{f_s}\left(\sigma_y - \sigma_1 l_y^2 - \sigma_2 m_y^2\right)$$

$$\rho_z = \frac{1}{f_s}\left(\sigma_{zf} - \sigma_{1f} l_z^2 - \sigma_{2f} m_z^2\right) \text{ or } \rho_z = \frac{1}{f_s}\left(\sigma_z - \sigma_1 l_z^2 - \sigma_2 m_z^2\right)$$

as appropriate where $\sigma_{xf} = \sigma_x - f_c$; $\sigma_{yf} = \sigma_y - f_c$ and $\sigma_{zf} = \sigma_z - f_c$

Positive values of the reinforcement ratios indicate compressive steel ratio and negative values indicate tensile steel ratio. The total reinforcement ratio obtained should be $\rho = |\rho_x| + |\rho_y| + |\rho_z|$.

(12) Determine the next l_y under the same pre-determined l_x and m_x through steps (4) to (11) from the previously determined l_y which yield a feasible solution for reinforcement, if such l_y exists.

(13) After the exhaustion of l_y to $\sqrt{m_x^2 + n_x^2}$, determine the next set of l_x and m_x by (2) and obtain another set of total reinforcement ratio, if in existence, through steps (2) to (12).

(14) Take the reinforcement ratios of the three directions when the total ratio is minimum among all sets of l_x and m_x tried. Refinement for a better solution may be carried out between calculated values, giving almost minimum total steel ratios.

For an efficient solution, the foregoing can be programmed for searching for optimal solutions. The authors have done so by using visual basic of excel. The following Worked Example 10.6 table contains solutions for five cases with concrete strength at $f_c = 20.25$ N/mm^2 and reinforcement strength at $f_y = 435$ N/mm^2. For the stresses and reinforcement ratios, a positive sign indicates tension and a negative sign indicates compression. The stresses σ_1, σ_2 and σ_3 are the final principal stresses after the addition of reinforcement.

Worked Example 10.6

	Cases				
	1	**2**	**3**	**4**	**5**
σ_x(kPa)	−22000	−8000	−8000	8000	−6000
σ_y(kPa)	−18000	−24000	24000	9000	10000
σ_z(kPa)	−20000	−26000	20000	6000	−18000
τ_{xy}(kPa)	2500	2500	2500	2500	2500
τ_{xz}(kPa)	2000	2000	2000	2000	2000
τ_{yz}(kPa)	3000	3000	3000	3000	3000
l_x	−0.200	0.900	0.150	0.450	0.800
m_x	−0.829	−0.335	−0.188	−0.043	−0.199
n_x	0.522	0.279	0.971	0.892	0.566
l_y	−0.403	−0.344	0.953	0.698	0.581
m_y	0.555	−0.153	0.290	0.640	0.493
n_y	0.727	0.926	−0.091	−0.321	−0.648
l_z	0.893	0.268	0.264	0.557	0.150
m_z	0.065	0.930	−0.938	−0.768	−0.847
n_z	0.445	0.253	−0.223	−0.318	−0.510
σ_1(kPa)	−11176	−9772	−20093	−7967	−6853
σ_2(kPa)	−13231	−5721	−6826	−129	−6989
σ_3(kPa)	−20250	−20250	0	0	0
ρ_x(%)	−1.595	0.490	−1.998	2.210	−0.307
ρ_y(%)	−0.320	−1.225	1.192	2.974	3.220
ρ_z(%)	−0.075	−0.683	2.894	1.964	−2.949
ρ_{total}(%)	1.989	2.398	6.085	7.148	6.477

10.12 ALTERNATIVE APPROACH BY CHENG

Cheng has worked on two special structures where more than one million solid elements are used to model the system with staged construction as shown in Figure 10.18. The use of solid elements instead of classical structural elements is required due to the shape and dimensions of the various members. From the computed nodal

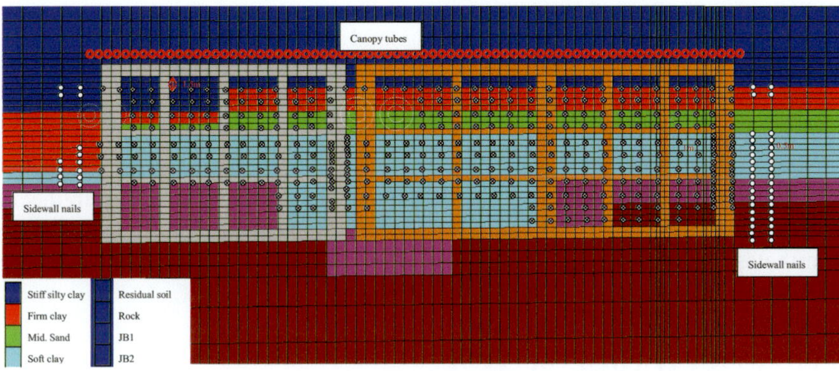

(a) Analysis of a jacked-in tunnel at Australia by Cheng et al. (2024)

(b) Analysis of a deep excavation in clay with barrette, diaphragm, capping beams and other structural elements

FIGURE 10.18 Two projects with more than one million solid elements used to model geotechnical materials and structural elements. (a) Analysis of a jacked-in tunnel in Australia by Cheng et al. (2024). (b) Analysis of a deep excavation in clay with barrette, diaphragm, capping beams and other structural elements.

stress at any section, the mean normal stress is computed, which is multiplied with the sectional area to give the axial stress on that section. Based on the normal stress at the two extreme fibres, the mean stress is determined so that a positive and negative equivalent bending stress at the two extreme fibres are determined. Based on the positive and negative stress blocks, the equivalent bending moment across this section is determined. At the centre of this section, the shear stress is determined. Based on the classical beam theory, the shear force at this section is related to the maximum shear stress at the centre; hence, the shear force at this section is determined. Since the equivalent axial, bending, and shear are determined, the reinforcement can be designed in accordance with the classical reinforced concrete design method. This approach is not commonly used in practice, but it is a practical method when the structure under consideration is difficult to define the type of structural members. The advantage of this method is that the final outputs to the engineers are the bending moment, axial force, and shear force for which most of the engineers are familiar, while the effects of finite sizes or changing sizing can be considered easily. Nevertheless, the recovery of these internal forces are approximate, as this approach still relies on the classical beam/column theory.

10.13 CONCLUDING REMARKS

Approaches for reinforced concrete design for (1) in-plane load for membrane element; (2) out-of-plane bending for plate element; (3) in-plane load plus out-of-plane for shell element; and (4) 3-dimensional stresses for solid element have been discussed in this chapter. They are applicable for stresses in "nodes" analyzed mostly nowadays by the finite element method. These design approaches enable a designer to carry out point-to-point design on reinforced concrete structures which is particularly useful in locations where section design is not appropriate due to invalidity in the assumption of "plane remains plane." More realistic and accurate designs can then be achieved.

REFERENCES

British Standard Institution BSI (2023) *Eurocode 2: Concrete – Specifications, performance, production and conformity BSEN1992-1-1:2023+A2:2021.* British Standard Institution.

Brondum-Nielsen, T. (1974) Optimum Design of Reinforced Concrete Shells and Slabs, Report No. R 44. Technical University of Denmark, Copenhagen, Denmark.

Cheng Y.M., Law C.W. and Liu L.L. (2024) Analysis, design and construction of foundations, 2nd edition, CRC Press.

Clark, L.A. (1976) The Provisions of Tension and Compression Reinforcement to Resist In-Plane Forces. Magazine of Concrete Research, 28(94), 3–12.

Johansen, K.W. (1962) Yield-Line Theory. London: Cement and Concrete Association.

Law, C.W., Cheng, Y.M. and Su, R.K.L. (2007) Approach for Reinforcement Design in Reinforced Concrete Structures Based on 3-Dimensional Stress Field. The HKIE Transactions, 14(2), 9–18.

Subedi, N.K. (1975) Design of Reinforced Concrete Sections Subjected to Membrane Forces. The Structural Engineer, 53(7), 155–162.

Wood, R.H. (1968) The Reinforcement of Slabs in Accordance with a Pre-Determined Field of Moments. Concrete, 2, 69–76.

11 Reinforced Concrete Detailing

11.1 INTRODUCTION

Proper reinforced concrete detailing is to ensure that reinforcement bar (rebar) is correctly placed and fixed so that the rebar can perform its designed functions effectively and efficiently. The most important or ultimate function of the rebar is the development and enhancement of strength of the reinforced concrete structure to resist various loads under good distribution for achieving stability and safety. However, although the rebar is essential in providing strength to the structure, it is nevertheless also a potential cause for deterioration of the structure by its rusting. So proper placing and fixing of it is necessary for protection against various environmental stresses. Thus, proper reinforced concrete detailing is indispensable for durability. In addition, proper reinforced concrete detailing can also facilitate construction, promoting quality of the concrete work which is also crucial for the good performance of the structure.

To these ends, various codes provide requirements in reinforced concrete detailing according to the standard of work being aimed at, under balance against various factors including cost, durability, environmental conditions, available construction techniques and workmanship level. The requirements are different to some extent among these codes. For ease of implementation, the requirements are often in prescriptive form with minimum calculations, e.g., maximum or minimum percentages.

The authors do not intend to give comprehensive reinforced concrete detailing requirements on various codes in this chapter. Instead, the fundamental principles and implications are discussed under various topics. Reference to the reputable codes will be drawn for illustration as necessary.

11.2 MINIMUM AND MAXIMUM REBAR PERCENTAGES

Determination of rebar contents in a structural member primarily depends on structural design calculations. Nevertheless, codes of practice always stipulate minimum percentages for various members. The main purpose is to ensure certain minimum strength in the structure, in particular tensile strength as the concrete is weak in tension. Such minimum strength serves to cover loads which may not have been considered in general design, such as shrinkage and creep, temperature and other loads unanticipated during the design stage. Minimum rebar provision also serves to promote ductility (e.g., compressive rebar in beams) and minimize concrete cracking. So, durability can be promoted.

DOI: 10.1201/9781003636380-11

The main purpose of maximum rebar percentage, however, is to avoid rebar congestion in the structure which may lead to the formation of voids (honeycombing) and therefore weak sections which would instead reduce strength and adversely affect durability. In addition, overprovision of rebar in some cases would also impair ductility, e.g., overprovision of tensile rebar in a beam will lead to premature concrete crushing prior to rebar yielding which is an unpreferable brittle failure mode. Another example is the limitation of shear rebar in a beam as provided in all codes of practice. It is because excessive provision of shear rebar will also lead to overstressing of the diagonal concrete strut (as discussed in Chapter 9, Figure 9.16) prior to the yielding of the shear rebar. This is not favoured as the former failure is brittle and the latter is ductile. Nevertheless, the minimization of rebar congestion should also be associated with bar spacing, bar diameter and maximum aggregate sizes. These should also be considered.

As a convenient provision, codes of practice often specify certain minimum or maximum percentages under various situations. Some of them are in simple figures, e.g., 0.1% (each face) of cross-sectional area in wall which is loaded by in-plane compressive load as specified by BSEN1992-1-1:2023 (2023). Others may depend on other parameters, e.g., when the wall carries in-plane normal and shear stresses, BSEN1992-1-1:2023 (2023) requires the minimum reinforcement to be $0.25A_c f_{ctm} / f_{yk}$ where A_c is the concrete cross-sectional area, f_{ctm} is the mean tensile strength of concrete and f_{yk} is the characteristic strength of reinforcement. Nevertheless, most of the codes do allow alternative proposals by justification.

11.3 ANCHORAGE – BOND AND LAP, TRANSVERSE REINFORCEMENTS AT LAPS

Bond:

For effective functioning of the rebar in concrete to provide tensile and compressive strengths, it must be adequately anchored or bonded in the concrete. For design purposes, the bond strength in concrete has to exceed its tensile strength so that the rebar will fail by tension instead of bond slippage. This mode of failure is preferred because this type of bond failure is brittle while the rebar failure by tension is ductile.

The bond strength of a rebar in concrete comprises (1) chemical adhesion; (2) friction resistance; and (3) mechanical interlock in a deformed bar. A deformed bar is one with a ribbed or corrugated surface. Chemical adhesion is the weakest component and would be broken at small slips of bars. After adhesion breaks, friction develops due to the surface roughness of the rebar, which is enhanced by the deformations (ribs/slugs) of the surface of the rebar in case of deformed bar. The greatest contribution to bond strength in deformed bar is the mechanical interlock between the ribs and the concrete, as illustrated in Figure 11.1. In the figure, the rebar under tension or compression will create bearing stresses on the concrete by its ribs, which may lead to the crushing of the concrete. In addition, as the bearing stresses have components in the radial outward directions, circumferential tension will also be developed in the concrete around the rebar, which may crack the concrete and subsequently

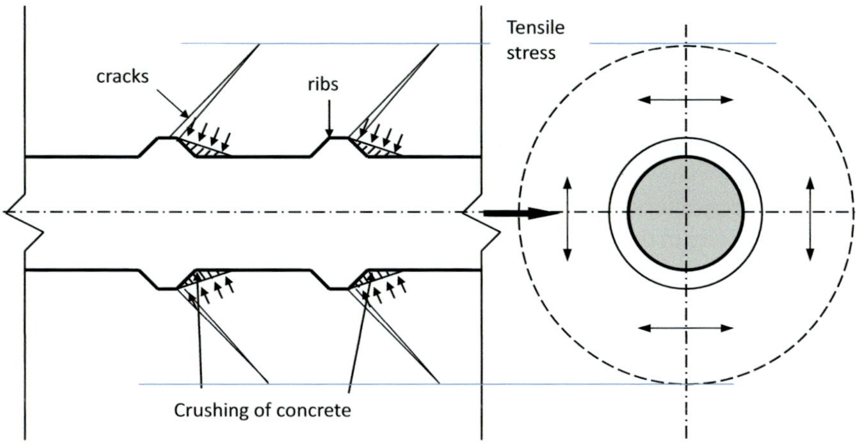

FIGURE 11.1 Failure of bond of deformed bar by concrete crushing.

reduce bond strength. The cracking would be more severe in thin concrete cover to the rebar. Transverse rebars in the direction of the tensile stress may be required to reduce cracking, which is not only a strength problem but also a durability problem. Summing up, higher concrete strength, greater concrete cover, and confinement of concrete by transverse rebars are favourable for good bonding.

The CEB FIP Model Code 2010 (2012) has a concise description of the bond behaviour in concrete briefly reproduced as follows with reference to its figure 6.1-1 and table 6.1-1 extracted as in Figure 11.2.

The CEB FIP bond failure model comprises several stages according to the mode of failure. For pull-out failure, bond stress first ascends non-linearly with slip in which the ribs penetrate into the mortar matrix leading to local crushing and concrete micro-cracking. When the bond stress reaches the maximum, it can sustain by remaining constant for certain further slip if concrete is under good confinement

Table 6.1-1: ***Parameters defining the mean bond stress-slip relationship of ribbed bars (acc. to Eqs. (6.1-1) to (6.1-4))***

	1	2	3	4	5	6
	Pull-Out (PO)		**Splitting (SP)**			
	$\varepsilon_s < \varepsilon_{s,y}$		$\varepsilon_s < \varepsilon_{s,y}$			
	Good bond cond.	**All other bond cond.**	**Good bond cond.**		**All other bond cond.**	
			unconfined	**stirrups**	**unconfined**	**stirrups**
τ_{bmax}	$2.5\sqrt{f_{cm}}$	$1.25\sqrt{f_{cm}}$	$7.0\cdot\left(\frac{f_{cm}}{25}\right)^{0.25}$	$8.0\cdot\left(\frac{f_{cm}}{25}\right)^{0.25}$	$5.0\cdot\left(\frac{f_{cm}}{25}\right)^{0.25}$	$5.5\cdot\left(\frac{f_{cm}}{25}\right)^{0.25}$
s_1	1.0 mm	1.8 mm	$s(\tau_{bmax})$	$s(\tau_{bmax})$	$s(\tau_{bmax})$	$s(\tau_{bmax})$
s_2	2.0 mm	3.6 mm	s_1	s_1	s_1	s_1
s_3	c_{clear} 1)	c_{clear} 1)	$1.2s_1$	$0.5c_{clear}$ 1)	$1.2s_1$	$0.5c_{clear}$ 1)
α	0.4	0.4	0.4	0.4	0.4	0.4
τ_{bf}	$0.40\tau_{bmax}$	$0.40\tau_{bmax}$	0	$0.4\tau_{bmax}$	0	$0.4\tau_{bmax}$

1) c_{clear} is the clear distance between ribs

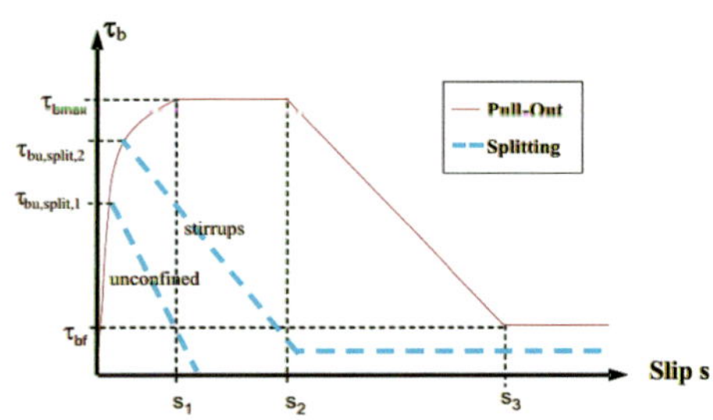

Figure 6.1-1: *Analytical bond stress-slip relationship (monotonic loading). $\tau_{bu,split,1}$ and $\tau_{bu,split,2}$ denote peak local bond resistance in the absence ($K_{tr}=0$) and presence respectively of confining stirrups*

FIGURE 11.2 Table 6.1-1 and figure 6.1-1 extracted from fib (2012).

during which advanced crushing and shearing off of the concrete between the ribs takes place. The post-peak linear descending branch shows a reduction of bond strength when the concrete "corbels" to the ribs are being sheared off. For the splitting failure, the model is reduced to a non-linear ascending and simplified linear descending branch.

Despite the many factors affecting bond strength, codes of practice used to express the required bond length, l_b of a rebar in terms of a multiple of the bar diameter for simplicity in design, under the assumption of constant ultimate bond strength τ_b along the circumference of the rebar and other nominal parameters.

$$\tau_b l_b \pi\phi = \frac{1}{4}\pi\phi^2 f_{yk} \Rightarrow l_b = \frac{f_{yk}}{4\tau_b}\phi \tag{11.1}$$

where ϕ is the rebar diameter f_{yk} is the ultimate axial stress of the rebar.

The former British Code BS8110-1:1997 (1997) adopts this approach and set $\tau_b = \beta\sqrt{f_{cu}}$ (in N/mm²) where β is a constant which is equal to 0.5 for high-ribbed bar and f_{cu} is the characteristic cube strength of concrete.

τ_b is, however, also related to other parameters such as concrete cover, rebar spacing and confinement. It even varies with the method of concrete casting which affects concrete compactness. The use of simple multiples to rebar diameter for determination of bond length by codes is generally under assumed and likely conservative parameters. The Eurocode BSEN1992-1-1:2023 (2023) lists in its Table 11.1 extracted as Table 11.1 in this book for required anchorage length under the assumptions of rebar being (i) ribbed bar $\phi \leq 32$ mm; indented bar $\phi \leq 14$ mm; (ii) stressed to 435 MPa; (iii) concrete cover $\geq 1.5\phi$. The Eurocode uses the term "Anchorage length" instead of bond length.

The equation 11.3 of the Eurocode based on which the values of its Table 11.1 are derived is reproduced as follows:

TABLE 11.1
Table 11.1 of BS EN 1992-1-1:2023

ϕ [mm]	Anchorage length l_{bd}/ϕ							
	f_{ck}							
	20	25	30	35	40	45	50	60
≤12	47	42	38	36	33	31	30	27
14	50	44	41	38	35	33	31	29
16	52	46	42	39	37	35	33	30
20	56	50	46	42	40	37	35	32
25	60	54	49	46	43	40	38	35
28	63	56	51	47	44	42	40	36
32	65	58	53	49	46	44	41	38

Note: The values of Table 11.1 (NDP) are derived from Formula (11.3).

$$l_{bd} = k_{lb}.k_{cp}.\phi.\left(\frac{\sigma_{sd}}{435}\right)^{n_\sigma}.\left(\frac{25}{f_{ck}}\right)^{\frac{1}{2}}.\left(\frac{\phi}{20}\right)^{\frac{1}{3}}.\left(\frac{1.5\phi}{c_d}\right)^{\frac{1}{2}} \geq 10\phi \qquad (11.2)$$

where l_{bd} is the anchorage length,

$k_{lb} = 50$for persistent and transient design situations with $n_\sigma = 3/2$; and
$k_{lb} = 35$for accidental situations with $n_\sigma = 3/2$;
k_{cp} is the coefficient accounting for casting effects on bond conditions with values :

$k_{cp} = 1.0$ for bars with good bond conditions comprising (i) bar with inclination 45° to 90° to the horizontal; and (ii) bars with an inclination less than 45° to the horizontal which are up to 300 mm from the bottom of the formwork or at least 300 mm from the free surface during concreting;

$k_{cp} = 1.2$ for poor bond conditions and for all bars used in slipform construction unless it is shown that the vertical bars cannot move during casting;

$k_{cp} = 1.4$ for all bars executed under bentonite or similar slurries unless data is available for the specific slurry to be used.

σ_{sd} is the design stress of the rebar;
f_{ck} is the characteristic strength of concrete;
$c_d = \min\{0.5c_s; c_x; c_y; 3.75\phi\}$ where c_s is the clear distance between parallel bars; c_x and c_y are the concrete covers to rebar measured in X and Y directions respectively at bend. The 3.75ϕ as the "maximum" value c_d can attain is the threshold where the cover effect is limited, i.e., any cover in excess of 3.75ϕ can only be taken as 3.75ϕ.

The equation factor $\frac{1.5\phi}{c_d}$ accounts for the effect of concrete cover and rebar spacing.

Worked Example 11.1
Check default value of Equation 11.2

Consider $f_{ck} = 45$; and $\phi = 32$; $k_{lb} = 50$for persistent and transient design situations with $n_\sigma = 3/2$; $k_{cp} = 1.0$ for bars with good bond conditions; $\sigma_{sd} = 435$; $c_d = c_y = 1.5\phi$. (The last two items are default values) Substituting into Equation 11.2, $l_{bd} = 43.6\phi$ which agrees with Table 11.1.

Nevertheless, Eurocode allows reduction of anchorage length by confinement reinforcement crossing the potential splitting surface shown in Figure 11.3 reproduced from Figure 11.5 of the Eurocode.

To introduce the effect of confinement, the parameter c_d in Equation 11.2 is replaced by

$$c_{d,conf} = \min\left\{c_x; c_y + 25\frac{\phi_t^2}{s_t}; \frac{c_s}{2}; 3.75\phi\right\} + \Delta c_d \leq 6\phi \qquad (11.3)$$

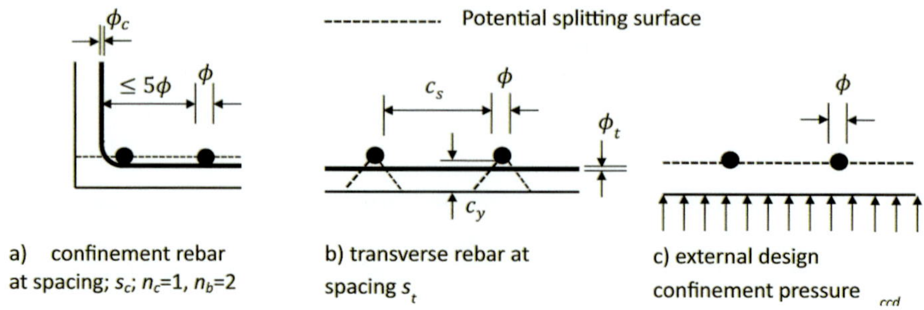

FIGURE 11.3 Confinement and transverse rebar to minimize concrete splitting.

where $\Delta c_d = \left(70\rho_{conf} + 12\sigma_{ccd} / \sqrt{f_{ck}}\right)\phi$;

ρ_{conf} is the ratio of the confinement reinforcement referred to the diameter of the bar to be anchored or spliced that $\rho_{conf} = \dfrac{n_c.\pi.\phi_c^2}{4.n_b.\phi.s_c}$;

ϕ is the diameter of the bar to be spliced or anchored;

ϕ_c is the diameter of the confinement reinforcement;

n_c is the number of legs of the confinement reinforcement crossing the potential splitting surface;

n_b is the number of anchored bars or pairs of lapped bars in the potential splitting failure surface;

s_c is the spacing of the confinement reinforcement along the bar to be anchored;

ϕ_t is the diameter of the transverse reinforcement between the bar to be anchored and the free surface (see Figure 11.3 b);

s_t is the spacing of the transverse reinforcement along the bar to be anchored;

σ_{ccd} is the design value of the mean compressive stress perpendicular to the potential splitting failure (see Figure 11.3 c).

Worked Example 11.2

Calculate the bond length of the side and central rebars as shown in Figure 11.4. First calculate the bond length without consideration of confinement.

Using Equation 11.2 $l_{bd} = k_{lb}.k_{cp}.\phi.\left(\dfrac{\sigma_{sd}}{435}\right)^{n_\sigma}.\left(\dfrac{25}{f_{ck}}\right)^{\frac{1}{2}}.\left(\dfrac{\phi}{20}\right)^{\frac{1}{3}}.\left(\dfrac{1.5\phi}{c_d}\right)^{\frac{1}{2}} \geq 10\phi$

For the corner bar, $c_s = 150 - 32 = 118\,\text{mm}$:

$$c_d = \min\{0.5c_s; c_x; c_y; 3.75\phi\} = \min\{0.5\times118; 60; 60; 3.75\times32\} = 59\,\text{mm}$$

$$l_{bd} = 50\times1\times32\times\left(\frac{435}{435}\right)^1\times\left(\frac{25}{45}\right)^{\frac{1}{2}}\times\left(\frac{32}{20}\right)^{\frac{1}{3}}\times\left(\frac{1.5\times32}{59}\right)^{\frac{1}{2}} = 1258\,\text{mm}$$

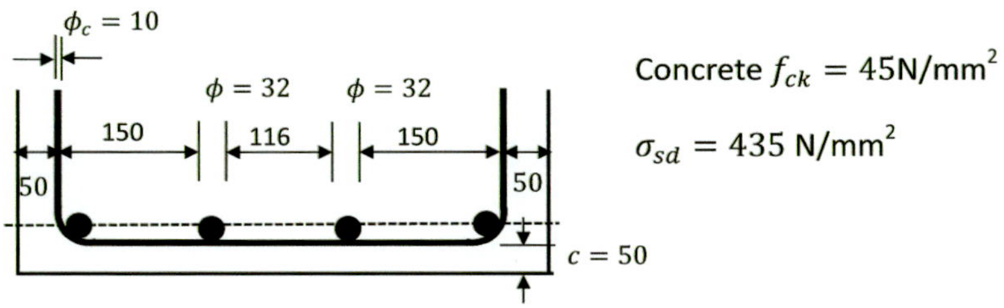

FIGURE 11.4 Worked Example 11.1.

For the central bar:

$$c_d = \min\{0.5c_s; c_x; c_y; 3.75\phi\} = \min\{0.5\times116; 60; 60; 3.75\times32\} = 58\,\text{mm}$$

$$l_{bd} = 50\times1\times32\times\left(\frac{435}{435}\right)^1\times\left(\frac{25}{45}\right)^{\frac{1}{2}}\times\left(\frac{32}{20}\right)^{\frac{1}{3}}\times\left(\frac{1.5\times32}{58}\right)^{\frac{1}{2}} = 1269\,\text{mm}.$$

Now take confinement into consideration:

$$\Delta c_d = \left(70\rho_{conf} + 12\sigma_{ccd}/\sqrt{f_{ck}}\right)\phi \text{ and } \rho_{conf} = \frac{n_c.\pi.\phi_c^{\ 2}}{4.n_b.\phi.s_c}$$

For the corner bar, $\rho_{conf} = \dfrac{n_c.\pi.\phi_c^{\ 2}}{4.n_b.\phi.s_c} = \dfrac{2.\pi.10^2}{4\times4\times32\times150} = 0.00818;$

$$\sigma_{ccd} = 0$$

$$\Delta c_d = \left(70\rho_{conf} + 12\sigma_{ccd}/\sqrt{f_{ck}}\right)\phi = \left(70\times0.00818 + 12\times0/\sqrt{45}\right)32 = 18.33$$

$$c_{d,conf} = \min\left\{c_x; c_y + 25\frac{\phi_t^{\ 2}}{s_t}; \frac{c_s}{2}; 3.75\phi\right\} + \Delta c_d \le 6\phi$$

$$= \min\left\{60; 60 + 25\times\frac{10^2}{150}; \frac{118}{2}; 3.75\times32\right\} + 18.33 = 59 + 18.33 = 77$$

$$l_{bd} = 50\times1\times32\times\left(\frac{435}{435}\right)^1\times\left(\frac{25}{45}\right)^{\frac{1}{2}}\times\left(\frac{32}{20}\right)^{\frac{1}{3}}\times\left(\frac{1.5\times32}{77}\right)^{\frac{1}{2}} = 1101\,\text{mm}$$

For the central bar, $\rho_{conf} = \dfrac{n_c.\pi.\phi_c^{\ 2}}{4.n_b.\phi.s_c} = \dfrac{2.\pi.10^2}{4\times4\times32\times150} = 0.00818;$

$$\sigma_{ccd} = 0$$

$$\Delta c_d = \left(70\rho_{conf} + 12\sigma_{ccd}/\sqrt{f_{ck}}\right)\phi = \left(70\times0.00818 + 12\times0/\sqrt{45}\right)32 = 18.33$$

$$c_{d,conf} = \min\left\{c_x; c_y + 25\frac{\phi_t^2}{s_t}; \frac{c_s}{2}; 3.75\phi\right\} + \Delta c_d \leq 6\phi$$

$$= \min\left\{210; 60 + 25\times\frac{10^2}{150}; \frac{118}{2}; 3.75\times 32\right\} + 18.33 = 59 + 18.33 = 77$$

$$l_{bd} = 50\times 1\times 32\times\left(\frac{435}{435}\right)^1\times\left(\frac{25}{45}\right)^{\frac{1}{2}}\times\left(\frac{32}{20}\right)^{\frac{1}{3}}\times\left(\frac{1.5\times 32}{77}\right)^{\frac{1}{2}} = 1101\text{mm}$$

So taking confinement into account can reduce the bond length by some 13%.

Similarly, ACI318M-19 (2019) states in its equation (25.4.2.4a) for the calculation of development length ℓ_d (i.e., bond length) in mm.

$$\ell_d = \frac{f_y}{1.1\lambda\sqrt{f_c'}}\frac{\psi_t\psi_e\psi_s\psi_g}{\left(\dfrac{c_b + K_{tr}}{d_b}\right)} \tag{11.4}$$

where f_y is the specified yield strength of rebar in MPa;

f_c' is the specified concrete compressive strength in MPa;

λ is the modification factor to reflect the reduced mechanical properties of lightweight concrete relative to normal weight concrete of the same compressive strength. $\lambda = 1$ for normal weight concrete;

ψ_t, ψ_e, ψ_s and ψ_g are modification factors for casting location in tension, rebar coating, rebar size and rebar grade respectively.

c_b is the lesser of: (a) the distance from the centre of a bar or wire to the nearest concrete surface, and (b) one-half the centre-to-centre spacing of bars or wires being developed, mm.

$K_{tr} = \dfrac{40A_{tr}}{sn}$ with A_{tr} is total cross-sectional area of all transverse rebar within spacing s that crosses the potential plane of splitting through the rebar being developed, mm^2.

s is the centre-to-centre spacing of rebar.

n is the number of rebar being developed or lap spliced along the plane of splitting.

d_b is the rebar diameter in mm.

$(c_b + K_{tr})/d_b$ is the confinement term and the ACI Code limits it to 2.5. It can be seen that this term is similar to c_d and Δc_d in the Eurocode discussed in the foregoing, playing the same role as cover plus confinement.

Equation 11.4 effectively relates ℓ_d as a multiple of d_b. It also shows that ℓ_d is directly proportional to f_y and inversely to $\sqrt{f_c'}$ which is a measure of the concrete's tensile strength or bond strength. In addition, c_b as the measure of concrete cover or bar-to-bar spacing and the confinement stress K_{tr} decreases ℓ_d.

Lap:

While bonding of rebar in concrete is the anchorage of a single rebar in concrete to develop the rebar's full strength, lap refers to "overlapping" of two rebars in concrete.

In a lap splice, two bars overlap and the force of a bar is transferred to another beyond its end, thus achieving continuity of the axial force of one rebar to another. A simplified transfer of rebar force can be illustrated as follows, each bar with peak at the start of the lap and zero at its end, in contrast to that of bond which may assume uniformity along the rebar.

As the presence of the two rebars reduces the effective contact area with concrete for each bar, lap length therefore generally has to be longer than bond length. The Eurocode BSEN1992-1-1:2023 (2023) requires determination of lap length to be based on Equations 11.2 and 11.3 with at least 1.2 times the bond length and 15 times of the rebar diameter, whichever is the greater. For calculation, BSEN1992-1-1:2023 (2023) requires the replacement of c_s replaced by the sum of clear bar spacing on both sides of a rebar under consideration.

The ACI318-25, by Equation 11.4 listed above, requires a longer bond length by c_b being zero or a small value in lapping of bars.

Transverse Reinforcement at Laps:
As discussed and illustrated by Figure 11.1, circumferential tension around a stressed rebar will be incurred, which tends to crack the concrete, as shown in Figure 11.6.

The tensile stress at the concrete cover can be large enough to crack the concrete, which is undesirable. An effective means to reduce cracking is to add transverse rebar to intercept the cracking, as shown in Figure 11.7. It has also been discussed in the foregoing that the reduction of the possibility of cracking will strengthen the concrete for the provision of bond strength.

Irrespective of the necessity in the increase of bond strength, the Eurocode BSEN1992-1-1:2023 (2023) requires provision of transverse reinforcement with a total cross-sectional area not less than the cross-sectional area of one longitudinal bar within a lap. The transverse rebars are to be placed at 0.3 times of the lapped length at both ends.

In addition, as lapping of bars is a location of potential weakness, it should be avoided in "critical sections" where the local internal forces are strong, e.g., beam and column ends. So, although it is convenient to lap column bars at floor levels which are the column ends, some codes do specify that lapping be moved to the mid-floor level, especially for ductility design.

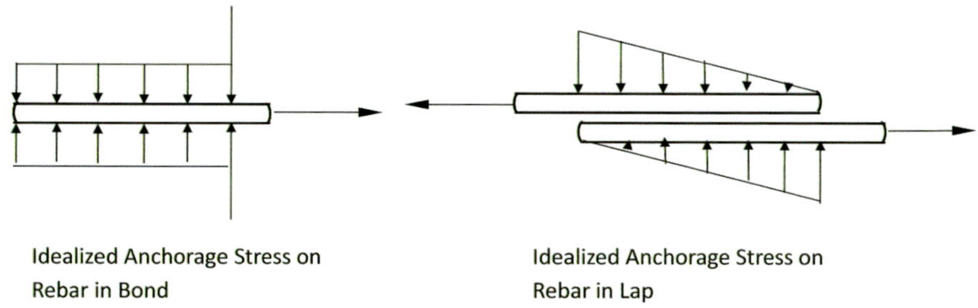

FIGURE 11.5 Idealized stress profiles on rebars in bond and lap.

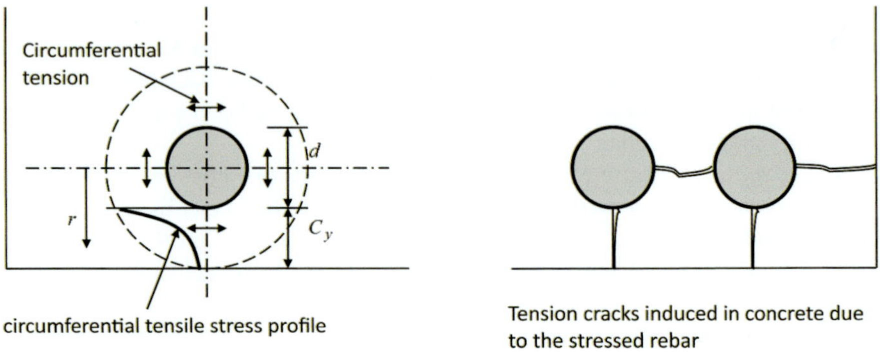

FIGURE 11.6 Tensile stress induced on concrete due to bonds by rebar.

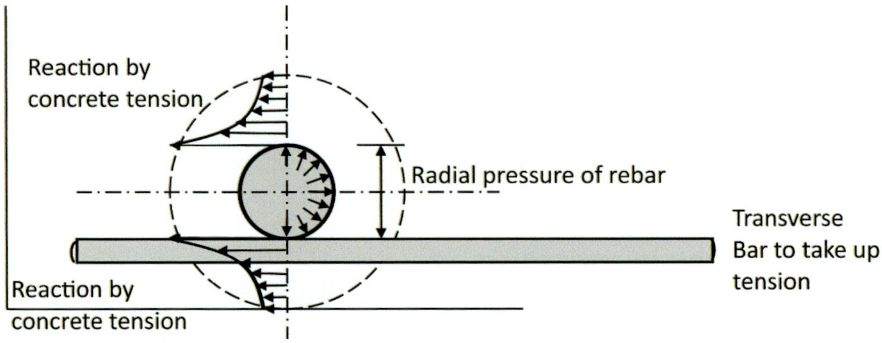

FIGURE 11.7 Transverse bar to resist concrete tensile stress due to bond.

11.4 REBAR BENDING RADIUS, BAR BEARING STRESS

Unavoidably, rebar has to be bent for fixing to various angles in the concrete structure. However, to preserve material integrity, the bend radius has to be limited; otherwise, excessive lengthening at the outer surface would occur, leading to significant overstressing or even cracking of the rebar. The excessive bar deformation will also to some extent exhaust the ductility of the material which is originally intended to be provided in design. So various codes set lower limits for the bend radius of a rebar, generally a multiple of 3 or 4 times its rebar diameter for the inner radius of the bend. For example, the Eurocode BSEN1992-1-1:2023 (2023) and BSEN1992-1-1:2004 (2004) specify the bend diameter (known as mandrel diameter) as 4 times and 7 times the rebar diameter for rebar diameters not greater than 16 mm and greater than 16 mm respectively. The American Code ACI318M-19 (2019) specifies minimum bend diameters to be 6 to 20 times the rebar diameter for different ranges of rebar diameter. Generally, the allowed minimum bend radius increases with bar diameter, and mild steel can tolerate a smaller diameter than high tensile steel.

In addition, when the rebar is tensioned, it exerts radial pressure against the concrete inside the bend. The pressure increases with decrease of the bend radius, which

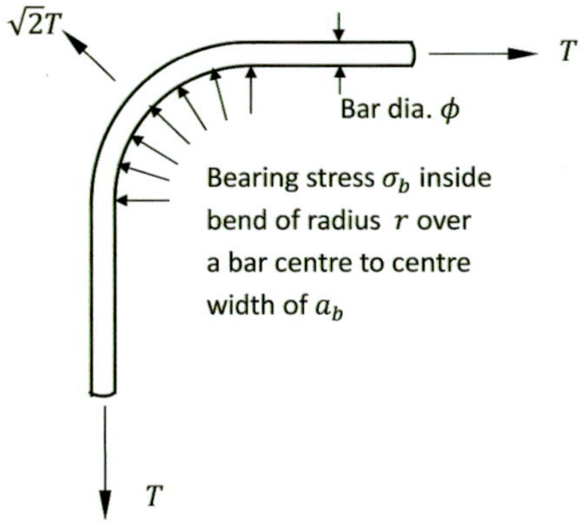

For equilibrium, the radial bearing stress has to balance a resultant force of $\sqrt{2}T$.

$$\therefore \int_0^{\pi/2} \sigma_b r\phi d\theta \cos\left(\frac{\pi}{4}-\theta\right) = \sqrt{2}T$$

$$\Rightarrow \sigma_b r\phi \int_0^{\pi/2} \cos\left(\frac{\pi}{4}-\theta\right) d\theta = \sqrt{2}T$$

$$\Rightarrow \sigma_b r\phi \int_0^{\pi/2} \left[\cos\frac{\pi}{4}\cos\theta + \sin\frac{\pi}{4}\sin\theta\right] d\theta = \sqrt{2}T$$

$$\Rightarrow \sigma_b r\phi \frac{1}{\sqrt{2}} \times 2 = \sqrt{2}T \Rightarrow \sigma_b = \frac{T}{r\phi}$$

FIGURE 11.8 Simplified model for estimation of bearing stress inside rebar bend.

could potentially crush the concrete and weaken the bond. The former British Code BS8110-1:1997 (1997) provides a formula for the calculation of bearing stress inside a standard right-angle bend based on a simple model as shown in Figure 11.8, under the assumption that the bearing stress is radial and constant over the bend.

The calculated bearing stress is used to check the bearing capacity of concrete. BS8110-1:1997 (1997) limits the bearing stress as shown in Equation 11.5.

$$\sigma_b = \frac{T}{r\phi} \le \frac{2f_{cu}}{1+2\dfrac{\phi}{a_b}} \tag{11.5}$$

where f_{cu} is the characteristic cube strength of concrete.

From Equation 11.5, it is obvious that the bearing pressure increases with decrease in rebar diameter. So, this is another controlling criterion on the bend radius of rebar. Nevertheless, the Eurocode BSEN1992-1-1:2023 (2023) 11.3(3) allows the checking of concrete bearing stress to be omitted if the bend complies with its Figure 11.6 reproduced in Figure 11.9 as follows, and under conditions (i) $f_{yk} \le 500$ MPa, and (ii) $f_{ck} \ge 25$ MPa; (iii) bar clearance $c_x \ge 1.5\phi$ from an edge parallel to the bend.

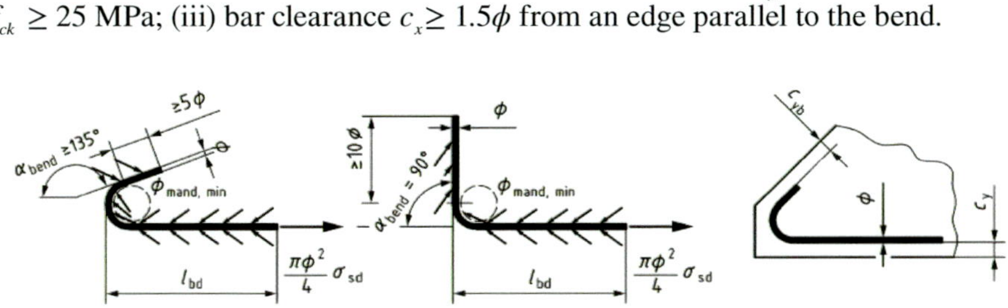

a) standard hook

b) standard 90° bend anchorages

c) definition of nominal cover $c_d = \min\{c_s/2; c_x; c_y; c_{yb}\}$ (definition of c_s and c_x, see Figure 11.3 c))

FIGURE 11.9 Anchorage with standard hook and bend in tension, figure 11.6 of BSEN1992-1-1:2023.

As a bend offers greater resistance to the tension of the bar at its far end by the concrete bearing, some codes allow the reduction of bond length in the presence of a bend. BSEN1992-1-1:2023 (2023) 11.4.4, permits tension bond length to be reduced by 15ϕ for bend with a bent angle not less than 90°, with the minimum value of 10ϕ.

11.5 CONCRETE COVER AND SPACING OF REINFORCEMENT BARS

As discussed in Section 11.3, concrete cover and spacing of rebar are controlling factors on bond strength.

In addition, concrete cover is also essential in enhancing durability as it acts as a protective layer to rebar against (i) environmental corrosion such as rusting, carbonation, chloride, and sulphate attack; and (ii) fire attack as has been discussed in Chapter 8.

Adequate of spacing for rebar is also important for concreting for minimization of risk in honeycombing.

11.6 PREVENTION OF "BUCKLING" OF COMPRESSION BARS

A rebar under strong compression may "buckle" outwards to the exterior if the lateral restraint by embedding concrete and transverse rebars (in the form of links) are not adequate. Nevertheless, as explained in Chapter 2, the phenomenon of buckling is generally not correct. The rebar actually fails by excessive P-delta effect due to the "initial imperfections" inherent in it which create high eccentric moments under high compressive loads. Such failures can easily take place under strong cyclic loads due to seismicity.

Concrete cover can restrain rebar from buckling to certain extent. However, the more effective means is by transverse rebar in the form of "tie" (or link, stirrup). The link serves to reduce the "effective length" of the rebar so that its lateral movements will be reduced. Requirements in terms of diameters and spacing of such ties are often specified in codes of practice, especially in areas of high compressive stresses to reduce the risk of buckling, although the more important function of these ties are "confinement" which will be discussed in the next section.

11.7 CONFINEMENT BY LATERAL TIES – ENDS IN COLUMNS, BOUNDARY ELEMENTS IN WALLS

The effect of confinement by lateral ties to enhance bond strength of rebar in concrete has been discussed in the previous sections. Nevertheless, another important phenomenon is the increase in concrete strength and ductility by these ties. Such increases are especially important at beam ends, column ends, and boundary elements in walls which are highly stressed.

Generally, when the links in the transverse direction of the concrete section are stressed, they will create lateral compression onto the concrete as illustrated in Figure 11.10. The stress can be quantified under the assumption that they are stressed

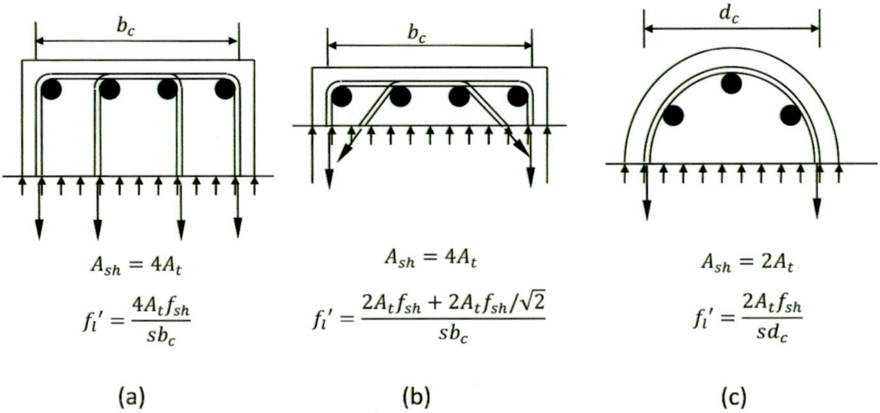

FIGURE 11.10 Full confining stresses provided by different arrangements of transverse reinforcement.

to their ultimate design values. In the figure, A_{sh} is the total cross-sectional area of the links, A_t is the cross-sectional area of a single leg of link, f_{sh} is the ultimate strength of the links, f_l is the lateral pressure created by the links, s is the vertical spacing of the ties.

The increase in concrete strength can be illustrated by Figure 11.11 which is well recognized by many researchers including Mander et al. (1988). In the figure, not only has the peak strength of the concrete increased, there is also a significant increase in concrete strength at the "post-peak" portion, indicating that the concrete

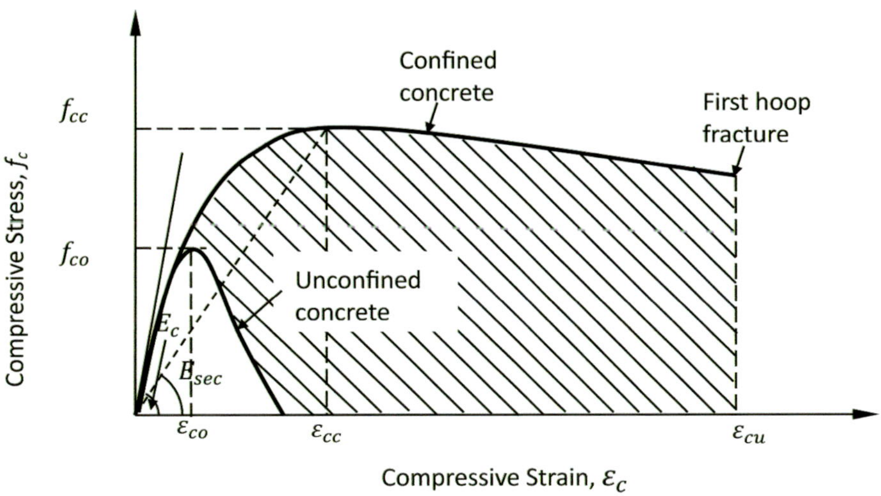

FIGURE 11.11 Stress-strain model proposed for monotonic loading of confined and unconfined concrete.

can tolerate much higher strain (i.e., deformation) after it has attained its maximum stress, which is the ductility behaviour favoured by designers.

The reason why lateral confinement can increase concrete strength is because the lateral confinement restrains concrete from lateral dilation, which is a sign of failure in compression by longitudinal splitting, as illustrated in Figure 11.12.

Nevertheless, Mander et al. (1988) considers that the laterally confining stress $\left(f_l^{'}\right)$ if provided by links in rectangular section as shown in Figure 11.10 is not uniform within the cross-section. Due to the arching effect he assumes, there are areas (within the parabolic areas between the links as shown in Figure 11.13) not stressed and the effect of the confinement has to be "discounted" by eliminating these areas.

Considering that only the shaded part is effective under the lateral confined stress, the confined stresses derived in Figure 10.13 have to be multiplied by the following factor $k_{e,rect}$ for discount.

$$k_{e,rect} = \left(1 - \frac{1}{b_c d_c} \Sigma \frac{(w')^2}{6}\right)\left(1 - \frac{s'}{2b_c}\right)\left(1 - \frac{s'}{2d_c}\right) \div (1 - \rho_{cc}) \tag{11.6}$$

where ρ_{cc} is the ratio of the longitudinal bars and other symbols are defined in Figure 11.12.

For circular sections, as the arching effect shown in Figure 11.10(c) does not exist, the term $\Sigma \frac{(w')^2}{6}$ in Equation 11.6 does not apply; the $k_{e,cir}$ factor will only be

$$k_{e,cir} = \left(1 - \frac{s'}{2b_c}\right)\left(1 - \frac{s'}{2d_c}\right) \div (1 - \rho_{cc}) \tag{11.7}$$

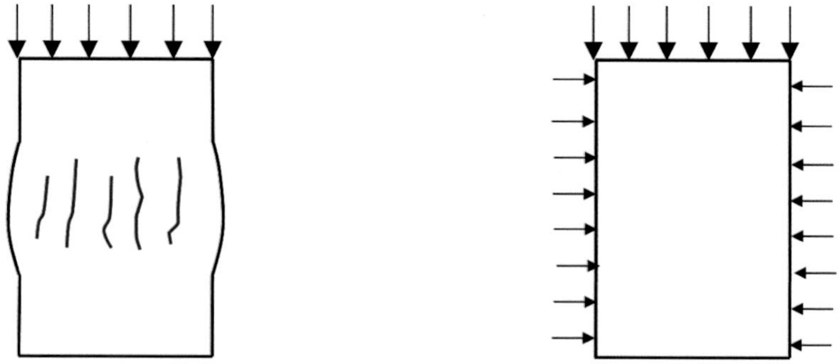

a) Failure by compression forming longitudinal splitting cracks

b) Lateral confinement stress help reduce risk of splitting cracks

FIGURE 11.12 Lateral confinement stress help reduce the risk of splitting cracks when concrete is under compression.

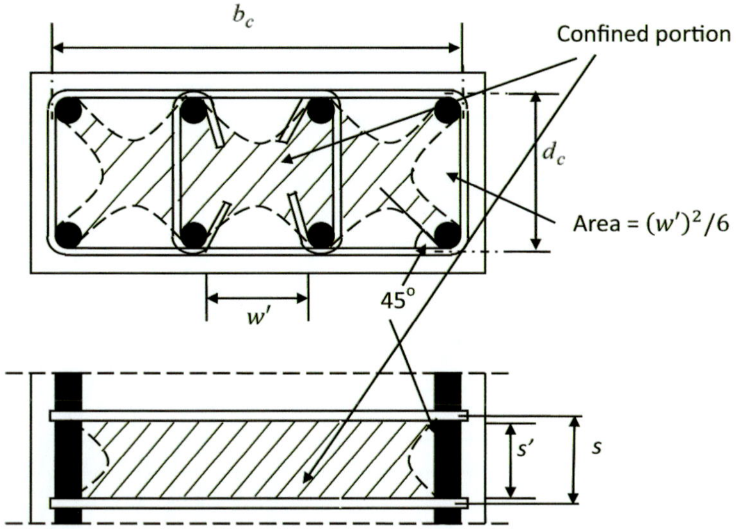

FIGURE 11.13 Illustration of effectively confined core for rectangular hoop reinforcement.

For rectangular section, Mander et al. (1988) proposes that under triaxial compression with equal effective stress in all directions, the concrete strength will be increased from f_{co}'under no confinement to f_{cc}' by the following equation:

$$f_{cc}' = f_{co}'\left(-1.254 + 2.254\sqrt{1 + \frac{7.94 f_l'}{f_{co}'}} - 2\frac{f_l'}{f_{co}'}\right) \quad (11.8)$$

If the lateral confined stresses derived in Figure 11.9 in the two directions for a rectangular section are not identical, Mander et al. (1988) provides a chart for the estimation of the factor f_{cc}'/f_{co}' as extracted in Figure 11.14.

In Figure 11.14, under confinement stress ratios at $f_{l2}'/f_{co}' = 0.17$ and $f_{l1}'/f_{co}' = 0.09$, the confined strength ratio f_{cc}'/f_{co}' increase is 1.65.

The Eurocode BSEN1992-1-1:2023 (2023) has incorporated the approach of Mander et al. (1988) in the determination of lateral confinement stress. However, it takes a more conservative approach in the determination of increase in concrete strength by adopting the lesser value in the two directions. The concrete strength is by simple formulae as listed below.

$\Delta f_{cd} = 4.\sigma_{c2d}$ for $\sigma_{c2d} \le 0.6 f_{cd}$

$$\Delta f_{cd} = 3.5.\sigma_{c2d}^{3/4}.f_{cd}^{1/4} \quad \text{for} \quad \sigma_{c2d} > 0.6 f_{cd} \quad (11.9)$$

where f_{cd} is the design compressive strength of concrete; Δf_{cd} is the increase of f_{cd}; σ_{c2d} is the lateral confinement stress.

To achieve good confinement of concrete by single leg links, it is a good practice to use links with end hooks of bent angles not less than 135°. Otherwise, the end bend

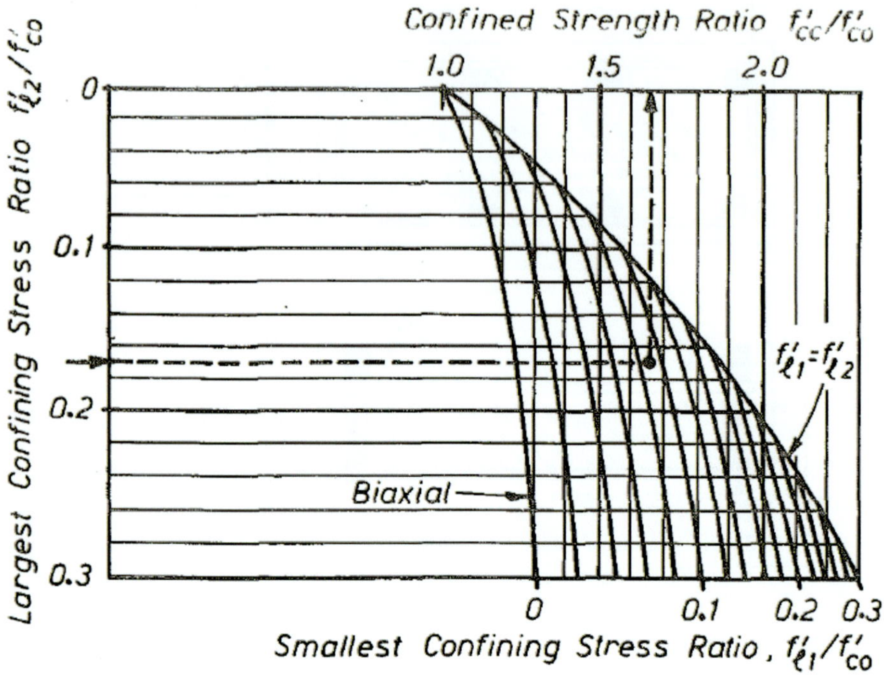

FIGURE 11.14 Extract from Mander et al. (1988) figure 4.

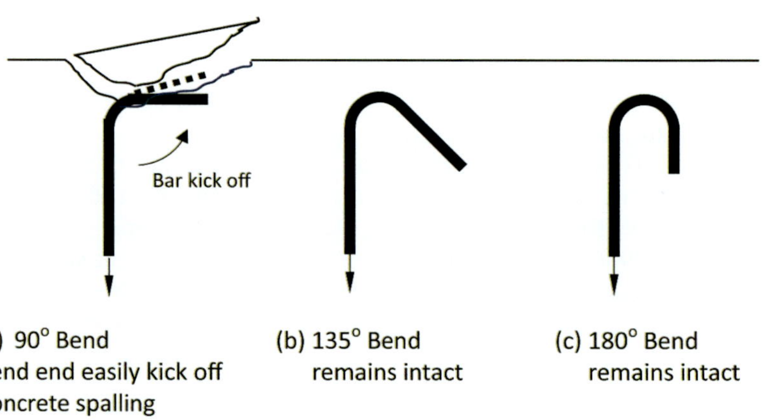

FIGURE 11.15 Performance of links at different bend angles.

can easily "kick off" under tension, which causes concrete spalling as illustrated in Figure 11.15. Such requirements are common in ensuring ductility. In addition, such links can also perform better in restraining vertical bars from buckling as they are more difficult to "kick off."

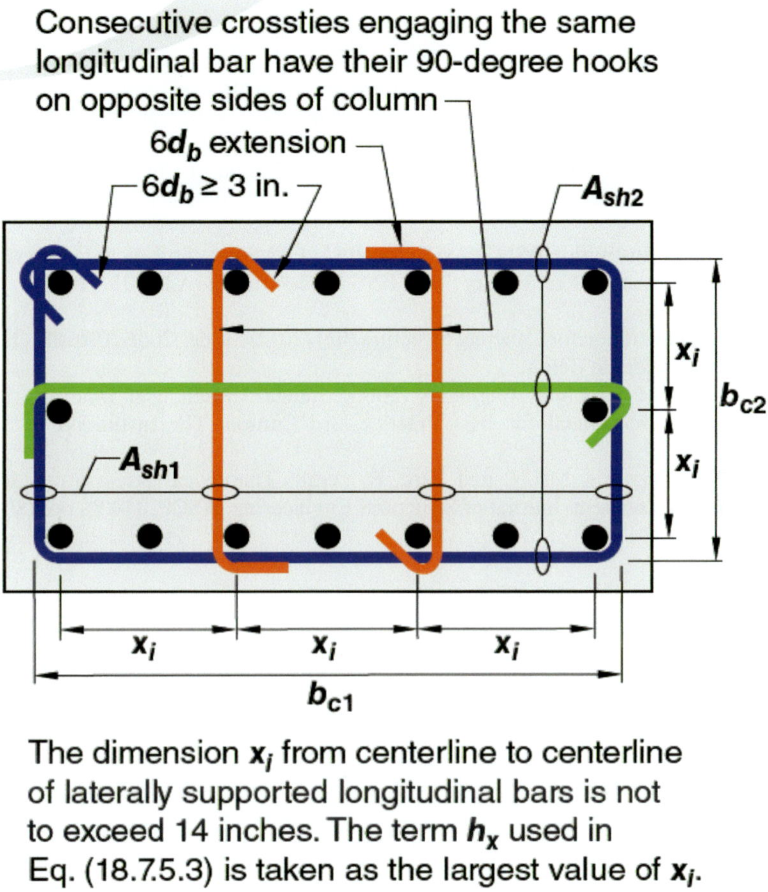

FIGURE 11.16 Extract from ACI318-25.

However, links with angle bends greater than 90° are difficult (or time-consuming) to install. ACI 318M-19 allows provision of alternate link arrangement as shown in its figure R18.7.5.2 extracted in Figure 11.16 of this book. By this configuration, a certain degree of confinement and vertical rebar restraints can be ensured in opposite faces of the section.

11.8 CONCLUSION

An account of reinforced concrete detailing has been outlined in this chapter, with emphasis on the working principles. Provisions and requirements from codes are quoted for illustration as appropriate. More discussion about the detailing of reinforcement is also given by the Institution of Structural Engineers (2006).

REFERENCES

ACI318M-19 (2019) Commentary on Building Code Requirements for Structural Concrete (ACI 318R-19), Reported by ACI Committee 318. American Concrete Institute.

British Standard Institution (1997) BS 8110-1: Structural Use of Concrete. British Standard Institution.

British Standard Institution BSI (2004) Eurocode 2: Design of Concrete Structures – Part 1-1: General Rules and Rules for Buildings BS EN 1992-1-1:2004. British Standard Institution.

British Standard Institution BSI (2023) Eurocode 2: Concrete – Specifications, Performance, Production and Conformity BS EN 1992-1-1:2023+A2:2021. British Standard Institution.

International Federation for Concrete Structure (fib) (2012) Model Code 2010. fib – Fédération Internationale du Béton.

The Institution of Structural Engineers (2006) Standard Method of Detailing Structural Concrete: A Manual for Best Practice, 3rd Edition. The Institution of Structural Engineers.

Mander, J.B., Priestley, M.J.N. and Park, R. (1988) Theoretical Stress-Strain Model for Confined Concrete. Journal of Structural Engineering, ASCE, 114(8), pp. 1804–1825.

12 Precast Concrete Design and Construction in Building

12.1 INTRODUCTION

Precast concrete is a construction method where concrete elements are cast, cured, and manufactured in a controlled factory environment before delivery to the construction site for assembly. This approach offers many advantages over conventional cast-in-place concrete. The advantages include improved quality control as the precast components are constructed in factory-controlled conditions, which can also lead to high efficiency of production with the least wastage. Thus, sustainability is also enhanced. Economy can generally be achieved if under large-scale production. In addition, the reduced long-term maintenance by the good quality of work also helps reduce costs. While the site construction can eliminate casting of concrete, faster overall construction speed can often be achieved as only mechanical maneuvering and assembly of precast components are required on site. This type of construction is also particularly favoured in countries with a shortage of skilled labour in building construction.

However, precast construction also presents design and logistical challenges. Structural connections between precast components and in situ concrete require careful engineering to ensure adequate stiffness, ductility, and durability. The structural connections often exhibit reduced rigidity, robustness, and ductility, which can affect overall structural performance if not properly addressed. Additionally, handling, transportation, storage, and on-site assembly demand meticulous planning, especially in congested or hard-to-access sites.

Despite these challenges, precast concrete construction has gained popularity in many countries due to its quality, efficiency, and sustainability benefits. The method has evolved from producing individual structural elements (such as beams, slabs, and stairs) to entire modular units, including bathrooms, kitchens, and even complete living spaces, now widely known as modular integrated construction (MiC). The Housing Authority of Hong Kong, whose role is to provide public housing to the local residents at large, has been extensively exploring, developing, and using precast concrete construction in the past decades. Its work has essentially drafted the history of the evolution of precast concrete construction as demonstrated in a slide from one of its presentation by Wong (2021) which is extracted in Figure 12.1. In the figure, various precast components in typical living units are displayed.

DOI: 10.1201/9781003636380-12

Significant research and innovation continue to explore and improve by addressing the limitations of precast systems. Cheng has adopted precast concrete truss plank for a footbridge construction in Hong Kong and has carried out relevant research works (Cheng and Fan 1993, 1994). More discussions about precast concrete structures are given by Steinle et al. (2019)

This chapter explores key aspects of precast concrete design and construction in buildings, discussing best practices, challenges, and advancements in the field.

12.2 MANUFACTURE OF PRECAST CONCRETE UNITS

Except for the very heavy structures such as transfer plate or pile cap, nearly all concrete structural elements, which include slab, façade, staircase, beam, column, and wall can be constructed as precast units. Historically, precast concrete construction started from the light and minor elements such as façade, staircases, etc. and gradually evolved to more important structural elements. Nowadays, precast units also include beams, columns, walls, and assemblies of them. Whether a structural element can be "precast" depends on its size and weight as limited by the power of transportation and lifting.

The structural elements are cast mostly in steel or fibre glass moulds which can be reused many times in the factory. Also, these moulds offer precise dimensions with smooth surfaces. In addition, grout loss is reduced to a minimum. Reinforcing bars can also be cut and bent precisely in the factory or formed into pre-fabricated cages with spot welds. Steam curing is sometimes employed for the quick gain of strength for early lifting. Walls can be cast horizontally, which is better than the vertical in

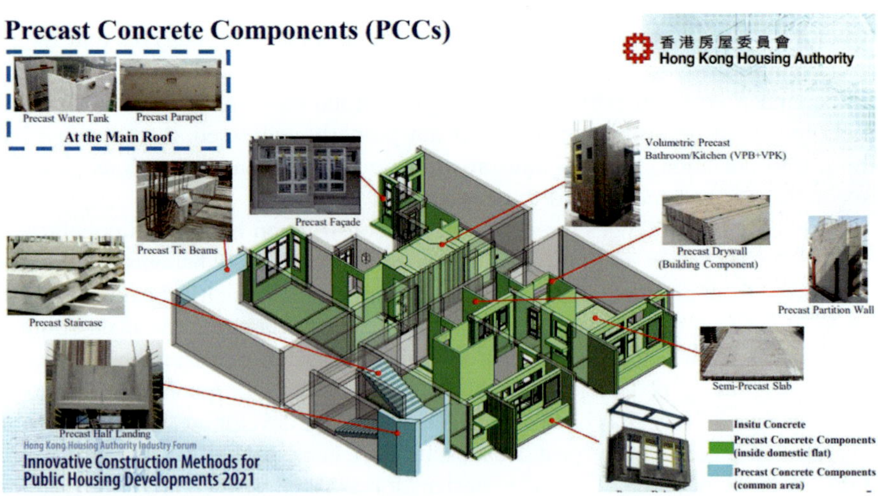

FIGURE 12.1 Use of precast concrete components in buildings by the Hong Kong housing authority.

situ casting as the problem of segregation and non-uniformity due to gravity can be reduced.

So, under factory conditions, workmanship and quality control are generally better than those on site.

12.3 DESIGN CONSIDERATIONS

Design considerations in precast concrete construction for member design are similar to those of in situ concrete. The Eurocode BSEN1992-1-1:2023 (2023) is applicable to both in situ and precast concrete, with some additional rules for the latter including connection design, provisions for heat curing, creep, and shrinkage for concrete, etc. The Eurocode provides approaches with formulae to modify concrete strengths due to heat curing and differential creep and shrinkage. Similarly, the American Code ACI318-25 (2025) incorporates provisions and requirements for precast concrete as parts of the code.

12.3.1 Connections

In terms of design, the main differences in structural behaviour between cast-in-place and precast concrete buildings in stiffness, robustness, and ductility are attributed by the structural connections. While the in situ connections with full lapping of bars (monolithic) are best in ensuring structural continuity and rigid joint formation, the precast concrete connections depend very much on the design configuration and construction workmanship to achieve similar performance. Of course, there are exceptions where the connection is intentionally designed as pinned or semi-rigid (up to a certain level of bending capacity) instead of a full monolithic connection.

The following types of structural connections are discussed, with respect to the formation of monolithic connections or pinned connections:

(1) Structural Connections in Beams

A typical connection of precast concrete beam to the in situ concrete structure is shown in Figure 12.2. The precast beams may be left with the top portion un-concreted but with transverse reinforcements exposed for anchoring of top bars and shear resistance. Rigid joint connection can generally be assumed in such a configuration.

Pinned connection can also be designed with the full precast concrete beams resting on column and wall supports generally through the use of corbel or half joint as shown in Figure 12.3. Dowell bars may be used at the supports for fixing purposes.

Under the pinned connection, moment-resisting frames by beams on column/wall cannot be formed. The lateral stability of the building has to rely on other structural elements, which may be stiff core walls or other frames with beams under rigid connections with the column/wall structural elements.

(2) Connection in Slab

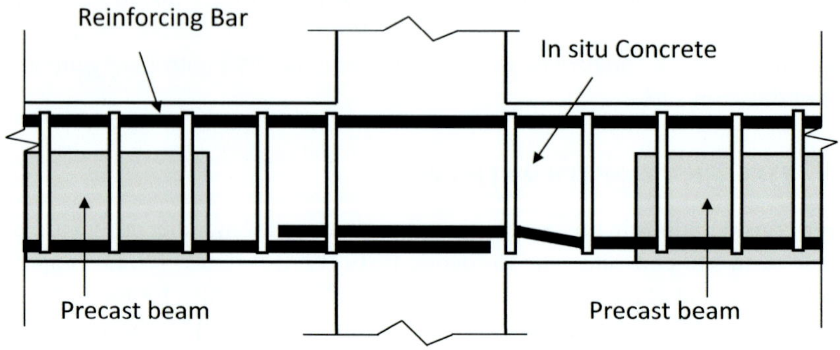

FIGURE 12.2 A typical monolithic connection for precast concrete beam.

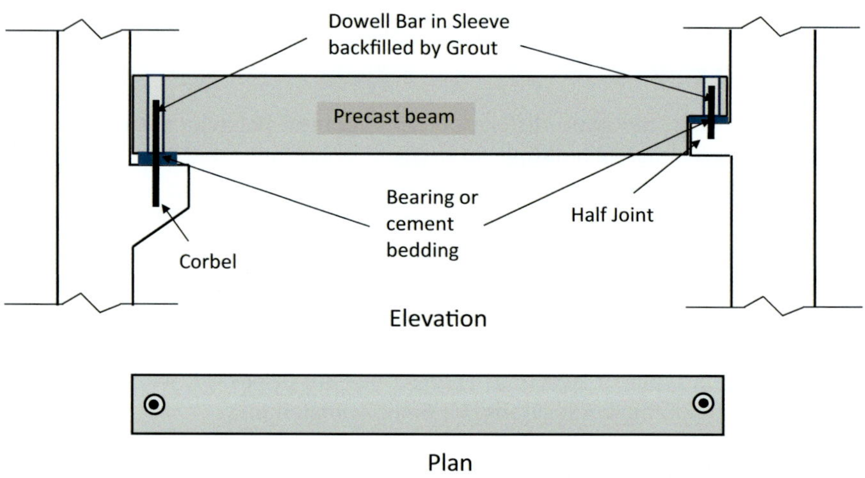

FIGURE 12.3 A typical pinned connection for precast concrete beam.

The joints are similar to that of beam as shown in Figure 12.4. However, the precast portion may form only half of the slab thickness, which may also act as formwork for the in situ portion. So, this may be termed as semi-precast slab construction.

As similar to beams, full depth slabs can also be supported on half joints or corbels on beams or walls with similar details as shown in Figure 12.3.

(3) Connections (Vertical) in Precast Columns and Walls

The connection of precast columns and walls at successive floors is achieved by lapping longitudinal reinforcing bars. A widely used technique involves embedding long sleeves in the upper column or wall segment. These sleeves are designed to

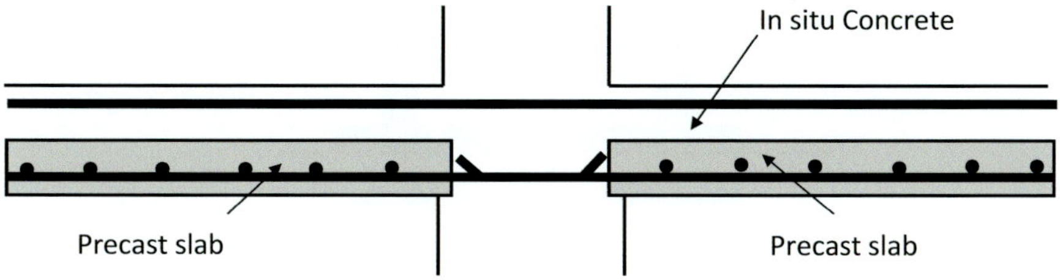

FIGURE 12.4 Typical precast slab construction.

receive the protruding bars from the lower segment. Once the elements are positioned, the sleeves are filled with grout to complete the structural joint, as illustrated in Figure 12.5. To streamline construction, the sleeves are frequently connected by ducts, enabling all grouting to be performed in one continuous operation.

Instead of lapping longitudinal bars in sleeves filled with grout, the grouted splice sleeve is a newer technique that improves upon the lapping of reinforcing bars in left-in sleeves in concrete by grouting as described previously. The grouted splice sleeve involves a steel tube that accepts two reinforcing bars, one inserted into each end. Once the bars are in place, the tube is filled with high-strength grout to lock them in position (Figure 12.6). A major benefit is that it requires a much shorter lap length (as short as 250 mm) compared to conventional lapping, which is enabled by the high-strength grout.

This product functions as a mechanical coupler but is more suited for precast concrete, where it is impossible to screw the longitudinal bars into a threaded coupler. In

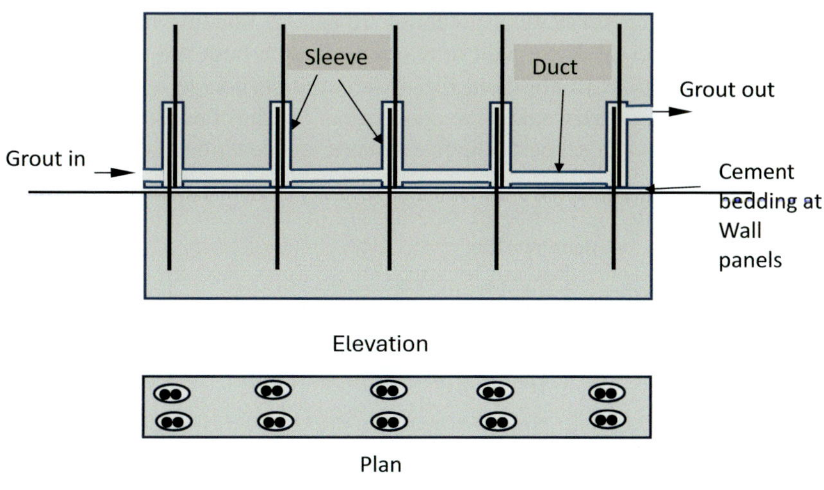

FIGURE 12.5 Connection between walls in successive floor.

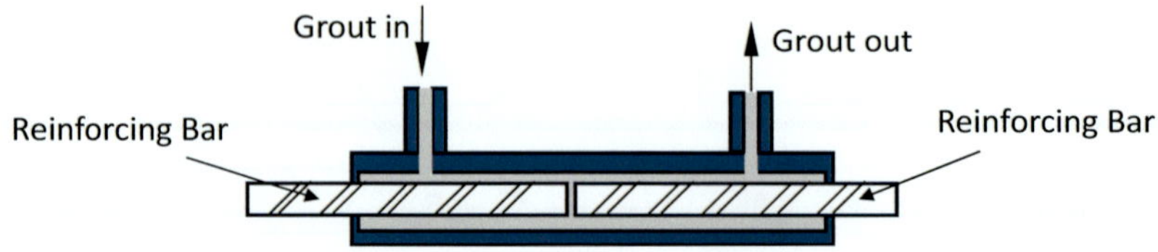

FIGURE 12.6 Splice sleeve in splicing reinforcing bars.

addition, splice sleeves can be designed to withstand repeated stresses of an earthquake, allowing them to be used in seismic design.

(4) Joints (Horizontal) in Walls

Very often, it is favoured to structurally integrate a wall of a modular unit (e.g., a volumetric precast bathroom or kitchen or even a living unit) with an in situ wall. Figure 12.7 shows the integration.

In Figure 12.7, the zig-zag form of the precast wall panels serves two purposes: (i) increase the stiffness of the precast unit as required during handling; and (ii) increase friction with the in situ wall. In addition, the shear hoops act as shear connectors to connect the precast panel to the in situ wall.

If two precast wall panels each from a modular unit are to be joined together, the details as shown in Figure 12.8 can be used.

An extract from another slide of Wong (2021) shows the joining of walls by which the details in Figures 12.7 and 12.8 are applicable (Figure 12.9).

Summing up, joints in precast concrete buildings are generally weaker than those in in situ buildings due to the presence of "cold joints," which are unavoidable in precast concrete construction if the connections are not well designed and constructed. The cold joints, being surfaces of potential slippage if bonding between the grout and the precast concrete is not well developed, weaken both the strength and ductility of the connection. In addition, the differential shrinkage of concrete creates stresses that would further weaken the connection. So, the connections are locations of potential weakness in precast concrete structures. Nevertheless, these joints can

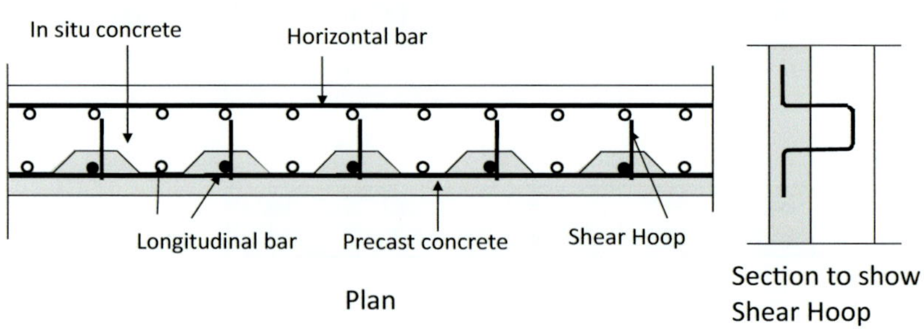

FIGURE 12.7 Precast wall panel joining an in situ wall.

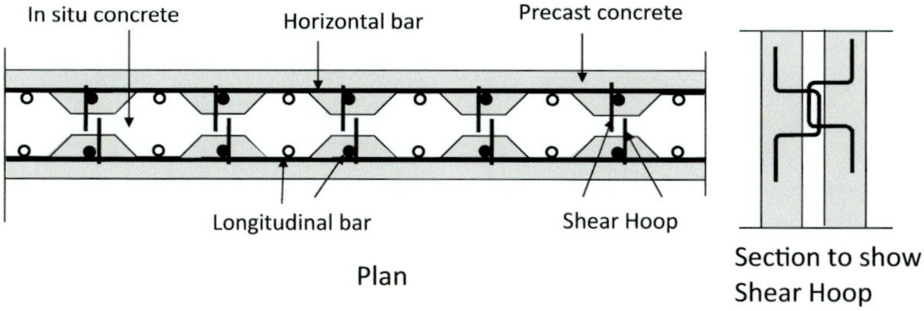

FIGURE 12.8 Sandwich wall for joining two precast wall panels.

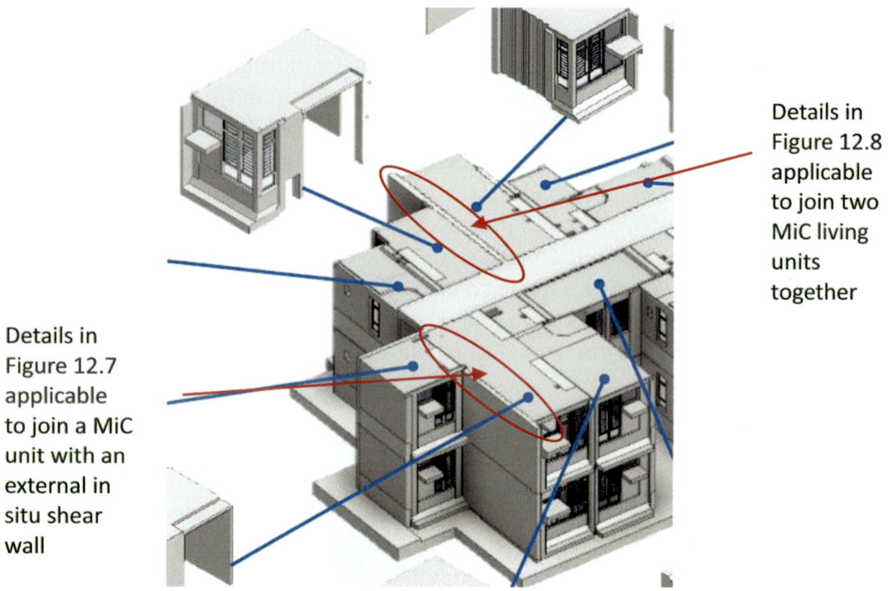

FIGURE 12.9 Extract from Wong (2021) showing applicability of Figures 12.7 and 12.8.

be strengthened by the addition of shear links. ACI318-25 (2025) provides a general approach. By this approach, reinforcing bars in tension across a potential shear failure face can exert a force normal to the surface, which would provide friction for load transfer, as shown in Figure 12.10 extracted from figure R22.9.4.3a and b of ACI318-25 (2025).

The shear capacity of the interface where the reinforcing bars are in tension can be taken as the friction V_n can be calculated by the equation (22.9.4.3) of ACI318-25 (2025) reproduced in Equation 12.1:

$$V_n = A_{vf} f_y \left(\mu \sin\alpha + \cos\alpha \right) + \mu N_u \tag{12.1}$$

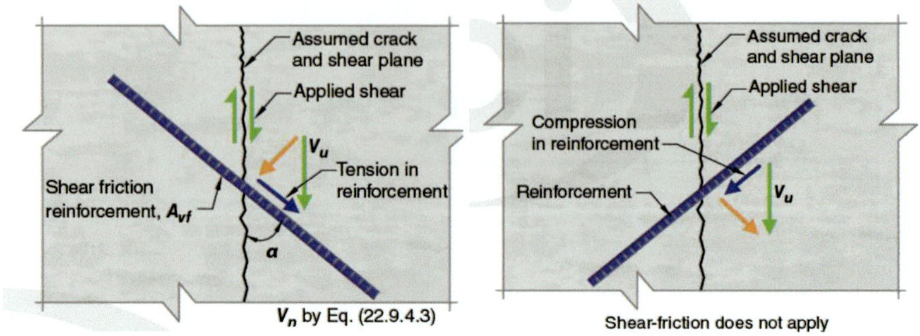

FIGURE 12.10 Extracts from ACI318-25 (2025) figure R22.9.4.3a and b.

where A_{vf} and f_y are the cross-sectional area and specified yield strength of reinforcing bars respectively; μ is the coefficient of friction being 1.4 for monolithic normal weight concrete, 1 for concrete placed on clean and intentionally roughed surface, 0.6 for concrete placed on clean and unintentionally roughed surface; N_u is the minimum factored compressive force concurrently acting with V_n. The equation is on the assumption that the reinforcing bars are fully stressed to f_y and provides the resistance directly by their components along the surface and friction due to their normal reaction. The normal compressive force N_u, if in existence would also be counted to providing friction. On the contrary, if the reinforcing bars are in compression, the shear friction would not exist. The Code of Practice for Precast Concrete Construction 2016 (2016) of Hong Kong provides a similar formula for resisting shear force V as

$$V = 0.6F_b \tan\alpha_f \tag{12.2}$$

where $F_b = 0.87f_yA_s$ with f_y as the yield strength and A_s as the steel cross-sectional area. $\tan\alpha_f$ plays the role of coefficient of friction ranging from 0.7 (smooth surface) to 1.4 (roughened or castellated surface without continuous insitu strips across the ends of connection) to 1.7 (roughened or castellated surface with continuous insitu strips across the ends of connection) depending on surface condition.

The shear capacity so arrived at can then be checked and converted to shear strength in terms of force per unit area by dividing the shear force by the respective areas with the reinforcing bar area A_s. The shear strength can subsequently be checked against the designed shear stress v at the interface which can be calculated by the first principle as

$$v = \frac{VQ}{zI} \tag{12.3}$$

where V is the design shear force, Q is the first moment of area to the interface under consideration, z is the width of the section and I is the second moment of area of the section.

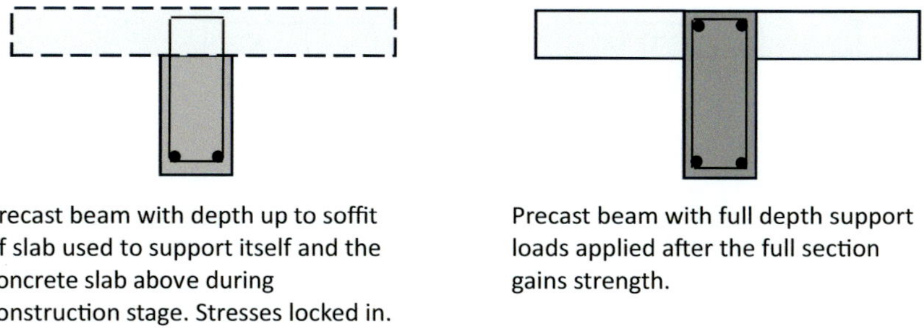

FIGURE 12.11 Illustration of staged application of load with respect to staged completion of beam sections.

12.3.2 Locked-in Stress

Precast components are often installed before completing construction to its full section. In these cases, an incomplete section is often used to support its self-weight and associated dead load as illustrated in Figure 12.11. Later, the completed section must support additional loads, such as superimposed dead, live, and wind loads which are applied after the full section gains strength.

The phenomenon differs from cast-in-place concrete, where the full structural section supports all loads from the outset. Consequently, precast elements are often subject to "locked-in" stresses. The following worked example demonstrates this design consideration.

Worked Example 12.1

A precast concrete beam of full section of 400(w) by 600(d) of clear span 10 m is first formed with precast section 400(w) by 450(d). It is used to support its own weight $0.4 \times 0.45 \times 24 = 4.32\text{kN/m}$, dead load from the adjacent slab of thickness 200 mm and 3 m tributary width of $3 \times 0.15 \times 24 = 10.8\text{kN/m}$. Finally, the beam has to carry superimposed dead load of 4.5 kN/m and live load 7.5 kN/m. The beam is designed as simply supported. The concrete mix is C35/45, i.e., $f_{ck} = 35\text{N/mm}^2$. Reinforcing bar $f_{yk} = 500\text{N/mm}^2$. The design code is BSEN1992-1-1:2023 (2023). Section RC design reference is made to Moseley et al. (2012).

Stage 1: When the beam has to support its precast section of 400(w) by 450(d), weight of wet concrete of the slab together with formwork load of 1.5 kN/m, construction load of 2.25 kN/m. The mid-span design moment is

$$M_{1,temp} = \frac{1}{8}\left[1.35 \times (4.32 + 10.8 + 1.5) + 1.5 \times 2.25\right] \times 10^2 = 322.65\text{kNm}$$

$$d = h - 50 - 16 = 450 - 66 = 384\text{mm}$$

$$\eta = 1,\ \alpha_{cc} = 0.85,\ \gamma_s = 1.5$$

$$K = \frac{M_{Ed}\left(= M_{1,temp}\right)}{f_{ck}bd^2} = \frac{322.65 \times 10^6}{35 \times 400 \times 384^2} = 0.1563 < 0.167, \text{ singly reinforced.}$$

$$z = d\left[0.5 + \sqrt{0.25 - \frac{0.5K}{\eta\alpha_{cc}/\gamma_s}}\right] = 384\left[0.5 + \sqrt{0.25 - \frac{K}{1.134}}\right] = 320.61\text{mm}$$

$$A_{s1,temp,req} = \frac{M_{Ed}}{0.87 f_{yk} z} = 2313\text{mm}^2, \text{(provide 3H32)}$$

Ignoring the construction load and the formwork which are removable, the design moment is M_1 becomes 255.15kNm, following the above steps, $A_{s1,req} = 1745\text{mm}^2$ which is required for the permanent design. This moment of 255.15kNm is "locked-in" the beam.

Stage 2 : When the slab hardens and the full beam section becomes 400(w) by 600(d).

While the own weights of slabs and beams have been resisted under the provision of 2011 mm², the full section is only required to resist the superimposed dead load and the live load which creates a design moment of

$$M_2 = \frac{1}{8}\left[1.35\times 4.5 + 1.5\times 7.5\right]\times 10^2 = 216.56\text{kNm}$$

$$d = h - 50 - 16 = 600 - 66 = 534\text{mm}$$

$$\eta = 1,\ \alpha_{cc} = 0.85,\ \gamma_s = 1.5$$

$$K = \frac{M_{Ed}\left(= M_2\right)}{f_{ck} bd^2} = \frac{216.56\times 10^6}{35\times 400\times 534^2} = 0.054 < 0.167, \text{ singly reinforced.}$$

$$z = d\left[0.5 + \sqrt{0.25 - \frac{0.5K}{\eta\alpha_{cc}/\gamma_s}}\right] = 534\left[0.5 + \sqrt{0.25 - \frac{K}{1.134}}\right] = 507.1\text{mm}$$

The neutral axis depth is $(534 - 507.1)\times 2 = 53.8\text{mm}$ which would not encroach into the compression zone of the previous precast beam.

$$A_{s2,req} = \frac{M_{Ed}}{0.87 f_{yk} z} = 982\text{mm}^2.$$

Adding to $A_{s1,req}$, total $A_{s,req} = 1745 + 982 = 2727\text{mm}^2$.

(Provide 4H32 where $A_{s,req} = 3217\text{mm}^2$)
Compared with the in situ scenario, the total moment =

$$M_{insitu} = \frac{1}{8}\left[1.35\times\left(4.32 + 10.8 + 4.5\right) + 1.5\times 7.5\right]\times 10^2 = 471.71\text{kNm}$$

$$= M_1 + M_2 = 255.15 + 216.56 = 471.71\text{kNm}$$

Following the above procedure, $A_{s,in-situ} = 2303\text{mm}^2 < A_{s,req} = 2727\text{mm}^2$.

So, it would be an under-design of reinforcement content if the construction sequence is ignored.

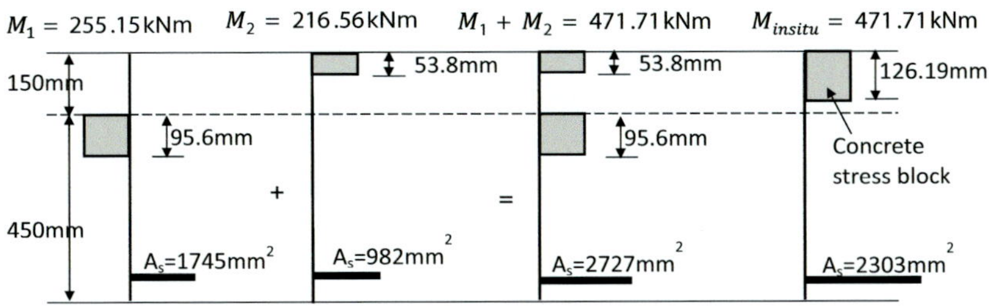

FIGURE 12.12 Worked Example 12.1 summary.

The sectional design of the stages and that of in situ are summarized in Figure 12.12.

Worked Example 12.2

In this worked example, a centre support is added in Stage 1 of Worked Example 12.1. The Stage 1 analysis would be that of the 400(w) by 450(d) precast beam on 3 supports and the RC design is carried out accordingly. The reaction on the centre support, R_m would also be obtained. In Stage 2, the centre support is removed when the slab concrete hardens and gains strength. The removal is equivalent to adding a downward force equal to R_m which would nullify the upward reaction. So, the analysis of Stage 2 is to add the loads of superimposed dead load and live load together as UDL with centre load R_m. Reinforcement is then designed to the full section accordingly. The bending moment diagram of the beam for the individual cases is as shown in Figure 12.13.

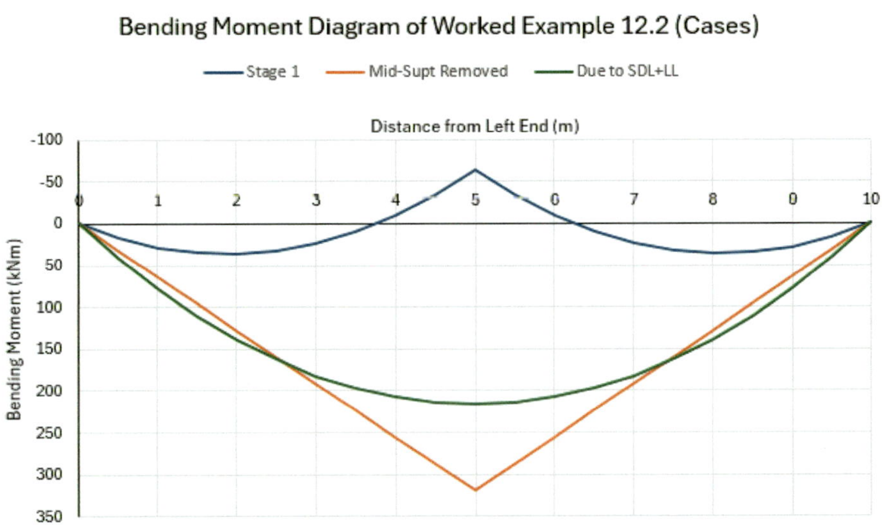

FIGURE 12.13 Bending moment diagrams – individual cases – Worked Example 12.2.

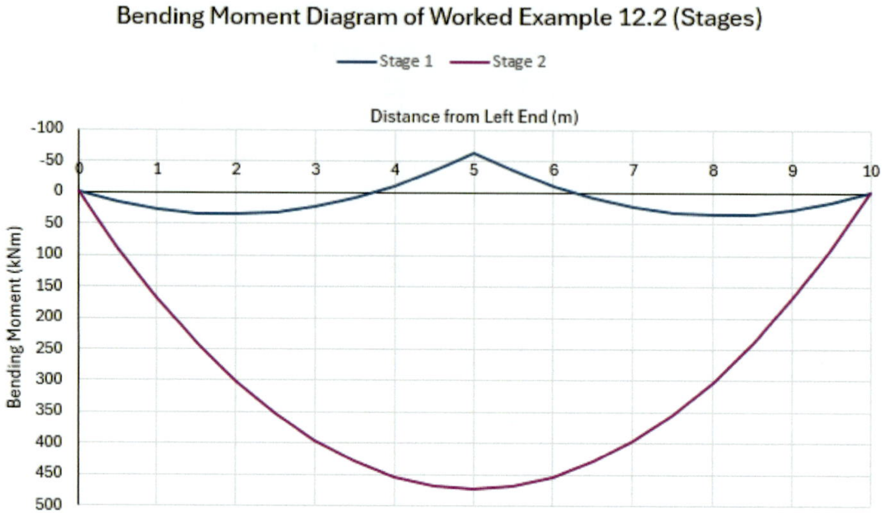

FIGURE 12.14 Bending moment diagrams – stages – Worked Example 12.2.

Summing up the cases for Stage 2 (final stage), the bending moment diagrams for Stage 1 and Stage 2 are finalized as in Figure 12.14.

Stage 2 bending moment is same as the in situ scenario for design by the full section. So, as far as the sagging moment at the mid-span of the beam is concerned, it is equal to that due to the full load on the beam. Nevertheless, to resist the temporary hogging moment at Stage 1, reinforcing bars near the top of the precast part of the beam is required. Readers may perform design as similar to Worked Example 12.1.

12.3.3 Robustness and Stability

The following related to the robustness and stability of the precast concrete buildings are discussed:

(i) Load paths

Load paths in structures are the routes through which vertical (gravity) and horizontal (lateral) loads such as those from wind or seismic activity are transferred from a building's superstructure down to its foundation. In both precast and cast-in-situ concrete construction, this typically involves slabs transferring loads to beams, which then transfer them to columns or walls, and finally to the foundation.

A critical feature of monolithic, cast-in-place structures is the development of multiple, redundant load paths. Forces from any single member can travel through several routes to the foundation, distributed according to the relative stiffness in each path. This redundancy is a key safety benefit. Should one path fail, the loads will be redistributed to others which carry the load. As such, the risk of disproportionate collapse as a measure of robustness of a building is reduced.

In contrast, the structural performance of precast concrete buildings is often not as robust in this regard. Due to their inherently weaker connections (as outlined in Section 12.2.1), precast concrete structures typically have fewer and less effective load paths, particularly for resisting lateral loads. For connections designed as pinned or weak, frames will be less effective at resisting lateral forces and overturning moments. Consequently, if a primary load path fails in a precast building, there are fewer redundant systems available to take up the excess load. This contrasts with the redundant nature of monolithic cast-in-place frames, which can better accommodate unexpected joint failures.

FIB (2008) discusses mechanism in the transference of loads through different types of connections in precast concrete buildings and suggests good practices for improvement. It suggests providing an alternate load path if practical. If not, the structure should be designed so that damages due to local failure should be kept through strengthening and the use of ties.

(ii) Ties

The tying systems to achieve robustness of buildings as specified for in situ buildings are applicable to precast concrete buildings. Generally, similar to other codes, BSEN 1991-1-1:2023 (2023) specifies four types of ties: (1) peripheral ties; (2) internal tie; (3) horizontal column and wall tie; and (4) vertical column and wall tie, as discussed in Section 5.4 of Chapter 5. The codes specify either minimum quantities of reinforcements directly or forces to be provided by these ties at their yield strengths. The requirements are identical in in situ buildings and precast concrete buildings.

Normally, the requirements of ties in in situ buildings can often be inherently satisfied by requirements in general reinforcement detailing. In contrast, precast concrete buildings, characterized by discontinuities between discrete components and cast-in situ connections, sometimes require these ties to be explicitly designed and added as additional reinforcements.

(iii) Minimum Lateral Load

As similar to the provisions of ties, the requirements of minimum lateral loads for ensuring robustness as described in 5.4 of Chapter 5 are applicable to both in situ and precast concrete buildings. Such minimum lateral load may have some bearing on the design of precast concrete buildings which may be inherently less robust than in situ buildings.

12.3.4 Stiffness

Stiffness of a concrete building is significantly affected by the rigidities of the connections of its members. The stiffness of a moment frame will be reduced if the connections are changed from rigid to semi-rigid or pinned. In case the rigid joint assumption is in doubt for a connection by limited restraints of member movements, particularly those involving an assembly of welded plates, grouted dowels or bolted brackets, the majority of which contain precast-to-precast or precast-to-*in situ* interfaces, it may have to be designed as a pinned or semi-rigid connection. Furthermore, plastic hinges may be more readily formed in the connections where precast and in situ members join, which would lead to greater deflections at high levels of stresses. Consequently, the stiffness of the building structure is reduced. In addition, the floor slabs as diaphragms in transmitting lateral loads also affect the overall stiffness of the building. The in-plane rigidity of the floor diaphragm can be significantly reduced under the use of discrete precast floor slab structures with ineffective connections. So, there are greater challenges in the modelling of the precast concrete structure for analysis.

12.3.5 Ductility

The requirements for ductility in precast concrete buildings are identical to that of in situ buildings. As the main differences between precast concrete buildings and in situ buildings lie mainly in the connections, the discussion focuses on the ductility provisions and requirements for connections.

As discussed in Chapter 8, the ductility requirements at connections can be summarized as

(i) Confinement of concrete core by transverse reinforcement and connections is to be avoided in the critical zone as illustrated in Figure 12.15 – The concept of the critical zone in the beam has been discussed in Chapter 8.
(ii) Long reinforcement bond length to avoid bond failure of reinforcement prior to yielding – Bond length determination based on over-strength of longitudinal bars.
(iii) Avoidance of lapping of reinforcement bars in critical regions (with high bending moment) – Lapping of reinforcement bars is not to take place at beam column joints.

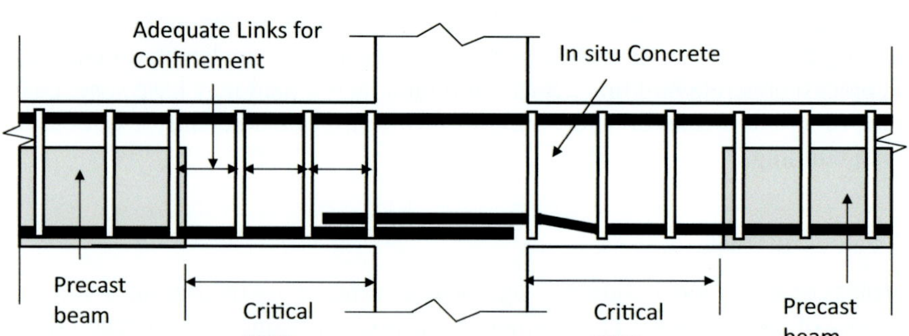

FIGURE 12.15 Adequate links for confinement purposes in precast beam.

(iv) Tension bars in the beam not to exceed 2.5% while the compression bar be not less than half of that of beam.
(v) Preferential failure mode such that beam failure takes place before column – Moment of resistance at beam ends not to exceed 80% of that of the supporting columns to fulfil the strong column weak beam requirement.

12.3.6 Connection Layout and Design for Lifting

As connections in precast concrete represent structural weaknesses, they should be minimized, and to be located in areas of low stress. In addition, constraints imposed by transportation and site lifting power, particularly for large, 3-dimensional elements like MiC modules, limit both element size and connection layout. Consequently, careful planning is essential to optimize the design and resolve these competing demands.

Locations of lifting hooks for the precast units should be designed and the structural capacities of the precast units checked against the loads induced during lifting. Reinforcements should be added if necessary.

12.4 CONSTRUCTION/INSTALLATION CONSIDERATIONS

12.4.1 Production Management in Factory

A quality assurance scheme should be set up by which quality control is to be implemented for the critical activities in the production factory, particularly on

(i) Concrete mix design, procedures of concreting, curing (including steam curing for early strength gain procedures);
(ii) Precise dimensions of the formwork and placing of reinforcements.
(iii) Testing of concrete and reinforcements to ensure adequacy in strength.
(iv) Where applicable, the finishing and fittings on the precast concrete units.
(v) All precast concrete units shall be traceable throughout the production, transportation, installation and post-installation stages. These are best achieved through the use of RFID (Radio Frequency Identification), which is an electronic tag buried in the precast concrete unit. It carries the unit's information in relation to the details of production, quality, etc.

12.4.2 Delivery/Site Storage

Precast units have to be well protected during transportation. Any extra stresses induced on the structure shall be checked and be additionally reinforced if necessary. Extra bracing for stability shall be required for members that are slender in the temporary stage. Figure 12.16 shows a part of a precast water tank being braced for stability during transportation as extracted from Wong (2021).

Delivery route and site storage shall be well planned with regard to the constraints in the delivery route and site storage area. Where necessary, temporary or permanent

FIGURE 12.16 A part of a precast water tank being braced during transportation.

structures shall be used for temporary support of the precast units, with additional reinforcements to the permanent structures as required. An example is to strengthen the large podium structure (if available on the site) to store the precast units.

12.4.3 Lifting and Installation

Safety and precision are important during lifting and installation

(i) The lifting appliances (mostly tower cranes) must have the adequate capacity to lift the precast unit.
(ii) Bracing to the precast units should be considered for stability during lifting for slender structure.
(iii) The precast concrete unit shall be checked for the stresses induced in it during lifting. Where required, a horizontal frame or beam may be used as shown in Figure 12.17 as extracted from the Hong Kong Precast Concrete Code 2016 (2016) to eliminate horizontal stresses in the unit.
(iv) Skilled labour is to be employed in the site installation process including levelling, grouting, bolting or welding for the connection formations so that the precast concrete units are precisely installed with adequate strength in the connections with others.

12.4.4 Post-Erection

The following post-erection measures are discussed:

(i) Strict control on the timing and procedure of dismantling the temporary supports has to be exercised to ensure safety.
(ii) Joint sealing at the joints between panels for the achievement of water tightness. Finishing shall be added.

FIGURE 12.17 Horizontal frame to nullify horizontal stresses in the precast unit during lifting.

(iii) The connections shall be inspected and tested in accordance with the regulations and the project specifications by verification of the grouting, welding and bolting as applicable.
(iv) The work should be well documented with as-built drawings.

12.5 CONCLUSION

As discussed, precast concrete offers significant advantages in the construction industry including fast speed of construction, high quality control, improvement of on-site safety, and environmental benefits through reduced waste and lesser future maintenance. However, the economy and feasibility in a particular building work depend on factors such as construction quantity, ease of access, and storage capacity of the Site. Availability of skilled labour is also a factor to be considered. In addition, larger structural sizes are often resulted with reduced robustness and stiffness in general. So, challenges include the higher initial cost, the need for specialized connections, and transportation limits for large or volumetric elements. Nevertheless, despite these challenges, precast concrete construction remains an attractive method, especially when combined with techniques like modular integrated construction (MiC), to enhance productivity and quality in building projects. There is a lot of room for improvement and future development of the construction method.

REFERENCES

American Concrete Institute ACI Committee 318 (2025) Building Code Requirements for Structural Concrete ACI 318M-25. American Concrete Institute.

British Standard Institution BSI (2023) Eurocode 2: Concrete – Specifications, Performance, Production and Conformity BS EN 1992-1-1:2023+A2:2021. British Standard Institution.
Cheng, Y.M. and Fan, Y. (1993) Modelling of Reinforcement in Concrete and Reinforcement Confinement Coefficient. Finite Elements in Analysis and Design, 13, 271–284.
Cheng, Y.M. and Fan, Y. (1994) Laboratory and Numerical Tests on Concrete Plank Composite with Reinforcement Truss. *Journal of Structural Engineering*, 20, 207–216.
Fib (CEB-FIP) (2008) Bulletin 43: Structural Connections for Precast Concrete Buildings. Fédération Internationale du Béton.
The Hong Kong SAR Government Buildings Department (2016) Code of Practice for Precast Concrete Construction 2016. Buildings Department, Hong Kong SAR Government.
Mosley, W.H., Bungey, J. and Hulse, R. (2012) Reinforced Concrete Design to Eurocode 2. Palgrave Macmillan.
Steinle, A., Bachmann, H. and Tillmann, M. (2019) Precast Concrete Structures, 2nd Edition. Ernst & Sohn, John Wiley.
Wong, W.H.R. (2021) Precast Concrete Construction and MiC for Public Housing Developments. Hong Kong Housing Authority.

13 An Illustrative Example for Structural Analysis and Design of a Building

13.1 INTRODUCTION

To elaborate the materials covered in Chapters 2 to 12 and to demonstrate the practical application of them, the analysis and design of a hypothetical 30 storeys high-rise reinforced concrete office building of height 128 m are performed and some details are contained in this chapter. The procedures comprise structural scheming, computer modelling, determination of loads and applications in the computer model. Load combinations are also carried out. Reinforced concrete design has also been carried out on selected structural members. The following Eurocodes are used in the analysis and design:

(i) BSEN1991-1-1:2025 for gravity loads assessment.
(ii) BSEN1992-1-4:2005 for wind load assessment.
(iii) BSEN1998-1:2004 for seismic action assessment.
(iv) BSEN1990:2023 for load combination
(v) BSEN1992-1:2023 for reinforced concrete design.

13.2 GENERAL DESCRIPTIONS OF THE BUILDING AND BASIC DESIGN PARAMETERS

The plan dimensions of the hypothetical building are 32 m by 24 m. The building height is 128 m. There are 30 storeys, with an upper floor lift machine room. The floor plan of the typical floor is shown in Figure 13.1. The building will be referred to as "the Building" hereafter.

The Building is situated at a sea front with relatively strong wind loads. The seismicity of the location is medium. The basic design parameters are listed as follow:

(i) Construction materials: Reinforced concrete of grade C35/45;
(ii) Basic wind speed (10 minutes wind speed at 10 m height): 25 m/sec;
(iii) Peak ground acceleration : 0.15 g.
(iv) Low seismicity. Type 2 elastic spectrum to be chosen.
(v) Ductility design taken as low, i.e., DCL.

DOI: 10.1201/9781003636380-13

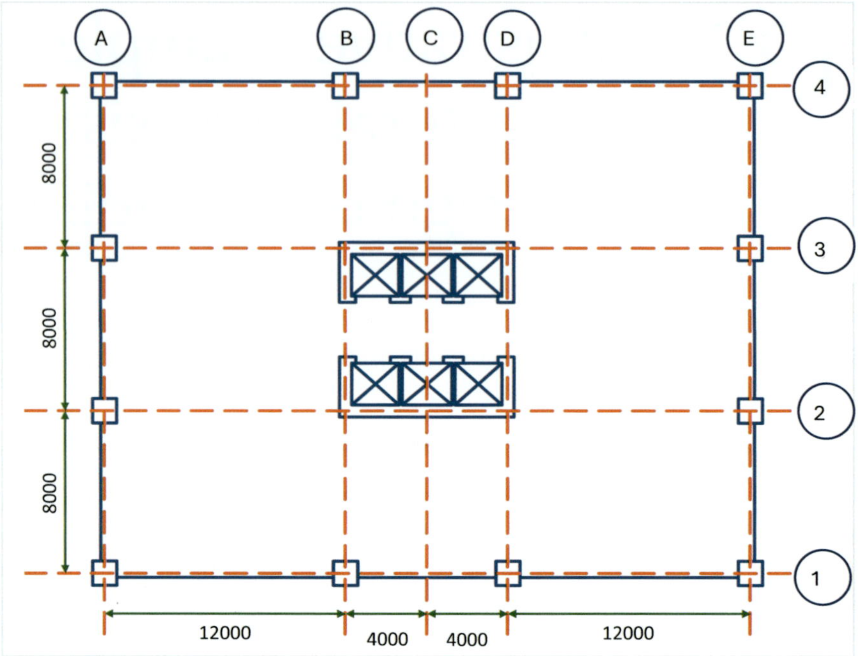

FIGURE 13.1 Floor plan of a typical floor of the Building.

(vi) Ground comprises deep deposits of loose-to-medium cohesionless soil (ground type D to BSEN1998-1:2004) for seismicity consideration.

(vii) Design for snow load is not required.

13.3 STRUCTURAL SCHEMING AND PRELIMINARY DETERMINATION OF STRUCTURAL SIZES

The central core wall and the 12 nos. of columns along the perimeter of the Building form the structural supports for gravity, wind and seismic loads. As coupling by beam members between the central core wall and the columns is required to form moment resisting frames which are essential to resist the lateral wind and seismic loads, beams of considerable stiffness are required. So, the beam-slab system is considered as the most appropriate and economical floor system in resisting both gravity and lateral loads.

As the Building is not very tall, by experience beam-column frames coupled with the central core wall should be adequate to resist the imposed gravity and lateral wind and seismic loads. Strong building structural systems such as framed tube or outriggers are not necessary.

The "framing plans" for typical floors are determined as shown in Figure 13.2, with structural sizes determined by experience and simple calculations. A computer

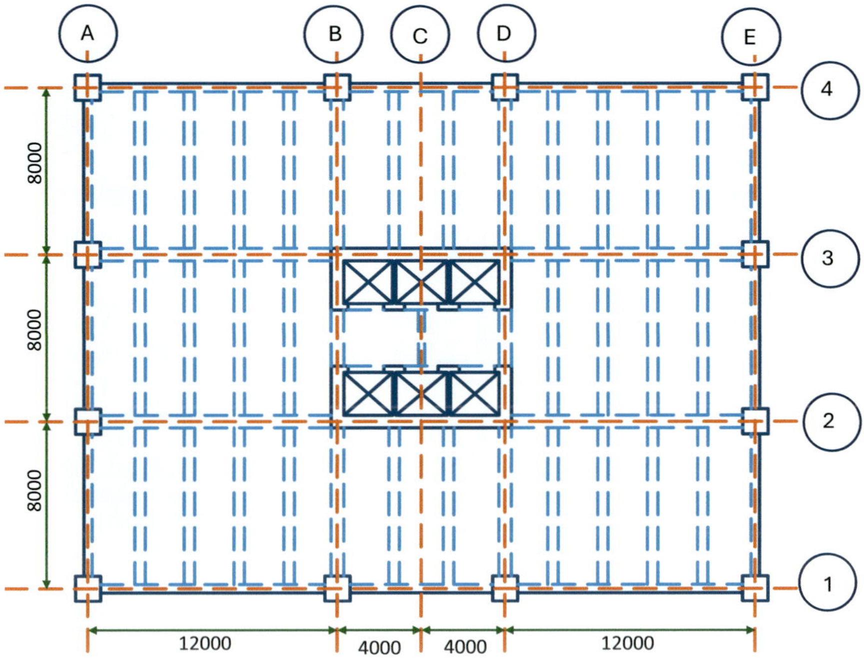

FIGURE 13.2 Framing plan of a typical floor of the Building.

model is built accordingly. Subsequent optimization can easily be done with the formed computer model by revising the structural sizes as necessary.

The structural sizes used are summarized as follows, all with concrete grade C35/45:

(i) The sub-walls in the core wall are generally 600 mm thick except for the cross walls between the lift shafts which are 150 mm and the flange wall at the lift door openings which are 300 mm thick;
(ii) All columns are 1200 mm by 1200 mm
(iii) All perimeter beams are 400 mm(w) by 1200 mm(d)
(iv) All internal main beams linking columns and walls are 600 mm(w) by 800 mm(d)
(v) All secondary beams are 500 mm(w) by 500 mm(d)
(vi) All slabs are 150 mm thick.

13.4 FORMULATION OF THE COMPUTER MODEL

The computer model is formed as 3-dimensional one by commercial software. The beams and columns are formed by line elements while the slabs and walls are formed by "shell" as planar elements, each with six degrees of freedom. The computer model is schematically shown in Figure 13.3.

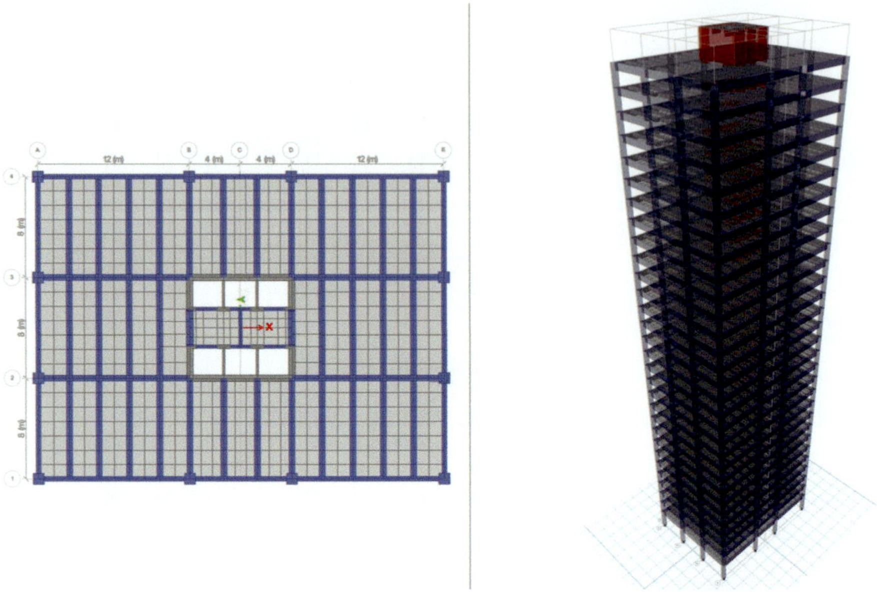

FIGURE 13.3 Computer model of the Building.

The rigid diaphragm assumption is adopted for the floor structures, which is well justified. Wind loads at each floor are applied at the centre of the rigid diaphragm due to symmetry.

The columns are assumed to be on pinned supports at their foundations.

13.5 DETERMINATION OF GRAVITY LOAD TO BSEN1991-1:2025

In accordance with the Eurocode, the gravity loads imposed on the Building are determined as follows :

- (i) Own weight of the structure is based on unit weight of concrete at 24 kN/m^3.
- (ii) The superimposed dead load (SDL) contributed by finishing and services is taken as 2.4 kN/m^2 for the office floors. Screeding as SDL is taken as 4.2 kN/m^2 for the roof.
- (iii) Dead weight of façade along the Building's perimeter is calculated to be 6 kN/m. Roof parapet is 3.96 kN/m.
- (iv) Office is included in Category B in the Eurocode BSEN1991-1:2025 (2025). Live load is 3 kN/m^2 accordingly with an addition of 0.35 kN/m^2 for partition, giving a total load of 3.35 kN/m^2. Live load reduction to area is not applicable because of the small tributary areas.
- (v) Live load for lift machine room is taken as 25 kN/m^2.
- (vi) Roof live load is taken as 3 kN/m^2, as Category B in the Eurocode BSEN1991-1:2025 (2025).

The loads are applied to the respective floor slabs of the computer model as uniformly distributed loads.

13.6 DYNAMIC PROPERTIES OF THE BUILDING

The dynamic properties of the Building is analyzed by the software up to 30 no. of modes under the mass calculated by Dead Load + 0.3 Live Load where dead load comprises the structure's own weight and superimposed dead load. The Modal Participating Mass Ratios of the first 30 modes are listed as in Table 13.1. The number of modes is so selected that the summed mass ratios of the movements exceed 95%.

From Table 13.1, it can be seen that the first mode, second mode, and third mode are respectively translation in X, translation in Y and on-plan rotation of natural frequencies 0.2496 Hz, 0.2573 Hz and 0.3681 Hz respectively.

13.7 DETERMINATION OF WIND LOAD

The determination of wind load is in accordance with BSEN 1991-1-4:2005 (2005). The clauses, equations quoted in the following are based on the Eurocode.

13.7.1 Design Parameters

13.7.1.1 Fundamental Value of the Basic Wind Velocity, $v_{b,0}$

In accordance with 4.2(1)P of the Eurocode, the fundamental value of the basic wind velocity, $v_{b,0}$, is the characteristic 10 minutes mean wind velocity, irrespective of wind direction and time of year, at 10 m above ground level in open country terrain with low vegetation such as grass and isolated obstacles with separations of at least 20 obstacle heights. $v_{b,0}$ should be obtained from the National Annex which is to be specified by the individual nation.

Take an example, the UK National Annex to the Eurocode contains a map in its figure NA.1 with contours showing $v_{b,map}$ which is the fundamental basic wind velocity before altitude correction by an altitude factor c_{alt} is applied. c_{alt} is given as follows:

$$c_{alt} = 1 + 0.001A \text{ for } z \le 10\text{m}$$

$$c_{alt} = 1 + 0.001A(10/z)^{0.2} \text{ for } z > 10\text{m}$$

where A is the altitude of the site above mean sea level and z can be taken as $0.6h$ with h as the Building height in accordance with figure 6.1 of the Eurocode.

So, if the Building is in London where $v_{b,map} = 21.5\text{m/sec}$ and $A = 24\text{msl}$. As the Building height $h = 127\text{m}$ so $z = 76.2\text{m}$,

$$c_{alt} = 1 + 0.001A(10/z)^{0.2} = 1 + 0.001 \times 24(10/76.2)^{0.2} = 1.016$$

$$v_{b,0} = v_{b,map}c_{alt} = 21.5 \times 1.016 = 21.8\text{m/sec}.$$

Other nations may specify $v_{b,0}$ in some other manners.

For demonstration purpose, $v_{b,0}$ is taken as 25 m/sec in this illustrative example.

TABLE 13.1
Summary of Modal Participating Mass Ratios of the Building

Case	Mode	Period	Frequency	UX	UY	UZ	RX	RY	RZ
□	□	sec	Hz	□	□	□	□	□	□
Modal	1	4.006	0.2496	0.6995	0	0	0	0.3117	0
Modal	2	3.887	0.2573	0	0.7303	0	0.2809	0	1.113E-06
Modal	3	2.717	0.3681	0	7.932E-07	0	6.132E-07	0	0.8258
Modal	4	1.106	0.9042	0	0.1594	0	0.4237	0	0
Modal	5	1.022	0.9785	0.1543	0	0	0	0.3324	0
Modal	6	0.885	1.1299	0	5.569E-07	0	1.554E-06	0	0.0978
Modal	7	0.554	1.8051	0	0.0463	0	0.1034	0	1.093E-06
Modal	8	0.509	1.9646	0	1.605E-06	0	3.493E-06	0	0.0348
Modal	9	0.448	2.2321	0.061	0	0	0	0.1208	0
Modal	10	0.36	2.7778	0	0.0224	0	0.0674	0	3.712E-06
Modal	11	0.348	2.8736	0	5.051E-06	0	1.533E-05	0	0.0171
Modal	12	0.259	3.8610	0	0.0124	0	0.0345	0	0.0002
Modal	13	0.258	3.8760	2.156E-06	0.0002	0	0.0006	5.54E-06	0.0093
Modal	14	0.257	3.8911	0.0326	0	0	0	0.084	5.209E-07
Modal	15	0.202	4.9505	0	9.367E-06	0	2.901E-05	0	0.0055
Modal	16	0.199	5.0251	0	0.0081	0	0.0252	0	6.515E-06
Modal	17	0.17	5.8824	0.0191	0	0	0	0.0507	0
Modal	18	0.164	6.0976	0	1.593E-06	0	4.722E-06	0	0.0033
Modal	19	0.158	6.3291	0	0.0055	0	0.0162	0	1.031E-06
Modal	20	0.137	7.2993	0	6.035E-07	0	1.883E-06	0	0.0021
Modal	21	0.129	7.7519	0	0.0039	0	0.0123	0	0
Modal	22	0.124	8.0645	0.0117	0	0	0	0.0338	0
Modal	23	0.117	8.5470	0	0	0	9.697E-07	0	0.0013

Modal	24	0.108	9.2593	0	0.0028	0	0.0086	0	0
Modal	25	0.102	9.8039	0	0	0	6.913E-07	0	0.0009
Modal	26	0.097	10.3093	0.0074	0	0	0	0.0216	0
Modal	27	0.092	10.8696	0	0.0021	0	0.0066	0	0
Modal	28	0.089	11.2360	0	0	0	7.468E-07	0	0.0006
Modal	29	0.08	12.5000	0	0.0015	0	0.0048	0	3.768E-06
Modal	30	0.079	12.6582	0	0.00001375	0	4.235E-05	0	0.0004
Sum				0.9855	0.995	0	0.9843	0.955	0.9991

13.7.1.2 Basic Wind Velocity, v_b

By Equation 4.1 of the Eurocode, the basic wind velocity is

$$v_b = c_{dir} c_{season} v_{b,0}$$

where c_{dir} is the directional factor for various directions and c_{season} is the seasonal factor. Both are to be specified in the National Annex. The Eurocode, however, recommends 1 for both factors. So

$v_b = c_{dir} c_{season} v_{b,0} = 25 \times 1 \times 1 = 25\,\text{m/sec}$

13.7.1.3 Mean Wind Velocity $v_m(z)$, z Is the Height above Ground

By Equation 4.3 of the Eurocode, the mean wind velocity for a specified height z is

$$v_m(z) = c_r(z) c_o(z) v_b$$

where $c_r(z)$ is the roughness factor, given in Section 4.3.2 of the Eurocode as

$c_r(z) = k_r \ln\left(\frac{z}{z_0}\right)$ for $z_{min} \le z \le z_{max}$

$c_r(z) = c_r(z_{min})$ for $z \le z_{min}$

in which $z_0 = 0.003$ $z_{min} = 1\text{m}$ for open sea of the site and $z_{0,II} = 0.05$ in accordance with table 4.1 of the Eurocode.

So under Equation 4.5 of the Eurocode $k_r = 0.19\left(\frac{z_0}{z_{0,II}}\right)^{0.07} = 0.19\left(\frac{0.003}{0.05}\right)^{0.07} = 0.156$

$c_o(z)$ is the orography factor, taken as 1 as the Building is on flat ground

Combining

$$v_m(z) = c_r(z) c_o(z) v_b = 0.156 \times \ln\left(\frac{z}{0.003}\right) \times 25 = 3.9 \times \ln\left(\frac{z}{0.003}\right)$$

This is the mean wind velocity profile along the height of the Building.

13.7.1.4 Turbulence Intensity $I_v(z)$

By Equation 4.6 of the Eurocode, to determine the gust wind speed, the standard deviation is $\sigma_v = k_r v_b k_1$ where $k_r = 0.156$ and $k_1 = 1$ (recommended by the Eurocode), so $\sigma_v = 0.156 \times 25 \times 1 = 3.9\,\text{m/sec}$.

By Equation 4.7 of the Eurocode, the turbulence intensity

$I_v(z) = \frac{\sigma_v}{v_m(z)} = \frac{k_1}{c_o(z) \ln(z / z_s)}$ for $z_{min} \le z \le z_{max}$

$I_v(z) = I_v(z_{min})$ for $z \le z_{min}$

13.7.1.5 Peak Velocity Pressure $q_p(z)$

By Equation 4.8 of the Eurocode, the peak velocity pressure at height z is

$$q_p(z) = \left(1 + 7I_v(z)\right)\frac{1}{2}\rho v_m^{\,2}(z)$$

$$q_p(z) = \left(1 + 7\frac{1}{1 \times \ln\left(\dfrac{z}{0.003}\right)}\right)\frac{1}{2} \times 1.25 \times \left[3.9 \times \ln\left(\frac{z}{0.003}\right)\right]^2$$

This equation will be used to determine the $q_p(z)$ at respective heights.

13.7.1.6 Structural Factor $c_s c_d$

In accordance with c_s and c_d are combined together into a single factor $c_s c_d$ for determination.

The basic equation for $c_s c_d$ as listed in Equation 6.1 of the Eurocode is

$$c_s c_d = \frac{1 + 2.k_p.I_v(z_s).\sqrt{B^2 + R^2}}{1 + 7.I_v(z_s)}$$

The various parameters are $c_s c_d$ determined as follows:

$z_s = 0.6h = 76.8\text{m}$ (By figure 6.1 of the Eurocode)

$z_0 = 0.003$ (By Table 4.1 of the Eurocode for exposure to open sea)

$$v_m(z_s) = 3.9ln\left(\frac{76.8}{0.003}\right) = 39.586\,\text{m/sec}$$

By Section B.1 of the Eurocode for z_t, L_t and α

$z_t = 200\,\text{m}$, $L_t = 300\,\text{m}$

$$\alpha = 0.67 + 0.05\ln(z_0) = 0.67 + 0.05\ln(0.003) = 0.3795$$

$$L(z_s) = L_t.\left(\frac{z_s}{z_t}\right)^{\alpha} = 208.62\text{m (By Equation B.1 of the Eurocode)}$$

$$B^2 = \frac{1}{1 + 0.9.\left(\dfrac{b+h}{L(z_s)}\right)^{0.63}} = \frac{1}{1 + 0.9.\left(\dfrac{24+128}{208.62}\right)^{0.63}} = 0.5756$$ (By Equation B.3 of the Eurocode)

$\delta = 0.1$ (By table F.2 of the Eurocode)

$$f_L\left(z_s,n_{1,x}\right)=\frac{n.L\left(z_s\right)}{v_m\left(z_s\right)}=\frac{0.2496\times 208.62}{39.586}=1.3154$$ (By Section B.1(2) of the Eurocode)

$$S_L\left(z_s,n_{1,x}\right)=\frac{n.S_v\left(z_s,n\right)}{\sigma_v^{\ 2}}=\frac{6.8.f_L\left(z_s,n\right)}{\left(1+10.2f_L\left(z_s,n\right)\right)^{5/3}}=\frac{6.8\times 1.3154}{\left(1+10.2\times 1.3154\right)^{5/3}}=0.1047$$

(By Equation B.2 of the Eurocode)

By Section B.2(6) of the Eurocode for η_h, η_b, $R_h\left(\eta_h\right)$ and $R_b\left(\eta_b\right)$

$$\eta_h=\frac{4.6.h}{L\left(z_s\right)}f_L\left(z_s,n_{1,x}\right)=\frac{4.6\times 128}{208.62}\times 1.3154=3.7125$$

$$\eta_b=\frac{4.6.b}{L\left(z_s\right)}f_L\left(z_s,n_{1,x}\right)=\frac{4.6\times 24}{208.62}\times 1.3125=0.6961$$

$$R_h\left(\eta_h\right)=\frac{1}{\eta_h}-\frac{1}{2.\eta_h^{\ 2}}\left(1-e^{-2.\eta_h}\right)=0.2331$$

$$R_b\left(\eta_b\right)=\frac{1}{\eta_b}-\frac{1}{2.\eta_b^{\ 2}}\left(1-e^{-2.\eta_b}\right)=0.6612$$

$$R^2=\frac{\pi^2}{2\delta}S_L\left(z_s,n_{1,x}\right).R_h\left(\eta_h\right).R_b\left(\eta_b\right)=0.7966$$ (By Equation B.6 of the Eurocode)

$$v=n_{1,x}\sqrt{\frac{R^2}{B^2+R^2}}=0.1902$$ (By Equation B.5 of the Eurocode)

$$k_p=\sqrt{2\ln\left(v.T\right)}+\frac{0.6}{\sqrt{2\ln\left(v.T\right)}}=3.273>3$$ where $T=600\text{sec}$ (By Equation B.4 of the Eurocode)

$k_1=1$, $c_o\left(z_s\right)=1$ where c_o is the orography factor taken as 1 in this example

$$I_v\left(z_s\right)=\frac{k_1}{c_o\left(z_s\right).\ln\left(z_s/z_0\right)}=\frac{1}{1.\ln\left(76.8/0.003\right)}=0.0985$$ (By Equation 4.7) of the Eurocode)

$$c_sc_d=\frac{1+2.k_p.I_v\left(z_s\right).\sqrt{B^2+R^2}}{1+7.I_v\left(z_s\right)}=\frac{1+2\times 3.331.\sqrt{0.576^2+0.404^2}}{1+7\times 0.0985}=1.039$$

(X-direction)

The calculations are repeated for the Y-direction for natural frequency 0.2573 Hz and exposed width $b=32$, $c_sc_d=1.016$ (Y-direction)

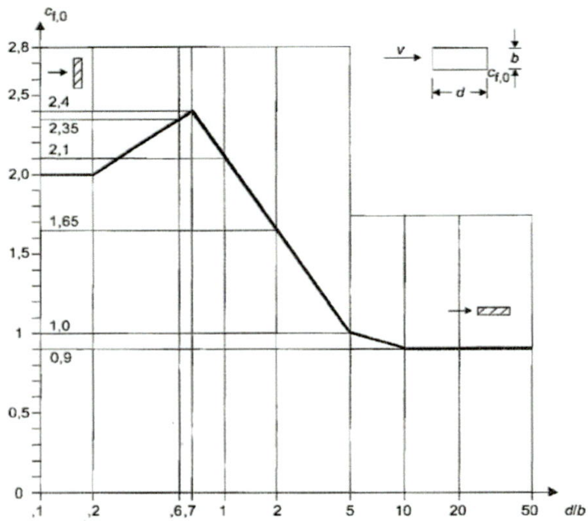

Figure 7.23 — Force coefficients $c_{f,0}$ of rectangular sections with sharp corners and without free end flow

FIGURE 13.4 Extract of figure 7.23 of the Eurocode.

13.7.1.7 Force Coefficient

The force coefficients in the two directions can be calculated in accordance with figure 7.23 of the Eurocode extracted as Figure 13.4 as follows. In the figure, d/b is in logarithm scale.

From figure 7.23 of the Eurocode,

In X-direction, $b = 24\text{m}$, $d = 32\text{m}$, $d/b = 1.333$.

$$C_{fx} = 2.4 + \frac{\left(log_{10}1.333 - log_{10}0.7\right)}{\left(log_{10}5 - log_{10}0.7\right)} \times \left(1 - 2.4\right) = 1.941$$

In Y-direction, $b = 32\text{m}$, $d = 24\text{m}$, $d/b = 0.75$.

$$C_{fy} = 2.4 + \frac{\left(log_{10}0.75 - log_{10}0.7\right)}{\left(log_{10}5 - log_{10}0.7\right)} \times \left(1 - 2.4\right) = 2.351$$

13.7.2 Total Wind Force

By Equation 5.4 of the Eurocode (2005), the total wind force on the Building is

$$F_w = c_s c_d \sum C_f q_p\left(z\right) A_{ref}$$

where F_w is the resultant wind force

A_{ref} is generally the height between the upper mid-storey height and the lower mid-storey height multiplied by the exposed width, b. A 1.1 m height parapet is assumed at the roof floor, which incurs additional wind load on the structure.

The Eurocode states in its 7.1.2(1) the following:

If instantaneous fluctuations of wind over surfaces can give rise to significant asymmetry of loading and the structural form is likely to be sensitive to such loading (e.g., torsion in nominally symmetric single-core buildings) then their effect should be taken into account.

As the Eurocode has not explicitly quantified the criteria for exemption from the torsion check, the criterion from the American Load Code ASCE7-22 is used. The criterion is that a building is considered torsionally regular if it can achieve no greater deflection at any point of any storey under 25% reduced direct wind load plus an additional torsion due to an eccentricity of 15% of the exposed width due to the reduced direct load than that under the unreduced full wind load without torsion. Using the wind load arrived at in the foregoing and performing the Building's check for torsional regularity, the deflections of 25% reduced load plus torsion and full load without torsion are checked in the X and Y directions with the following results:

X-direction: 107.59 mm versus 128.94 mm
Y-direction: 166.68 mm versus 181.66 mm

So the Building is "torsionally regular" according to the definition of ASCE7-22 (2022) with no need to check for the additional torsion.

The wind loads on the floors are determined and tabulated as Table 13.2

13.8 DETERMINATION OF SEISMIC EFFECT

The response spectrum method is chosen for the assessment of the seismic effect. The code used is the Eurocode BSEN1998-1:2004 (2004) for determination of the response spectrum.

By the Eurocode BSEN1998-1:2004 (2004), the ground acceleration of a building shall be determined by the nation based on its locality. As a hypothetical building, the ground acceleration is chosen to be $a_g = 0.15g$ in a region of moderate seismicity. Other parameters to the Eurocode chosen are as follows:

(i) The elastic spectrum chosen is Type 2 as the Building is not situated in a location of high seismicity.
(ii) The ground type chosen is Type D as the Site has deep deposits of loose-to-medium cohesionless soil with SPTN < 15 (table 3.1 of the Eurocode)
(iii) The ductility class of the Building is chosen to be Low (DCL). The behaviour factor $q = 1.5$ (table 6.1).
(iv) The Importance class is II in accordance with table 4.3 of the Eurocode, so importance factor $\gamma_I = 1$.

TABLE 13.2
Summary of Wind Shears in X and Y directions

	Height (m)	A_{ref}-X (m^2)	A_{ref}-Y (m^2)	v_m (m/s)	$q_p(z)$ (kPa)	$c_sc_dxq_p(z)$ Cf_x	F_{wx} (kN)	$c_sc_dyq_p(z)$ Cf_y	F_{wy} (kN)
UR	128	20	20	41.579	1.790	3.613	72.264	4.292	85.845
Roof	123	120.8	154.4	41.423	1.779	3.592	433.863	4.267	658.759
Storey 29	119	96	128	41.294	1.770	3.574	343.074	4.245	543.400
Storey 28	115	96	128	41.161	1.761	3.555	341.301	4.223	540.593
Storey 27	111	96	128	41.023	1.752	3.536	339.470	4.201	537.693
Storey 26	107	96	128	40.880	1.742	3.516	337.577	4.177	534.694
Storey 25	103	96	128	40.731	1.732	3.496	335.617	4.153	531.590
Storey 24	99	96	128	40.577	1.721	3.475	333.585	4.128	528.371
Storey 23	95	96	128	40.416	1.710	3.453	331.475	4.102	525.029
Storey 22	91	96	128	40.248	1.699	3.430	329.281	4.075	521.555
Storey 21	87	96	128	40.073	1.687	3.406	326.996	4.046	517.935
Storey 20	83	96	128	39.889	1.675	3.381	324.611	4.017	514.158
Storey 19	79	96	128	39.697	1.662	3.355	322.118	3.986	510.208
Storey 18	75	96	128	39.494	1.649	3.328	319.504	3.954	506.068
Storey 17	71	96	128	39.280	1.635	3.300	316.758	3.920	501.718
Storey 16	67	96	128	39.054	1.620	3.269	313.864	3.884	497.135
Storey 15	63	96	128	38.814	1.604	3.238	310.806	3.846	492.292
Storey 14	59	96	128	38.558	1.587	3.204	307.563	3.806	487.155
Storey 13	55	96	128	38.284	1.569	3.168	304.109	3.763	481.684
Storey 12	51	96	128	37.990	1.550	3.129	300.415	3.717	475.833
Storey 11	47	96	128	37.671	1.530	3.088	296.443	3.668	469.541
Storey 10	43	96	128	37.324	1.508	3.043	292.145	3.615	462.734
Storey 9	39	96	128	36.944	1.483	2.994	287.461	3.557	455.314

(Continued)

TABLE 13.2 (CONTINUED)
Summary of Wind Shears in X and Y directions

	Height (m)	$A_{ref.}X$ (m^2)	$A_{ref.}Y$ (m^2)	v_m (m/s)	$q_p(z)$ (kPa)	$c_sc_dxq_p(z)$ Cf_x	F_{wx} (kN)	$c_sc_dyq_p(z)$ Cf_y	F_{wy} (kN)
Storey 8	35	96	128	36.522	1.457	2.941	282.310	3.493	447.156
Storey 7	31	96	128	36.048	1.427	2.881	276.585	3.423	438.087
Storey 6	27	96	128	35.509	1.394	2.814	270.134	3.343	427.869
Storey 5	23	96	128	34.884	1.356	2.737	262.734	3.251	416.149
Storey 4	19	96	128	34.139	1.311	2.646	254.041	3.144	402.381
Storey 3	15	96	128	33.217	1.256	2.536	243.472	3.013	385.640
Storey 2	11	96	128	32.007	1.186	2.395	229.917	2.845	364.169
Storey 1	7	132	176	30.245	1.088	2.196	289.846	2.608	459.093
GRD	0	84	112	0	0	0	0	0	0
						Sum	9329.339	Sum	14719.848

TABLE 13.3
Extract of Table 3.3 of BSEN1998-1:2004 (2004)

Ground type	S	T_B (s)	T_C (s)	T_D (s)
A	1,0	0,05	0,25	1,2
B	1,35	0,05	0,25	1,2
C	1,5	0,10	0,25	1,2
D	1,8	0,10	0,30	1,2
E	1,6	0,05	0,25	1,2

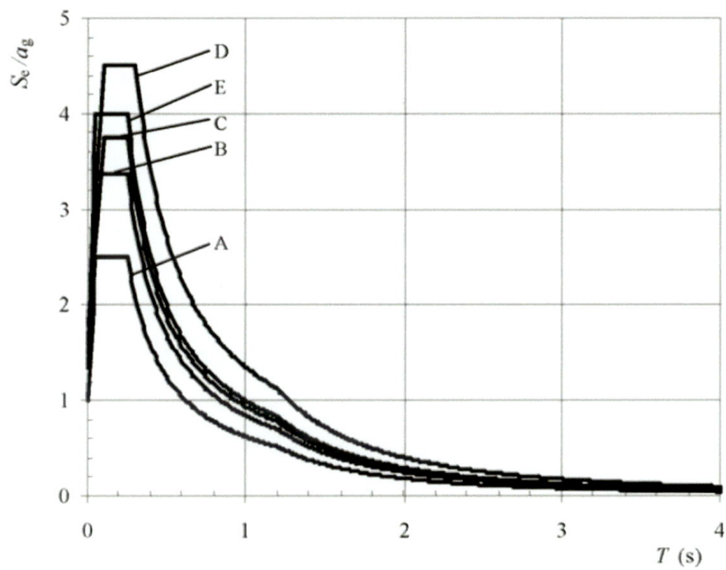

FIGURE 13.5 Extract of figure 3.3 from the Eurocode BSEN1998-1:2004.

With these determined parameters, the response spectrum can be determined as follows, as extracted from a software:

By table 3.3 and figure 3.3 of BSEN1998-1:2004 (2004) extracted as Table 13.3 and Figure 13.5 of this book as follows:

The following parameters are determined in accordance with table 3.3 of the Eurocode:

5% damping is adopted. $\eta = 1$ for 5% viscous damping by section 3.2.2.2 of the Eurocode.

Soil Factor, $S = 1.8$; $T_B = 0.1\text{sec}$, $T_C = 0.3\text{sec}$; $T_D = 1.2\text{sec}$ as shown in Figure 13.6.

Behaviour factor $q = 1.5$ in accordance with section 5.3.3 of the Eurocode for DCL

The importance class of the Building is II. By table 4.3 of the Eurocode, the importance factor $\gamma_I = 1$.

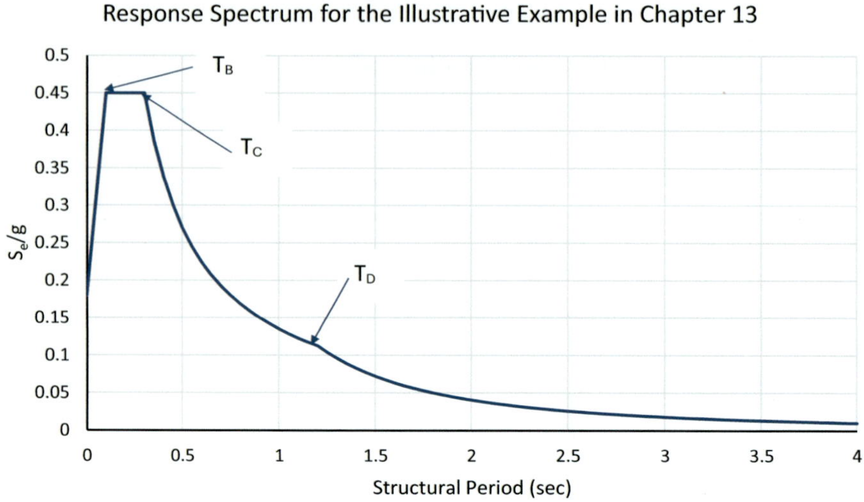

FIGURE 13.6 Response spectrum for analysis of the Building.

The response spectrum is then constructed in accordance with section 3.2.2.2 of the Eurocode as shown in Figure 13.6.

$$0 \le T \le T_B : S_e(T) = a_g.S.\left[1 + \frac{T}{T_B}(\eta 2.5 - 1)\right]\frac{\gamma_I}{q}$$

$$= 0.15g \times 1.8 \times \left[1 + \frac{T}{0.1} \times 1.5\right]\frac{1}{1.5} = 0.18(1 + 15T)g;$$

$$T_B \le T \le T_C : S_e(T) = a_g.S.2.5\eta\frac{\gamma_I}{q} = 0.15g \times 1.8 \times 2.5 \times 1 \times \frac{1}{1.5} = 0.45g$$

$$T_C \le T \le T_D : S_e(T) = a_g.S.2.5\eta\frac{T_C}{T}\frac{\gamma_I}{q} = 0.15g \times 1.8 \times 2.5 \times 1\frac{0.3}{T}\frac{1}{1.5} = \frac{0.135}{T}g$$

$$T_D \le T \le 4s : S_e(T) = a_g.S.2.5\eta\frac{T_C T_D}{T}\frac{\gamma_I}{q} = 0.15g$$

$$= 0.15g \times 1.8 \times 2.5 \times 1\frac{0.3 \times 1.2}{T^2}\frac{1}{1.5} = \frac{0.162}{T^2}g$$

Lateral seismic loads will be determined in accordance with the method with the spectral acceleration determined by Figure 13.6. An outline of the method with an illustrative worked examples are contained in Chapter 5 and Appendix E. Due to the complexity of the structures of the Building as many modes are involved together with their mode shapes, the seismic loads are determined by the software engaged in the analysis of the Building which are automatically used in the member forces analysis for design purposes.

13.9 LOAD COMBINATION

With reference to Section 5.12 in Chapter 5, the following load combinations in accordance with BSEN1990:2023 are formulated:

(i) $\gamma_g G_k : 1.35G_k$ (No variable action);

(ii) $\gamma_g G_k + \gamma_q Q_k : 1.35G_k + 1.5Q_k$ (live load as the only variable action);

Wind load as the only variable action: $\gamma_g G_k + \gamma_q Q_w$

(iii) $1.35G_k + 1.5Q_{wx}$

(iv) $1.35G_k + 1.5Q_{wy}$

(v) $1.35G_k - 1.5Q_{wx}$

(vi) $1.35G_k - 1.5Q_{wy}$

Live load as the leading variable action + wind load $\gamma_g G_k + \gamma_q Q_k + \gamma_q \psi_{0,w} Q_w$

(vii) $1.35G_k + 1.5Q_k + 1.5\times 0.6Q_{wx} = 1.35G_k + 1.5Q_k + 0.9Q_{wx}$

(viii) $1.35G_k + 1.5Q_k + 1.5\times 0.6Q_{wy} = 1.35G_k + 1.5Q_k + 0.9Q_{wy}$

(ix) $1.35G_k + 1.5Q_k - 1.5\times 0.6Q_{wy} = 1.35G_k + 1.5Q_k - 0.9Q_{wx}$

(x) $1.35G_k + 1.5Q_k - 1.5\times 0.6Q_{wy} = 1.35G_k + 1.5Q_k - 0.9Q_{wy}$

Wind load as the leading variable action + live load $\gamma_g G_k + \gamma_q \psi_{0,s} Q_k + \psi_{0,w} \gamma_q Q_w$

(xi) $1.35G_k + 1.5\times 0.7Q_k + 1.5Q_{wx} = 1.35G_k + 1.05Q_k + 1.5Q_{wx}$;

(xii) $1.35G_k + 1.5\times 0.7Q_k + 1.5Q_{wy} = 1.35G_k + 1.05Q_k + 1.5Q_{wy}$;

(xiii) $1.35G_k + 1.5\times 0.7Q_k - 1.5Q_{wx} = 1.35G_k + 1.05Q_k - 1.5Q_{wx}$;

(xiv) $1.35G_k + 1.5\times 0.7Q_k - 1.5Q_{wy} = 1.35G_k + 1.05Q_k - 1.5Q_{wy}$; ;

Lower bound of permanent action and wind load as the only variable action $\gamma_{g,inf} G_k + \gamma_q Q_w$

(xv) $1.0G_k + 1.5Q_{wx}$;

(xvi) $1.0G_k + 1.5Q_{wy}$;

(xvii) $1.0G_k - 1.5Q_{wx}$;

(xviii) $1.0G_k - 1.5Q_{wy}$;

Seismic effect $G_k + A_{Ed,ULS} + \psi_{2,k} Q_k + \psi_{2,w} Q_w$ ($\psi_2 = 0$ in accordance with table A.1.7 of the Eurocode for wind load, so $\psi_{2,w} = 0$). Load combination by square root of sum of squares, i.e., $\sqrt{A_{Ed,ULSX}{}^2 + A_{Ed,ULSY}{}^2}$ for $A_{Ed,ULS}$ is chosen which will be automatically done by the software.

(xix) $G_k + A_{Ed,ULS} + 0.3Q_k + 0 \times Q_w = G_k + A_{Ed,ULS} + 0.3Q_k$.

(xx) $G_k - A_{Ed,ULS} + 0.3Q_k + 0 \times Q_w = G_k - A_{Ed,ULS} + 0.3Q_k$

The 20 load combinations are inputted into the software and enveloped for reinforcement design. An enveloped load case is also inputted which can provide enveloped values (maximum and minimum) of the 20 combinations for ease of design.

13.10 RESULTS – LATERAL DEFLECTION DUE TO WIND

Gravity loads, wind loads and parameters for determination of seismic effects obtained in the previous sections are inputted into the computer model and analyzed. The deflection profiles of the Building for wind load in the X and Y directions are shown in Figure 13.7.

The deflections at main roof level are 129 mm and 181 mm respectively. The maximum ratio of deflection to building height is $181/128000 = 1:707$ which is considered satisfactory.

13.11 REINFORCEMENT DESIGN

The reinforcement design is based on BSEN1992-1-1:2023 with concrete grade C35/45. The stress-strain curve of the concrete has first to be defined by (8.4) of the Eurocode as follows:

$$\sigma_{cd} = \begin{cases} f_{cd}\left[1-\left(1-\dfrac{\varepsilon_{cd}}{\varepsilon_{c2}}\right)^2\right] & \text{where } 0 \le \varepsilon_{cd} \le 0.002 \\ f_{cd} & \text{where } \varepsilon_{c2} < \varepsilon_{cd} \le \varepsilon_{cu} \end{cases}$$

where f_{cd} is the design value of concrete compressive strength which can be determined as $f_{cd} = 19.83\,\text{N/mm}^2$ according to (5.3) of BSEN1992-1-1:2023; ε_{cd} is the strain of concrete which is assumed to be linearly distributed across the concrete surface. $\varepsilon_{c2} = 0.002$ and $\varepsilon_{cu} = 0.0035$. Accordingly, the stress distribution is rectangular when the strain exceeds 0.002 and parabolic otherwise, which is termed "parabola-rectangle stress distribution."

Alternatively, the Eurocode also allows a simplified "rectangular stress distribution" extending for a depth of 0.8 times the neutral axis depth from the extreme fibre. The two stress distributions are shown by figure 8.2 of the Eurocode extracted as in Figure 13.8 in this chapter.

As demonstrations, the rectangular stress distribution will be used for beam design and the parabola-rectangular distribution will be used for column and wall design in this illustrative example.

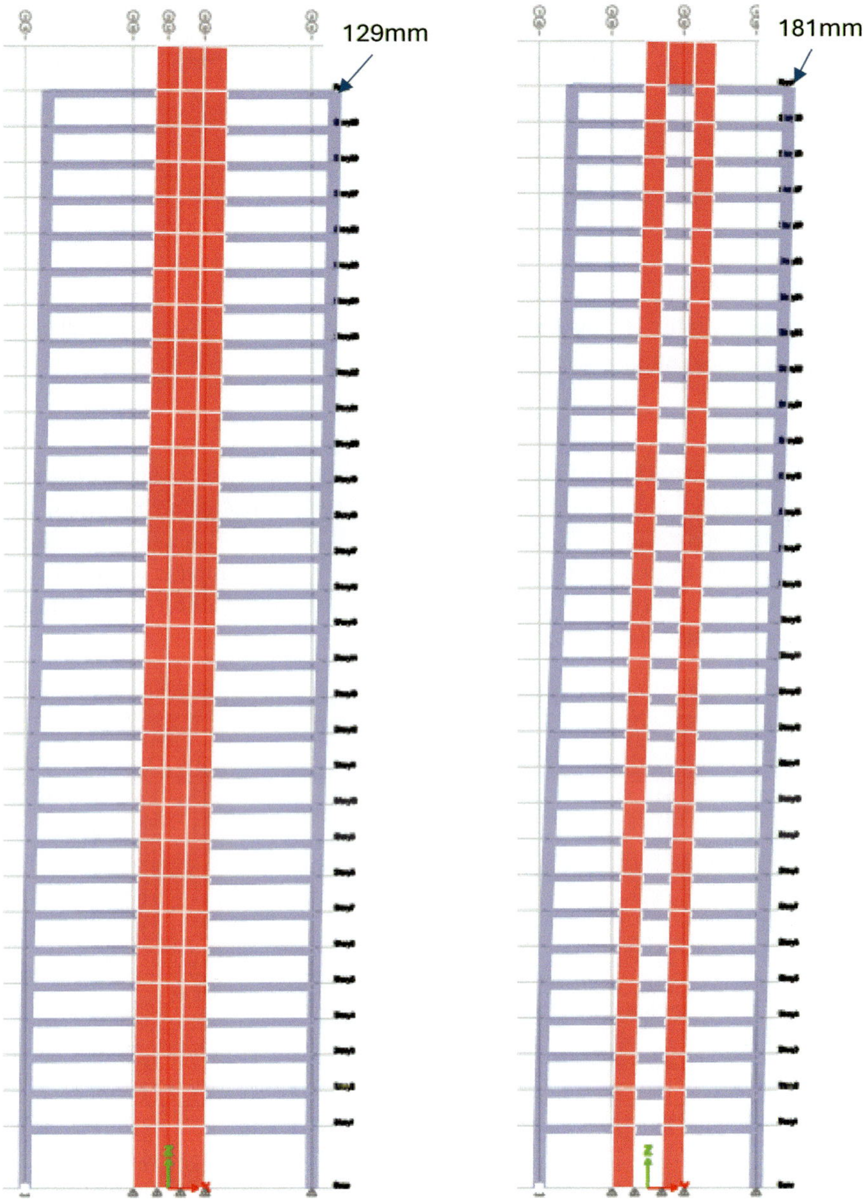

(a) Wind Deflection Profile at Grid 2 for wind in X-direction

(b) Wind Deflection Profile at Grid B for wind in Y-direction

FIGURE 13.7 Deflection profiles due to wind load.

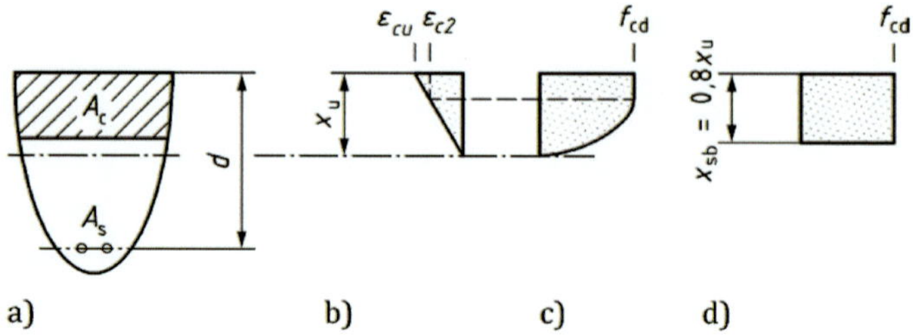

a) cross-section

b) assumed strain distribution

c) parabola-rectangle stress distribution

d) rectangular stress distribution

FIGURE 13.8 Extract from figure 8.2 of BSEN1992-1-1:2023.

The rc section designs of a beam, a column and a core wall at a floor are designed as follows with respect to the loads determined by the computer analysis. Figure 13.9 shows the locations of the selected beam, column and wall for demonstration of reinforced concrete section design.

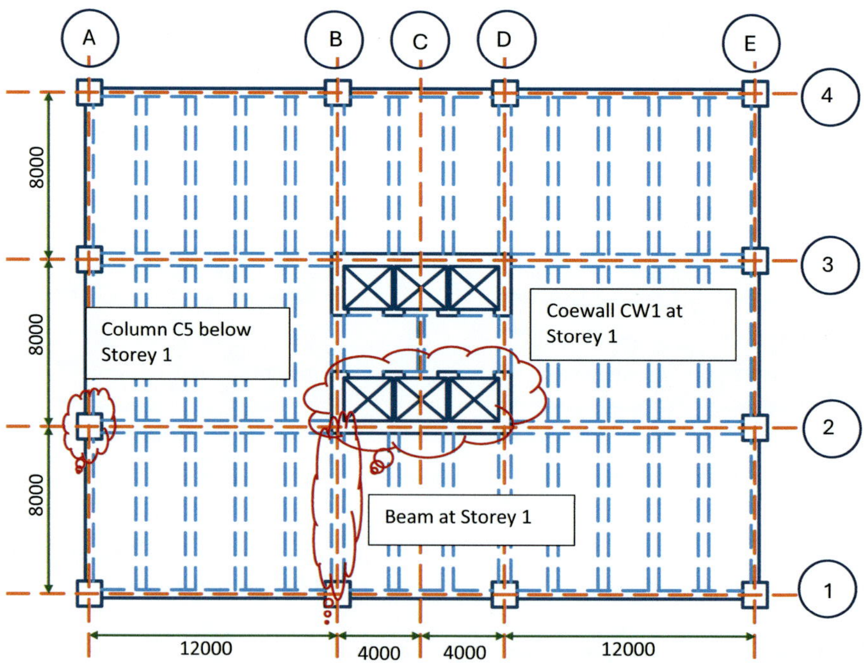

FIGURE 13.9 Location of selected beam, column core wall for demonstration of reinforced concrete section design.

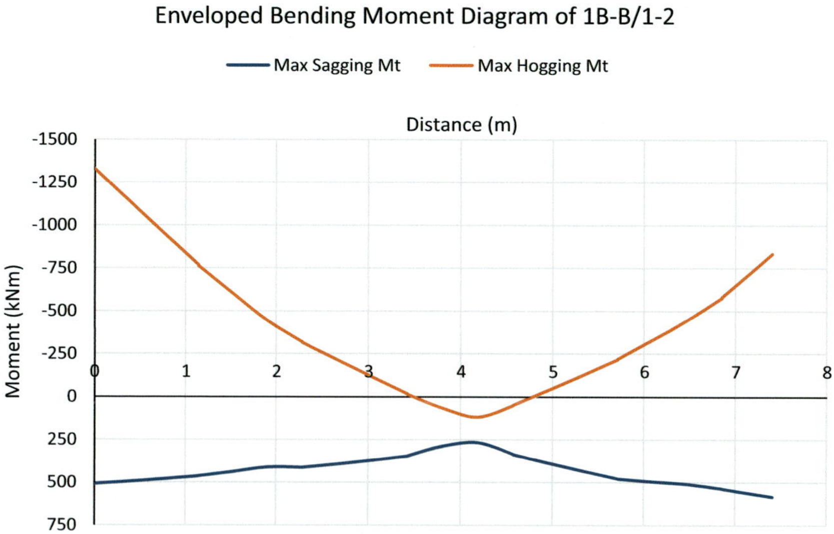

FIGURE 13.10 Enveloped bending moment diagram for beam 1B-B/1-2.

13.11.1 Beam RC Design – Beam 1B-B/1-2 in Storey 1

Readers may refer to Chapter 9 for the reinforced concrete section design for beam by BSEN1992-1-1:2023.

The enveloped bending moment and shear force diagrams of the beam 1B-B/1-2 are extracted from the analytical results and plotted in Figures 13.10 and 13.11. It

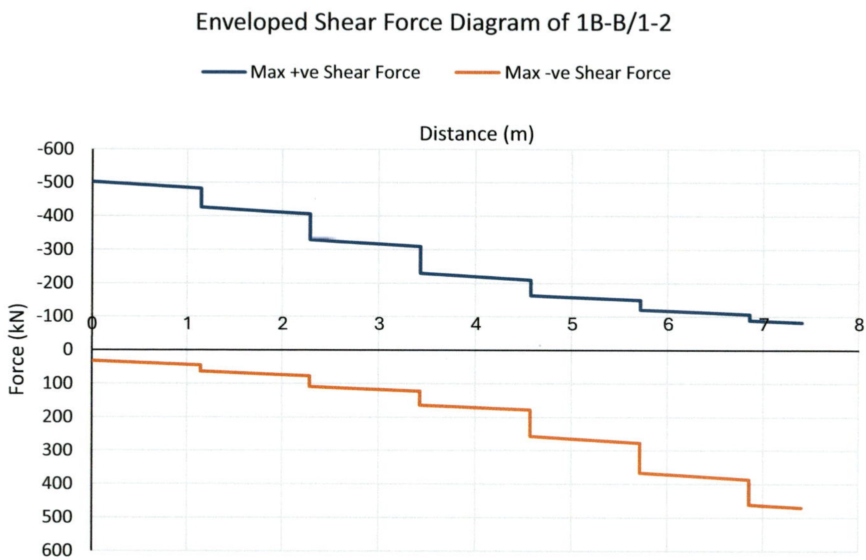

FIGURE 13.11 Enveloped shear force diagram for beam 1B-B/1-2.

can readily be seen from the bending moment diagram that the beam is under strong wind load as compared with the gravity load such that there are reversals of bending moments at its ends.

Design parameters:

$f_{ck} = 35\text{N/mm}^2$; $f_{cd} = 19.83\,\text{N/mm}^2$ (By (5.3) of BSEN1992-1-1:2023)

$f_{yk} = 500\text{N/mm}^2$; $b = 600\text{mm}$; $h = 800\text{mm}$ $c = 40\text{mm}$;

$b = 800 - 40 - 10 - 16 = 734\,\text{mm}$ (assume link of 10 mm and longitudinal of H32).

The summary of reinforced concrete section design to bending moment and shear is tabulated as follows in Table 13.4. The last column contains relevant equations no. of the Eurocode as relevant. The design to bending is based on the "rectangular stress distribution" of concrete.

By the foregoing design calculations, the reinforcement of the beam is sketched as follows in Figure 13.12.

Shear resistance of the beam is in accordance with Equation 8.27 of BSEN1992-1-1:2023 which has been reproduced in Equation 9.16 in Chapter 9. It has to be greater than $\tau_{Rdc,min}$ in Equation 8.20 of the Eurocode.

13.11.2 Column RC Design – Column A2 (Intersection of Grids A and 2) at Ground Level

As there are strong interactions between axial loads and the biaxial moments, using enveloped results involving the design of maximum values of them from different load cases at the same time would lead to significant over-design. So, the design would be based on the 20 load combinations listed in Section 13.8.

The column design will be based on the "parabola-rectangle stress distribution" of concrete under the approach described in Sections 9.4.2 and 9.4.3 of Chapter 9 for a column section under axial load and biaxial moments. Briefly speaking, the moments of resistance of the column section with pre-determined reinforcements are worked out for y and z-directions by varying neutral axis depths in each direction so that the axial load capacity of the section balances the external load. Then, Equation 8.2 of BSEN1992-1-1:2023 as $\left(\dfrac{|M_{Edz}|}{M_{Rdz,N}}\right)^{\alpha_N} + \left(\dfrac{|M_{Edy}|}{M_{Rdy,N}}\right)^{\alpha_N} \leq 1.0$ is checked.

The check is done by spreadsheet in Excel Visual Basic drafted by the authors.

Table 13.5 summarizes the design of the column with pre-determined reinforcement as shown in Figure 13.13.

From Table 13.5, the ratios $N_{Ed} / N_{Ed,0}$ and $\left(\dfrac{|M_{Edz}|}{M_{Rdz,N}}\right)^{\alpha_N} + \left(\dfrac{|M_{Edy}|}{M_{Rdy,N}}\right)^{\alpha_N}$ are less than unity. The column section is therefore adequate to resist all the 20 load combinations.

Axial load – moment capacity curves of the column section are also plotted for reference (Figure 13.14).

TABLE 13.4
Summary of reinforced section design of Beam 1B-B/1-2

	Supt at Column B1	Mid-Span	Supt at Corewall	Reference to BSEN1992-1-1:2023
Mt (hogging) (kNm)	1130.86		834.82	
$K = M / f_{ck} bd^2 =$	0.100		0.074	
$z / d = 0.5 + \sqrt{0.25 - K / 1.134} =$	0.902		0.930	
$z =$	662.300		682.647	
$A_{st,req} = M / 0.87 f_{yk} z$ (mm^2) =	3925.230		2811.301	
$A_{st,pro}$ (mm^2)=	4021.25		3217	
Steel Provided	(5H32)		(4H32)	
Mt (sagging) (kNm)	509.49	267.57	583.45	
$K = M / f_{ck} bd^2 =$	0.045	0.025	0.052	
$z / d = 0.5 + \sqrt{0.25 - K / 1.134} =$	0.959	0.977	0.952	Moseley et al. (2012)
$z =$	703.592	717.456	698.947	
$A_{st,req} = M / 0.87 f_{yk} z$ (mm^2) =	1664.659	857.340	1918.979	
$A_{st,pro}$ (mm^2) =	4021.25	3217	3217	
Steel Provided	(4H32)	(4H32)	(4H32)	
Shear V_{Ed} (kN)	502.17		470.66	
$\tau_{Ed} = V_{Ed} / bz$ (N/mm^2) =	1.264		1.149	equation 8.18
$\gamma_V =$	1.4		1.4	table 4.3
$\rho_l = A_{st,pro} / bd =$	0.009		0.007	
d_{dg} (max aggregate size) (mm) =	20		20	
$\tau_{Rdc} = \frac{0.66}{\gamma_V}\left(100 \rho_l f_{ck} \frac{d_{dg}}{d}\right)^{(1/3)} =$	0.411		0.411	equation 8.27
$\tau_{Rdc,min} = \frac{11}{\gamma_V}\sqrt{\frac{f_{ck}}{f_{yk}}\frac{d_{dg}}{d}} =$	0.343		0.343	equation 8.20
$\tau_{Rdc} - \tau_{Rdc,min} =$	0.921		0.739	
$(A_{sw} / s)_{req} = \frac{b(\tau_{Rdc} - \tau_{Rdc,min})}{f_{yk}} =$	1.024		0.886	equations 8.42 and 8.43
Provided Links	H10-4legs-200		H10-4legs-200	
$(A_{sw} / s)_{prov}$ (mm)	1.57		1.57	

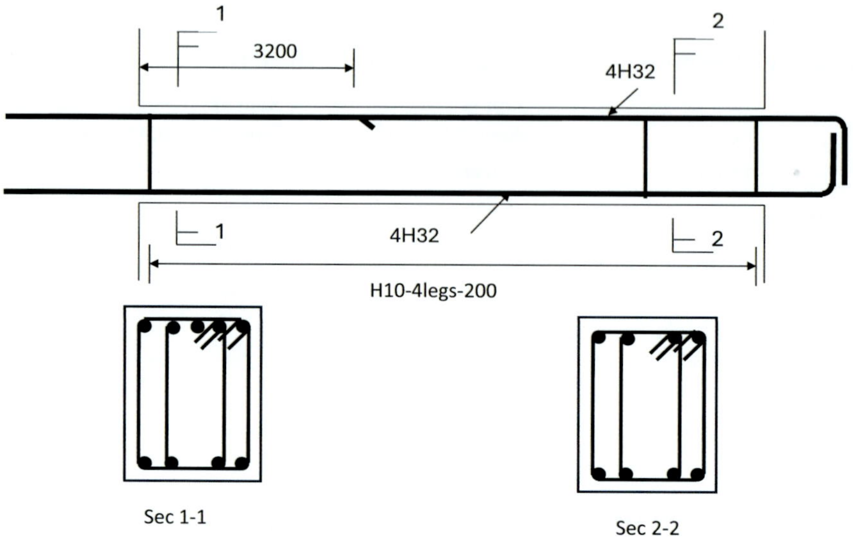

FIGURE 13.12 Reinforcement details of beam 1B-B/1-2.

13.11.3 Core Wall RC Design – Core Wall CW1 at Storey 1

As similar to column, the reinforcement design of core wall CW1 is done by designing to the 20 load combination cases instead of the enveloped case.

In the computer model, the core wall is an assembly of individual wall panels simulated as shell elements. For simplicity of design, the forces of the individual panels of the core wall are lumped together into total forces and moments by the computer software. The lumped forces and moments are then checked on the whole core wall section with pre-determined reinforcement. Figure 13.15 shows the proposed reinforcement arrangement.

The reinforcement design check is in accordance with the approach described in section 9.4.5 of Chapter 9 by an Excel spreadsheet drafted by the authors in Visual Basic. The proposed reinforced section comprising 202 nos. of H32 is shown in Figure 13.15.

Under $f_{cd} = 19.83$ N/mm^2, summary of the Sectional Properties and Axial Load Capacities are as follows:

Cross-sectional area = 9315000 mm^2
Steel percentage = 1.74%
Centroid X-coordinates = 0 mm
Centroid X-coordinates = −3272.746 mm
Maximum Axial Compression Capacity $N_{Rd,c}$ = 259091.41 kN
Maximum Tensile Compression Capacity $N_{Rd,t}$ = 70669.25 kN

TABLE 13.5

Reinforced Concrete Design for Column A2 at Ground Floor, d_{nx} and d_{ny} Are Neutral Axis Depths Tried that the Column Section Would Balance N_{Ed}. Other Symbols Such As N_{Ed} Are the Usual Ones Adopted in the Eurocode

Load Combination Cases	N_{Ed} (kN)	$M_{Ed,y}$ (kNm)	$M_{Ed,z}$ (kNm)	d_{ny} (mm)	$M_{Rdy,N}$ (kNm)	d_{nz} (mm)	$M_{Rdz,N}$ (kNm)	N_{Ed} / N_{Ed0}	α_N	Eqn(8.2)
$1.35G_k$	28787	109.9	524.56	1089.5	5551.5	1102.8	5849.6	0.6771	1.4809	0.0311
$1.35G_k+1.5Q_k$	38218	16.636	503.3	1585.9	1907.6	1619.5	2023.5	0.8989	1.8315	0.0784
$1.35G_k+1.5Q_k+0.9Q_{wx}$	35977	17.984	680.78	1400.2	2826.1	1420.4	3001.7	0.8462	1.7436	0.0754
$1.35G_k+1.5Q_k+0.9Q_{wy}$	36944	132.57	495.51	1469.3	2434.5	1493.6	2587.8	0.8689	1.7815	0.0582
$1.35G_k+1.5Q_k-0.9Q_{wx}$	40460	15.288	325.82	1943.6	942.72	2004.3	1007.7	0.9516	1.9193	0.1149
$1.35G_k+1.5Q_k-0.9Q_{wy}$	39493	99.303	511.1	1754.1	1359.1	1799.2	1447.4	0.9288	1.8814	0.1484
$1.35G_k+1.5Q_{wx}$	27982	14.409	678.96	1062.3	5804.5	1075.1	6123.9	0.6581	1.4651	0.04
$1.35G_k+1.5Q_{wy}$	35453	9.9147	87.355	1367.4	3036.5	1386.3	3221.1	0.8338	1.7231	0.002
$1.35G_k-1.5Q_{wx}$	29593	205.39	370.16	1117.6	5283.7	1131	5563.5	0.696	1.4967	0.0251
$1.35G_k-1.5Q_{wy}$	33841	181.07	396.15	1282	3676.8	1297.8	3888.5	0.7959	1.6599	0.0293
$1.35G_k+1.05Q_k+1.5Q_{wx}$	32532	17.54	763.06	1225.3	4191.9	1239.3	4420.4	0.7651	1.6086	0.0594
$1.35G_k+1.05Q_k+1.5Q_{wy}$	34144	208.53	454.26	1296.5	3557.3	1312.8	3764.3	0.803	1.6717	0.0379
$1.35G_k+1.05Q_k-1.5Q_{wx}$	38392	177.94	480.25	1605.3	1833.2	1640.1	1945.6	0.903	1.8383	0.0901
$1.35G_k+1.05Q_k-1.5Q_{wy}$	40003	13.047	171.46	1844.8	1138.9	1897.1	1214.4	0.9409	1.9014	0.0244
$G_k+1.5Q_{wx}$	17623	10.437	553.82	758.05	8253.3	766.79	8894.2	0.4145	1.2621	0.0303
$G_k+1.5Q_{wy}$	25094	5.9427	37.784	968.63	6622.4	979.99	7021.6	0.5902	1.4085	0.0007
$G_k-1.5Q_{wx}$	19234	201.42	245.02	799.46	7940.8	808.21	8531.5	0.4524	1.2936	0.0187
$G_k-1.5Q_{wy}$	23483	185.04	271.01	919.26	7021.2	929.75	7472.6	0.5523	1.3769	0.0171
$G_k+0.3Q_k+Ad$	18914	129.91	548.3	791.05	8004.2	799.8	8604.7	0.4448	1.2874	0.0339
$G_k+0.3Q_k-Ad$	26167	111.9	11.423	1002.7	6334.9	1014.6	6702.6	0.6154	1.4295	0.0032
							Max	0.9516	Max	0.1484

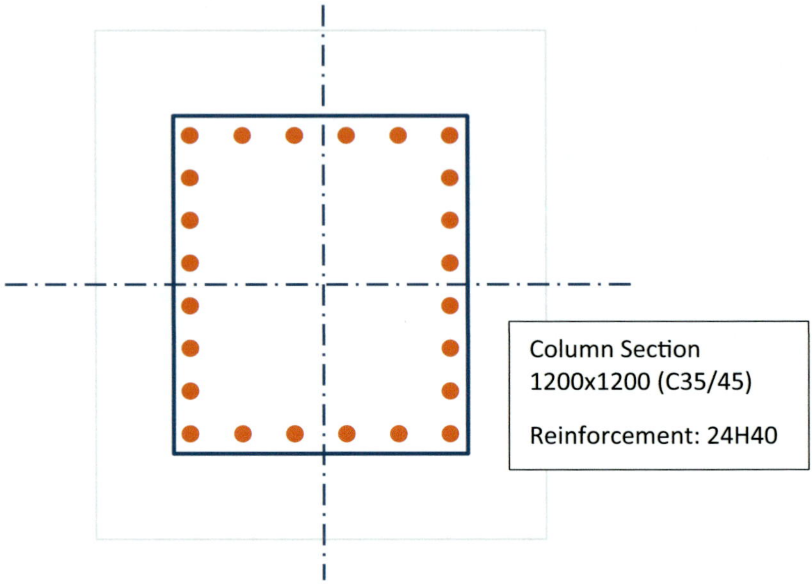

FIGURE 13.13 Reinforcement section of column A2 at ground floor.

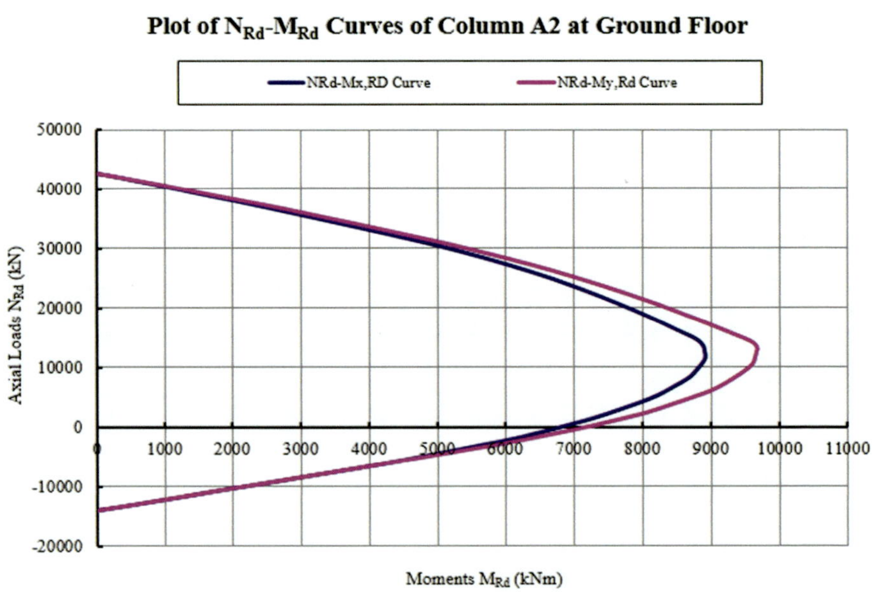

FIGURE 13.14 Axial load moment – capacity curves for column A2 at ground floor.

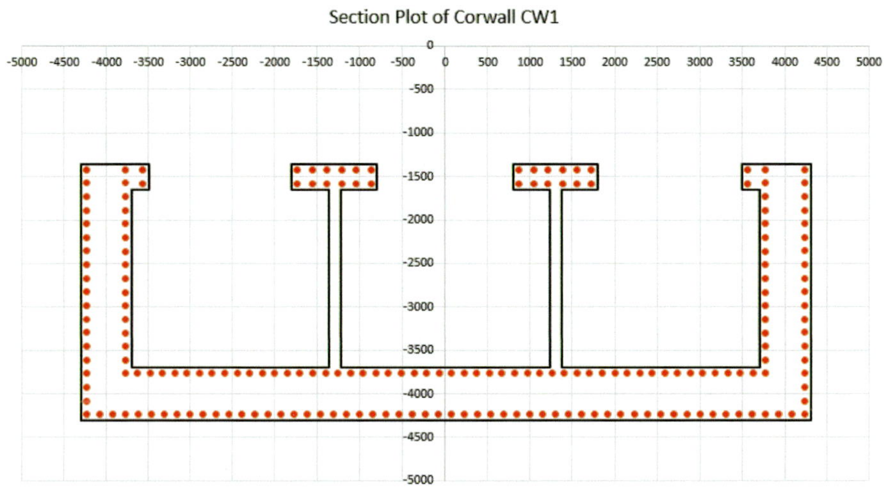

FIGURE 13.15 Reinforced concrete section of CW1 at storey 1.

The results of the 20 load combination cases which have been doubled by checking the wall at the top and bottom levels of ground floor to form 40 cases are checked for adequacy in Table 13.6. In the table, the following symbols are defined as in addition to those generally defined in the Eurocode such as N_{Ed}:

(i) $M_{Ed\text{-}res}$: Resultant of the applied moment
(ii) $\phi_{applied\text{-}mt}$: orientation of the resultant of the applied moment;
(iii) $\phi_{NA\text{-}final}$: orientation of the final neutral axis of the section to balance the external axial load and moments;
(iv) x_{final}: final neutral axis depth;
(v) $M_{Rd\text{-}res}$: Resultant of the moment of resistance of the Section to balance the external axial load and moments;
(vi) $\phi_{Mt\text{-}strength}$: orientation of the resultant of the moment of resistance of the section under $\phi_{NA\text{-}final}$ and x_{final}.

In terms of utilization, the maximum ratio of $N_{Ed} / N_{Rd,c}$ is 0.78 and $M_{Ed,res} / M_{Rd,res}$ is 0.61. So, the section is adequate.

The capacity curves of the core wall section in terms of biaxial moments at various constant axial loads are plotted as in Figure 13.16 for reference. Users may estimate from these curves for any set of given axial load and biaxial moments.

13.12 CONCLUDING REMARKS

The hypothetical Building has been checked to comply with strength and deformation requirements. Nevertheless, there is room for structural size reductions as there are quite some margins to the deflection limit generally set at 1 in 500. Generally,

TABLE 13.6
Reinforced Concrete Design for Core Wall CW1 at Storey 1

Case	N_{Ed} (kN)	M_{Edy} (kNm)	M_{Edz} (kNm)	$M_{Ed,res}$ (kNm)	$\phi_{applied\text{-}mt}$	$\phi_{NA\text{-}final}$	x_{final}(mm)	$M_{Rd,res}$ (kNm)	$\phi_{Mt\text{-}strength}$	$N_{iteration}$	Result
1	130188.38	−1412.366	172.4724	1422.8573	3.0200783	3.1292592	2880.9591	91547.82	3.0097588	2	OK
2	131554.91	940.8315	170.8844	956.22455	0.1796725	0.0329014	1423.3865	136525.72	0.1796831	2	OK
3	156948.02	−1849.717	172.4189	1857.7354	3.0486476	3.1354129	3071.2857	72925.259	3.0634429	2	OK
4	158314.55	1232.0768	170.6799	1243.8428	0.1376541	0.0240987	1920.5167	123383.2	0.1353041	2	OK
5	156948.04	−1849.72	79077.779	79099.41	1.5941832	2.0332887	7702.1956	288985.98	1.5939035	6	OK
6	158314.57	1232.0724	92062.995	92071.239	1.5574142	1.621438	7529.1438	295732.75	1.5574317	1	OK
7	125899.93	1127.3329	172.1275	1140.3979	0.1515154	0.0266134	1279.6246	136376.19	0.1474327	2	OK
8	127266.46	18839.292	170.2006	18840.061	0.0090341	0.0020895	1173.9239	135954	0.0119542	1	OK
9	156948.01	−1849.714	−78732.94	78754.666	4.6888998	4.2490915	7701.2333	288952.45	4.689177	6	OK
10	158314.53	1232.0812	−91721.64	91729.91	4.725821	4.6624074	7527.6701	295744.22	4.7258025	1	OK
11	187996.12	−4826.767	172.7103	4829.8558	3.1058261	3.1391203	3389.5505	50741.782	3.1111687	1	OK
12	189362.64	−16375.14	171.1592	16376.033	3.1311407	3.1428492	3401.7575	49735.471	3.1583876	1	OK
13	130188.41	−1412.371	131681.41	131688.98	1.5815216	1.7710424	6523.4681	342601.24	1.5815476	3	OK
14	131554.94	940.8241	153324.74	153327.63	1.5646603	1.5923777	6302.1559	345520.92	1.5647288	1	OK
15	130188.35	−1412.36	−131336.5	131344.06	4.7016356	4.5118372	6523.7743	342591.9	4.7016092	3	OK
16	131554.87	940.8388	−152983	152985.87	4.7185389	4.6909549	6301.8596	345524.6	4.718471	1	OK
17	78441.557	3549.3842	171.9867	3553.5486	0.0484175	0.0058511	630.37176	119559.36	0.0718378	1	OK
18	79808.084	30286.191	170.0856	30286.668	0.0056159	−0.001069	613.17622	120043.11	6.2704673	1	OK
19	181935.2	−6374.115	172.9581	6376.4613	3.1144649	3.1410082	3301.1202	55095.858	3.1338578	1	OK
20	183301.73	−28404.53	171.6833	28405.047	3.1355485	3.1426903	3320.775	54119.143	3.1561357	1	OK
21	148920.16	−1718.517	131681.37	131692.58	1.5838462	1.8916501	7400.747	308728.23	1.5838832	4	OK
22	150286.69	1144.6959	153324.6	153328.87	1.5633306	1.6337718	7171.0448	311454.55	1.5633225	1	OK
23	97173.309	3243.2382	171.9493	3247.7932	0.0529682	0.007332	767.22186	128286.65	0.0628724	1	OK

24	98539.835	30490.062	159.9424	30490.536	0.0055736	8.776E−05	739.22902	128768.07	0.0007476	1	OK
25	200666.95	−6680.261	172.9206	6682.4989	3.1157131	3.1413065	3568.4074	41634.283	3.1377607	1	OK
26	202033.48	−28200.66	171.5401	28201.178	3.1355099	3.1427266	3596.8936	40643.998	3.1567935	1	OK
27	148920.1	−1718.506	−131336.5	131347.74	4.699305	4.3910833	7400.8699	308717.28	4.6992678	4	OK
28	150286.63	1144.7106	−152983.1	152987.4	4.7198714	4.6496241	7170.6308	311458	4.7198799	1	OK
29	87668.975	−951.0936	131625.08	131628.51	1.578022	1.6121633	4712.5121	398144	1.5780556	1	OK
30	88589.195	633.5492	153268.93	153270.24	1.5666628	1.4493668	4730.582	397038.15	1.5667074	2	OK
31	87668.91	−951.0821	−131392.8	131396.23	4.7051506	4.6707989	4713.0003	398139.89	4.7051167	1	OK
32	88589.13	633.5639	−153038.8	153040.1	4.7165288	4.8339053	4730.6244	397036.77	4.7164843	2	OK
33	35922.12	4010.6618	115.6574	4012.3291	0.0288295	3.26E−05	404.92295	95574.162	0.0005286	1	OK
34	36842.34	29978.916	114.2748	29979.133	0.0038118	−0.000816	412.2136	96121.73	6.2701104	1	OK
35	139415.76	−5912.838	116.6287	5913.9877	3.1218706	3.1419819	2899.2492	85055.558	3.1462618	1	OK
36	140335.98	−28711.8	115.8726	28712.036	3.137557	3.142264	2907.9567	84433.253	3.1497243	1	OK
37	63425.587	6400.96	116423.49	116599.32	1.5158717	0.8612884	3448.9224	377433.55	1.5162134	9	OK
38	64345.807	18367.71	137660.62	138880.59	1.4381523	0.5500564	2815.8156	339276.35	1.4393089	12	OK
39	121643.08	−8462.173	−116191.2	116498.97	4.6396876	3.9465939	5860.2191	327847.55	4.6399096	9	OK
40	122563.3	−16994.69	−137430.5	138477.34	4.5893534	3.734325	5270.144	301283.37	4.5897496	11	OK

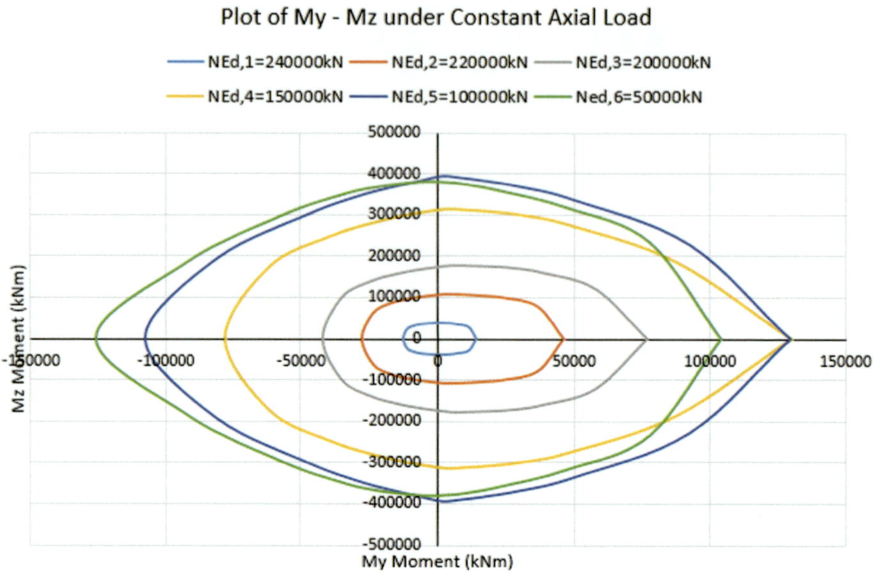

FIGURE 13.16 Plots of My-Mz moments of resistance of CW1 under different axial loads.

the beam, column and wall sizes in the upper floors can be reduced. Readers may set up the model accordingly and try reductions to achieve more optimal structural arrangement through the steps stipulated in the foregoing.

REFERENCES

British Standard Institution (2004) Eurocode 2: Design of Concrete Structures – Part 1-1: General Rules and Rules for Buildings BS EN 1992-1-1:2004. British Standard Institution.

British Standard Institution (2004) Eurocode BS EN 1998-1:2004 Design of Structures Against Earthquake Resistance – Part 1: General Rules, Seismic Actions and Rules for Buildings. British Standard Institution.

British Standard Institution (2005) Eurocode BS EN 1991-1-4:2005 Actions on Structures – General Actions: Wind Load. British Standard Institution.

British Standard Institution (2023) *Eurocode BS EN 1990:2023 Eurocode Basis of Structural Design*. British Standard Institution.

British Standard Institution (2023) Eurocode BS EN 1992-1-1:2023 Eurocode 2: Design of Concrete Structures. British Standard Institution.

British Standard Institution (2023) *Eurocode 2: Concrete – Specifications, Performance, Production and Conformity BSEN1992-1-1:2023+A2:2023.* British Standard Institution.

British Standard Institution (2025) BS EN 1991-1-1:2025 Actions on Structures Eurocode 1 – Actions on structures Part 1-1: Specific weight of materials, self-weight of construction works and imposed loads for buildings.

Mosley, W.H., Bungey, J. and Hulse, R. (2012) Reinforced Concrete Design to Eurocode 2. Palgrave Macmillan.

14 Introduction to Building Foundation Design

Choices, Types, and Design Principles

14.1 GENERAL INTRODUCTION

Foundations are structural elements that transfer loads from the superstructure to the underlying soil or rock. A structure may be supported on individual foundations or on larger foundations. In general, foundations can be classified into two broad categories: shallow foundations and deep foundations. Shallow foundations transmit the structural loads to the soils immediately beneath the foundation, which spreads the applied load over a larger area so that the induced stresses and settlements are within the tolerable limits. There are two major types of shallow foundations: (1) footing or raft foundation with depth/breadth ≤ 1.0; (2) deep foundations which transmit the applied load to deeper soils or rock. In the past, empirical rules and rules of thumb combined with good judgment can usually produce acceptable foundation design as long as they are applied to structures and soil/rock conditions similar to those in the past. Such an approach is, however, no longer valid for the modern structures with heavy loadings and may be irregular and sometimes strange architectural designs. In many projects, both types of foundations can be adopted at different locations, which are largely controlled by the loads from the superstructure.

A foundation system must satisfy the following requirements:

(1) Structural requirement – the foundation must be able to take up the applied loads with acceptable movements; that is, both the ultimate limit state and serviceability limit state must be satisfied in the design of the foundation. In general, the serviceability limit state is more difficult to satisfy than the strength of the foundation, and most of the failures are also related to settlement, tilting and relative settlement, while a total collapse of the foundation system (strength) is less common.

(2) Geotechnical requirement – the design and construction of the foundation must consider the soil and rock properties, underground water, underground caverns, adjacent structures and other factors. There have been many reported construction problems in various poor ground conditions in Hong Kong, and this requirement can be challenging in some cases.

DOI: 10.1201/9781003636380-14

(3) Construction requirement – the choice of the foundation system can be greatly controlled by the availability of construction plants, construction materials and government policies. In Hong Kong, the noise restriction from the government has nearly eliminated the use of driven piles while the safety regulation has practically banned the use of hand-dug caisson.

The general considerations for the design and construction of a foundation are:

(1) Cost of construction – engineers should pay the effort to minimize the construction which is the most critical cost of a project. Some engineers, however, tend to accept lower design fees and produce more conservative designs with less time and effort for design. The authors do not view this phenomenon as healthy, unless it is urgent to carry out the real construction works with insufficient time for design.
(2) Time of construction – a poor design without sufficient site investigation and time for design can cause serious delays and VO in a project, which is commonly found in many large-scale projects in Hong Kong. The rise in time and cost due to corrections and VO can be tremendous for many major projects in Hong Kong and many other countries.
(3) Effect on the surrounding – the construction of the Mass Transit Railway concourses and tunnels have induced significant ground and building settlement in Hong Kong. Besides that, there are also many projects with water leakage through retaining walls, ground settlement and tilting of buildings, ground subsidence, noise and environmental pollution. In the authors' view, some of these problems are difficult to avoid unless a major increase in time and cost of construction can be allowed. Nevertheless, the engineers should try their best to minimize these problems, and the incorporation of a suitable monitoring system, either the traditional or the modern internet of things (IOT) monitoring system is highly recommended for those major construction projects.
(4) Health and safety – compressed air, poison gas, PM2.5, noise and other problems have affected the health of many workers on site. More importantly, deaths of workers in the Hong Kong ZhuHai Macau Bridge projects and works in confined space are critical issues in the construction industry in Hong Kong. Currently, the Hong Kong Government has imposed the 4S (Smart site safety system) requirement in order to reduce the accidents and deaths on site.
(5) Social impact – for major construction projects, the time, cost and quality of construction are major social or even political issues.
(6) Environmental – noise, vibration and pollution (air, water, noise) have to be considered with care in a project. Due to the noise restriction, pile driving has nearly disappeared in Hong Kong.
(7) Manufacturability – the use of precast or MIC construction, high-strength steel and innovative materials/technology become a hot topic in Hong Kong, and the government is promoting the use of such techniques/materials wherever possible.

14.2 SHALLOW FOUNDATION (ULTIMATE LIMIT STATE)

A shallow foundation is usually within 2 to 3 metres below ground. Some textbooks and design codes, however, give some more explicit definitions. BS8004 and the Hong Kong Code of Practice for Foundations 2017 define shallow foundations with a founding depth of less than 3 m, while Bowles (1996) defines it with a depth over breadth ratio of less than one. A shallow foundation is an enlarged base (in plan area) for the support of columns or walls of the superstructure and spread the stresses of the columns or walls to a wider founding stratum without failure so as to achieve the ultimate (bearing capacity) and serviceability (settlement, angular distortion etc.) limit states. The determination of the geometry of the shallow foundation therefore depends on the loads of the columns and walls being supported and the strength and stiffness of the supporting soil or stratum.

Usually, settlement is the most critical issue for the design of a shallow foundation. If the ultimate capacity is approached (yield), settlement must be large. If settlement is small, the system will be far from yield. The control of bearing pressure can hence also control settlement, or vice versa. This approach is actually used for many shallow foundation designs in the past, when computers were not readily available to most of the engineers. Furthermore, so far as the authors are aware, there are very few ultimate limit state failures for shallow foundations, but there are vast numbers of settlement problems for them.

Shallow foundations are predominantly constructed of reinforced concrete as durability is an important design criterion, because the foundations are buried in soil where aggressive actions by corrosive materials can take place. In Hong Kong, the maximum concrete grade used for shallow foundation is typically limited to Grade 45 (f_{cu} = 50 N/mm^2). Since many buildings in Hong Kong are tall, bar sizes of 32 mm, 40 mm, and even 50 mm are commonly used in construction. The foundation is hence an expensive item in the overall construction cost, and savings in the foundation design can be an important issue.

The common types of shallow foundations are given in Figure 14.1. Many laboratory tests have been carried out on the bearing capacity and failure mechanisms of shallow foundations under different soil conditions and depths (see Bowles 1966). The bearing capacity and failure mechanisms of shallow foundations are also investigated by many researchers using different methods. The model test results, the theoretical studies of the basic governing equation as well as empirical correction factors are used to determine the bearing capacity, though many problems about these modification factors have been discussed by many researchers. The failure modes of a typical shallow foundation are classified as (1) General shear failure; (2) Local shear failure; and (3) Punching shear failure:

(1) General shear failure

This failure mode applies for dense and stiff soil with well-defined failure surfaces and zones as shown in Figure 14.2. The upheaval of soil adjacent to the shallow foundation is also observed in many practical cases (but not all). Even though the problem looks symmetrical, the actual failure mode is usually asymmetric with tilting of the footing at one side.

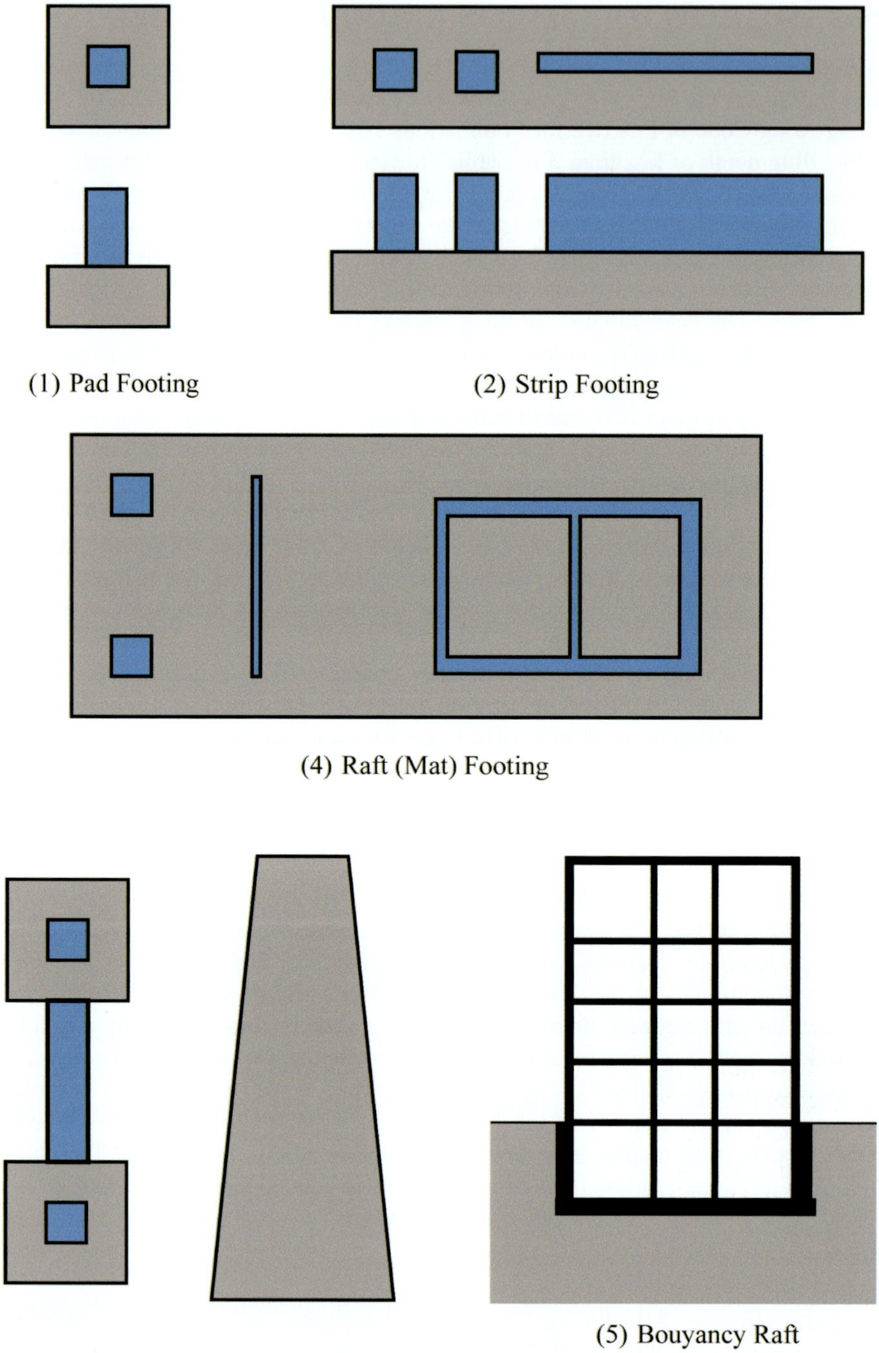

FIGURE 14.1 Different types of commonly used shallow foundations.

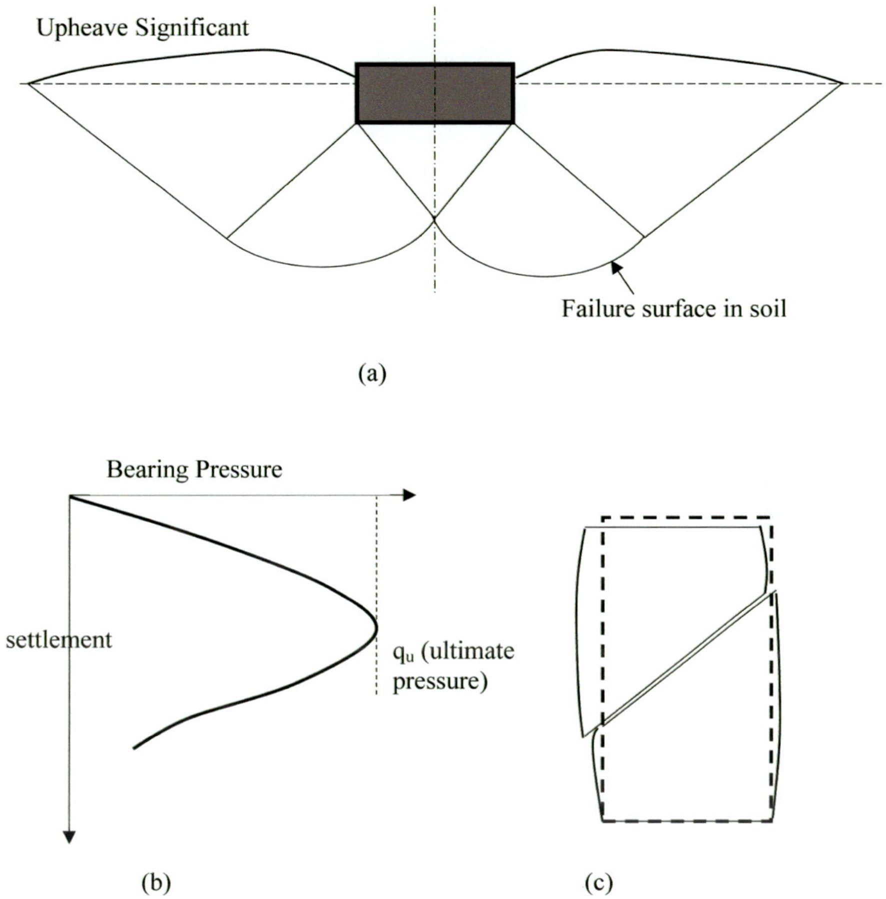

FIGURE 14.2 (a) General shear failure. (b) Settlement/pressure for general shear failure. (c) Failure mode for soil sample under general shear failure.

(2) Local shear failure
This failure mode is found for relatively loose or soft soil with significant compression of soil below the footing, which is shown in Figure 14.3. The failure surface in this mode is less well defined as compared with the general shear failure mode. Upheaval of soil around the footing is seldom observed with considerable footing settlement. A well-defined stress/settlement peak is not seen as compared with the previous failure mode.

(3) Punching shear failure
This failure mode applies to loose or soft soil at a greater depth with large displacement, as shown in Figure 14.4, and a clear and well-defined ultimate bearing capacity is difficult to be specified. Ground heaving is not observed in this failure mode, and this mode is usually associated with pile foundation but not for shallow foundation.

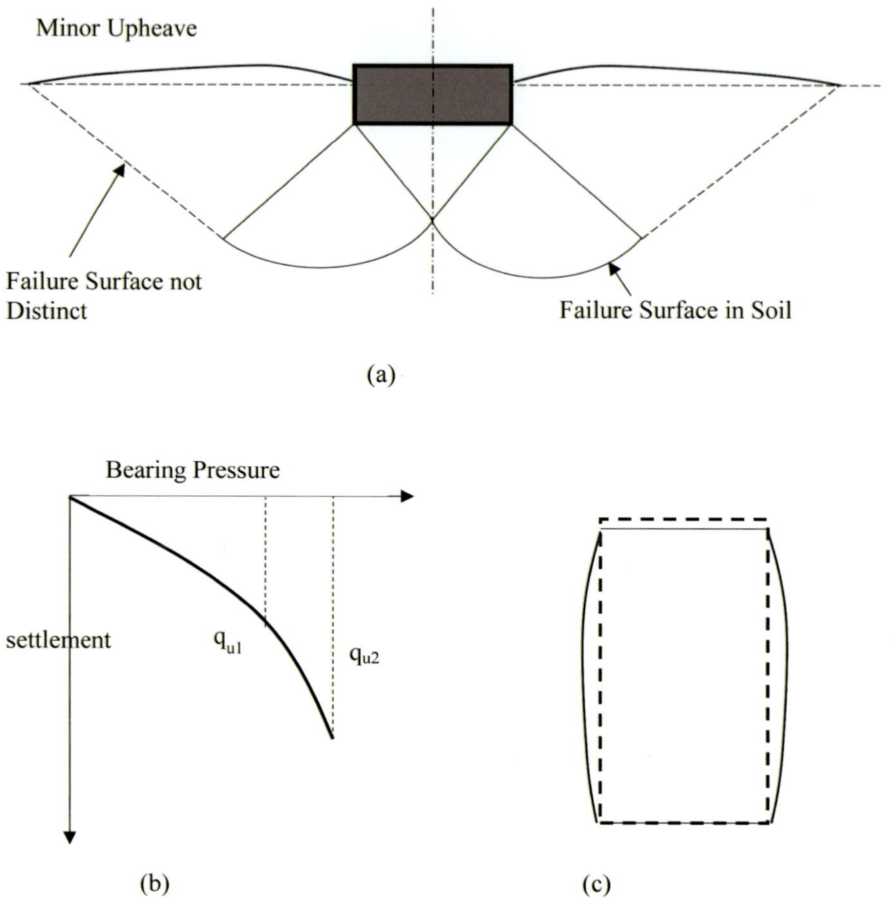

FIGURE 14.3 (a) Local shear failure. (b) Settlement/pressure relation for local shear failure. (c) Failure mode for soil sample under local shear failure.

Based on test results, Vesic (1973) concluded that the failure mode also depends on the depth of foundation and the relative density of the soil. An analytical solution is available for the general shear failure mode. For local shear failure mode where the soil compressibility is important, Terzaghi (1943) suggested reducing c' and ϕ' while Vesic (1973) proposed the rigidity index together with compressibility correction if necessary.

Failure of a shallow foundation can be disastrous, but is rare; hence, a safety factor of 3 (in some cases 2) is commonly adopted to control the applied pressure and the corresponding settlement. In fact, most failures of shallow foundations are due to excessive settlement, not bearing capacity failure.

The bearing capacity of a footing can also be obtained by plasticity theory. Consider the "general shear failure mode" in Figure 14.5, bearing capacity factors have been proposed by different engineers. The failure mechanism by most of them

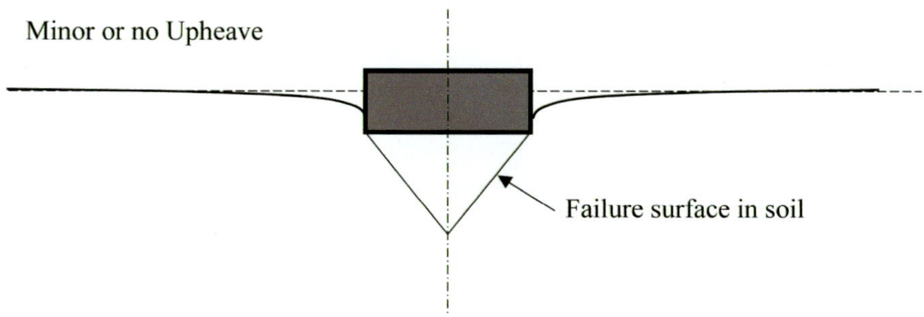

(a)

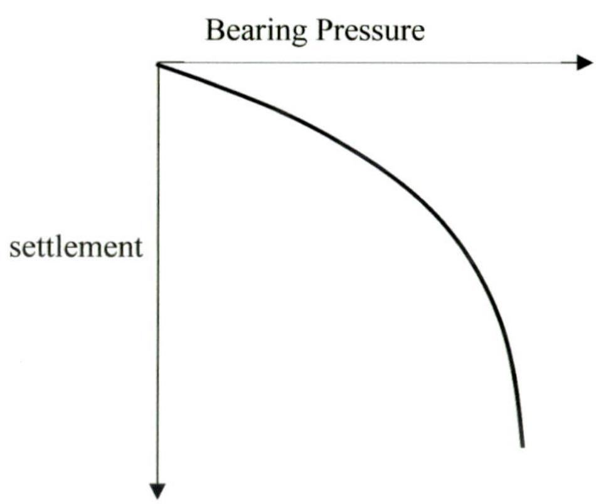

(b)

FIGURE 14.4 (a) Punching shear failure. (b) Settlement/pressure relation for punching shear failure.

is based on Figure 14.5. It should be noted that the inclination of the triangular wedge beneath the foundation is ϕ instead of $\pi/4+\phi/2$in Terzaghi's mechanism (1943), which is generally accepted to be not reasonable, particularly if ϕ is small. For a shallow foundation on ground surface with weightless soil with a surcharge, two bearing capacity factors are obtained by the classical limit equilibrium method. Consider Figure 14.5 for a plane strain condition (2-dimensional problem) on homogeneous and isotropic soil. At ultimate condition, the ultimate bearing capacity q_u push the soil mass F'FB to plastic equilibrium state (active Rankine zone). Angles AF'B and AFB are equal to $\pi/4+\phi/2$. Zones F'C'D' and FCD are passive Rankine zones which are pushed outwards due to horizontal stress generated by the Poisson ratio effect; hence, angles CFD and C'F'D' will be $\pi/4-\phi/2$. Zones F'B'C' and FBC are "transition zones" with rapid stresses and direction changes, which is commonly

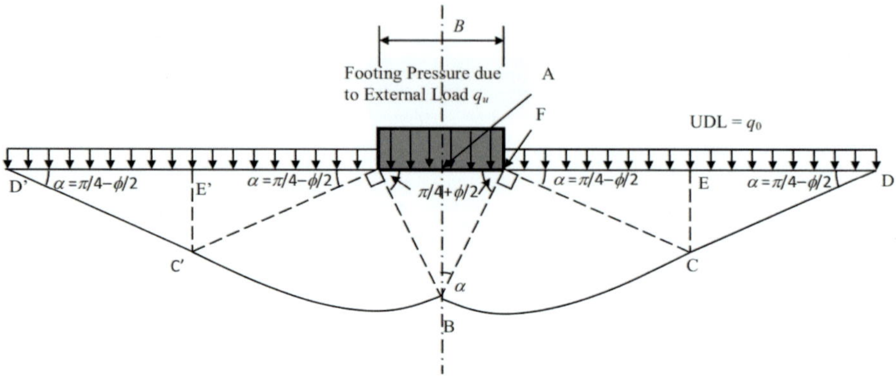

FIGURE 14.5 Failure mechanism for general shear failure of footing with zero unit weight of soil.

taken as logarithm spiral zone. This logarithm spiral curve can be derived from the governing plasticity differential equation if soil is weightless, else the logarithm spiral curve is only a good approximation.

Based on the failure mechanism in Figure 14.5 and consider the moment equilibrium of ABCEFA, the plane strain ultimate bearing capacity q_u can be derived by taking moment equilibrium about point F for weightless soil and given by Cheng et al. (2024).

$$q_u = q_0 N_q + cN_c \tag{14.1}$$

where $N_q = e^{\pi\tan\phi}\tan^2\left(\frac{\pi}{4}+\frac{\phi}{2}\right)$ $\quad N_c = \left(N_q - 1\right)\cot\phi$

If the weight of ABCEFA is considered, the eccentricity of the soil mass is required, which is tedious to determine. The derivation of the factor N_γ is complicated, and a simple and rigorous expression is not possible. Cheng et al. (2024) demonstrated that the failure mechanism in Figure 14.5 is valid only for the cohesive strength and surcharge terms but not for the self-weight term, but the difference in the bearing capacity from Figure 14.5 and the more rigorous failure mechanism from slip line analysis is small. Different researchers proposed different simplified formulas for N_γ based on laboratory tests or plasticity theory, and Equations 14.2, 14.3, and 14.4 are the most common expressions used by engineers. The N_γ expression by Vesic (1973) is used in the United States and Hong Kong. Actually, there are many expressions for N_γ in the literature by different researchers, and these expressions are developed using various simplifications, while a rigorous result is given by Cheng and Au (2005) and Cheng et al. (2024).

$$\text{Vesic} \quad N_\gamma = 2\left(N_q + 1\right)\tan\phi \tag{14.2}$$

$$\text{Hansen} \quad N_\gamma = 1.5 N_c \tan^2\phi \tag{14.3}$$

$$\text{Meyerhof} \quad N_\gamma = (N_q - 1)\tan 1.4\phi \tag{14.4}$$

Summing up, for a plane strain or infinite strip footing of width *B*:

$$q_u = 0.5\gamma_s B N_\gamma + q_0 N_q + c N_c \tag{14.5}$$

Plasticity based method gives the factor N_γ smaller than those by the classical Equations 14.2 to 14.4. The bearing capacity factors N_c, N_q for the Vesic, Meyerhof and Hansen (Bowles 1996) are the same, but not for the expression of N_γ. Nevertheless, the differences between Equations 14.2 and 14.4 are not major in most cases (Cheng and Au 2005). Equation 14.5 applies for an ideal 2-dimensional strip footing, but a real foundation differs from the failure mechanism in Figure 14.5 in:

(1) The actual problem is not a true 2-dimensional problem, as the length of the foundation is actually finite so that additional forces/energy are required to induce failure at the two ends (not considered in the plane strain analysis).
(2) Most of the shallow foundations (with few exceptions) are located below ground level. A virtual ground level is assumed at the foundation base for simplicity. The weight of soil above the virtual ground is considered as a surcharge term $q_0=\gamma D$ applied outside the foundation. This approach is only a simplification, as the failure surface will extend beyond the virtual ground level and additional shear resistance needs to be considered.
(3) Most of the loading on the shallow foundation will not be simple UDL, and horizontal forces and moments may act on the foundation.
(4) Non-homogeneous ground conditions may be encountered.

It should also be noted that if the resultant of the vertical load is eccentric to the footing with offsets from the centre of the footing by e_B and e_L along the dimension *B* and *L* respectively, the effective dimensions of the footing are reduced to $B' = B - 2e_B$ and $L' = L - 2e_L$. For general case, the corrected bearing capacity for a rectangular footing using the effective dimensions $L' \times B'$ is then

$$q_u = 0.5\gamma_s B N_\gamma \zeta_{\gamma s}\zeta_{\gamma i}\zeta_{\gamma t}\zeta_{\gamma g}\zeta_{\gamma d} + q_0 N_q \zeta_{qs}\zeta_{qi}\zeta_{qt}\zeta_{qg}\zeta_{qd} + c N_c \zeta_{cs}\zeta_{ci}\zeta_{ct}\zeta_{cg}\zeta_{cd} \tag{14.6}$$

The total ultimate vertical load of the footing is then given by

$$Q_u = q_u B' L' \tag{14.7}$$

Different modification factors have been proposed by Meyerhof and Hansen, which are based on model test results dated about 50 years ago (see Table 14.1 for the requirement in Hong Kong). Many researchers have questioned these modification factors that were established many years ago. In particular, the shape factor, which

TABLE 14.1
Bearing Capacity and Various Modification Factors Proposed by Vesic (1973) and Adopted in the GEO Guide (2006) in Hong Kong

Basic bearing capacity factors	$N_q = e^{\pi\tan\phi}\tan^2\left(\frac{\pi}{4}+\frac{\phi}{2}\right)$ $N_c = \left(N_q - 1\right)\cot\phi$ $N_\gamma = 2\left(N_q + 1\right)\tan\phi$
Shape factors (depending on $\frac{B}{L}$) or 3D corrections	$\zeta_{cs} = 1 + \frac{B}{L}\frac{N_q}{N_c}$ $\zeta_{\gamma s} = 1 - 0.4\frac{B}{L}$ $\zeta_{qs} = 1 + 0.4\frac{B}{L}\tan\phi$
Factors vertical load P with horizontal shear H	$\zeta_{ci} = \zeta_{qi} - \frac{1-\zeta_{qi}}{N_c\tan\phi}$ $\zeta_{\gamma i} = \left(1 - \frac{H}{P + B'L'c\cot\phi}\right)^{m_i+1}$ $\zeta_{qi} = \left(1 - \frac{H}{P + B'L'c \times \cot\phi}\right)^{m_i}$ $m_i = \frac{2 + \frac{B'}{L'}}{1 + \frac{B'}{L'}}$ for load inclination along B' $m_i = \frac{2 + \frac{L'}{B'}}{1 + \frac{L'}{B'}}$ for load inclination along L'
Factors for footing inclined at angle α to the Horizontal	$\zeta_{ct} = \zeta_{qt} - \frac{1-\zeta_{qt}}{N_c\tan\phi}$ $\zeta_{\gamma t} = \left(1 - \alpha \times \tan\phi\right)^2$ $\zeta_{qt} = \zeta_{\gamma t}$
Factors to modify sloping ground just adjacent to the edge of footing with an angle ω to the horizontal	$\zeta_{cg} = e^{-2\text{É}\tan\phi}$ $\zeta_{\gamma g} = \zeta_{qg}$ $\zeta_{qg} = \left(1 - \tan\text{É}\right)^2$ for $\omega \le 45^\circ$ $\zeta_{qg} = 0$ for $\omega > 45^\circ$
Factors for buried footing (not adopted in Hong Kong)	$\zeta_{cd} = 1 + 0.4k$ $\zeta_{qd} = 1 + 2k\tan f\left(1 - \sin\phi\right)^2$ $\zeta_{gd} = 1.0$ D is the depth below ground, k=D/B if it is less than or equal to 1.0, else $k=\tan^{-1}(D/B)$

is less than 1.0, deviates from theory as well as many laboratory tests (possibly due to inappropriate data analysis). Nevertheless, many engineers still use these simple formulae, which are on the safe side instead of tedious numerical analysis, as settlement is actually the most critical factor in foundation design and a high factor of safety for bearing capacity is usually provided; hence, refined analysis on bearing capacity appears to be not necessary. For non-homogeneous soil conditions, most engineers prefer to use the soil properties directly beneath the footing for the calculation, although some formulae are available in the literature for this case.

For cohesive soil, we can take $\phi = 0$, $N_q = 1$, $N_\gamma = 0$. N_c is indeterminate by direct substitution, but is given by $\pi + 2$ by taking the limit of $\left(N_q - 1\right)\cot\phi$ with $\phi \to 0$. The previous bearing capacity equation applies to foundations on good soil with general shear failure. For footing on compressible soil which is actually not suggested to be used, Terzaghi (1943) proposed to reduce c and tanϕ to 2/3 of their values, while Vesic (1973) suggested using the compressibility factors. In general, the Terzaghi (1943) approach is more popular among engineers.

Equation 14.5 is based on a smooth foundation base without base friction. For a rough base, Cheng and Li (2017) derived the exact expressions for N_c and N_q. For earthquake load, Cheng and Au (2005) gave the bearing capacity factors based on slip line analysis. Another commonly encountered problem is the unit weight of soil in Equation 14.5. If the water table is located well below the footing base or above the base, there is no doubt about the unit weight term. However, if the water table is located at a depth D_w (< B) below the footing base, this situation is usually approximated by an equivalent unit weight term as $\gamma' + \gamma_w * D_w / B$.

For bearing capacity problems on rock, the common practice is to adopt presumed values provided in building codes. Some engineers may adopt the SPT or CPT values of soil to assess the bearing capacity of footing, if sufficient previous database are available.

There are some rare cases where the finite element method is used for bearing capacity assessment. The numerical problem can be modelled by gradually increasing the uniform pressure until the ultimate condition is approached, and the ultimate condition is usually associated with a very large settlement. Cheng has found that many finite element programs fail to give a reasonable failure mechanism as well as the bearing capacity, which reflects the difficulty of non-linear analysis approaching yield. In Hong Kong, the use of the finite element method for bearing capacity determination is not allowed in routine design.

For illustration of the use of the bearing capacity equation, a problem as shown in Figure 14.6 is considered.

Effective width and Length of the Footing

$$B_{fe} = B_f - 2e_B = 5.4\text{m}$$
$$L_{fe} = L_f - 2e_L = 7.2\text{m}$$

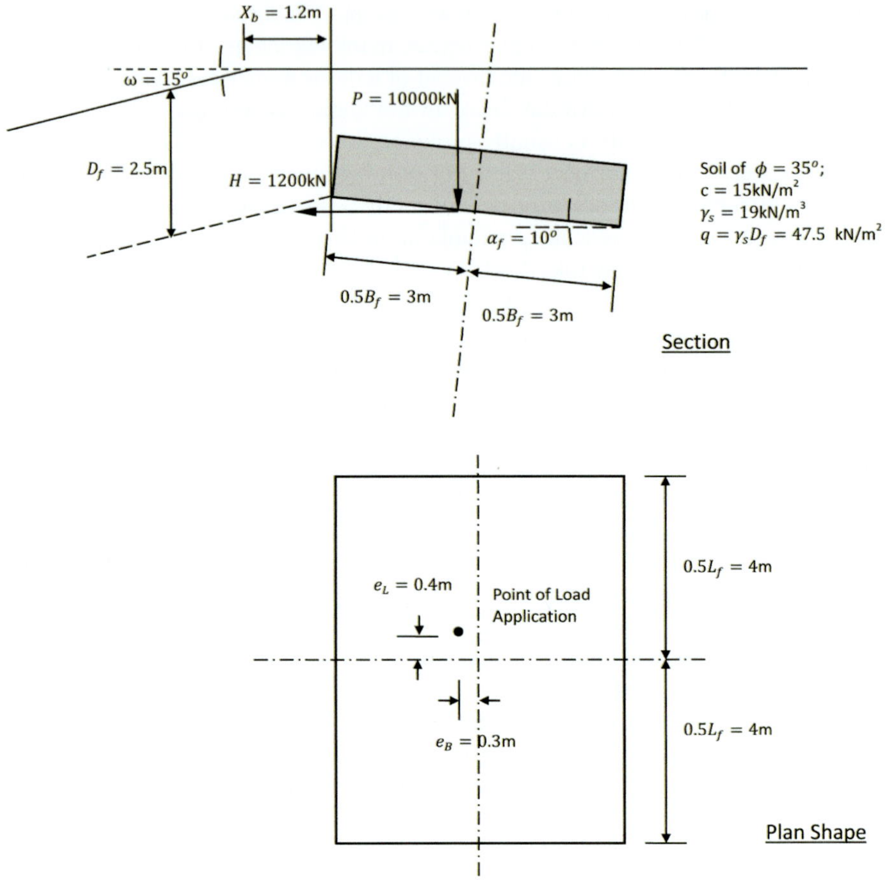

FIGURE 14.6 An illustration example for ultimate bearing capacity determination.

Bearing Capacity Factors

$$N_q = e^{\pi \tan\phi} \tan^2\left(\frac{\phi}{2} + 45^o\right) = 33.296$$

$$N_c = \left(N_c - 1\right)\cot\phi = 46.124$$

$$N_\gamma = 2\left(N_q + 1\right)\tan\phi = 48.029$$

Shape Factors

$$\zeta_{cs} = 1 + \frac{B_{fe}}{L_{fe}} \frac{N_q}{N_c} = 1.541$$

$$\zeta_{\gamma s} = 1 - 0.4\frac{B_{fe}}{L_{fe}} = 0.7$$

$$\zeta_{qs} = 1 + \frac{B_{fe}}{L_{fe}}\tan\phi = 1.525$$

Inclination Factors

$$m_i = \frac{2 + B_{fe} / L_{fe}}{1 + B_{fe} / L_{fe}} = 1.571$$

$$\zeta_{qi} = \left(1 - \frac{H}{P + B_{fe}L_{fe}c\cot\phi}\right)^{m_i} = 0.832$$

$$\zeta_{ci} = \zeta_{qi} - \frac{1 - \zeta_{qi}}{N_c \tan\phi} = 0.826$$

$$\zeta_{\gamma i} = \left(1 - \frac{H}{P + B_{fe}L_{fe}c\cot\phi}\right)^{m_i + 1} = 0.739$$

Tilt Factors

$$\zeta_{\gamma t} = \left(1 - \alpha_f \tan\phi\right)^2 = 0.768$$

$$\zeta_{qt} = \zeta_{\gamma t} = 0.768$$

$$\zeta_{ct} = \zeta_{qt} - \frac{1 - \zeta_{qt}}{N_c \tan\phi} = 0.761$$

Ground Sloping Factors

$$\zeta_{cg} = e^{-2\omega\tan\phi} = 0.687$$

$$\zeta_{qg} = \left(1 - \tan\omega\right)^2 = 0.536$$

$$\zeta_{\gamma g} = \zeta_{qg} = 0.536$$

Overall, the ultimate bearing capacity of the footing is

$$\begin{aligned} q_u &= cN_c\zeta_{cs}\zeta_{ci}\zeta_{ct}\zeta_{cg} + 0.5B_{fe}\gamma_s\zeta_{\gamma s}\zeta_{\gamma i}\zeta_{\gamma t}\zeta_{\gamma g} + qN_q\zeta_{qs}\zeta_{qi}\zeta_{qt}\zeta_{qg} \\ &= 460.868 + 10.933 + 825.844 = 1297.645\text{kN/m}^2 \end{aligned}$$

The bearing capacity equation and modification factors have been adopted in many national design codes, including the Hong Kong Foundation Code (2017), the Canadian Foundation Engineering Manual (2006), Eurocode EC7 (2000, 2004) and China codes. There is also a simple "allowable bearing capacity" approach according to the soil and rock, which is still used in many foundation design codes.

14.3 SHALLOW FOUNDATION (SERVICEABILITY LIMIT STATE)

Excessive foundation settlement will create adverse problems for the structure. Additional shear force and bending moment can be induced from the relative and/or absolute displacement (in some cases), and such additional internal forces are usually not considered in the design. Actually, there is virtually no bearing capacity problem for a normal foundation, as intolerable settlement and angular settlement of the superstructures will come much earlier than the bearing capacity failure. For the construction of Hong Kong Mass Transit Railway subways and tunnels, excessive settlement and serious cracks are found for those buildings on raft foundation in reclaimed areas. Extensive and expensive ground treatment has been carried out for the problems of induced settlements.

Settlement of a shallow footing is due to the deformation of the foundation soil or rock. Soil is usually not elastic even under normal conditions with small loading, but it is still treated as pseudo-elastic for ease of computation to estimate the "short term" settlement using the elastic theory. This approach is basically applicable for coarse-grained soil under small loading. However, the consolidation settlement for clay can last for months or years to completion, which is due to the expulsion of water in fine-grained impermeable soil. Furthermore, there is creep, another settlement which is important for clay soil. Creep is an important consideration for construction in reclaimed areas (a famous airport creep crisis in Japan). For the Hong Kong Chek Lap Kok airport first runway, which costs about 13 billion US dollars, is the most expensive airport construction in the world in the year 1997. The most critical factor for the extremely high construction cost is to eliminate creep by removing all the marine clay by dredgers before the reclamation. During the construction, half of the dredgers from different places in the world come to Hong Kong for this project! The construction cost of the third runway of the same airport is about 26 billion US dollars in 2020, which is due to the use of deep cement mixing (DCM) to strengthen the marine clay and to minimize the consolidation and creep settlement. There are other famous examples where creep and consolidation are important, which include the ZhuZhai-Hong Kong-Macau bridge project, Tower Pisa and others.

The basic solution to the equilibrium differential equation in X, Y, Z directions for a point load P on the surface of a semi-infinite, isotropic and homogeneous medium in Figure 14.7 is given by the Boussinesq Equations with the symbols defined by Figure 14.8:

$$\sigma_z = \frac{3P}{2\pi z^2}\left(\frac{1}{1+(r/z)^2}\right)^{2.5} \tag{14.8}$$

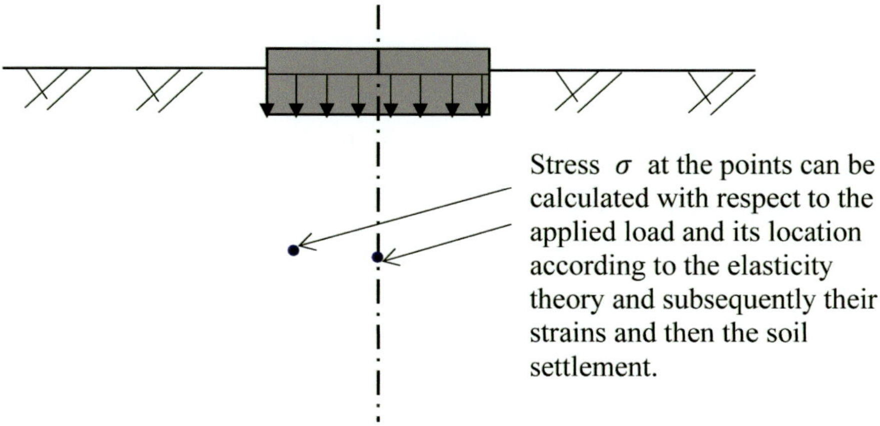

FIGURE 14.7 Estimation of settlement beneath footing.

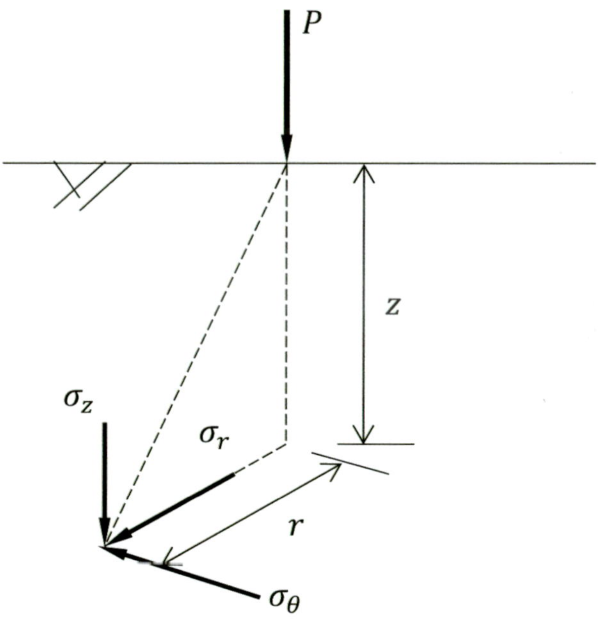

FIGURE 14.8 Boussinesq equation

$$\sigma_r = \frac{P}{2\pi}\left\{\frac{3r^2 z}{\left(r^2+z^2\right)^{2.5}} - \frac{1-2\mu}{r^2+z^2+z\sqrt{r^2+z^2}}\right\} \tag{14.9}$$

$$\sigma_\theta = \frac{-P}{2\pi}\left(1-2\mu\right)\left\{\frac{z}{\left(r^2+z^2\right)^{1.5}} - \frac{1}{r^2+z^2+z\sqrt{r^2+z^2}}\right\} \tag{14.10}$$

$$\tau_{rz} = \frac{3P}{2\pi}\left(\frac{rz^2}{\left(r^2+z^2\right)^{2.5}}\right) \tag{14.11}$$

$\tau_{r\theta} = \tau_{z\theta} = 0$ using this axis system while τ_{rz} is the shear stress defined with the rz plane.

If the soil is isotropic which means $E_x = E_y = E_z = E$ and $\mu_x = \mu_y = \mu$, the vertical strain is:

$$\varepsilon_z = \frac{1}{E}\left[\sigma_z - \mu\left(\sigma_x + \sigma_y\right)\right] \tag{14.12}$$

The vertical strain at coordinate (x, y, z) due to a finite plan size load on ground surface is

$$\varepsilon_z = \int \varepsilon_z q dA \tag{14.13}$$

where q is the uniform pressure defined for an infinitesimal area dA, which can be varying over the footing.

For a BxL rectangular footing (B less than L) under a uniformly distributed load p with a rock or non-deformable stratum at depth H as measured from the base of the foundation (not ground surface), the total settlement at a corner is

$$\Delta H = \int_0^H \int_0^B \int_0^L \varepsilon_z dxdydz \tag{14.14}$$

Although ε_z is based on Boussinesq Equations which applies for an elastic half space, it is still approximately applicable even if a hard stratum exists at a depth H.

For Equation 14.14, the mathematics can be simplified as

$$\Delta H = \int_0^\infty \int_0^B \int_0^L \varepsilon_z dxdydz - \int_H^\infty \int_0^B \int_0^L \varepsilon_z dxdydz \tag{14.15}$$

The analytical solution for Equation 14.15 for the settlement at the corner of a rectangular footing is

$$\Delta H = qB\frac{1-\mu^2}{E_s}\left(I_1 + \frac{1-2\mu}{1-\mu}I_2\right)I_f \tag{14.16}$$

where E_s is the Young's modulus of soil, μ is the Poisson's ratio

$$I_1 = \frac{1}{\pi}\left[M\ln\frac{\left(1+\sqrt{M^2+1}\right)\sqrt{M^2+N^2}}{M\left(1+\sqrt{M^2+N^2+1}\right)} + \ln\frac{\left(M+\sqrt{M^2+1}\right)\sqrt{1+N^2}}{1+\sqrt{M^2+N^2+1}}\right]$$

$$I_2 = \frac{N}{2\pi}\tan^{-1}\left(\frac{M}{N\sqrt{M^2+N^2+1}}\right)$$

where $M = L/B$, $N = H/B$. I_f is the depth factor to account for the vertical load not applied on the ground surface (see Cheng et al. 2024). Many engineers simply take I_f to be 1.0 as a conservative design, unless the depth of foundation D is appreciable.

In using the elastic settlement equation, it should be noted that the **effective loading** instead of total load is adopted by many engineers in the calculation, that is $q-\gamma D$, provided that q refers to the total pressure beneath the foundation. This is based on the rationale that the original overburden stress γD will not generate any elastic settlement. This approach is adopted in all textbooks and design codes that the authors are aware of. However, a problem in Hong Kong has caused the authors to rethink about this problem. For a building with a raft foundation supporting an average load of 120 kPa, after demolition and re-construction works, the average loading on the raft footing will be about 160 kPa. What should be the design load for settlement calculation? Since the concept of effective loading has a major impact in this case, the authors view that 160 kPa should be used in the design calculation! The rationale behind this is that after demolition of the superstructure and existing raft footing, the soil will rebound to some extent, and the upheaval will be removed to the designed level for the new construction works. The new 160 kPa is imposed on this new configuration, and the settlement will be induced by this 160 kPa. However, using the unloading/reloading Young's modulus of soil until the previous maximum overburden pressure is exceeded, and the applied loading is located at a certain depth below the ground surface is a more realistic phenomenon. This should be the actual situation that is happening on site, and the full applied pressure should be adopted in the design calculation, but using an unloading/reloading modulus.

For foundation on multi-layered soil, some engineers adopt the use of equivalent Young's modulus of soil, which is given by $E_{eq} = \Sigma E_iH_i/\Sigma H_i$. If the footing stiffness is considered, there are only minor changes to the settlement, as the settlement is mainly controlled by the settlement of the foundation soil. Hence, the use of a rigid footing cannot really reduce the amount of settlement, but will greatly reduce the differential settlement. For rigid foundation, settlement is about 0.93×settlement at centre of flexible foundation. For the Young's modulus of soil, many engineers rely on the statistical relation with the SPT or CPT results, and an additional factor of safety is sometimes incorporated into the design codes in some countries. Some typical values are 10000 kN/m^2 for fill, 10000–15000 kN/m^2 for alluvium, 15000 to 20000 kN/m^2 for completely decomposed granite in Hong Kong, where a factor of safety of about 2 has been applied. For the Poisson's ratio, some typical values adopted by the engineers are 0.3–0.4 for N equals 4–10, 0.2–0.35 for N equals 11 to 30 and 0.15–0.3 for N above 30.

Example: A 2 m × 4 m footing at 1.5 m below ground is loaded with a total UDL of 200 kPa. Determine the maximum differential elastic settlement along the long direction if the soil properties are E = 15000 kPa, μ = 0.2 and bedrock is present at a depth of 7 m below ground, and the unit weight of soil = 19 kN/m^3.

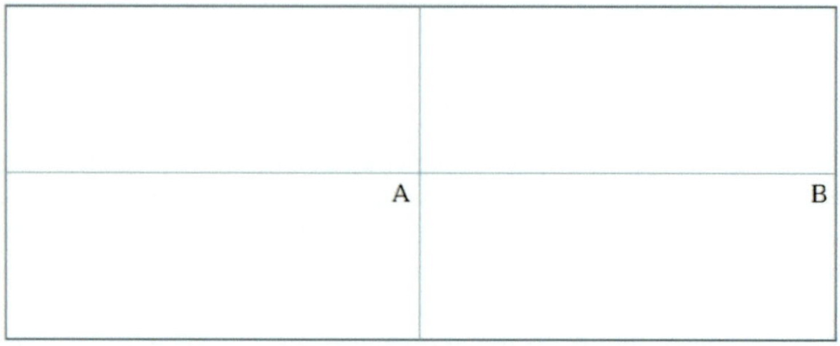

FIGURE 14.9 Calculation of the differential settlement for a simple footing.

Solution (Figure 14.9):

The effective imposed loading q = 200 – 1.5 × 19 = 171.7 kPa (classical approach)

For point A, divide the rectangle into 4 smaller 1 m × 2 m rectangles, M = 2/1, N = 5.5/1 (note that 6.5 m is used as H, which is the depth from the bottom of the footing to bedrock), D/B = 0.75

$$I_1 = 0.546\ I_2 = 0.0535,\ I_f = 0.73, \text{ hence}$$

$$\Delta H_A = 4 \times 171.5 \times 1 \times \frac{1-0.2^2}{15000}\left(0.546 + \frac{1-2\times0.2}{1-0.2}\times 0.0535\right)\times 0.75 \times 1000 = 19.28$$

mm (note that the result has been multiplied by 4 when using the smaller rectangle for calculation)

For point B, divide the rectangle into 2 smaller 1 m × 4 m rectangles, M = 4/1, N = 5.5/1

$$I_1 = 0.584\ I_2 = 0.0923,\ I_f = 0.73,$$

$$\Delta H_{AB} = 2 \times 171.5 \times 1 \times \frac{1-0.2^2}{15000}\left(0.584 + \frac{1-2\times0.2}{1-0.2}\times 0.0923\right)\times 0.75 \times 1000 = 10.73$$

mm. Hence $\Delta H_{AB} = 10.73\text{mm}$ (note that the result has been multiplied by 2)

The differential settlement is the distance between two points against the difference in the settlement, which is 1 in 2000/(19.28-10.73), or 1 in 235.

It should be noted that the Hong Kong Foundation code (2017) and Euro Code 7 (2004) limit the differential settlement to 1 in 500, and similar or slightly different values are used in other countries. The maximum allowable settlement of the foundation is less than 30 mm in the Hong Kong Foundation Code (or 50 mm in the Euro Code 7). In general, the maximum settlement is a less important factor in the foundation design, as the differential settlement will control the design in many cases. More importantly, the maximum settlement is important only for those cases where the utilities entering into a building cannot tolerate the settlement. Cheng has observed the effects of differential settlement on many old buildings (on raft foundation) around the Wan Chai area in Hong Kong which were built more than 50 years ago. Many of the walls, columns and beams are seriously cracked, with a crack width wide enough for a finger. Such cracking is generated from the additional bending moment and shear force arising from the differential settlement, which have not been considered in the design 50 years ago. Recently, there are also many differential settlement problems for the buildings in China, due to the improper foundation construction. Cheng has also observed some settlements for new buildings at Wan Chai and Sheung Wan around the MTR construction. The maximum building settlements amount to 250 mm to 300 mm, yet there is not any cracking at the buildings which sit on piles. This is due to the fact that these buildings undergo rigid body movement during the settlement. As long as the utilities can tolerate the settlement, there are no major practical problems for these buildings.

Additional example for readers: Determine the differential settlement along the long and short direction for a 2 × 4 m footing on ground subjected to a UDL of 180 kPa. The soil parameters are E = 15000 kPa and μ = 0.2. (Rock is assumed to be at 10 m below ground)

Answer: centre settlement 30.59/short side 18.07/long side 21.19 mm, hence 1 in 159.7 along long direction and 1 in 106.4 along short direction.

Besides the "short term" or elastic settlement, there is a relatively "long-term settlement" in fine-grained soil of low permeability to water. The process is termed "consolidation" by which the volume of a mass of fully saturated soil under pressure (exerted from a foundation) is reduced slowly due to expulsion of water from its voids to the exterior through the soil grains of low permeability. The process often takes time in terms of years to completion until the excess pore water pressure set up by the increase in total stress has completely dissipated. Currently, the use of preloading with the use of wick drains is commonly adopted to accelerate the consolidation process.

For the 1-dimensional consolidation theory, both the compression index C_c and the coefficient of compression m_v are used, and there are some common wrong concepts among the engineers in using these two parameters. For a saturated clay stratum with initial void ratio e_0. Under applied pressure, there will be a change of void ratio of Δe, and the vertical strain is expressed as :

$$\varepsilon_z = \frac{\Delta e}{1+e_0} \tag{14.17}$$

The vertical strain ε_z is

$$\varepsilon_z = \frac{\Delta H}{H} = \frac{\Delta e}{1+e_0} \tag{14.18}$$

Denote $\Delta e = e_0 - e$, with e the void ratio after consolidation. A typical relation of e with effective vertical stress σ' for a saturated clay is shown in Figure 14.10. The compression is termed "virgin" if the current stress on the clay soil is the maximum stress, and this condition is termed "normally consolidated." The $e - \log\sigma'$ relation is a good linear relation for the virgin compression line. If the stress is then reduced,

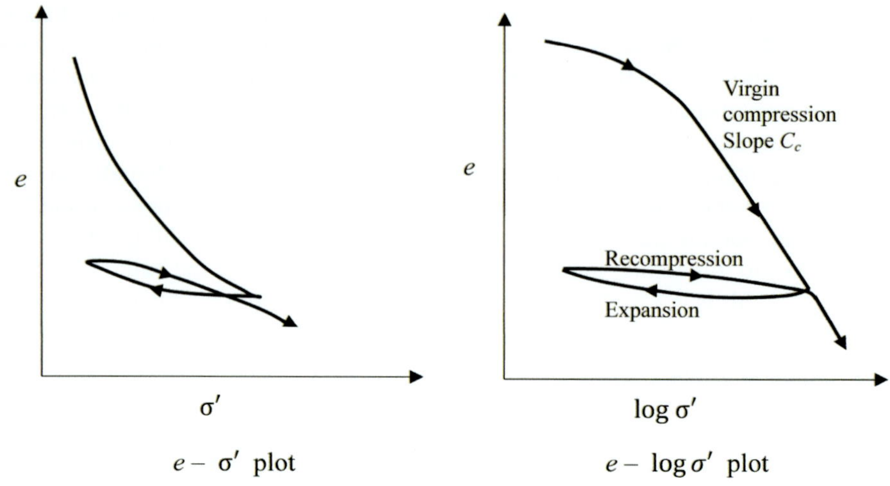

FIGURE 14.10 Typical relation of void ratio of clay with effective stress.

the soil will slightly expand while the void ratio can only recover a small amount of the original value. If the stress is increased again, then the void ratio will return to the previous value at approximately the same previous stress level. The compression and reloading path as shown in Figure 14.10 (hysteresis) is usually taken as a straight line for simplicity.

By Figure 14.10, two parameters can be defined :

(i) The coefficient of compressibility m_v

$$m_v = \frac{1}{1+e_0}\left(\frac{e_0 - e_1}{\sigma_1' - \sigma_0'}\right) \tag{14.19}$$

where σ_0' is the initial effective stress, e_0 is the initial void ratio, e_1 is the void ratio at the increased stress of σ_1'. m_v relates the volumetric change with the increase in effective stress. m_v is however not really a material constant, though many engineers have mistaken it to be constant in the consolidation process.

(ii) The "compression index" C_c

$$C_c = \frac{e_0 - e_1}{log\left(\sigma_1' / \sigma_0'\right)} \tag{14.20}$$

It is usually a good constant for the straight line portion of the $e - \log\sigma'$ plot in Figure 14.10. For the unloading/reloading part, the "expansion index" C_e can be defined. The use of C_c and C_e are basically constants as compared with the volume compressibility m_v.

Settlement of soil due to consolidation is

$$S_c = \int \varepsilon_z dz \tag{14.21}$$

$$\varepsilon_z = \frac{\Delta e}{1+e_0} = \frac{e_0 - e_1}{1+e_0}$$

If the stress increase is constant throughout the clay soil with depth H_c, Equation 14.21 becomes $S_c = \varepsilon_z H_c$. Consolidation settlement for normally and over consolidated clay is given as:

$$S_c = \frac{C_c H_c}{1+e_0} \log \frac{\sigma_1'}{\sigma_0'} \tag{14.22a}$$

$$S_c = \frac{C_e H_c}{1+e_0} \log \frac{\sigma_c'}{\sigma_0'} + \frac{C_c H_c}{1+e_0} \log \frac{\sigma_1'}{\sigma_c'} \tag{14.22b}$$

where σ_σ' is the maximum overburden stress for over-consolidated soil. Under additional increase in stress $\Delta\sigma$, $\sigma_1' = \sigma_0' + \Delta\sigma$. If the calculation is based on m_v which can be defined as $m_v = \frac{\Delta\varepsilon_v}{\Delta\sigma'}$ the settlement is

$$S_c = \int_0^H \varepsilon_v dz = \int_0^H m_v \Delta\sigma' dz \tag{14.23}$$

If $\Delta\sigma'$ is constant, or approximately constant with depth

$$S_c = m_v \Delta\sigma' H \tag{14.24}$$

This equation is commonly used by the engineers, but the settlement of clay is more complicated as the induced stress usually changes with depths. Under this condition, the average σ_0' and σ_1' of the whole clay stratum is used for computation if the thickness is limited. A better method is to divide the clay stratum into a series of layers, and the mean stress in each layer is used for the calculation of the settlement for each layer. The total settlement will be the sum of the individual settlement of all the layers. To determine $\Delta\sigma$, the simple load spread methods of: 1 (horizontal) and 2 (vertical) can be used if the thickness of clay is limited. Otherwise, the Boussinesq equation will be applied. It should be noted that the volume compressibility is not really constant during the consolidation process, so this concept must be used with care.

The worked example in Figure 14.11 illustrates the calculation of the consolidation settlement of the normally consolidated clay at the centre axis of the footing.

Worked Example 14.1 (with reference to Figure 14.11)

(i) Given: the clay is Normally Consolidated Clay. The average initial overburden stress σ_0' and average σ_1' are determined.

$$\sigma_0' = 3 \times 16.5 + 1.2 \times (17.5 - 9.8) + 1.25 \times (16 - 9.8) = 66.49\text{kN/m}^2;$$

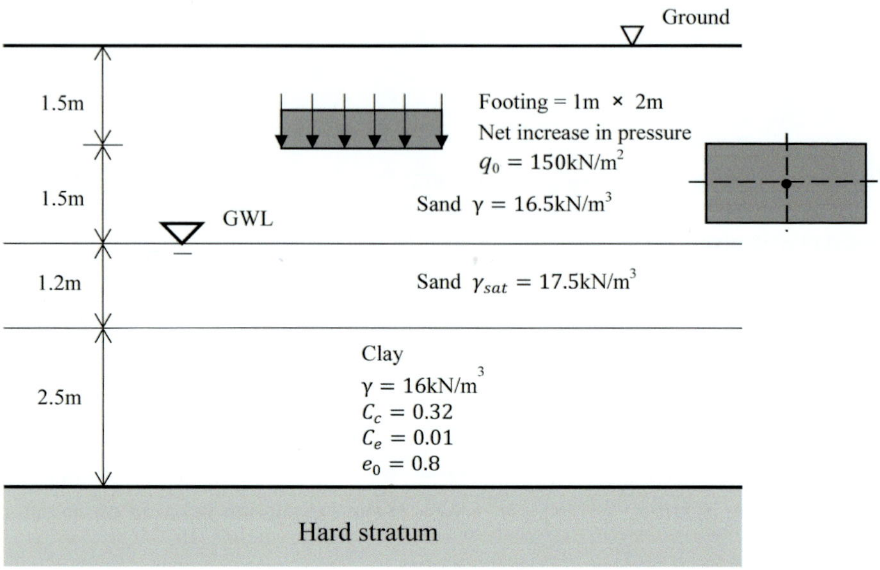

FIGURE 14.11 Settlement calculation for normally consolidated and over-consolidated clay.

For the average $\Delta\sigma$ due to the footing load, the stress increase at the top, mid-depth and bottom of the clay layer is determined by dividing the footing divided into four equal parts ($B = 0.5$m, $L = 1$m). The stress factor I_σ for each quadrant is :

So at top z = 2.7m, $m = \dfrac{B}{z} = 0.1852;\ n = \dfrac{L}{z} = 0.3704,\ I_\sigma = 0.0287$

At mid-depth, z = 3.95m $m = \dfrac{B}{z} = 0.1266;\ n = \dfrac{L}{z} = 0.2532,\ I_\sigma = 0.0143$

At bottom, z = 5.2m $m = \dfrac{B}{z} = 0.1266;\ n = \dfrac{L}{z} = 0.2532,\ I_\sigma = 0.0085$

The average value I_σ is obtained by the trapezoidal rule (or Simpson rule) as

$$I_{\sigma,ave1} = \frac{1}{4}(0.0287 + 2\times 0.0143 + 0.0085) = 0.0165$$

Or by Simpson Rule $I_{\sigma,ave2} = \dfrac{1}{6}(0.0287 + 4\times 0.0143 + 0.0085) = 0.0158$

Choosing $I_{\sigma,ave2} = 0.0158$, $\Delta\sigma' = 0.0158\times 150\times 4 = 9.48\text{kN/m}^2$;

So the consolidation settlement is

$$S_c = \frac{C_c H_c}{1+e_0}\log\frac{\sigma_1'}{\sigma_0'} = \frac{0.32\times 2.5}{1+0.8}\log\frac{66.94+9.48}{66.94} = 0.0256\text{m} = 25.6\text{ mm}$$

Given: the Clay is over-consolidated with $\sigma_c' = 70$ kN/m², the consolidation settlement from the pressure 66.49 kN/m² to $\sigma_c' = 70$ kN/m² is

$$S_{c1} = \frac{C_e H_c}{1+e_0}\log\frac{\sigma_c'}{\sigma_0'} = \frac{0.1\times 2.5}{1+0.8}\log\frac{70}{66.94} = 0.0027 \text{ m}$$

The settlement after the preconsolidation pressure is

$$S_{c2} = \frac{C_c H_c}{1+e_0}\log\frac{\sigma_1'}{\sigma_c'} = \frac{0.32\times 2.5}{1+0.8}\log\frac{76.42}{70} = 0.0169 \text{ m}$$

So the total settlement is

$$S_{c1} + S_{c2} = 0.0196\text{m} = 19.6 \text{ mm}$$

In the past, many engineers adopted the volume compressibility m_v in the settlement calculation. It appears that the main reason is due to the calculation of the degree of consolidation. As discussed by Cheng et al. (2024), this approach is not satisfactory in that the volume compressibility has to be assumed to be constant in the consolidation process which is, however, not true. The volume compressibility as determined from laboratory tests is actually a stress-dependent parameter, and it can change by several times during reclamation. It is not surprising that the measured pore pressure and settlement on site usually cannot match with the calculation in Hong Kong where most of the consolidation settlements are determined by the use of m_v. Cheng et al. (2024) recommend the use of the compression index (which is practically a constant parameter) for all the consolidation calculations, and the degree of consolidation has to be determined from a computer analysis. Instead of using those complicated finite element program for such one-dimensional analysis, Cheng has developed a computer code for the degree of consolidation based on 1D large displacement finite element consolidation analysis which is simple to use. This more refined approach is, however, not commonly adopted in Hong Kong.

Vibro-compaction of sand fill or variations such as the use of stone columns to improve the characteristics of the marine clay have been adopted in Hong Kong. The use of deep cement mixing (DCM) is adopted for the third Hong Kong Airport runway project where the in situ marine clay will remain in place and stabilized by high pressure cement grouting, following the major shift in the cellular cofferdam (maximum 6 m) associated with the Hong Kong ZhuHai Macau Bridge project which is induced by the marine clay consolidation. For the artificial island in the Hong Kong ZhuHai Macau Bridge project, the use of vacuum preloading, lowering of the water table and preloading in the form of surcharge are used for accelerating the consolidation process.

For sandy soil, the effect of creep is usually not important, though it is also considered in some countries. For clay, besides the consolidation, creep or secondary consolidation can also be a critical problem. Under unchanged loading and practical the end of the consolidation process, the settlement will still keep on increasing with time. It must be noted that creep starts at the beginning of the consolidation process, and various means are used to separate these processes. However, many engineers prefer to separate creep from the consolidation process for ease of calculation, and the estimation of creep starts at the end of the primary consolidation process. The accurate determination of creep is very difficult, as the test for the creep parameters can be extremely time-consuming. Furthermore, the creep process is actually not a linear process. A classical creep equation is given by Equation 14.25

$$S_c = \frac{C_s H}{1+e_p} log\left(\frac{t_2}{t_1}\right) \tag{14.25}$$

Where e_p is the void ratio at the end of primary consolidation, C_s is the creep (secondary consolidation) factor, t_1 and t_2 are the time for primary consolidation and the time under consideration. Many engineers have criticized this equation, as if t_2 tends to infinity, the creep settlement will tend to infinity. Creep must stabilize at some time, and this is the basis for many advanced creep models where the creep strain rate will decrease with time. Even so, many engineers still adopt this equation because t_2, which will be the expected lifespan of the construction work, will actually be finite. Furthermore, due to the limitations of the laboratory tests, parameters for more advanced constitutive models are not easily calibrated. From a practical point of view, there is no major harm in using this equation. The most important point in the determination of creep is not the equation form, but the parameters for creep. Accurate parameters for creep are difficult and time-consuming to determine. Advanced creep models such as those by Cheng et al. (1996) and many others are hence not commonly adopted in actual practice. Due to the high uncertainty of creep, the use of complete dredging or deep cement mixing (DCM), as shown in Figure 14.12 are both for the airport reclamation construction in Hong Kong, for which both methods of construction are expensive.

14.4 USE OF SAND DRAIN/WICK DRAIN/VACUUM PRELOADING TO ACCELERATE CONSOLIDATION

Besides the use of surcharge, sand drains or wick drains are also commonly used in many reclamation projects to accelerate the primary consolidation. The configuration for the drains is usually arranged in a triangular or square pattern. Wick drain (Figure 14.13) and geodrains are fast in construction and cheaper than classical sand drain, but classical sand drain is still used a lot without the need to import the wick drains. For rectangular wick drain, the equivalent radius used for the calculation is given by $(b + t)/\pi$. It is also a common practice to use additional surcharge together with wick drain to accelerate the consolidation process. Vacuum preloading is also a common practice in applying surcharge in China. 1 atmospheric pressure is close to 10 m height of water. Assuming a vacuum efficiency of 60% for the preloading, vacuum preloading is approximately equivalent to 3 m height of soil, and vacuum preloading can save transportation cost and is good enough for many problems.

FIGURE 14.12 Deep cement mixing for airport reclamation project in Hong Kong.

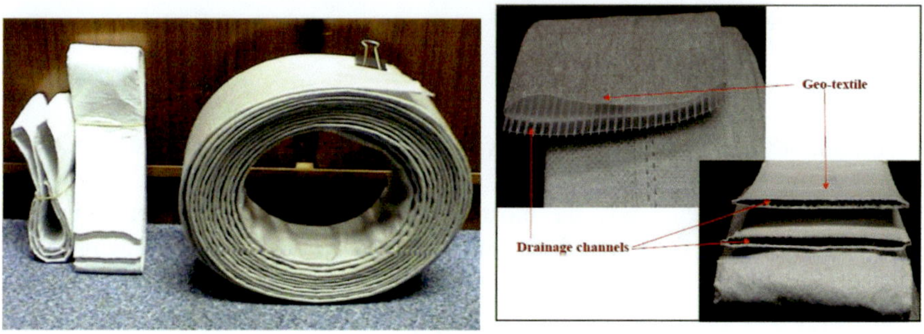

FIGURE 14.13 Examples of some wick drain.

Common methods for construction of sand drain/wick drain:

(1) Sand drain with sand – use of lightweight rotary drill with wash boring or auger drilling method to form the sand drain
(2) Sand drain is usually formed by sand bag – forming good quality but large diameter and cheap sand drain is not easy due to necking during installation.

FIGURE 14.14 Installation of wick drains.

Sand bag made of geosynthetic with good strength with a diameter of 7–12 cm can be used.

(3) Wick drain – currently the most common method which is installed by driving using a mandrel which can be retracted (Figures 14.14 and 14.15).

14.5 USE OF FOUNDATION CODES

Foundation settlement is important as it affects the performance of the superstructure, both in the ultimate and serviceability limit states. For the determination of shallow foundation settlements, the common national and international codes of practice for foundations either list the formulae as derived and discussed in the foregoing or ask the user to base their calculations on engineering principles. In some codes, empirical values for the parameters in the formulae such as Young's modulus of soil as related to SPTN values may be given. Or more generally, the codes would

FIGURE 14.15 Use of wick drain to accelerate consolidation. (a) Wick drains on site. (b) Water coming out from wick drain.

simply ask the user to refer to laboratory or site tests. A brief discussion on the following codes is given below:

The Eurocode EC7 (6.6.2) has defined three types of settlement which are (i) immediate; (ii) consolidation; and (iii) creep. The clause allows commonly recognized methods for evaluation of the settlements. The Code also provides sample methods as specified in its Annex F. In the Annex, though detailed formulae are not given, basically the approaches for the three kinds of settlements are as indicated in the foregoing of this chapter. And yet it acknowledges that the calculations are not regarded as accurate, but only approximate indications.

The old British Code BS8004–1986 (2.1.2.3) only lists the different types of settlement as similar to EC7 with a discussion of the nature but without an approach on the quantification of the settlement. The new version BS8004 – 2015 simply makes reference to EC7.

The Hong Kong Code of Practice for Foundation 2017 lists the formulae for settlement determination of coarse- and fine-grained soil and that of secondary settlement, which are identical to the foregoing of this chapter.

The codes often stress that settlement and relative settlement should not incur detrimental effects to the structures, both in the ultimate and serviceability limit states. Some codes do not recommend limiting values while some do. The Eurocode EC7 lists limiting values of structural deformation and foundation movement for normal routine structures in its Annex H which mainly comprises

(i) Acceptable absolute settlement can be up to 50 mm;
(ii) Acceptable relative rotations (defined as relative settlements between two points divided by the distance between the two points) generally not to exceed 1/500 for "sagging mode," which should be halved for "hogging mode."

EC7 does not specify the loads (dead or live or wind) in giving rise to the settlement to which these limits apply. Presumably, the loads should be dead load + live load.

The Code of Practice for Foundations 2017 (Hong Kong) 2.3.2(2) lists the following limits in general for checking of buildings or structures not particularly sensitive to movement:

(i) Absolute settlement of 30 mm due to dead + live loads.
(ii) Differential settlement/distance apart not to exceed 1:500 dead to dead + live load.
(iii) Angular rotation is again defined as differential settlement/distance apart not to exceed 1:500 due to wind or other transient loads.

For reclamation, the normally accepted residual settlement parameters adopted for general reclamations in Hong Kong are as follows:

(1) Maximum Residual Settlement at 50 years after completion of the defects and maintenance liability period – 500 mm.
(2) Maximum Differential Settlement at 50 years after completion of the defects and maintenance liability period – 1:300.

By the Hong Kong Foundation Code, (ii) and (iii) can be considered separately. From the practical point of view, the code allows exemption from checking (i) if the foundation is resting on rock. Exceedance of the above limits would, however, require the user to carry out assessment not only the structures, but also on adjacent structures. The use of such criteria has a limitation of accumulation effect, which is also not clearly specified in nearly all the design codes. Consider a case when the construction has induced about 25 mm settlement after completion of the works. If a new construction starts immediately after the previous work and induces another 25 mm, should the new construction be considered acceptable? Based on the code, the second construction has complied perfectly with the requirement, but the area has actually settled by a total of 50 mm over a relatively short period. The situation will be further amplified if there is a third or fourth construction work nearby. This problem has occurred for a settlement problem at Tin Shui Wai in Hong Kong, where the actual total settlement is very significant and has induced some problems to the surrounding area. Many design codes have not laid down clear guidelines about this problem, so the problem of accumulation effect can be critical in some cases.

14.6 MONITORING OF SETTLEMENT

There are various uncertainties with the consolidation/settlement of loose fill and soft marine clay, and many engineering problems have occurred in different projects in Hong Kong and many other countries. Even with great care in the analysis and design, it is still advisable to monitor the movement, stress/strain, pore water pressure, tilting, vibration, air quality, weather, and various issues during construction, as well as long-term monitoring of the construction. For example, the settlement of

FIGURE 14.16 GPS/GNSS monitoring of ground movement for Hong Kong ZhuHai Macau bridge project.

the fill zone in the Hong Kong Link of the Hong Kong ZhuHai Macau Bridge, the third Airport runway, and other large-scale construction projects. Cheng is responsible for the monitoring works of some of these projects. For example, for the monitoring of the third Airport runway project in Hong Kong, monitoring works in the form of inclinometer, magnetic probe extensometer, vibrating wire piezometer (VWP), standpipe, settlement markers, deformation monitoring system, vibration, tile plate, and others are used with different monitoring frequencies. For the Hong Kong Link reclamation, GPS/GNSS is used by Cheng for monitoring the ground movement as shown in Figure 14.16.

For IOT monitoring which is now commonly used in construction, a platform is usually required by most engineers (there are only limited cases where API for data transfer are provided to the engineers but not a platform). A sample view for the platform developed by Cheng is shown in Figure 14.17.

In Hong Kong, the settlement and similar issues are usually responded to through an AAA monitoring plan as:

Alert level

(1) RC to notify and submit a written report to RSE/RGE.
(2) RSE/RGE to assess the effect of movements and predict future movements.

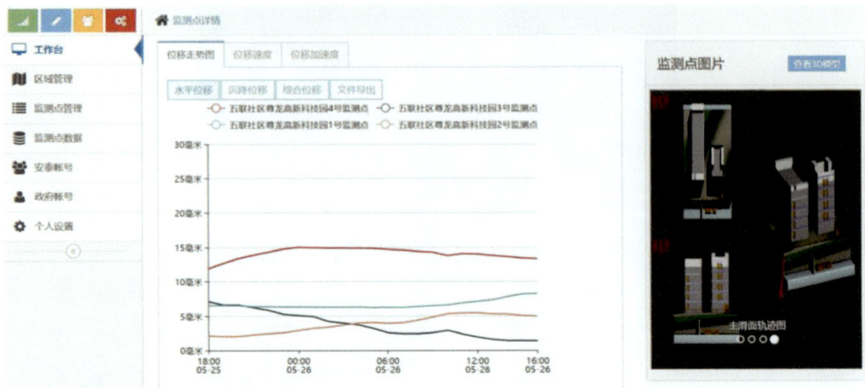

FIGURE 14.17 Platform for monitoring of settlement and groundwater table for a building in Shenzhen, China.

(3) RSE/RGE to agree with RC on the suitable action plan which may include the installation of additional check points and/or increasing monitoring frequency, and the remedial/mitigating measures to be taken upon reaching the "Alarm level."

Alarm level

(1) RC to notify and submit updated report to RSE/RGE and notify the BD immediately.
(2) RSE/RGE and RC to conduct a joint site inspection to determine if any construction activities should be temporarily suspended.
(3) RC to implement the necessary remedial/mitigating measures in accordance with the agreed action plan.
(4) RSE/RGE and RC to discuss the instrument response and review the effectiveness of the response action.
(5) RSE/RGE to agree with RC on the emergency plan detailing the measures to be taken upon reaching "Action Level."
(6) Any construction activities may be suspended if the response action has been implemented and on the advice of the RSE/RGE as necessary.
(7) RSE/RGE to review the method statement of piling works to determine whether modification to the construction methods is required to prevent Action Level from being reached.

Action level

(1) All works that will cause ground movement are to be ceased.
(2) RC to notify and carry out a joint site inspection with the RSE/RGE immediately. The BD and the relevant parties should be notified immediately.

TABLE 14.2
Triggering Levels as Used in Hong Kong

Items	Alert	Alarm	Action
Ground settlement	12 mm	18 mm	25 mm
Services settlement/angular rotation	12 mm or 1:600	18 mm or 1 in 450	25 mm or 1 in 300
Building tilting	1:1000	1:750	1:500

(3) RC to implement the necessary emergency measures in accordance with the agreed emergency plan.
(4) RC to submit an incident report to RSE/RGE and the BD detailing the full history of movements and remedial/emergency measures implemented.
(5) RSE/RGE to review the incident and agree with RC on further remedial and preventive measures to enable resumption of the suspended works.
(6) Construction activities should not be resumed until the necessary remedial and preventive measures have been completed to the satisfaction of the BD.
(7) If the trigger values and response action are revised, the amended plans should be submitted to the BD for approval. The suspended construction activities should not be resumed until the amended plans are approved by the BA and the consent is given.

The triggering levels that are adopted in the Hong Kong Foundation Code 2017 is given in Table 14.2. It should also be noted that under the GEO Publication 1/2023, a 5-level alarm system is suggested.

14.7 INTRODUCTION TO PILE FOUNDATION

Piles are usually classified by the material type, method of installation, or load transfer method. For load transfer, a pile can be classified as end-bearing piles or friction piles according to the relative proportion between the skin friction and end-bearing. It should be noted that many end-bearing piles actually function as friction pile under working load condition. There are many types of piles in the world, and piles can be classified according to different criteria:

(1) End-bearing and friction pile – This classification is based on the load transfer of the piles. If the main pile capacity is derived from friction along its shaft, it is termed as "friction pile." If the main pile capacity is derived from its end-bearing capacity on a hard stratum, it is termed as "end-bearing pile." It should be noted that the classification is mainly used for design purposes, as a pile's actual geotechnical capacity is a combination of both friction and end-bearing, and the ratio between the shaft friction and end-bearing capacity will vary with the applied loading. Many end-bearing

piles actually function as friction pile under the working load condition, which is revealed by many static load tests in Hong Kong and other places.

(2) Percussive pile and non-percussive pile – The classification is based on the method of installation. A percussive pile (driven pile) is installed by percussive driving. A non-percussive pile is normally installed by forming a hole in the ground (bored pile) by methods such as auger or rotary cutting after which (i) a pre-formed pile such as precast concrete pile or steel pile is placed; or (ii) wet concrete is poured into the hole to form the pile.

(3) Displacement pile, small displacement pile and non-displacement pile – This classification is based on the amount of displaced soil during installation. Displacement (or large displacement) pile involves relatively large soil displacement, and this type of pile is usually solid pile or hollow pile with closed ends. A small displacement pile creates relatively small displacement of soil during the installation, which is commonly steel H or hollow section piles where the cross-sectional area is small. Non-displacement pile is formed by boring into the ground before the pile installation. The significance of soil displacement dictates the friction along the pile shaft and densification/looseness of the surrounding soil.

(4) By materials:

Timber pile – seldom used nowadays, but many old structures are supported on timber piles.

Concrete pile – usually in the form of precast piles (reinforced and more commonly prestressed) or cast-in situ piles.

Steel pile – mostly tubular or H-pile, but piles in other shapes are sometimes used.

Composite pile – a combination of two or more structural materials, and most commonly tubular steel pile with concrete or cast-in situ concrete pile with steel section.

(5) Mode of loading – Axially loaded, transverse or laterally loaded, moment resisting, or combinations of these loadings under different load cases.

(6) Shape – Square (solid or hollow), octagonal (solid or hollow), circular (solid or hollow), fluted, H, pipe and others.

Miscellaneous – piles may have a constant cross-section or be tapered, and on some occasions, may have different sizes along their lengths. Many special piles are used in China to suit different ground conditions and to reduce the cost of construction. Lime columns and stone columns are some low-capacity piles that are sometimes used to support light structures, while CFG may be used for medium loading capacity purposes. Lime columns have been used in the racecourse project in Hong Kong, while stone columns are sometimes used in China. Caisson is another form of pile that is still used in limited countries, despite the high fatal rate records in the past.

14.8 INSTALLATION OF PILES

Piles can either be prefabricated as a steel section pile or precast concrete pile prior to installation into the ground or formed by cast-in situ concrete. There are many

(a) Steel H pile

(b) pipe pile

(c) Precast pretensioned concrete pile

FIGURE 14.18 Some typical prefabricated driven piles. (a) Steel H-pile. (b) Pipe pile. (c) Precast pretensioned concrete pile.

types of piles in the world, and this section only discusses the more commonly used piles in construction.

14.8.1 Prefabricated Piles

Prefabricated pile (Figure 14.18) can be installed into the ground by percussive driving, and the pile capacity comprises the shaft friction and the end-bearing at its tip, or a combination of both capacities in most cases. Prefabricated pile is driven by a heavy hammer dropped from height. During the impact, energy is transferred from the kinetic energy of the hammer to the energy dissipation along the shaft and the base. To confirm the pile capacity, temporary compressions of pile and soil and permanent set of the pile are usually measured. H and tubular piles are the most popular steel piles in construction, but other shapes (usually high capacity) are also available. The use of steel tubular piles is popular in many earthquake zones, as the section modulus of the pile is the same for all directions of earthquake. There are also cases where a hole is bored with the insertion of prefabricated pile, but this is not a common practice.

Besides steel piles, precast concrete piles are also commonly used (Figure 14.18c). In the past, reinforced concrete piles were used for low-capacity purposes, and the shape of the reinforced concrete pile is usually square. Such low-capacity reinforced concrete piles are not popular now, but are still used in some countries where the load capacity requirement is not high. The main advantage of such piles is the low investment cost and the flexibility in production and transportation. Since the capacity of the reinforced concrete piles is low, they are now largely replaced by the precast prestressed concrete piles in many countries. The precast prestressed spun concrete pile (Figure 14.18c) with a diameter of 400 mm to 600 mm is manufactured by using high-grade concrete (cube strength up to 78 MPa) with prestressed tendons embedded in the pile. The allowable capacity ranges from 1690 kN to 3500 kN per pile, and is commonly used for many building works in Hong Kong, Macau, and China. Spinning of the pile shaft is involved during the manufacturing of the pile. The prestressing force by the tendons serves to resist the transient tensions that may be induced during pile driving. It should be noted that the prestressing force doesn't reduce the ultimate pile capacity, as the applied load will shorten the pile and hence reduce the prestressing force in the pile. In Hong Kong, each pile is normally 12 m in length, and the connection between piles is achieved by welding the steel joint plates at the ends of each pile section. Steel conical pile shoes are usually installed at pile ends for pile driving. The common sizes of this pile type include 500 mm diameter with 100 mm thick walls and 500 mm diameter with 125 mm thick walls. The maximum compressive capacities of these two types of piles are 2300 kN and 2700 kN respectively. Since the displacement for this pile is large, the driving sequence should be carefully arranged. Normally, driving is commenced for piles in the central portion of the site, and driving progresses in an outward direction towards the peripheral of the site. This type of pile is not suitable for sites where underground boulders and other obstructions exist, since the piles may crack under excessive hard driving, which has been found in some piling projects.

There are other types of driving type piles, which are not commonly used nowadays, which include the Franki pile, Vibro pile, pressure pile and others. These piles are seldom used in Hong Kong these days, as the engineers prefer to use larger diameter bore piles instead of more smaller diameter piles. While percussive driving may induce undesirable sound pollution and ground vibrations, a pile may be silently jacked into the ground usually by high hydraulic pressure. Pile jacking is less efficient than percussive driving and is generally used when percussive piling is not allowed.

14.8.2 Socketed H-pile

The prefabricated pile can be installed in a pre-bored hole formed by pre-boring or augering. Shaft friction is usually less significant for this type of pile, and the pile capacity is usually derived from the end-bearing with a hard stratum or the bond stress with the rock socket (Figure 14.19a,b). For the former case, the H-pile rests on rock, but this is less common in Hong Kong. In Hong Kong, piles socketed into rock are more common. A socketed H-pile is formed by pre-boring a hole through the soil stratum using ODEX drill in the past and concentric driving system recently and then into hard rock. The H-section steel pile is then installed into the hole and cement grout is injected into the hole which acts as a medium in bonding the pile to the rock by which the pile derives its geotechnical bearing capacity. It should be noted that while socketed H-pile is common in Hong Kong, it is seldom adopted in

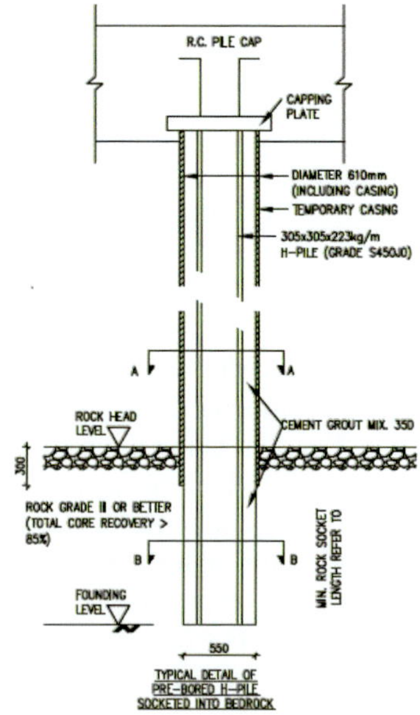

(a) typical socket H pile

(c) the use of permanent corrugated steel casing in very poor soil

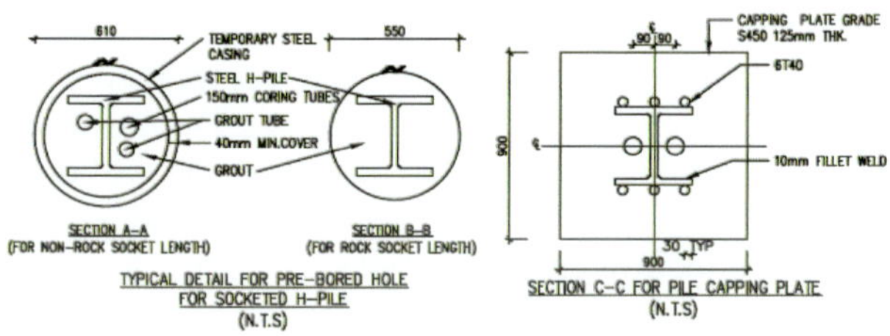

(b) Section A-A and B-B for Fig.14.19a

FIGURE 14.19 Socket H-pile construction. (a) Typical socket H-pile. (b) Section A-A and B-B for Figure 14.19a. (c) The use of permanent corrugated steel casing in very poor soil.

other countries. The reason behind the adoption in Hong Kong is to avoid piles resting on underground boulders as well as providing pile capacity with a higher level of confidence (at the expense of high cost).

Steel piles (H or other shapes) are usually installed by driving, but the socketed H-pile is installed by a different method as described in the previous paragraph. Shear connectors are commonly installed along the socket H-pile shaft to increase the bonding between the cement grout and steel, and many field tests in Hong Kong have proved that the steel/cement bond stress is greatly increased by the use of shear connectors. The major difference between the socketed H-pile and another pile type called minipile is that the steel casing is normally extracted along with grouting while the steel casing remains in the ground for minipile after installation. Another difference is that the socketed H-pile can be designed for bending, which is not possible for the minipile. Different sizes of steel casings and steel sections are available for this pile type. One of the most common steel casing sizes is the 610 mm diameter hollow section, which is sufficient to cater for all different sizes of the H-sections with adequate cover.

Typical method statement for socketed H-pile construction in Hong Kong:

(1) Drilling with the first temporary casing (commonly 610 mm diameter), alignment is checked during the installation
(2) Drilling Works by Concentric driving system or ODEX Method, where the ODEX drill bit cuts ahead of the steel casing with a diameter slightly larger than the casing. It should be noted that the ODEX system is not allowed to be used now, as there have been several problem cases which were caused by this system. For the ODEX system, the casing can be hammered down easily and continuously. High pressure air is applied through the flushing holes at the front of the drilling bit to push the soil up. Once the designed depth is reached, the drill bit is drilled in reverse order. For the concentric driving system, a ring bit system will be used to install 610 mm O/D casing with minimum disturbance to the surroundings. The casing is welded to the shoe to seal around the end of the casing and to prevent air from passing outside the casing. The casing and the casing shoe are carried down the hole as a whole at the same time. The casing shoe and casing, however, do not rotate to minimize the disturbance to the surroundings. During the drilling, only air is used as the flushing medium, and no dewatering is allowed. The maximum operating air pressure is usually limited to 17 bars. Upon reaching the rock head, the pilot drill bit will be replaced by the "down the hole" hammer (activated by compressed air) for drilling the rock socket. Down the hole hammer is faster than rotary in cutting the rock socket.
(3) Preparation of steel H-pile – shear studs and steel bars are commonly welded to the pile to enhance the bond stress.
(4) Splicing of temporary casing/H-pile. Ultrasonic or liquid penetration test can be used for the weld testing.
(5) Obtaining rock sample after completion of drilling
(6) Preparation for grouting – a grout mix used in Hong Kong comprises 900 kg cement, 378 kg water (W/C ratio 0.42), Cebex 100 4.08 kg for non-shrinkage purposes per 0.648 m^3 of grout. The tests for the grout include flow cone efflux test, bleeding and free expansion test, and cube strength test.
(7) Air-lifting to clean the hole before grouting
(8) Cement grouting for the socketed pile. The minimum 28 days strength for the cement grout is typically 35 MPa, and the water cement ratio is usually limited to 0.45 max.
(9) Extraction of temporary steel casing by vibrating hammer or hydraulic jack in some cases
(10) Grout Testing Work

14.8.3 Minipile

A minipile usually comprises a few (4 to 7 nos., but a high number is not recommended) steel reinforcing bars in a drill hole (not exceeding 400 mm) filled by grout. Its bearing capacity is usually derived from the bond stress with the rock socket. Due to its small size, the pile capacity is comparatively low. However, as the machinery required for construction is small and the speed of construction is fast, the pile is suitable for the construction of buildings in sites where accesses are difficult. In addition, due to the limited load carrying capacity (1374 kN for 4T50 pile) of the pile and the requirement of minimum spacing (the greater of 750 mm and twice of the outer diameter), the pile type can normally carry loads from buildings of up to 18 storeys.

For a minipile, 4 or more number of 50 mm diameter steel bars are installed in a pre-bored pile (Figure 14.20) through soil to rock and the hole is then grouted by cement grout to achieve bonding with the rock for derivation of the bearing capacity. Other than the conventional reinforcement bar, 67.5 mm high-strength DSI bar with a strength of 840 N/mm^2 has been used to replace the conventional steel bar to form a super-minipile. The strength of the DSI bar can hence greatly enhance the pile capacity of the minipile. The more important application of such DSI bar is to reduce the number of bars in a minipile so that the load sharing among the bars can

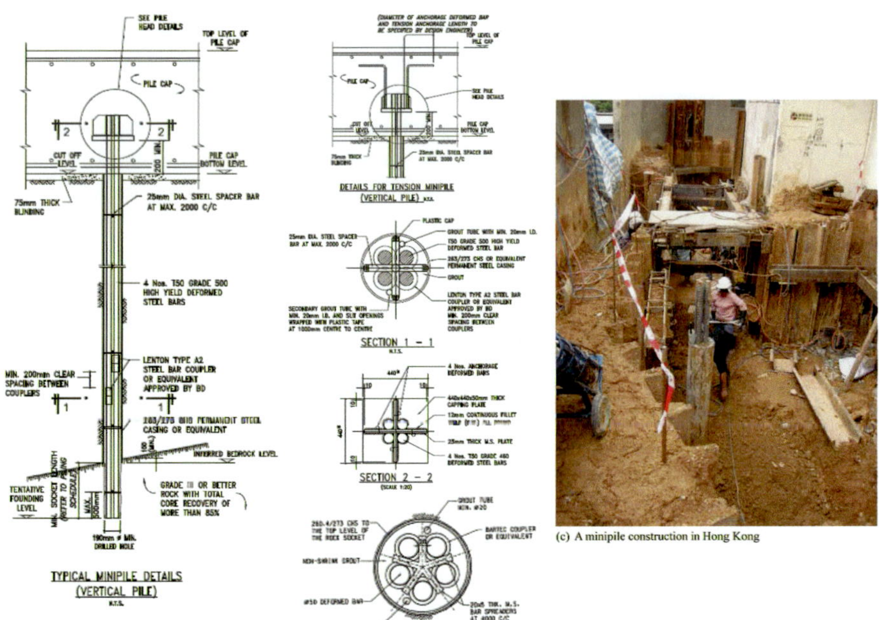

(a) Typical minipile details in Hong Kong

(b) Details for minipile bar arrangement

(c) A minipile construction in Hong Kong

FIGURE 14.20 Typical details for minipile in Hong Kong. (a) Typical minipile details in Hong Kong. (b) Details for minipile bar arrangement. (c) A minipile construction in Hong Kong.

be more even. The main disadvantage of such application is the cost, as DSI bar is several times more expensive than conventional steel bar.

Typical method statement for minipile construction in Hong Kong:

(1) Set up drilling rig for minipile, and a working platform may be required for construction on slopes
(2) Drilling with alignment checking of minipile
(3) Splicing of Steel Casing
(4) Collecting rock sample at founding level
(5) Air-lifting to clean the pile after completion of drilling
(6) Prefabrication of steel bars on ground
(7) Installation of steel bars
(8) Cement grouting

14.8.4 Bored Pile

A cast-in situ concrete pile is often formed by pre-boring a hole in the ground down to the design level, and the hole is then concreted after the reinforcement cage installation. The bored pile size can range from 400 mm to 3500 mm diameter. By definition to the British and Hong Kong standards, pile diameters above 750 mm are termed "large diameter bored pile," otherwise it is termed "small diameter bored pile." A bored pile can derive its bearing from friction along its pile shaft and/or end-bearing.

A bored pile is usually constructed by driving a steel casing into the ground and the soil within the casing is removed by the use of grab or other means. Reverse circulation drilling (RCD) technique is now commonly used for many large diameter piles, and typical RCD used for bored pile construction is shown in Figure 14.21c. RCD uses the self-weight and rotary motion of the drill to crush hard materials. Crushed materials are then retrieved by circulation of fluids within the drillhole. This method effectively reduces the noise and vibration during the construction of rock sockets. It is also possible to bore the hole using drilling fluid without a casing, but there are also some reported unsuccessful cases. Concreting is then performed after the installation of the reinforcement cage with the casing gradually withdrawn from the ground. Bell-out as shown in Figure 14.22 is commonly used, which is based on the allowable stress on concrete being greater than that for rock with $A_s f_{cu} \geq A_b f_{rock}$.

Koden test is an acoustic device originally used in checking overbreak of diaphragm wall trench (details will be covered in later section of this chapter). The application is now extended to checking verticality and bell-out of bored piles. Ultrasonic Koden test is usually used to check the integrity of the bell-out for the bored pile. Since the diameter of the bored pile is usually large in Hong Kong, pre-drilling to confirm the founding level (instead of just relying on the site investigation) is practically mandatory nowadays.

(a) use of grab

(b) use of auger

(c) Use of RCD (common in Hong Kong)

FIGURE 14.21 Different types of plants used for bored pile construction. (a) Use of grab. (b) Use of auger. (c) Use of RCD (common in Hong Kong).

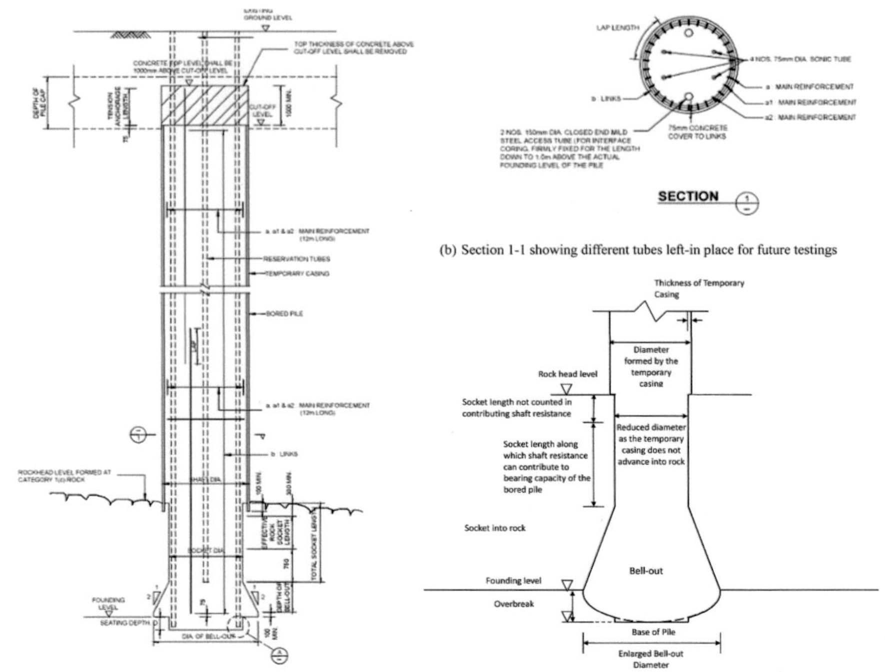

(a) Typical bored pile details with bell-out

(b) Section 1-1 showing different tubes left-in place for future testings

(c) bell-out with socket

FIGURE 14.22 Typical details for large diameter bored pile with bell-out. (a) Typical bored pile details with bell-out. (b) Section 1-1 showing different tubes left in place for future testing. (c) Bell-out with socket.

Some other considerations during bored pile construction include:

(1) It should be noted that the finished concrete top level shall be at least 750 mm above the top cut-off level. The top 750 mm concrete for the bored pile is usually weak due to inadequate compaction by self-weight of the tremie concrete and will therefore be removed to expose the pile with good-quality concrete to the pile cap.
(2) If a permanent liner is required (usually for very poor soil which will collapse during casing extraction), the gap between the temporary and permanent liner shall be backfilled with soil.
(3) During construction, alignment and verticality of each bored pile shall be checked frequently. If the deviation exceeds the allowable limit, remedial actions must be carried out as soon as possible to rectify the problems.
(4) In case of obstruction, the use of a backhoe or hand-held machine may be used if the obstruction is located at a shallow depth. If the obstruction is found at depth, chiselling by rock chisels or under-reaming is commonly used.

(5) Sonic access tubes are installed for most of the large diameter bored piles in Hong Kong to assess the quality of the concrete. The sonic tubes, which are capped with screw-on steel watertight shoes, are usually fixed to the reinforcement cage and project 300 mm above the cut-off level of the pile.
(6) In case if there is a thin layer of sediment (segregation) at the pile-rock interface, remedial pressure grouting is usually carried out. The grout shall be non-shrink cement grout with a minimum grouting pressure of 25 bars, which is maintained for a minimum of 5 minutes. The sediment or defective concrete shall be cleaned by high pressure water with a minimum jetting pressure of 200 bars prior to pressure grouting.

For pile foundation in marble areas with sinkholes, a modern technique is to use high water pressure to crack the vertical faces of the bored pile or barrette, followed by the injection of grout through the formed cracks. The grouting thus formed will significantly enhance the shaft friction of the pile. This type of pile is called "shaft grouted barrette" or "shaft grouted bored pile," which will be discussed later, and this approach is popular in some marble areas in Hong Kong.

The typical procedures for shaft grouting are:

(1) Crack the fresh concrete cover using a double packer and water within 24 hours of concrete casting.
(2) Carry out shaft grouting for each manchette from bottom to top. The refusal grouting pressure is around 50 bars in Hong Kong.

If the intake cannot be achieved on some manchettes, the target intake immediately above, below or on its side should be increased.

It should be noted that tremie concreting is required for the construction of bored piles or barrettes. A reduction factor of 0.8 on the concrete cube strength is commonly used for the design of these piles in Hong Kong.

For illustration, the complete procedures for a large diameter bored pile construction in Hong Kong is given in Figure 14.23. In general, the temporary reinforcement, which is very heavy, will be temporarily suspended on the casing before the insertion and connection of the next reinforcement cage by bolts. A similar method is used for lowering the reinforcement cage in diaphragm wall construction.

A typical bored piles construction sequence using casing for a project in Hong Kong is given below:

(1) Set out pile location in accordance with construction drawings and carry out the predrilling work to determine the rock head level and founding level of each bored pile. Initially, predrilling is carried out to verify the location of the rock head level by means of wash boring up to the rock head level. Then, a rock core is usually retrieved by a triple-tube barrel in Hong Kong. The rock core should be in Grade III or better and at least 5 m (or socketed length) with total core recovery of more than 85%. Point Load Test (PLT) is carried out to verify whether the rock strength is enough (minimum 1

FIGURE 14.23 Typical procedures for the construction of a large diameter bored pile. (a) Setting out of bored pile position. (b) Mobilization of hydraulic oscillator. (c) Setting up the temporary steel casing. (d) Removal of soil by grab. (e) Connection of temporary steel casing. (f) Use of chisel to break boulder. (g) Cutter to cut in rock. (h) A reamer to form the bell-out. (i) Reverse circulation to remove spoil. (j) Cleaning of pile base. (k) Lowering of steel cage. (l) Tremie concreting of bored pile.

MPa for I50). For the test, at least one diameter length of the rock core is chosen for in situ testing; otherwise, the irregular shape should be tested in the laboratory due to test specimen manufacture.

(2) Set out pile location in accordance with construction drawings and reference points for commencement of bored pile construction. Allow time for RE checking prior to pitching of temporary casing.
(3) All monitoring points shown on the piling layout plan will be installed prior to the commencement of piling work. All monitoring points together with the existing one will be maintained throughout the Contract. The frequency of monitoring will be in accordance with the approved plan.
(4) Pitch temporary steel casing at the pile location and drive the casing down to the rockhead level of bedrock or hard stratum with a heavy-duty hydraulic oscillator/rotator. The steel casing shall form a watertight seal to prevent the collapse of soil into the unlined shaft at all times.
(5) Excavate within steel casing with hammer grab working by a crane. Casings are coupled together by means of welding or bolting. The bottom casing has a thickened cutter shoe to facilitate penetration into hard ground.

(6) The excavated materials will be disposed of to the approved Government dumping area. The detailed disposal system and procedures are shown in the Environmental Management Plan.
(7) The surveyor will check the offset of the temporary casing after the first casing pitching and before air-lifting. Moreover, the verticality of the casing is frequently checked by a spirit level.
(8) In case of excessive out-of-tolerance of steel casing, the bored hole will be backfilled, then the casing will be re-installed into the correct location.
(9) A positive water head will be maintained within temporary steel casing throughout the course of excavation to ensure stability of the pile and prevent excess pore water pressure. In addition, prior to the installation of RCD, excavation level should always be kept at least 1 m above casing toe level.
(10) All underground obstructions or boulders, encountered during the course of pile excavation, shall be removed by grabbing or RCD.
(11) When casing is required to penetrate through rock prior to reaching bedrock level, reverse circulation drilling method will be adopted. The drilling rig will be securely attached to the top of the casing, either by hydraulic clamps or by special joints. A large diameter drill bit, attached to a heavy-duty drill string, is rotated by means of a power swivel located on the top of the rig. The cuttings will be air-lifted into a large mud tank for settlement and subsequent removal from site. Water will be returned to the pile shaft to balance any head of water in the ground. As drilling continues, additional drill strings will be connected until the required level is reached.
(12) Rock chip sample will be collected for inspection and verification of the rock head and founding level. Rock socket will be achieved by means of reverse circulation drill. The rock head, founding, and pile base level should be checked by the Resident Engineer.
(13) The extent of additional length for the formation of bell-out and the details of the bit will be based on the on site calibration record. The additional length will be included during the drilling of the rock socket.
(14) The rock chip will be collected at the inlet of the desilting tank from the discharged effluent of RCD.
(15) Bell-out base will then be formed by using a calibrated expandable drill bit mounted onto a reverse circulation drill rig to form the desired bell-out base.
(16) The size of an expandable drill bit must be checked and calibrated on the ground before the construction.
(17) The bell-out base is formed by applying the drilling motion coupled with the downward thrust to the drill rod from which the bell-out drill bit is expanded. This operation is continued until the downward displacement of the drill bit rod coincides with the value of the specific diameter of the bell-out base formed. Koden tests are commonly required to ensure the quality of the excavation and the bell-out.

(18) To ensure that the pile base is clean, tape water is filled instead of the circulated water. RCD with a drill bit is employed to keep the muddy water out of the pile by air-lifting. The muddy water then flows to a sedimentary tank before discharging. After changing the water, tremie pipes are set up for fine air-lifting. The purpose is to keep the base clean and without any debris. Water samples are then collected to inspect whether the base is clean. At least five spots for water sampling are taken. The pile depth can then be measured by a measuring tape with a steel rod at the front.
(19) The bored piles shall be constructed in accordance with the tolerances stated in the Specification.
(20) Monitoring of ground settlement and tilting to adjacent buildings should be carried out during bored pile construction, and the use of vibration check points and standpipe/piezometer may also be necessary.
(21) Each bored pile is required to core the Interface layer to verify the pile base is in exact contact with the rock profile without any debris/muddy soil.

Water-soluble polymers (partially hydrolysed polyacrylamides PHAs) long-chain molecules with very high molecular weight can be used to stabilize bored pile and diaphragm construction. The viscosity reduces the seepage losses to the ground and stabilizes the excavation. Such fluids have advantages such as a smaller site footprint, lower environmental impact and construction costs. In coarse soils, which are common in Hong Kong, polymer fluid may, however, experience excessive fluid loss, which can be overcome by adding bentonite, superabsorbent gel particles or proprietary fines that are dispersible in polymer fluids. The experience in Hong Kong shows that polymer fluid cannot be simply treated as bentonite slurries in construction. Polymer fluid (SlurryPro CDP) has been adopted for the West Rail Tuen Mun Station construction in 2001–2002 for the construction of nine rock-socketed bored piles with a diameter ranging from 1.5 m to 1.8 m and up to 51 m in length. The concentration of the polymer material was maintained at least 0.7 kg/m^3 to form a viscous solution, while the Marsh funnel viscosity should be over 65 seconds. Polymer fluids can be damaged by the high shear stresses from the use of centrifugal pumps; hence, a screw pump with lower shear stress was used to preserve the properties of the fluid. During construction, the polymer fluid was tested for pH, viscosity, density and sand content. Before the construction, the pH and viscosity were maintained between 9.5–11.5 and over 50s for Marsh funnel test, while the density and sand content were maintained below 1.1 g/cc and 3% respectively during construction.

14.8.5 PIP Pile

PIP pile has been popular in Hong Kong in the past decade as a low to medium capacity pile foundation. For example, the Jordan and Tsim Sha Tsui MTR stations and the Charter Bank basement in Hong Kong are constructed with the use of PIP piles. Some load capacity piles were also constructed using PIP piles in the past, but their use has basically been totally replaced by minipiles in recent years, due to the difficulty of maintaining good quality control for PIP construction. Apart from the use of a normal reinforcement cage, the use of a standard H-section as the reinforcement of PIP pile has been popular in the past (Jordan and Tsim Sha Tsui MTR station). Auger is commonly used for the construction of small diameter bored pile or PIP pile. The size of PIP pile is commonly in the order of 450–550 mm, as compared with the small diameter bored pile which has a diameter of 800–1000 mm in Hong Kong.

Typical process for PIP construction (similar to many other small diameter bored piles):

(1) Setting out for pile
(2) Drilling with auger
(3) Extension of auger drilling rod
(4) Drilling completed to the required depth
(5) Grouting of pile, preparation for prefabricated steel cage
(6) Flow cone test for grout
(7) Initial setting time test for grout
(8) Grout cube compression test
(9) Completion of grouting work
(10) Installation of steel cage into the pile shaft
(11) Surveying for the top level of steel bar
(12) Stabilizing the steel cage (in grout) by use of U-bolts
(13) Inspection of grout at the top level of the pile before backfill

Grouted piles in the form of PIP pile, CCP, mixed-in-place piles (CFG) have been used for some low-capacity works, and actually, such piles have been popular for footbridges and low-rise buildings in the past, and they are still popular in many countries, including China, but not in Hong Kong.

14.8.6 Barrette

Barrette is actually a pile that is constructed by the diaphragm wall technique. Besides the normal rectangular section, T section, L section, or X section are also commonly adopted. For some of the MTR stations in Hong Kong, the barrette serves as the foundation as well as columns in some MTR concourses. Hence, the connections between the barrette and the slabs have to be carefully designed. There is an interesting case in Hong Kong. For the construction of the Central Plaza in Hong Kong, diaphragm wall and bored pile are specified in the drawings. To avoid the mobilization of different plants, the contractor proposed to replace the bore pile with barrette with a major saving in the construction cost.

An interesting shaft grouted barrette construction in Hong Kong is given here. According to the ground investigation, the site is covered in succession with marine deposit, alluvium which is completely decomposed metssilstone then sedimentary. There is also a fault zone underneath the current site. Cavities infilled with soft materials with low N values are formed in the marble bedrock ranging from −15 to −130 mPD and the thickness of the cavity zone ranges from 10 m to 25 m. Groundwater is found at a depth of 2 m below ground. It is required to construct buildings up to 28 levels at such a site which is a challenge to the foundation design and construction. It is proposed to adopt shaft grouted barrette for the construction using Tube a' Manchette shaft grouting and base grouting through a flat jack. The top excavation is carried out by a grab, which is then followed by the use of hydro-fraise after a depth of 20 m. Forty-two millimeter diameter mild steel shaft grouting tubes are installed along the reinforcement cage. Shaft grouting is carried out 48 hours after concreting. Full-scale load tests are carried out to confirm the required shaft friction has been achieved. For the arrangement of the grouting tube for shaft grouting of a 2.8 × 0.8 m barrette, 3 number 42 mm diameter mild steel grouting tubes are provided outside the reinforcement cage along the long direction at each face, and one grouting tube is provided at each short side of the barrette. The fresh concrete cover is cracked with water by the use of a double packer within 24 hours of casting concrete. The shaft grouting tubes are flushed within 48 hours after concreting of the barrette, and the manchettes are cracked open with water. Grouting starts from the bottom of the grout tube and gradually moves upward. Based on previous experience, the target grout volume amounts to litres/m^2 of the grouted area. If the target grout volume cannot be achieved in one manchette, the manchette above and below that manchette will have additional compensated volume of grout. The grouting criterion for this barrette is grouting pressure greater than 30 bars and at least 250 litres of grout per 2 m length. The ultimate grouted shaft friction is about 4.5 N (kPa) for alluvium and 2.0 N for saprolite, while the ultimate end-bearing capacity is taken as 10 N (MPa) which is a commonly accepted figure in Hong Kong. A typical grout mix used in Hong Kong for 1 m^3 of grout consists of 1000 kg cement, 666 litres of water, 15 kg bentonite, 1.5 litres Bentocryl 86 and 4 litres Daracem 100.

According to the requirements of the Buildings Department in Hong Kong, 1% of the working barrette should be selected for load testing to twice the design load, and core drilling should be taken to the full length of the barrette for 5% of the working barrette. Sonic tests are also specified for barrette, similar to the bored pile.

The famous ICC building in Hong Kong is another example of the application of shaft grouted barrettes. ICC is located above a major fault zone, and there are difficulties in constructing good quality conventional end-bearing bored piles, particularly at the interface between the concrete and the end-bearing medium, which is likely to be affected by debris from sedimentation and the segregation of concrete in these deep piles. Furthermore, due to the variation in the bedrock profile, it is difficult and time-consuming to construct the bored piles in the presence of thick sand fill, alluvium, and the decomposed granite layers. The final choice involved a total of 240 shaft grouted barrettes with sizes generally 2.8 m × 1.5 m with 2 m clear spacing between pile edges within the footprint of a 76 m diameter cofferdam. The cofferdam is made up of 1500 mm thick shaft grouted diaphragm wall panels and constructed with a hydrofraise.

14.8.7 CFG Piles

Cement-Fly Ash-Gravel pile is a low-capacity pile based on stone column, which is used extensively in China and other countries. Stone column in soft soil is not reliable due to the low confining resistance. Cement and fly ash is mixed with gravel to form a more rigid column, which could be applied in most soil conditions. CFG pile has a high bond strength derived from macadam, gravel, sand, fly ash and a mix of cement and water, and forms a composite foundation with the soil cushion and the soil between the piles. Under different cement content and mix proportion, the strength of CFG mix lies between C15 to C25, which is between flexible and rigid pile. The top 0.5 m of the CFG pile is usually removed to expose the better pile material below for construction.

CFG pile is usually installed by three methods:

(1) The use of slurry to stabilize the excavation with tremie concreting which is suitable for many types of soft soil, sandy soil, fill, and cobbles.
(2) The use of helical drill where the mix is pumped through the pipe inside the drill to form the pile, which is also suitable for many types of ground conditions.
(3) The use of pipe driven by vibration, which is suitable for clayey, fine-grained soil, fills.

Some of the typical problems during CFG pile construction include:

(1) Clogging of the pipe during concreting.
(2) Necking, debris/gas at pile bottom or even cracking of pile.
(3) Collapse of holes.

Besides working as a low-capacity pile, CFG pile is also used for cases like bridge approaches where abutment pile will be subjected to lateral stress from soil movements. Another application is the treatment of soft embankment foundation soils in mountainous area with difficult access and a lack of other construction materials or heavy machines. CFG pile composite foundation will be a low cost effective solution under such condition.

Steel reinforcing bars are usually not required for the pile, and the industrial waste-fly ash is used as an admixture for the concrete. The bearing capacity of the soil between piles is mobilized by providing an appropriate thickness of mattress layer between the piles head and the bottom of the supported structure. Because of the stress concentration at the pile top, piles will penetrate into the cushion layer; thus, the load is partly transferred through the negative side friction developed at the upper part of the pile. Many authors studied the effect of the cushion thickness and the engineering properties on the load redistribution mechanism. The results indicated that beyond a certain mattress thickness, the cushion effect will become constant.

In most CFG pile, lean concrete is used as a more popular filling material. The advantages of this pile include:

- CFG can be applied in many soil conditions, such as peat soil, soft clay, fat clay, sandy soil, complicated layered soil, or hard soil conditions.
- Different materials such as cement-fly ash-gravel, cement-lime-gravel, cement-sand, and others can be used to form this low-cost pile,
- CFG could form a column from low stiffness column (similar to DCM column) to high stiffness column, which means that the capacity of CFG pile can range over a wide spectrum to fulfil the project requirement.
- CFG pile is fast in construction with low machinery requirements, and can be more cost-saving compared with other piles.

Some of the precautions in CFG construction:

(1) Low strain echo tests (7 days) together with load tests (28 days) are commonly used to assess the pile integrity. Core drilling may be necessary in some cases as the quality control.
(2) The verticality of CFG pile is limited by the capacity of the rotary auger. In general, the control of verticality can be a problem.
(3) The amount of debris material at the pile bottom is difficult to control.
(4) The slump of the concrete is limited to 160–200 mm in general.
(5) The speed of withdrawal of the drill is typically controlled within 2–3 m/min.

CFG pile derives its capacity mainly from the shaft friction. With reference to a CFG pile design in Beijing, the ultimate shaft friction adopted ranged from 15 to 45 kPa, and the unit friction will vary with the soil type.

14.8.8 Shaft Grouted Piles

Extremely deep bedrock can pose a construction challenge to the conventional large diameter bored pile on rock and the pre-bored H-piles socketed into rock. If friction piles are used without relying on the pile end-bearing capacity, most of the friction piles have limited pile capacities which will limit the development of the superstructures. The use of shaft grouting to enhance the shaft frictional capacity has been introduced to Hong Kong since the 1990s for friction bored piles and barrettes in weak saprolites. This technique has recently been extended to areas with meta-siltstone/sandstone formation underlying by marble bedrock beyond 150 m below ground. The full-scale load test for the KCRC West Rail Phase I reported by Plumbridge et al. (2000) demonstrated that the measured shaft friction capacities for shaft grouted barrettes and bored piles in completely decomposed granite (CDG) and volcanic (CDV) could increase up to two or three-fold when compared with those without shaft grouting. The use of shaft grouted barrettes in MTRC Kowloon Station Mega Tower Development (Chan et al. 2004) and the private residential developments at Tung Chung Designated Area (Sze et al. 2007) also demonstrated enhancement of shaft friction by the shaft grouting technique in sandy alluvium, completely decomposed rhyolite (CDR), CDG and diamict deposit (DD). Lui et al. (1993) also reported significant increase in the ultimate shaft capacity for small diameter pressure grouting Pakt-in-Place (PIP) pile. This type of pile is then adopted in a series of projects in Yuen Long (barrettes), Hollywood Road (mini piles), Seymour Road (mini piles), Wong Nai Chung Road (pre-bored H-piles) in Hong Kong. The famous 484 m high ICC building in Hong Kong also sits on shafted grouted barrettes. In general, Tube a' Manchette method is used to ensure good shaft grouting, and one of the major advantages of this grouting method is the possibility of multiple grouting.

Based on some load test results in Hong Kong, it is generally found that there is very little loading transferred to the pile base for shafted grouted piles. The ultimate shaft friction can be mobilized in colluvium or similar type of soil or the top of the CDG zone, but is not fully mobilized for the lower portion of the CDG zone. In general, the shaft friction capacity can be increased by several times by shaft grouting, but the percentage of increase for stiffer soil is generally lower than that for the less stiff soil.

Shaft grouting technique has been used to pre-bored H-piles in Hong Kong. The allowable shaft for shaft grouted pre-bore H-pile is typically taken as 0 for fill or similar. For CDG, it can be taken as 1.6 N with a capping value of 64 kPa, while higher values are actually measured in different projects. Based on the test results in Hong Kong, a shear displacement of 3–7 mm is sufficient to mobilize the maximum shear stress. Some of the recommendations from the Structural Engineering Committee on shafted grouted large diameter bored piles include (1) the ultimate resistance in CDG can be taken as 4.5 N but limited to 130 kPa (working load); (2) the allowable load under tension is taken as half the allowable load under compression; (3) proof load tests are necessary under some cases.

14.8.9 Hand-dug Caisson

Hand-dug caisson is invented in Hong Kong which is used as a replacement for expensive bore pile construction. Hand-dug caisson is commonly used at sites with steeply sloping terrain, difficult access or insufficient working platform. A timber portal is usually set up on the ground surface with a female working at the top to support the worker down below the excavation (Figure 14.24). It is common that the male who works underground is the husband of the female who works on the ground surface, as deaths and injuries due to electric shock and falling objects have happened many times in Hong Kong. In this construction, each caisson ring overlaps with the subsequent ring by 75 mm. The diameter of the caisson can range from 0.8 m to several metres (not common). For the caisson ring, it will be subjected to axial compression due to the arch action. The use of concrete without reinforcement is a common practice, but circular reinforcement is commonly used for those large diameter caissons. This technique has been banned in Hong Kong, but it is still practised

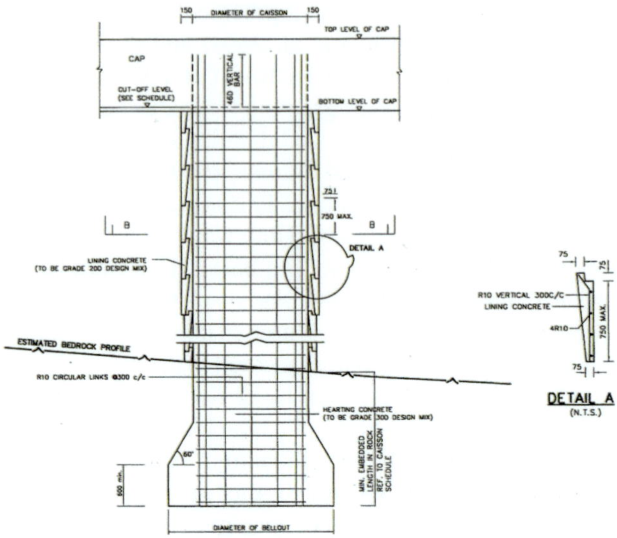

(a) Typical drawings for hand-dug caisson

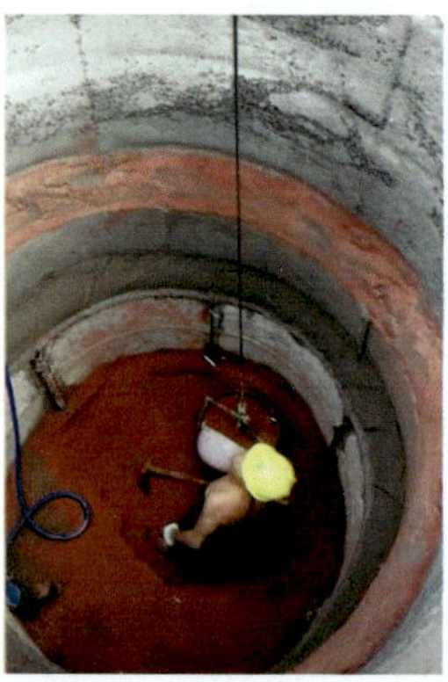

(b) Photo for hand-dug caisson construction

(c) Set up of working platform on ground surface for caisson construction

FIGURE 14.24 Hand-dug caisson construction in Hong Kong. (a) Typical drawings for hand-dug caisson. (b) Photo for hand-dug caisson construction. (c) Set up of working platform on ground surface for caisson construction.

in Macau and China. The design of hand-dug caisson is similar to large diameter bored piles, where only the end-bearing capacity is considered.

14.9 DESIGN OF PILE – VERTICAL LOAD

The design of a pile for the vertical load should consider both the structural capacity and the geotechnical requirements. In general, the structural capacity and hence the pile type is chosen first, which is followed by the geotechnical design for the length of the pile and the requirement for bell-out or socket length. The design of the pile is largely controlled by the local building code and practice, and there is not a universal method of design. The examples below illustrate the design practice used in Hong Kong which is based on a consideration of both the pile capacity and pile settlement, and the design practice can be used for reference in other countries.

14.9.1 Steel Pile by Driving or Jacking/Bored and Socketed

The maximum loading capacities (axial load) of steel piles are given by:

$$\text{(Installed by Driving) } P = 0.3 f_y A_p$$
$$\text{(Installed by Pre-boring or Jacking) } P = 0.5\, f_y A_p$$

where f_y is the yield strength of the steel and A_p is the cross-sectional area of the pile. A lower factor of 0.3 is used for pile installed by driving is to reduce bearing capacity of the driven pile so as to minimize the use of heavy hammer and high drop height which can subsequently lead to high stresses during pile driving (dynamic load is different from static load, and the axial stress can be magnified during driving). Pile capacity is increased by 25% if wind load is considered. For combined bending and axial load, the working stress can be increased by 50%. The minimum yield stresses of steel should be determined from the "design strength" of steelwork in accordance with table 3.2 of the Code of Practice for the Structural Use of Steel 2011 (CoPSteel2011) or BSEN10025. The maximum allowable average bond stress between grout and steel H-pile shall be 0.5 MPa for grout having minimum cube strength of 30 MPa at 28 days and 0.45 MPa for grout having minimum cube strength of 25 MPa at 28 days.

Thus, for the H-Section S450 305 × 305 × 223 where the cross-sectional area is 284 cm^2, the yield stress is 430 MPa (as the flange and web thicknesses respectively at 30.4 mm and 30.3 mm), the axial load carrying capacity for a driven H-pile without wind is $0.3 \times 30 \times 8400 \times 0^{-3} = 3676$ kN (add 25% with wind). Currently, Grade 450 JO class 1 complying with BS EN 10025:2004 is commonly adopted in Hong Kong. The design load carrying capacity of a socketted H-pile S450 305 × 305 × 223 H-Section of area 284 cm^2 without wind is $0.5 \times 30 \times 8400 \times 0^{-3} = 6127$ kN. It should be noted that due to the static load test of the pile which is loaded to twice the design load, to avoid buckling of the pile during the static load test which has happened but is rare, some engineers limit the ratio to 0.49 instead of 0.5 in the pile capacity determination.

To fully utilize the bearing capacity affected by the cross-section, the minimum socket length into Category 1(c) rock or better in accordance with table 2.1 of the Hong Kong Code of Practice for Foundation is determined as follows (suppose diameter of the bored hole in the rock is 550 mm and the presumed bond or friction between rock and grout according to table 2.2 of the Code of Practice for Foundation is 700 kPa)

$$0.55\pi \times 700 = 6127 \Rightarrow L = 5.066\text{m}$$

Design example for Bearing Capacity of Rock-Socketed H-Pile

Design Condition:

Diameter of pile = 0.55 m
Socket Length ls = 4.5 m (Assume)
Grade of rock = II
Type of Steel H-Pile = 305 × 305 × 223
Sectional Area of H-Pile A = 28500 mm^2
Perimeter of Pile L = 1727.88 mm

Material Strength

Grade of grout = 30 MPa
Reduction Factor for Grout Ra = 0.85
Yield stress of steel H-pile = 415 MPa
Allowable stress/Yield stress = 0.5
Allowable bond stress between rock and grout = 700 kPa (under compression)
Allowable bond stress between rock and grout = 350 kPa (under tension)
Allowable bond stress between steel and grout=600 kPa (under compression)
Allowable bond stress between steel and grout = 600 kPa (under tension)

Calculation of Bearing Capacity

a) Bearing Capacity of Steel H-Pile = $0.5 \times f_y A_p$ = 5913.75 kN
Assume provide 16 no. of H-pile, Total load capacity of H-pile = 5913.75 x 16 = 94620 kN
Assume N.S.F up to 20% of the pile capacity = 18924 kN
Design Load = Column Load + N.S.F = 60000 + 18924
= 78924 kN < 94620 kN OK !

14.9.2 Small Diameter Bored Pile

A small diameter bored piles is a bored pile having a diameter not exceeding 750 mm. It is formed by boring a casing into the ground and subsequently filling the hole with concrete or grout. As the pile derives its bearing largely from skin friction, pressure grouting and pressurizing the concrete are favoured. Small diameter bored pile is less popular in Hong Kong due to its low capacity, which would result in a large number of piles for the Hong Kong buildings, which are generally tall or very tall. Small diameter bored piles are, however, still very popular in other countries.

The following design principles should be adopted in general:

(1) The geotechnical capacity of a small diameter bored pile is usually derived from shaft friction, but the combined use of end-bearing and shaft friction in soil or rock can also be used.

(2) The contribution of friction from any fill or marine deposit layer, if it exists, should usually not be included in design unless it can be proved that there is no possibility of future consolidation.
(3) Where pressure grouting or pressurizing the concrete is involved in construction, the empirical relationship of shaft friction of 4.8 N (kPa) is generally used for N values up to a maximum of 40, usually with a factor of safety of 3, as described in Clause 6.4.5.3 of GEO Publication No. 1/2006. Otherwise, smaller values should be used with reference to other publications or tests.
(4) Structural capacity of the small diameter bored pile should also be considered, and in case it is less than the geotechnical capacity, the structural capacity will control.
(5) The small diameter bored piles can be used to resist lateral loads even when they are not raking piles.

A sample design calculation of a small diameter bored pile with pressurizing concrete will be presented as follows. The design data of the pile are:

Pile Diameter = 508 mm, Shaft resistance taken as $\tau = 4.8$ N
Concrete strength = 20 MPa; Factor of safety = 3 (as skin friction is not tested);
No. of piles under the same pile cap: less than 4
Concrete mix of the pile is grade 20 where the permissible stress is 5 MPa.
Capacity of the pile in accordance with its structural provision is

$$P = \frac{1}{4} \times 508^2 \pi \times 5 \times 10^{-3} = 1013kN$$

Geotechnical capacity of the pile is determined in accordance with the empirical relations that shaft friction is τ=4.8 N and end-bearing is σ_b=5 N (pile terminated at soil). The SPTN values of the soil embedding the pile is given in Table 14.3, based on which the total shaft friction and end-bearing are calculated in Table 14.3, giving a sum of 1414 kN which is greater than that by the structural provision of 1013 kN. So the load carrying capacity of the pile is 1013 kN. For cast-in-place concrete foundations, the concrete strength should be reduced by 20% where groundwater is likely to be encountered during concreting or where concrete is placed underwater.

Since the structural capacity of the small diameter pile is only 1013 kN, the pile capacity is hence controlled by the structural capacity which is 1013 kN.

14.9.3 Large Diameter Bored Pile

Pile capacity of the large diameter bored pile is usually derived from end-bearing. As the concrete grade currently used are usually high, up to grade 45, giving a permissible compressive stress in the order of 12 MPa which is well in excess of the rock bearing pressure in the order of 3 to 10 MPa, bell-outs for enlargement of the end-bearing areas have been extensively used. The Hong Kong Foundation Code allows combined use of shaft resistance and end-bearing of rock in the determination of its geotechnical capacity under the following conditions:

(1) When bell-out is used, the determination of shaft friction is restricted to a height of one pile diameter;
(2) When bell-out is not used, the determination of shaft friction can be based on a height of two pile diameters.

A large diameter bored pile design for a project in Hong Kong is given below for reference:

Design Data:

Shaft Diameter = 3000 mm
Bell-out diameter = 3000 × 1.65 = 4950 mm
Thickness of the casing = 50 mm
Reduced diameter of the pile (within the socket above the bell-out) = 2900 mm
Allowable end-bearing pressure = 5000 kPa (for Category 1(c) rock)
Shaft resistance = 700 kPa (Under compression or transient tension)
Shaft resistance = 350 kPa (Under permanent tension)

Under compression

Effective socket length = 2.9 m (if bell-out capacity is used)
Effective socket length = 2 × 2.9 = 5.8 m (if bell-out bearing capacity is not used)
Shaft resistance of the pile (with bell-out) = 2.9 × 2.9 × π × 700 = 18495 kN
Shaft resistance of the pile (without bell-out) = 2.9 × 5.8 × π × 700 = 36989 kN
End-bearing capacity of the pile without bell-out = 2.9^2 × 0.25 × π × 5000 = 33026 kN
Capacity of the large diameter bored pile without bell-out = end-bearing without bell-out + shaft resistance of the pile = 36989 + 33026 = 70015 kN
End-bearing capacity of the pile with Bell-out = 4.95^2 × 0.25 × π × 5000 = 96211 kN
Under Condition (1), the geotechnical capacity of the pile is 96221 + 18495 = 114716 kN
Under Condition (2), the geotechnical capacity of the pile is 33026 + 36989 = 70015 kN
With consideration of wind effects, both capacities can be increased by 25%, which become respectively 143395 kN and 87519 kN.

Structural design of bored pile
Using concrete with grade 45, steel reinforcement for yield strength of 500 N/mm^2
The structural capacity of the pile with using 1% reinforcements.
Reduction factor of concrete strength due to groundwater = 0.8

Area of the pile = $0.25 \times 2900^2 \times \pi = 6605198.554$ mm^2

f_{cu} = 45 as the cube compressive strength in N/mm^2, which has to be reduced by 20% to cater for concreting under water. So effectively, f_{cu} = 45 x 0.8 = 36 MPa.

The ultimate structural capacity:

= Reduction factor due to groundwater water × Area of concrete × 0.45 × f cu

+ area of steel × 0.87 × yield strength

= $6605198.554 \times 0.45 \times 36 + 0.87 \times 6605198.554 \times 0.01 \times 500$

= 107004 + 28733 = 135737 kN

Taking an average partial load factor of 1.4, the allowable working load is 135737/1.4 = 96955 kN.

The structural capacity with 1% steel reinforcement is less than the geotechnical capacity with bell-out.

The structural capacity which is higher than the geotechnical capacity, and the allowable compression load of the pile is 79,521 kN without considering the wind effect and 99401 kN with considering the wind effect.

Lateral resistance check (taken out from a project in Hong Kong):

Design Data:

L= 39.32 m; z = 0 m

E_p= 23.97 kN/mm^2; I_p = 2.243176 m^4

T= 5.498 m

Design Deflection:

ULS: 25 mm

Check ULS:

L/T = 7.151692; z/T= 0 ⇒ $F_{\delta H}$ = 0.95

$H_u = F_{\delta H} \times HT^3/E_pI_p$ = 8513.998 kN

The maximum lateral force: 2833.202 kN < 8513 OK

$F_{\delta M}$ = 1.5

$M_u = F_{\delta M} * MxT^2/E_pI_p$ = 29646.31 kNm

The maximum banding moment: 12466.09 kN < 29463 OK

14.9.4 Minipile

Minipiles are now commonly used when the loading is not high, and there is a difficulty in mobilizing heavy boring machine or working with limited headroom. The most common minipile nowadays is the 219 mm diameter hollow section with four T50 bars, which will give a working capacity of 1400 kN. Larger capacity can be achieved by the use of larger minipile with the use of high yield thread bars (fy = 520 MPa). A minipile usually comprises a few (4 to 5, and more is not suggested) steel reinforcing bars in a drill hole (not exceeding 400 mm) filled by grout. Its bearing capacity is derived from the bond with rock. Due to its small size, the pile capacity is comparatively low, and it is not assumed to resist bending in the design. However, as the machinery required for construction is small and the speed of construction is fast, the pile is suitable for the construction of buildings in sites where accesses are difficult. In addition, due to the limited load carrying capacity (1374 kN for 4T50 pile) of the pile and the requirement of minimum spacing (the greater of 750 mm and twice of the outer diameter), the pile type can normally carry buildings up to 18 storeys.

The design principles of a minipile are briefly summarized as follows :

(1) The loading carrying capacity of a minipile is derived solely from the steel bars.
(2) The presumed allowable bond or friction between rock and concrete is given in table 2.2 of the Code.
(3) Values for design ultimate anchorage bond stress are given in table 8.3 of clause 8.4.4 of Code of Practice for Structural Use of Concrete 2004.
(4) The allowable buckling capacity of the minipile may be checked with considerations of lateral restraints from the grout, permanent steel casing, and the surrounding soil can be included.
(5) Minipile does not take bending or lateral force in the design. Lateral load will be taken up by raking minipile. No bending moment is allowed in all cases.
(6) Inclined minipiles are commonly used to take up the horizontal in a pile group.

Example:
Steel casing External Diameter = 219 mm; Steel casing Thickness = 5 mm

Rock hole Diameter = 190 mm;

4T50 Grade 460 High Yield Deformed Steel Bar; Design grout strength = 30 MPa;

Rock encountered: Category 1(c) rock, 700 kPa (presumed allowable friction between rock and concrete) is taken from table 2.2 of Code of Practice for Foundations 2004/2017;

Design of Load Carrying Capacity of the Pile (excluding casing and grout)

(i) Due to the structural strength of the reinforcing bars

$$P = 4\times 25^2 \times \pi \times 230 = 1806 \text{ kN}$$

(ii) To withstand the maximum load capacity of 1806 kN, the minimum socket grouted length should be $1806/(0.19\,\pi \times 700) = 4.32$ m
(iii) The bond strength of steel bars with grout needs also to be checked. As per clause 8.4.4 and table 8.3 of the Code of Practice for Structural Use of Concrete 2004, the allowable bond strength is 2.739 N/mm^2, in excess of the allowable friction between rock and concrete. So the bond between grout and the reinforcing bars is not a controlling design criterion.

The main advantage of minipile is that only relatively small mechanical plant is required for the construction; hence, they are suitable for sites in hilly terrain or sites with accessibility difficulties. Minipiles are hence popular for many low-rise developments in hilly terrain in Hong Kong.

TABLE 14.3
Typical Small Diameter Bored Pile Capacity Estimation

Depths Below round	Soil Description	Layer of Depth (m)	SPT N Value from G.I. Report	Design SPT N Value	Shaft Resistance (kN/m²)	Shaft Friction (kN)
0 m–6 m	Fill or marine deposit	–	–	–	Neglected	–
6 m–21 m	Completely Decomposed Granite	1.50	18	18	86.4	68.9
		1.50	25	25	120	95.7
		1.50	36	36	172.8	137.8
		1.50	48	40	192	153.2
		1.50	57	40	192	153.2
		1.50	62	40	192	153.2
		1.50	78	40	192	153.2
		1.50	91	40	192	153.2
		1.50	104	40	192	153.2
		1.50	118	40	192	153.2
			Allowable Friction resistance			1,374
			Allowable End-bearing resistance			40.5
			Total loading capacity of pile			1,414

14.10 GEOTECHNICAL DESIGN OF PILE

The analysis and design of a single pile for vertical load involve the determination of its load carrying capacity and settlement. Very often, load carrying capacity and settlement are not independent of each other. The load carrying capacity is often limited by pre-determined settlement instead of shear failure of the soil/rock mass or the structural material of the pile. Even if shear failure of the soil/rock mass is adopted, the failure load of the pile so determined is the "ultimate load" and the "allowable load" which is the characteristic or working load the pile is allowed to carry has to be the ultimate load divided by a factor of safety in the order of two to three.

Broadly speaking, the determination of a single pile's capacity for vertical load is by "static" and "dynamic" formulas. A single pile refers to one which theoretically is an isolated one where the effects of other piles are ignored. If effects from other piles are considered, it will be a pile group effect.

Static Formula for single pile:
A pile's bearing capacity can be considered to be made up of two components, which are friction along its pile and the resistance at its tip called "end-bearing"

$$P_p = P_s + P_b \tag{14.26}$$

where P_p is the total pile bearing capacity; P_s is the bearing due to pile friction; and P_b is the bearing capacity at end-bearing.

The shaft friction of the pile is generally determined by the Alpha (α-α) or the Beta (β-β) methods. The α-α method is generally applied to cohesive soil, which relates the friction of the pile shaft with soil to the undrained shear strength of the soil c_u by a factor α termed the "adhesion factor" so that the shear resistance is

$$f_s = \alpha c_u \tag{14.27}$$

And the contribution of the bearing by this friction is

$$P_s = \alpha c_u A \tag{14.28}$$

where A is the circumferential area of the pile. However, α is not a constant for a certain type of soil and some studies have shown that it may decrease with the increase of c_u. The choice of the parameter should be based on local statistical results.

Nevertheless, for cohesionless soil, the behaviour is different. The shaft friction is related to the effective stress of the soil σ_v'. The relationship is through a β factor. So this approach is known as β-approach.

$$f_s = \beta \sigma_v' \tag{14.29}$$

We may take a simple consideration by relating the normal stress of soil on the pile shaft to the vertical effective stress, say by K_o and then the normal stress by friction coefficient expressed as $\tan\delta$ and finally set $\beta = K_o \tan\delta$. While the determination of K_o and δ are empirical or experimental in nature, it is therefore convenient to combine them into a single factor of symbol β. Yet by considering $K_o = 1 - \sin\phi = 0.5$ based on a conservative value of $\phi = 30°$ and $\tan\delta = 0.5$ (as the coefficient of friction), it is often conservative to take $\beta = 0.5 \times 0.5 = 0.25$. Currently, β = 0.25 is commonly used by many engineers in Hong Kong in the design.

Obviously, the β value of a pile depends not only on the soil parameter but also on the method of pile installation. Some typical values for β as measured in Hong Kong are given in Table 14.4.

Nevertheless, the reported measurement of β values can range from 0.1 to 0.6. The Code of Practice for Foundations 2017 (Hong Kong) has limited β value to 0.25 in the absence of a more accurate assessment, and the factor is reduced to 0.2 for pile in tension in the absence of trial pile test.

In addition, due to the "arching effect" of the soil, the shaft friction does not increase indefinitely with the effective stresses. Tests by Vesic have shown that there is a critical depth beyond which the shaft friction becomes constant. The depth is usually taken as 20 pile diameters or the maximum allowable skin friction, whichever is reached first. The Code of Practice for Foundations 2017 (Hong Kong) has limited the ultimate skin friction to 120 kPa for tension pile. The axi-symmetric earth pressure for circular excavation by Cheng has also illustrated that the effective

TABLE 14.4
Typical β values for Skin Friction Estimation

Types of piles	Type of soils	β
Driven small displacement piles	Saprolites	0.1–0.4
	Loose to medium dense sand	0.1–0.5
Driven large displacement piles	Saprolites	0.8–1.2
	Loose to medium dense sand	0.2–1.5
Bored piles and barrettes	Saprolites	0.1–0.6
	Loose to medium dense sand	0.2–0.6
Shaft grouted bored piles or barrettes	Saprolites	0.2–1.2

overburden stress and hence the skin friction will achieve a maximum value beyond the critical depth.

Some engineers correlate the skin friction with the SPT. Based on limited studies in Hong Kong, the ultimate skin friction can be taken as 1.5 to 2.0 N in kPa, with a maximum value of 160 kPa for driven small diameter pile. For large driven displacement pile, the relation can be taken as 4.0 N, with a maximum value of 250 kPa. For other countries, different correlations should be applied. For barrette formed by grab, the ultimate skin friction can be taken as 1.2 N, with a maximum value of 200 kPa. For barrette formed by cutter, the ultimate skin friction can be taken as 0.8 N, with a maximum value of 200 kPa when shaft grouting is not used, else the values can be taken as 2.5 N and 200 kPa respectively. For normal bored pile, it is similar to the barrette, except that the ultimate skin friction should be taken as 2.1 N when the shaft grouting is applied. The ultimate end-bearing capacity in soil will be taken as 10 N, with a limit of 2000 kPa.

For some types of piles, for example, small diameter bored piles, there are also empirical approaches by relating the allowable skin friction directly to the SPT N values with limits to some values. The Code of Practice for Foundations 2017 (Hong Kong) specifies that the allowable skin friction of small diameter bored piles formed by continuous augering is to be in the order of 1.0 N to 1.6 N (in kPa) with a limit of 40 kPa.

The end-bearing component of the pile capacity often adopts a simple approach by specifying numerical values for the allowable or ultimate bearing capacity of the soil and rock as q_a or q_{ult} and the end-bearing capacity is simply the cross-sectional area multiplied by q_a or q_{ult} as

$$P_b = q_a A \tag{14.29}$$

where A is the cross-sectional area.

The q_a values are given in some codes including BS8004: 1986 and the Code of Practice for Foundations 2017 (Hong Kong) for rock and soil of different qualities. The values range from 10000 kPa for strong rock to 2000 kPa for weak. For soil, the values are 600 kPa to 50 kPa. In Hong Kong, the allowable end-bearing pressure is also commonly taken as 6.7 N for dry soil and 5 N (kPa) for wet soil.

However, from the principle of theoretical soil mechanics, failure mechanics similar to that of shallow foundations applies, by which the three bearing capacity factors N_q, N_c and N_γ can be applied. So the end-bearing capacity of the pile $Q_b$can be expressed as :

$$Q_b = \left(cN_c + qN_q + 0.5B\gamma N_\gamma\right)A_b \tag{14.30}$$

where A_b is the bearing area of the pile.

Of the three components, qN_q is the most important one as q equal to at least the soil weight down to the pile tip, which are generally of high values. The effect of the cohesive strength and γB terms are constant values that are much less than the surcharge term; hence, they are commonly neglected in the design.

Similar to N_q for shallow foundations, N_q for piles also depend on the ϕ value of soil. Different researchers, based on different approaches and assumptions, have produced different relations. The N_q by Berezantsev et al. (1961) appears to be popular in many countries, and generally, this factor is conservative compared with other published N_q values. This is probably the main reason for the popularity of the N_q by Berezantsev. A design figure for N_q is prepared by the authors, which is based on the correction of the original derivation by Berezantsev and is shown in Figure 14.25.

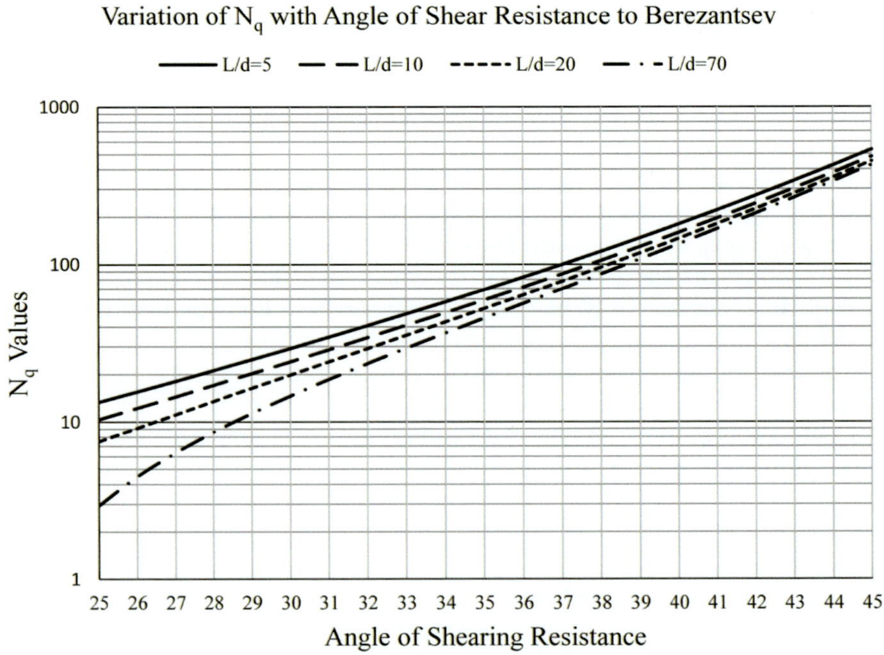

FIGURE 14.25 Design figure for N_q by Cheng et al. (2024).

Since pile foundation is typically controlled by settlement instead of capacity, there is no strong need to discuss other bearing capacity factors here. Interested readers can consult other references on the pile bearing capacity factors if necessary. In general, engineers in Hong Kong do not believe in these bearing capacity factors but rely on the use of other statistical relations such as SPT and presumed values for the preliminary design.

Piles are usually used in groups, and the pile capacity of a pile group is compared with a single pile. The pile capacity of a pile group is a complicated problem, and the problem heavily depends on the method of installation. For driven piles in loose to medium dense sand, the efficiency is usually greater than 1.0, as the densification and compaction effect will actually improve the soil properties. For driven pile in dense sand, it is possible that the driven process can actually loosen the structure of the soil, but generally this is not a critical effect. On the whole, it can be considered that the pile capacity of a pile group in sand is the sum of the individual pile capacity. On the other hand, if a pile under 1500 kN settles by 15 mm, will a pile group with 2 piles taking a loading of 3000 kN also settle by 15 mm? In general, this is not true, as the settlement of a pile will cause soil dragdown and hence settlement to the adjacent pile. From the consideration of the pile group settlement, the pile group capacity should be reduced. There are some empirical formulae in the literature for the pile group capacity, but the authors view that such empirical formulae are not believable unless they are supported by field measurements. In Hong Kong, a group reduction factor of 0.85 is commonly adopted by many engineers, but currently it is generally accepted that no reduction factor is required for pile group in sand. For Hong Kong, a group reduction factor is not required where:

(1) the centre-to-centre spacings are more than three times the perimeter of the piles or the circumscribed rectangles in the case of H-piles; or
(2) the load capacity of the piles is derived from end-bearing; or
(3) the piles are rock-socketed piles; or
(4) the piles are driven to refusal with the toes resting closely onto bedrock, according to the Code of Practice for Foundation; or
(5) the piles are driven to sound bearing strata, e.g., soil with SPT N value $\geq$ 200.

14.11 LATERAL LOAD ANALYSIS

In addition to the piles' primary use as axially loaded members to resist vertical loads, piles are sometimes required to resist lateral loads. The lateral loads mostly come from wind and soil, which often act on the pile heads. The lateral restraints are offered by the soil embedding the pile. Lateral shears and moments, and sometimes axial loads, will then be induced in the piles. The goals of designers are to determine deflections and stresses in the selected soil-pile system in order that they may be controlled within tolerable limits.

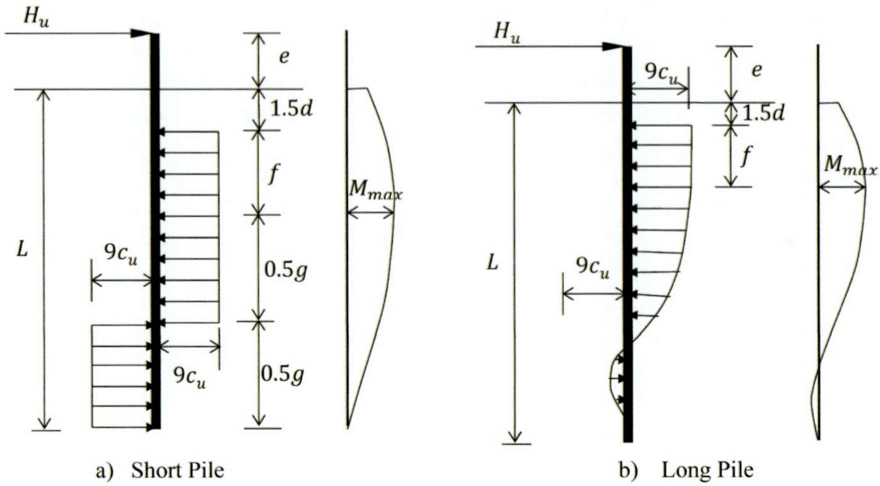

FIGURE 14.26 Ultimate lateral resistance of pile in cohesive soil.

14.11.1 Ultimate Analysis

Ultimate lateral load resistance of a pile can be originated from the resistance of soil. For cohesive soils, Broms (1964) takes the ultimate resistance as $9c_u$ where c_u is the undrained shear strength of the soil. The top $1.5d$ (where d is the pile diameter) of soil resistance is ignored, as illustrated for long and short piles in Figure 14.26.

By Figure 14.26(a) which is a short pile, the equilibrium of which is controlled by the soil resistance under the ultimate soil pressure distribution as shown,

$$H_u = 9c_u df \tag{14.31}$$

where H_u is the ultimate shear. As zero shear exists at $1.5d + f$ below ground, where the maximum moment in the pile will coexist, we may list

$$M_{max} = H_u(e + 1.5d + f) - H_u \times 0.5f = H_u(e + 1.5d + 0.5f) \tag{14.32}$$

Also, taking moment at zero shear location due to push and pull soil pressure below the location, a couple constituted by the soil pressure is also equal to M_{max}, i.e.,

$$M_{max} = 9c_u d \times 0.5g \times 0.5g = 2.25c_u dg^2 \tag{14.33}$$

In addition

$$L = 1.5d + f + g \tag{14.34}$$

So there are four Equations from 14.31 to 14.34 to solve H_u, M_{max}, f and g.

Equating Equations 14.32 and 14.33 with H_u substituting by Equation 14.31 and simplifying gives

$$4f(e+1.5d+0.5f)=g^2$$

Substituting Equation 14.34 to eliminate g

$$2f^2+(4e+6d)f=\left[(L-1.5d)-f\right]^2$$

$$\Rightarrow 3f^2+\left[(4e+6d)+2(L-1.5d)\right]f-(L-1.5d)^2=0$$

$$\Rightarrow 3f^2+(4e+2L+3d)f-(L-1.5d)^2=0 \tag{14.35}$$

Solving f by the above quadratic equation, H_u can be directly solved by Equation 14.31, together with M_{max} by Equation 14.33. If M_{max} exceeds the flexural capacity of the pile $M_p$at long length, the pressure profiles below the zero shear location will change as shown in Figure 14.26(b), M_p will then be taken into account as a limiting criterion for determination of H_u. So only Equations 14.31 and 14.32 need to be used with $M_{max}=M_p$, leaving H_u, f to be solved.

For cohesionless soil, the ultimate soil resistance is the passive pressure. Taking the arching effect of the soil into account, Broms (1964) takes the ultimate resistance as $3K_p\sigma'$. Some more updated results actually suggest replacing 3 by K_p. The ultimate lateral resistance of the pile will be under the combined consideration of the ultimate pressure of the soil and the ultimate bending capacity of the pile as illustrated in Figures 14.27.

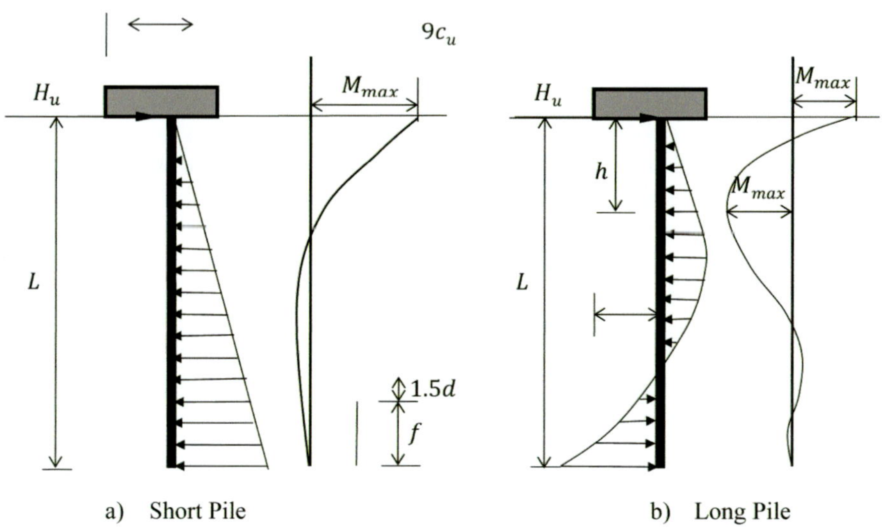

FIGURE 14.27 Ultimate lateral resistance of pile in cohesionless soil.

For the short pile as shown in Figure 14.27(a), the full passive soil resistance is mobilized which is equal to the ultimate shear of the pile S_{ult}

$$S_{ult} = 0.5\left(3K_p\sigma' dL\right) \tag{14.36}$$

The moment is the greatest at the pile head which is

$$M_{max} = 0.5\left(3K_p\sigma' dL\right)\left(\frac{2}{3}L\right) \tag{14.37}$$

and M_{max} is less than the flexural capacity of the pile.

However, if M_{max} is greater than the flexural capacity of the pile M_p, then the soil pressure distribution becomes that of Figure 14.27(b). Then the equation involving h which is the location of zero shear and the ultimate shear is set up as

$$S_{ult} = 0.5\left(3K_p\gamma hdL\right) \tag{14.38}$$

taking $\sigma' = \gamma h$ where γ is the unit weight of the soil with the addition of surcharge if any.

And the moment at depth h is $M_{max} = 0.5\left(3K_p\gamma hdh\right)\left(\frac{2}{3}h\right)$ and putting $M_{max} = M_p$ the flexural capacity of the pile which is a known value. So,

$$M_p = 0.5\left(3K_p\gamma hdh\right)\left(\frac{2}{3}h\right) \tag{14.39}$$

Solving h by Equation 14.39, S_{ult} can be calculated by Equation 14.38

If the pile head is at some depth below the ground so that σ' is not zero at pile head level, Equations 14.38 and 14.39 will be modified for determination of the S_{ult}.

The foregoing describes an approach for determination of ultimate shear of a pile. The allowable shear will then be the ultimate shear divided by a factor of safety in the order of 2 to 3. However, the controlling criterion for a pile resisting lateral load is often deflection which is normally a serviceability limit state. Approaches have been developed for estimation of the lateral deflections of piles.

Worked Example 14.2

A 305 × 305 × 180 Grade S460 floating pile embedded in cohesionless soil of ϕ'= 35° as shown. Its flexural capacity under the ultimate axial load is 300 kNm.

Case 1: The Pile is Pinned Connected at the Pile Cap (Figure 14.28)

Passive Pressure Coefficient $K_p = (1 + \sin\phi)/(1 - \sin\phi) = 3.69$

Width of the Pile D is 0.325 m

The ultimate soil lateral resistance according to Broms is 3 K_p σ. For rotational equilibrium of the pile, there must be a reversed direction of soil pressure at the pile tip. Assuming the reversed soil load is at a single pile tip level and taking moment about the tip level of the pile.

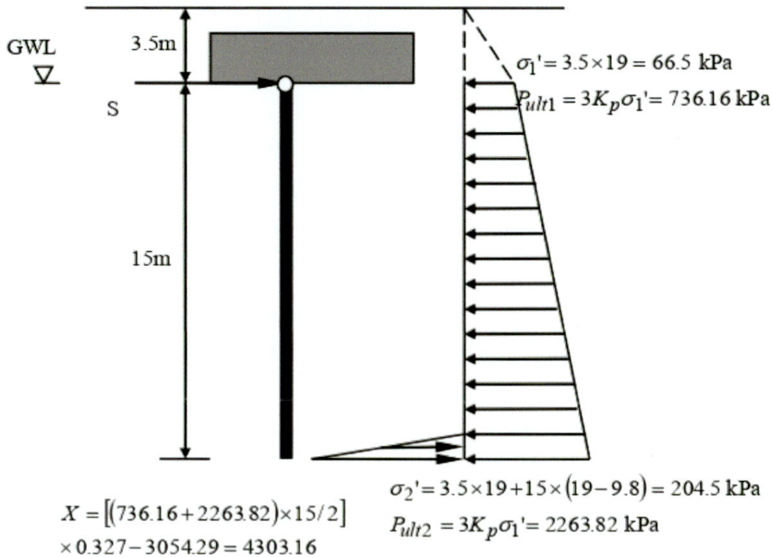

(a) Preliminary Check by Considering the Pile to be a "Short Pile"

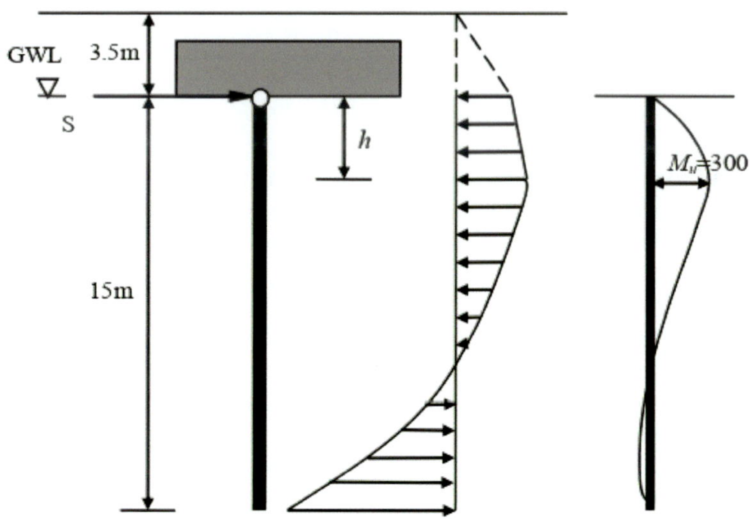

(b) Revised Check by Limiting to Flexural Strength of the Pile

FIGURE 14.28 Ultimate lateral capacity for case 1. (a) Preliminary check by considering the pile to be a "short pile." (b) Revised check by limiting to flexural strength of the pile.

S × 15 = [736.16 × 15 × 7.5 + 0.5 × 1527.66 × 15 x (15/3)] × 0.325 ⇒ S = 3035.6 kN

The level of zero shear is at h below the pile cap.

$$\left[736.16h+\frac{(2263.82-736.16)h}{15}\times\frac{h}{2}\right]\times 0.325-3035.61=0 \Rightarrow \text{h} = 8.123 \text{ m}$$

The maximum moment of the pile at the level 8.123 m below the pile cap is (736.16 × 8.123²/2 + 1527.66 × 8.123³/15/6) × 0.325 – 3035.61 × 8.123 = 13808 kNm > 300 kNm

As the ultimate moment is only 300 kNm, the level of maximum moment (zero shear) has to be raised and h re-calculated as in (b). At depth h, the ultimate lateral soil pressure is

$$736.16+\frac{(2263.82-736.16)h}{15}=736.16 + 101.844\, h$$

The lateral shear provided by soil from top level of pile to the new depth h is [736.16 + 101.844 h + 736.16] ÷ 2 × h x 0.325 = 16.5497 $h^2/2$ + 239.252 h.

As zero shear exists in this level, the lateral shear S at the top level of the pile is 16.5497h²/2 + 239.252 h. Listing the moment equation as follows and limiting it to 300 kNm

$$S\times h-\left(736.16\times h^2/2-101.844\times\frac{h}{2}\times\frac{h}{3}\right)\times 0.325=300$$

Solving the cubic equation, h = 1.485 m and S = 16.5497 h^2 + 239.252 h = 391.84

So the ultimate shear capacity is 391.84 kN

Case 2: The Pile is Rigidly Connected at the Pile Cap (Figure 14.29)

Assuming "short pile" by which the ultimate lateral soil resistance is mobilized, to achieve equilibrium, the total shear and moment at the top level of the pile are respectively

S = (736.16 + 2263.82)/2 × 15 × 0.325 = 7312 kN

M = [736.16 × 15 × 7.5 + 1527.66 × 15/2 x 10] × 0.325 = 27288 kNm > 300 kNm

As the ultimate moment is only 300 kNm, the level of zero moment has to be raised and h re-calculated as in (b). The assumption is based on that the maximum moment of 300 kNm occurs at some depths of the pile, as equal to that at the top level but in the reversed direction and at this level, the shear is zero. As the ultimate moment is only 300 kNm, the level of maximum moment (zero shear) has to be raised and h re-calculated as in (b). At depth h, the ultimate lateral soil pressure is

$$736.16+\frac{(2263.82-736.16)h}{15}= 736.16 + 101.844\, h$$

The lateral shear provided by soil from top level of pile to the new depth h is [736.16 + 101.844 h + 736.16] ÷ 2 × h × 0.325 = 16.5497 h^2 + 239.252 h

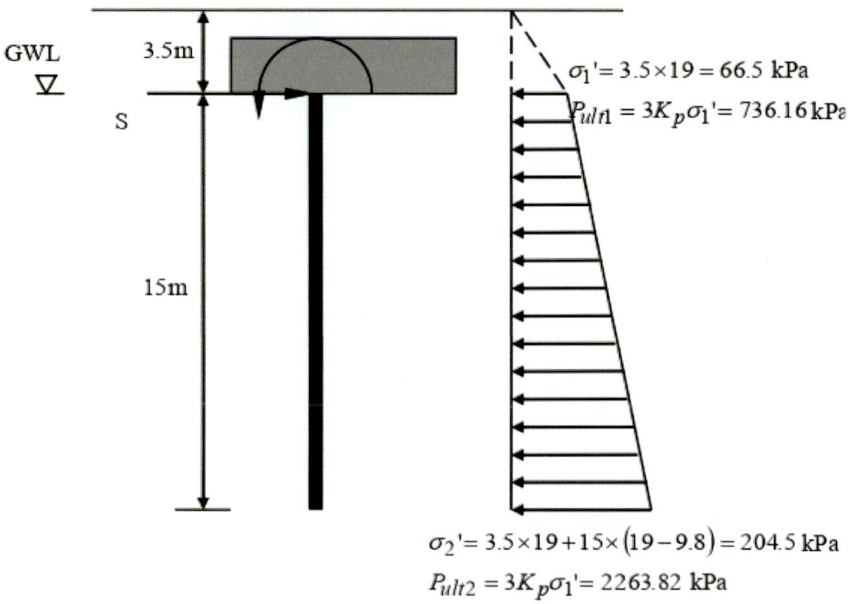

(a) Preliminary Check by Considering the Pile to be a "Short Pile"

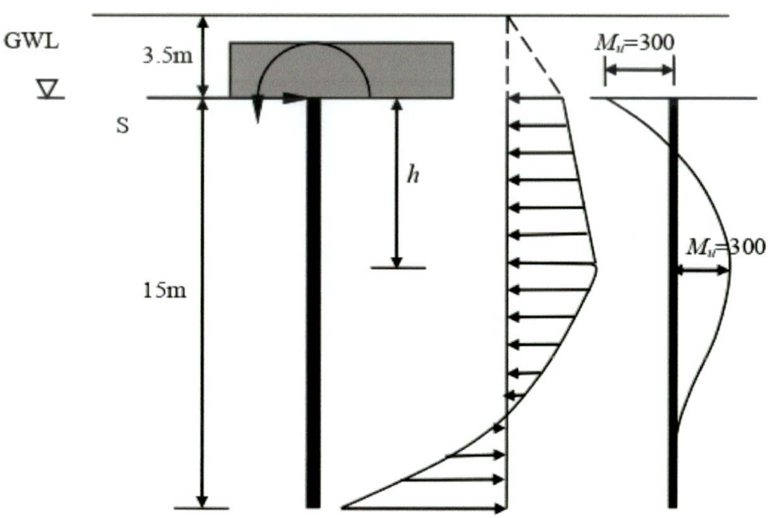

(b) Revised Check by Limiting to Moment Strength of the Pile

FIGURE 14.29 Ultimate lateral capacity for case 2. (a) Preliminary check by considering the pile to be a "short pile." (b) Revised check by limiting to moment strength of the pile.

As zero shear exists in this level, the lateral shear at the top level of the pile is also 6.5497 h^2 + 239.252 h. Listing the moment equation as follows and limiting it to 300 kNm

$$S \times h - \left(736.16 \times h^2 / 2 - 101.844h \times \frac{h}{2} \times \frac{h}{3}\right) \times 0.325 - 300 = 300$$

Solving the cubic equation, h=2053 mm and S = 561.09 kN which is the ultimate shear capacity.

14.11.2 Lateral Deflection of Pile

Poulos and Davis (1980) suggested the use of Mindlin's equations and their integrated forms for the lateral analysis of single piles and pile groups, which require the use of Young's modulus and Poisson's ratio of soil. Cheng has programmed such a method and found that numerical problems will occur if the number of segment is too large. Furthermore, over-interaction is obtained if this method is extended to pile group analysis. Very high stresses will be obtained at the pile heads which have to be dealt with by plastic analysis using iterative methods. In view of the various limitations, such a method is rarely used now.

To overcome the complication as mentioned above, an approach based on the assumption of "Winkler's springs" developed by Terzaghi (1955) is commonly used by many engineers. The method is known as the "horizontal subgrade reaction method." By the method, the soil restraint is idealized as a series of elastic springs (independent of each other and therefore called Winkler's springs) on the pile shaft. The spring value is denoted by k_h which is called "coefficient of horizontal subgrade reaction" and is related to the displacement as

$$p = k_h u \tag{14.40}$$

where p is the pressure and u is the horizontal displacement of the soil. k_h carries a physical meaning of stiffness which is the proportionality relating pressure and lateral displacement, i.e., the lateral pressure required to produce unit lateral displacement. In metric units, its unit is kN/m^2/m. For the simulation of a pile as a structural model for solution by the stiffness method, a beam element with a series of lateral point springs as supports can be constructed as shown in Figure 14.30. In the figure, the springs are discrete point springs which carry the value of $k_h B \Delta z$ where B is the width of the pile and ΔL is the tributary length of the pile the spring represents. Continuous supports can be assumed if the springs are closely spaced.

With B as the width of the pile, the load per unit length of the pile is $w = pB$. So we may list

$$w = k_h B u \tag{14.41}$$

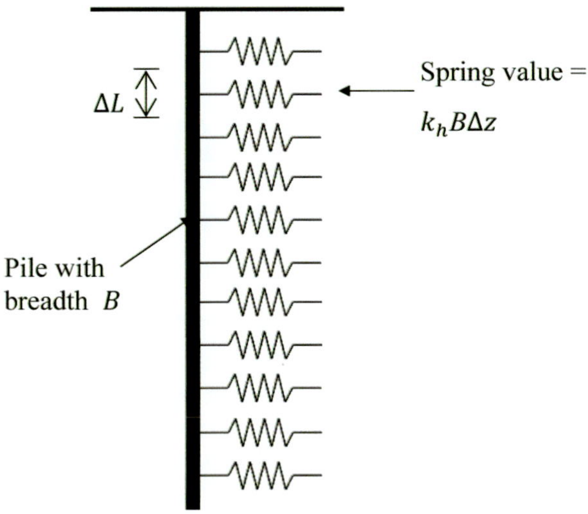

FIGURE 14.30 Computer model of a pile with lateral soil springs.

k_h depends on the stiffness of the soil and at the same time may depend also on the depth of soil z. A general relation is

$$k_h = k_L \left(\frac{z}{L}\right)^{\alpha} \tag{14.42}$$

where L is the level of the pile tip below ground, and k_L is the k_h value at $z = L$, α is a coefficient to be applied that depends on the type of soil. Generally, α is taken as 0 for cohesive soil so that the k_h value is constant along the pile shaft and α is taken as 1 for cohesionless soil, so that k_h varies linearly with depth.

So the "spring values" along the pile shaft can be defined under determined values of k_h. For cohesionless soil, k_h is expressed as

$$k_h = n_h \left(\frac{z}{B}\right) \tag{14.43}$$

where n_h is termed the "constant of horizontal subgrade reaction" and depends on the type of soil only. Terzaghi (1955) has produced values for different ranges of SPT N values. The values are reproduced with metric units in Table 14.5.

With the k_h determined either by Equation 14.42 or by Equation 14.43 for cohesive or cohesionless soil, simulation into the computer model as shown in Figure 14.30 can be carried out for analysis. Results comprising lateral deflections, internal forces of the pile and soil reactions can be obtained. It should be noted that this n_h is a derived soil property, as different from the Young's modulus which can be directly measured. If the lateral deflection is so large that the secondary effect due to additional moment induced from the vertical load and the lateral deflection is significant,

TABLE 14.5
Values of n_h for Cohesionless Soils

	Loose (SPT N Value = 4–10)	Medium (SPT N Value = 11–30)	Dense (SPT N Value = 31–50)
n_h for dry or moist sand	2200 kN/m²/m	6600 kN/m²/m	17600 kN/m²/m
n_h for submerged sand	1300 kN/m²/m	4400 kN/m²/m	10700 kN/m²/m

P-Y analysis should be considered (PY program by Cheng can be obtained from email request). For medium lateral deflection, the engineers in Hong Kong usually just add an additional moment due to the vertical load and the lateral deflection without further iteration.

14.12 STRUCTURAL ANALYSIS FOR PILE CAP AND RAFT FOUNDATION

Most of the pile caps and raft foundation are thick slabs with piles and superstructures (walls, columns and others). Furthermore, besides the support from piles, the soil underneath the cap can support part of the superstructure loadings. The head condition of the pile with cap can be pin-joint, fixed joint or semi-rigid joint. The problem is further complicated by the action of lateral load, finite sizes of piles and wall/column and the possible pile load transfer between piles. A rigorous analysis of the pile cap is difficult, and 3-dimensional finite element analysis is seldom adopted in practice. Currently, most of the engineers adopt a plate model for the cap, and the thick plate model with shear deformation is adopted in many finite element programs. Towards the use of the plate model as compared with true 3-dimensional analysis, Cheng has carried out comparative studies for several cases. It is found that the use of the plate model is a good approximation of the 3-dimensional model using brick element, hence the adoption of plate analysis (thick plate) is an acceptable method for engineering purposes. There are however various issues that should be considered:

(1) Finite size of piles and walls/columns – For piles and columns, many engineers treat them as point structures/loads which may be satisfactory if they are small in size. There are some cases where the moment and shear are very high so that the finite sizes of piles and columns are considered. For column loads, they can be distributed as uniformly distributed loads or trapezoidal loads (with moments). For piles, if the piles are modelled as vertical springs only, the vertical spring stiffnesses can be distributed over the pile areas. If the pile head condition carries moment, the

equivalent rotational spring can be distributed as many smaller rotational springs over the pile area.

(2) Superstructure stiffness – In general, the superstructure stiffnesses arising from columns/walls are neglected. If the stiffness from the superstructure has to be considered (partially) to reduce the differential settlement calculation, they are usually modelled as grillage beams for walls and rotational springs for columns, which can be defined easily in program PLATE.

(3) Thickness of cap – The thick plate model is usually adopted for the pile cap analysis, which may be adequate for normal purposes. To further reduce the deformation and internal forces calculations, some researchers propose the use of a brick element with a rotation degree of freedom. This approach appears to have a noticeable reduction in the computed internal forces, but there are difficulties in the reinforcement design so that it appears to be a rare practice among engineers.

(4) Piles under lateral load – Some finite element programs allow the user to define the cap with a pile. In general, the authors are not in favour of this approach, and many strange results have been reported to Cheng from his students in industry. Instead, the pile can be defined as a lateral rotational spring with the spring constant obtained separately from a pile lateral analysis. This approach has the advantage that the spring constant can be obtained easily in non-homogeneous conditions.

(5) Soil medium – Most of the pile caps are analyzed without the presence of soil medium for simplicity. In PLATE, the soil mass can be modelled as Winkler's support or the elastic continuum theory. For the load transfer between piles, many finite element programs fail to model this behaviour which may not be critical in general. This case can be modelled easily in PLATE by using the option "PILE" with or without pile interaction. Without pile interaction, each pile acts as an individual spring with the consideration of skin friction by Randolph's theory. Pile interaction is built into PLATE automatically by the "Pile interaction" button.

(6) The strut-and-tie model is now seldom adopted, except for those simple pile caps with only one column and three to four piles. On the other hand, the grillage model is still sometimes used by some engineers for the analysis and design of pile caps, due to the avoidance of numerical problems in many plate programs, simplicity in the design process. The grillage model, however, can never converge to the plate solution but can be a simple method to be used. For complicated problems with irregularly spaced columns and walls, it is very difficult to form a good grillage model.

(7) The twisting moment M_{xy} from plate analysis is still not appreciated by many engineers. The main reason appears to be the over-reliance on the use of computer programs without basic knowledge among many engineers. Without computer programs, many engineers can work it out correctly using Fourier analysis at the time when Cheng is still working as an engineer.

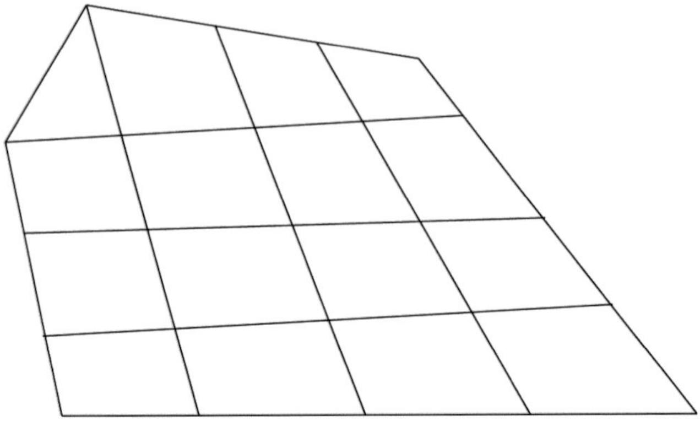

FIGURE 14.31 An irregular plan shaped raft footing being meshed into an assembly of plate elements to be analyzed by the finite element method.

Currently, most engineers have turned to the use of finite element plate analysis for raft foundation and pile cap analysis and design (limited use of the grillage method). With the development of the finite element method to formulate the stiffness matrix of a plate element, the structural analysis of a raft footing can be carried out by simulating the structure as an assembly of plate elements as shown in Figure 14.31.

Each of the plate elements has basically three degrees of freedom at each node which are the vertical settlement and out-of-plane rotations about two mutually perpendicular axes, say X and Y. The basic mathematical derivation of the shears, bending moments and twisting moments in plate bending are as follows based on Figure 14.32 and Figure 14.33. In Figures 14.32 and 14.33, the rectangular element of plan dimensions ∂x and ∂y is an infinitesimal element. By considering equilibrium and strain compatibility, we can get the following equations relating shears, bending and twisting moments and the strains by the plate bending theory.

$$\frac{\partial^4 w}{\partial x^4}+2\frac{\partial^4 w}{\partial x^2 \partial y^2}+\frac{\partial^4 w}{\partial y^4}=\frac{q}{D} \tag{14.44}$$

$$\frac{\partial^2 M_x}{\partial x^2}+2\frac{\partial^2 M_{xy}}{\partial x \partial y}+\frac{\partial^2 M_y}{\partial y^2}+q=0 \tag{14.45}$$

where q is the uniformly distributed load at the point (x, y) and $D=\frac{Et^3}{12(1-\mu^2)}$.

Equation 14.44 is termed the thin plate deflection equation, which should be modified for many pile cap and raft foundations where the thickness and shear deformation are not negligible. Most of the plate elements in commercial programs have a choice of thin/thick options for the users, but many plate elements in commercial programs suffer from the problem of distorted mesh shape, which is unavoidable for irregular problem. Cheng and Law (2008) developed a refined thick plate element which is good for thin plates as well, so that the engineers need not choose the thin/

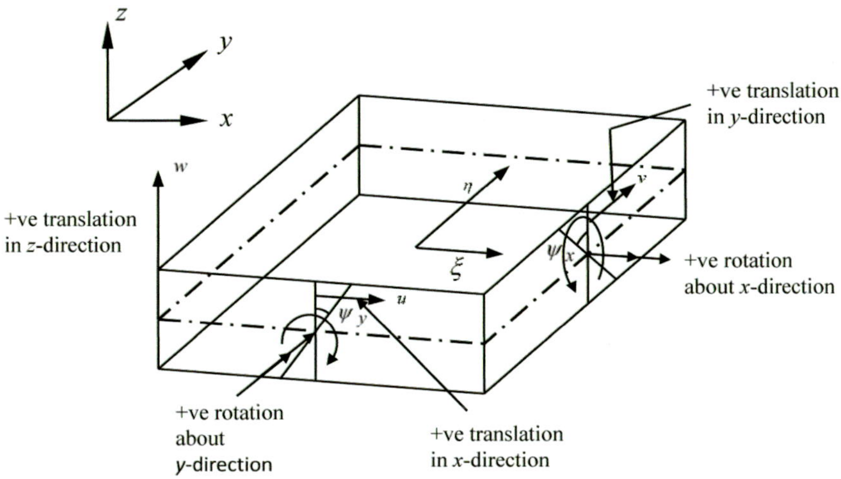

FIGURE 14.32 Plate bending theory for bending.

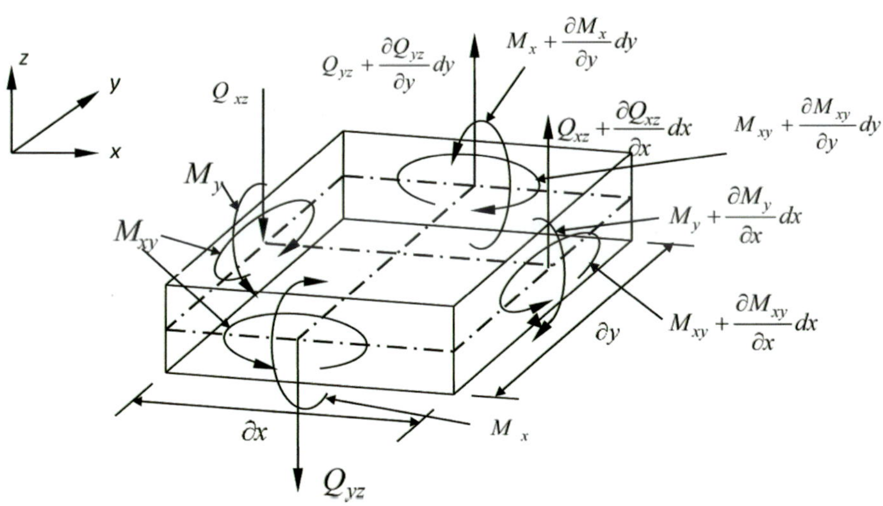

FIGURE 14.33 Plate bending theory for shear.

thick plate option. More importantly, the plate element by the authors is able to give good results even under distorted configurations, and the moment and shear are generally much more accurate than other plate elements under the same mesh.

For raft foundation, the support from soil/rock can be modelled by the use of Winkler's spring model or elastic support model. Consider the typical plate load test result as shown in Figure 14.34.

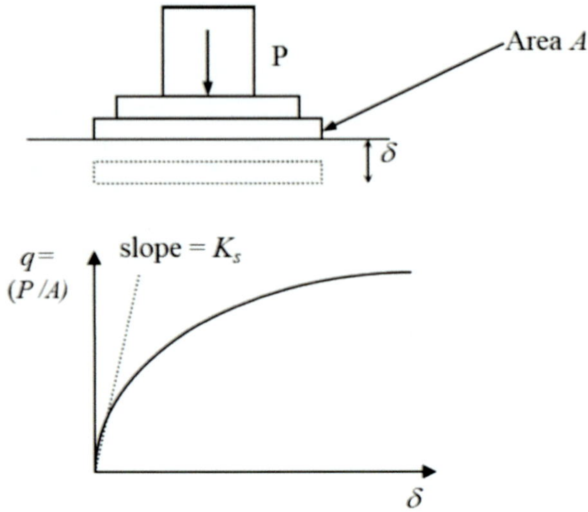

FIGURE 14.34 Results for a typical plate load test.

When the strain is small, q vs. δ can be taken as linear and

$$K_s = q/\delta \text{ (unit=kN/m}^3\text{)} \quad (14.46)$$

$$\text{Equation 14.46} \times: A\cdot K_s = q\cdot A/\delta = F/\delta$$

$$\text{or } \delta = F/A\,K_s \quad (14.47)$$

Hence $A\cdot K_s$ is conceptually similar to the classical 1-D spring, and $A\cdot K_s$ is taken as the spring constant. The response from the subgrade is hence a series of linear springs acting along the footing. K_s can be approximated as $0.65E/B(1-\mu^2)$ by Salvadurai or $E/B(1-\mu^2)$ by Vesic, or $\pi E/2B(1-\mu^2)\log_e(L/B)$. K_s is not really a constant parameter, as it is dependent on B according to the formulae by Vesic, Salvadurai or field tests; however, it is usually considered to be constant for simplicity, and Table 14.6 can be referenced if necessary.

TABLE 14.6
Some Suggested K_s by Terzaghi (1955)

Soil	K_s
Cu (100–200 kPa)	15–30 MN/m^3
Cu (200–400 kPa)	30–60 MN/m^3
Cu > 400 kPa	60 MN/m^3
Loose dry sand	5–20 MN/m^3
Medium dry sand	20–90 MN/m^3
Dense dry sand	90–300 MN/m^3
Wet loose sand	8 MN/m^3
Wet medium sand	25 MN/m^3
Wet dense sand	100 MN/m^3

Instead of assuming the settlement of any point support is only due to the load directly acting on it as a "Winkler's spring," the settlement of a point i is not only due to the point load P_i acting directly on it, but also due to loads acting on other points. The effect is first taken as elastic, by which the settlement is proportional to the load creating it. By means of the Boussinesq or Mindlin's Equations, the settlement of a point i due to a point load P_j at some distance apart can be expressed as

$$\delta_{ij} = I_{ij} P_j \tag{14.48}$$

where I_{ij} is the interaction coefficient between the point i and j that can be calculated based on the Boussinesq or Mindlin's Equations. From Boussinesq equation,

$$\mathrm{I}_{ij} = \frac{1-\mu^2}{\pi E_s r} \tag{14.49}$$

where r is the distance between points i and j. The use of Equation 14.49 will result in "interaction" of the supports and is considered the flexibility of the subgrade. The matrix of the flexibility coefficient [f] can then be established, which will give the equivalent stiffness matrix of the subgrade $[k_s]$ matrix inversion, and $[f]^{-1}= [k_s]$ is a full matrix that is superimposed to the stiffness of the beam or plate to give the overall stiffness matrix of the whole system, which will yield the displacements and internal forces under the prescribed loading. In the superposition of $[k_s]$ with the foundation stiffness, it must be noted that $[k_s]$ affects only the vertical displacement component during the assemblage of the global stiffness matrix. Cheng developed the program PLATE with this option, which is not available in most of the commercial plate analysis program. Cheng has further refined the analysis by adding a term radius of influence, beyond which the influence coefficient I_{ij} in Equation 14.49 is taken as 0. Some of the special features of the program PLATE by Cheng include:

(1) Unlike some other commercial programs, the user need not choose between the thick and thin plate options. The plate element by Cheng and Law (2008) has been verified to give very high accuracy for both thin and thick conditions.
(2) No grid is required! Only coordinates are used to define everything.
(3) Prescribed fixed or prescribed boundary conditions can be defined.
(4) Winkler, piles and elastic half space support conditions can be modelled.
(5) Contour lines for results of individual load case or load combination, result envelopes which are the maximum and minimum values out of all the load combinations can be viewed and generated. The user can also cut any section along any rotation to view the results.
(6) Reinforcement design to Code of Practice for Structural Use of Concrete 2004, Hong Kong. Wood-Armer design to the steel reinforcement is also available as an option.
(7) Uniform elastic half space ground support or stratified layers of soil ground support defined by boreholes are accepted. This is a unique feature not available in any other existing commercial program (Figure 14.35).

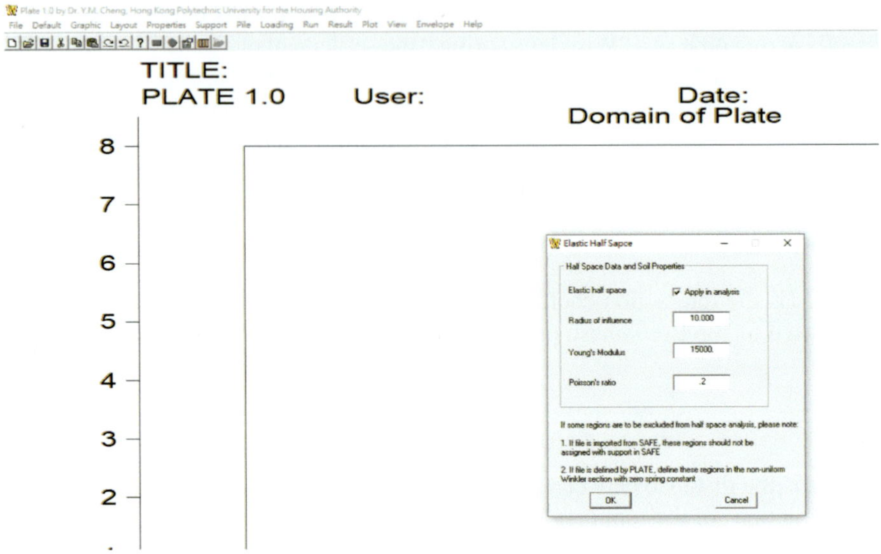

FIGURE 14.35 Radius of influence for elastic half space option.

(8) For pile cap on pile group, the interaction matrix is generated automatically. A unique feature of PLATE against other commercial program!
(9) Line moment is automatically transformed to point load internally.
(10) Reinforcement design at inclined axes is possible, a feature not available in other program !
(11) Punching shear is automatically checked in the design.
(12) Correct shear reinforcement design, which considers shear enhancement automatically. The design practice by engineers using Fx and Fy is not correct, and Vmax is used in PLATE for shear reinforcement design, with reduced shear reinforcement in general.
(13) Improved design strip bending moment, more economical, and avoids some strange mistakes in commercial program.
(14) A borehole option is available to define the 3-dimensional ground condition and soil parameters, which is also unique among the existing plate element programs (Figure 14.36).

Cheng has further incorporated the ring shear theory by Randolph (1987) into PLATE, and a pile can be modelled as an individual spring on a refined pile-settlement model as shown in Figure 14.37.

The refined PLATE program by Cheng enabled accurate and refined analysis with various practical engineering functions which are used by some engineers. For raft foundation analysis, both Winkler's spring, elastic half space and borehole options are available while the elimination of tension at the base of the raft foundation is possible. Based on some engineers' studies, the refined post-processing of the

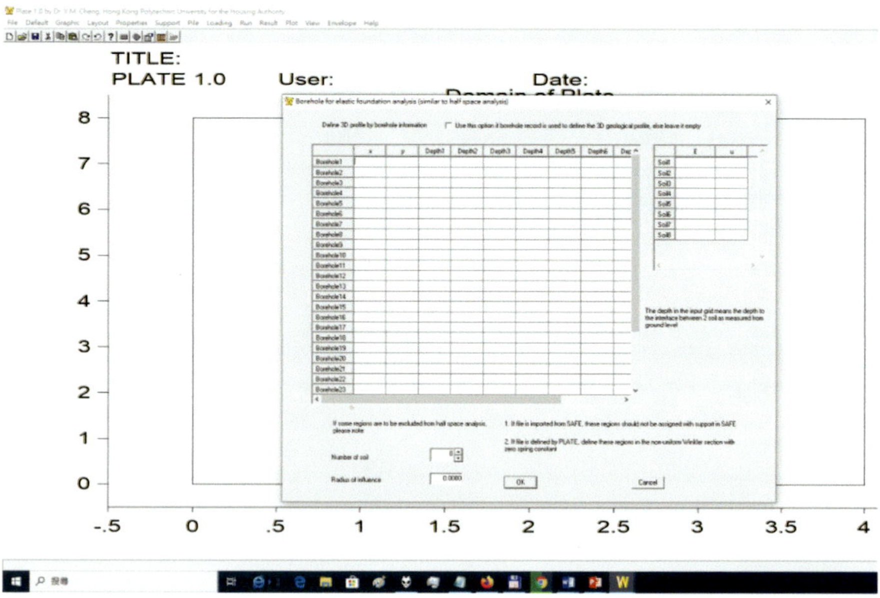

FIGURE 14.36 Borehole option in PLATE.

internal forces for reinforcement design can usually give smaller values as compared with some commercial programs, due to the use of more accurate and refined strip force determination. A sample program can be obtained from Cheng at natureymc @yahoo.com.hk.

14.13 Concluding Remarks

Building foundation design is a critical discipline that ensures safe and economical transfer of superstructure loads to the ground while limiting deformations to acceptable levels. This chapter has covered the core principles of **shallow** and **deep** foundations, with emphasis on the challenging ground conditions frequently encountered in urban areas like Hong Kong.

For **shallow foundations**, design is governed by:

(i) Ultimate limit state: bearing capacity calculated using plasticity-based equations with appropriate modification factors.
(ii) Serviceability limit state: usually the controlling criterion, requiring accurate prediction of elastic, consolidation, and creep settlements, as well as differential and angular distortions.

For **deep foundations** (especially piles), key aspects include load transfer mechanisms (shaft friction vs. end-bearing), empirical correlations with SPT N-values, rock socket bond stresses, installation effects, group interaction, negative skin friction, and lateral resistance. In Hong Kong, non-percussive systems, such as bored piles, socketed H-piles, shaft-grouted barrettes, and minipiles—dominate due to noise, vibration, safety, and adjacent-structure constraints

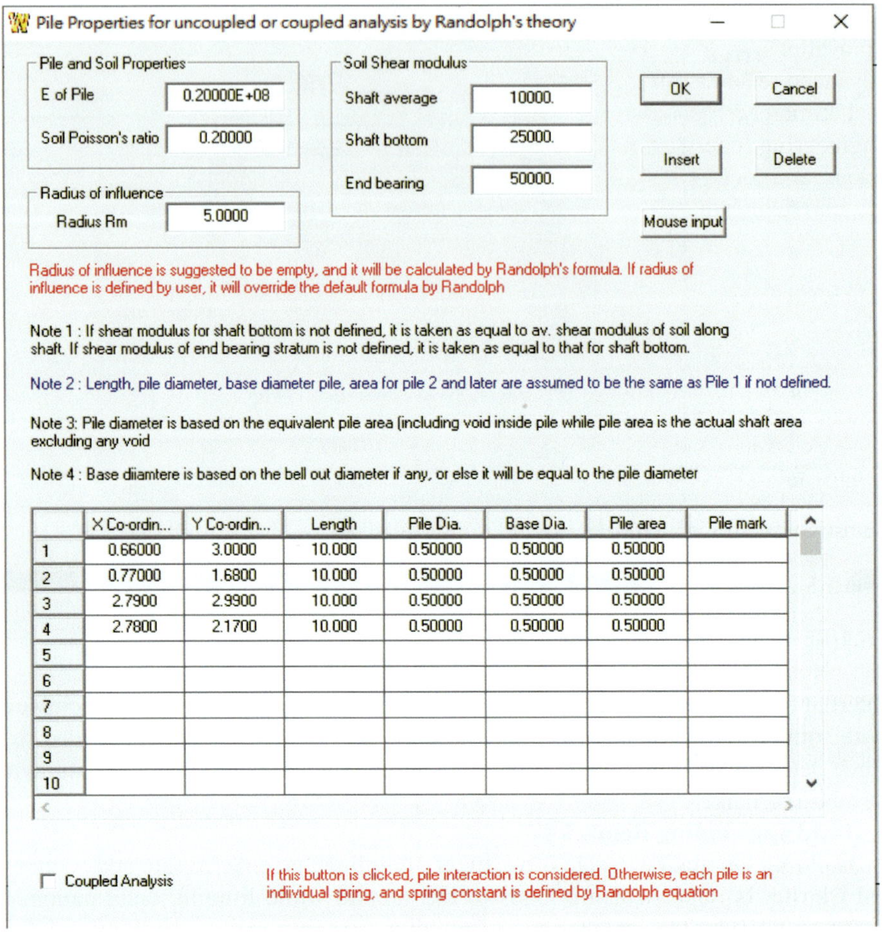

FIGURE 14.37 Pile option in program PLATE.

Successful foundation engineering requires:

(i) detailed site-specific geotechnical understanding,
(ii) high-quality investigation and testing,
(iii) realistic assessment of long-term soil-structure behaviour,
(iv) balanced consideration of cost, time, environmental impact, safety, and manufacturability,
(v) adherence to codes (e.g., Hong Kong Code of Practice for Foundations 2017) combined with sound judgment,
(vi) instrumentation and monitoring (including AAA triggers and IOT systems) to verify performance and enable prompt response.

Hong Kong's unique geology, dense urban setting, and strict regulations have driven innovations such as shaft grouting, pre-bored H-piles, deep cement mixing, and advanced reclamation techniques.

Ultimately, foundation design is a synthesis of geotechnical insight, structural engineering, construction practicality, regulatory compliance, economic considerations, and social responsibility, continuing to evolve with new technologies, materials, and field data.

REFERENCES

Berezantsev, V.G., Khristoforov, V.S. and Golubkov, V.N. (1961) *Load Bearing Capacity and Deformation of Piled Foundation*. Proceedings of the 5th International Conference on Soil Mechanics and Foundation Engineering, Vol. II, pp. 11–15.

Bowles, J.E. (1996) Foundation Analysis and Design, 5th Edition. The McGraw-Hill Companies, Inc.

Broms, B.B. (1964) Lateral Resistance of Piles in Cohesionless Soils. Journal of the Soil Mechanics and Foundations Division, ASCE, 90(SM3), 123–156.

British Standard Institution (2000) Eurocode 7: Geotechnical Design – Design Assisted by Field Testing. British Standard Institution.

British Standard Institution (2004) Eurocode 7: Geotechnical Design – General Rules. British Standard Institution.

Buildings Department (2017) Code of Practice for Foundations 2017. Buildings Department, Hong Kong SAR Government.

Canadian Geotechnical Society (2006) Canadian Foundation Engineering Manual, 4th Edition. Canadian Geotechnical Society.

Chan, G., Lui, J.Y.H., Lam, K., Yin, K.K., Law, C.W., Lau, R., Chan, A. and Hasle, R. (2004) *Shaft Grouted Friction Barrette Piles for a Super High-Rise Building*. Proceedings of New Perspectives in the Design and Construction of Foundation Structures, HKIE, pp. 83–98.

Cheng, Y.M. and Au, S.K. (2005) Slip Line Solution of Bearing Capacity Problems with Inclined Ground. Canadian Geotechnical Journal, 42, pp. 1232–1241.

Cheng, Y.M. and Law, C.W. (2008) Development of a New and Efficient Thick Plate Element. Structural Engineering and Mechanics, 29(3), 327–354.

Cheng, Y.M., Law, C.W. and Liu, L.L. (2024) Analysis, Design and Construction of Foundations, 2nd Edition. CRC Press.

Cheng, Y.M. and Li, N. (2017) Equivalence between Bearing Capacity, Lateral Earth Pressure and Slope Stability Problems by Slip-Line and Extremum Limit Equilibrium Methods. International Journal of Geomechanics, ASCE, 17(12), 04017113.

Cheng, Y.M., Lu, H.L. and Sun, J. (1996) Analysis of Nonlinear Creep in Geotechnical Engineering. Chinese Journal of Geotechnical Engineering, 18(5), 1–13.

Geotechnical Engineering Office (GEO) (2006) Foundation Design and Construction, GEO Publication No. 1/2006. Hong Kong SAR Government.

Lui, S.P.Y., Cheung, S.P.Y. and Chan, A.K.C. (1993) *Pressure Grouted Minipiles for a 12-Storey Residential Building at the Mid-Levels Scheduled Area in Hong Kong*. Proceedings of the Second International Conference on Soft Soil Engineering, Guangzhou, pp. 419–424.

Plumbridge, G.D., Littlechild, B.D., Hill, S.J. and Pratt, M. (2000) *Full-Scale Shaft Grouted Piles and Barrettes in Hong Kong – A First*. Proceedings of the Nineteenth Annual Seminar of the Geotechnical Division of the HKIE, Hong Kong, pp. 157–166.

Poulos, H.G. and Davis, E.H. (1980) Pile Foundation Analysis and Design. John Wiley & Sons.

Randolph, M.F. (1987) PIGLET: A Computer Program for the Analysis and Design of Pile Groups, Report No. GEO 87036. Department of Civil Engineering, University of Western Australia, Nedlands.

Sze, J.W.C., Lam, A.K.M., Pappin, J.W. and Chan, K.M. (2007) Design and Construction of Shaft-Grouted Friction Barrette in Tung Chung Designated Area. Proceedings of the HKIE Geotechnical Division Annual Seminar 2007, HKIE, pp. 299–304.

Terzaghi, K. (1943) Theoretical Soil Mechanics. John Wiley.

Terzaghi, K. (1955) Evaluation of Coefficient of Subgrade Reaction. Géotechnique, 5(4), 297–326.

Vesic, A.S. (1973) Analysis of Ultimate Loads for Shallow Foundations. Journal of the Soil Mechanics and Foundations Division, ASCE, 99(SM1), pp. 45–73.

Appendix A Worked Examples for Moment Distribution Method and Plane Frame Analysis by Hand Calculations

WORKED EXAMPLE A1 – THE CONTINUOUS BEAM COMPRISING 3 SPANS TB1–TB2–TB3 SHOWN IN FIGURE A1 IS TO BE ANALYZED BY THE SIMPLIFIED MOMENT DISTRIBUTION METHOD

Beam sizes are marked as depth x width in Figure A.A1.

All slabs are 150 mm thick with finishes 1.5 kN/m². So dead loads on slab is $0.15\times24.5+1.5=5.175\text{kN/m}^2$; live load is 2.5 kN/m².

Load of partition on TB3 is $0.1\times15\times(3-0.75)=3.375\text{kN/m}$.

The loading diagram is Figure A.A2, with magnitudes of loads determined in Table A1.

It is required to determine the fixed-end moments. They can be worked out by first principles such as the virtual work principle or the moment area method. A more detailed treatment about the moment distribution method is given by Volokh (2002). In commercial design, they can be based on formulae and in the pre-computer era by reference to design charts in references including Reynold's *Reinforced Concrete Designer's Handbook* (1988), the relevant charts are extracted in Figure A.A3(a) to A3(d). The formulae are also quoted.

By the formulae listed in Figures A3(a) to (d), the fixed-end moments of Worked Example A1 have been worked out as listed in the following Table A1 as LFEM (beam left end) and RFEM (beam right end), together with the free shears LSH (beam left end) and RSH (beam right end). The free shears are the beam end shears under simply supported conditions.

The stiffness of the beams is determined by $4I/L$ with the E value omitted as all structures are of taken as the same concrete grade. The column K values are $2(4I/H)$ where $H=3$.

The simplified moment distribution method is performed as follows, which follows a span-by-span distribution. The sign convention is taken as positive for clockwise moment and negative for anti-clockwise moment acting on a beam. During "carry over" involving moment carried over from one end to the other, the moment

DOI: 10.1201/9781003636380-15

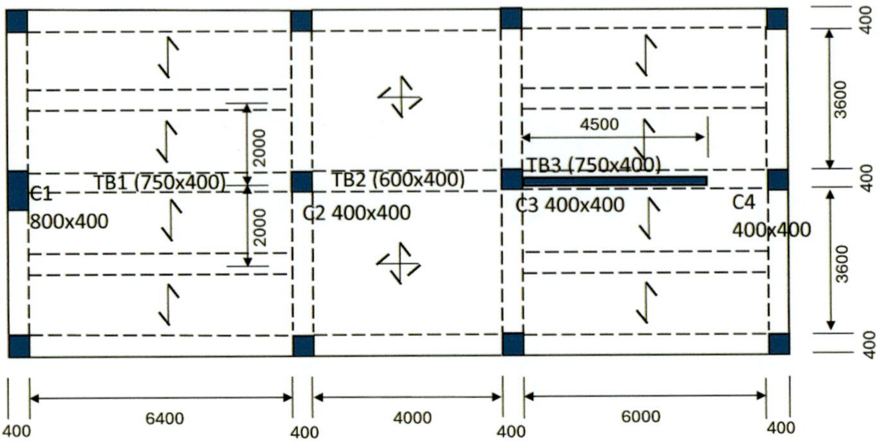

FIGURE A.A1 Framing plan of Worked Example A1.

direction does not change and therefore there is no sign change. Taking span TB2 dead load as an example, the following steps are taken, with reference to Table A2 in which distribution by the Simplified Moment Distribution Method is performed.

(i) The fixed-end moments (FEM) are first carried over to the other ends and summed. So for the left end where the right end moment is carried over to it, it is $-27.476\times0.3113/2-27.476=-31.752$. Similarly for the right end which is $27.476\times0.3192/2+27.476=31.861$

(ii) The left end moment is then distributed to the column and TB1 span by $-(-31.752\times0.2774)=8.808$ and $-(-31.752\times0.4034)=12.809$ respectively. The moment at TB2 left end is the sum of these two.

(iii) The moment of 12.809 at the right end of span TB1 is then carried over to the left end by $12.809/2$. It further distributes to the column C1 by $12.809/2\times0.579=3.708$.

(iv) The process is similar for the right side FEM of TB2 by $-31.861\times0.2706=-8.622$ to C3 and $-31.861\times0.4181=-13.321$ to the left end of TB3 and then $=(-13.321)/2\times0.3929=-2.617$ to the right end of TB3.

(v) For TB1 which is the left end span, the calculation process stops when the left end FEM distributes on the column junction. For the right side FEM, it will be distributed and carried over through the right two spans. The process is similar for the span TB3.

(vi) The final dead load moment at an end will be the sum of all spans.

(vii) The live load moment of the three spans will be distributed similarly to that of the dead load.

(viii) The maximum final live load moment at the left end of a beam is the sum of the moment due to live load on its span and all negative moments contributed by other spans, while the minimum live load moment is the sum of the

TABLE A1

Summary of Loads on Beams TB1–TB3

Beam Mark	DL/LL	Description		LSH (kN)	RSH (kN)	LFEM (kNm)	RFEM (kNm)
TB1	DL	OW 0.40 × 0.75 × 6.8 × 24.5 = 49.98 kN		60.180	60.180	−68.204	68.204
span = 6.8 m		Slab 5.175 × 1× 2× 6.8 = 70.38 kN					
	LL	Slab 2.5 × 1 × 2 × 6.8 = 34 kN		17.000	17.000	−19.267	19.267
TB2	DL	OW 0.40 × 0.6 × 6.8 × 24.5 = 25.87 2kN		12.936	12.936	−4.743	4.743
span = 4.4 m		Slab 5.175 × (4.4 + 0.4) ×2 ÷ 2 × 2 = 49.68 kN		24.840	24.840	−22.732	22.732
			Sum	37.776	37.776	−27.476	27.476
	LL	Slab 2.5 × (4.4 + 0.4) ×2 ÷ 2 ×2 = 24 kN		12.000	12.000	−10.982	10.982
TB3	DL	OW 0.40 × 0.75 × 6.4 × 24.5 = 47.04 kN		56.640	56.640	−60.416	60.416
span = 6.4 m		Slab 5.175 × 1 × 2 × 6.4 = 66.24 kN					
		Parti 3.375 × 4.5 = 15.1875 kN		9.374	5.814	−10.764	8.197
			Sum	66.014	62.454	−71.180	68.613
	LL	Slab 2.5 × 1× 2 × 6.4 = 32 kN		16.000	16.000	−17.067	17.067

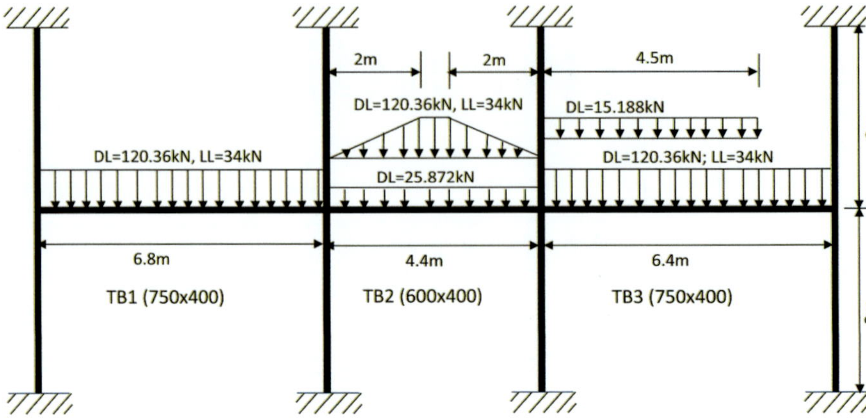

FIGURE A.A2 Loading diagram of TB1–TB2–TB3.

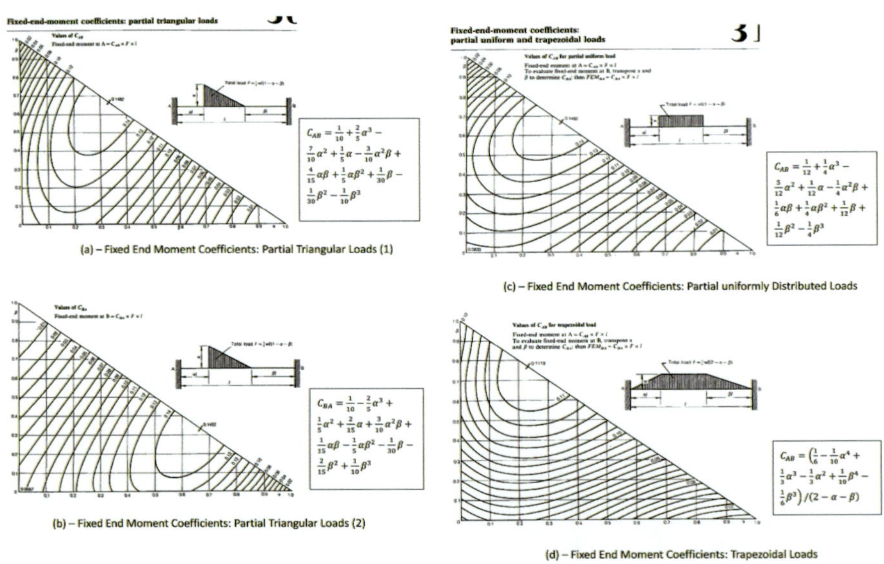

FIGURE A.A3 (a) Fixed-end moment coefficients: partial triangular loads (1). (b) Fixed-end moment coefficients: partial triangular loads (2). (c) Fixed-end moment coefficients: partial uniformly distributed loads. (d) Fixed-end moment coefficients: trapezoidal loads.

moment due to live load on its span and all positive moments contributed by other spans. The moment signs are reversed for the right end of a beam.

(ix) The dead load shears at the ends of each span are the sum of the free shears (calculated by assuming the span is simply supported) plus the elastic shear, which is the sum of the end moments divided by the span length to achieve

TABLE A2
Table for the Simplified Moment Distribution of the Continuous Beam (Approximate Method)

Section	C1	TB1		C2	TB2		C3	TB3		C4
	800 × 400	750 × 400		400 × 400	600 × 400		400 × 400	750 × 400		400 × 400
Span (m)		6.8			4.4			6.4		
K	0.011378	0.008272	0.008272	0.005689	0.006545	0.006545	0.005689	0.008789	0.008789	0.005689
DF	0.5790	0.4210	0.4034	0.2774	0.3192	0.3113	0.2706	0.4181	0.6071	0.3929
DLFEM		−68.204	68.204		−27.476	27.476		−71.180	68.613	
Dist		−13.757	14.357		−4.277	4.385		−20.827	14.880	
Sum		−81.961	82.561		−31.752	31.861		−92.007	83.493	
Span 1		−47.455	49.256	−22.902	−26.353	−9.075	3.566	5.509	1.082	
Span 2		3.708	12.809	8.808	−21.617	21.942	−8.621	−13.321	−2.617	
Span 3		−1.672	−5.777	−3.973	9.750	28.642	24.897	−53.539	32.804	
Sum		−45.420	56.288	−18.067	−38.221	41.509	19.841	−61.351	31.270	
FR Sh		60.180	60.180		37.776	37.776		66.014	62.454	
El Sh		−1.598	1.598		−0.747	0.747		4.700	−4.700	
Sum Sh		58.582	61.778		37.029	38.523		70.714	57.754	
LLFEM		−19.267	19.267		−10.982	10.982		−17.067	17.067	
Dist		−3.886	4.056		−1.709	1.753		−5.181	3.568	
Sum		−23.153	23.322		−12.691	12.735		−22.247	20.634	
Span 1		−13.405	13.914	−6.470	−7.444	−2.564	1.007	1.556	0.306	
Span 2		1.482	5.120	3.521	−8.640	8.770	−3.446	−5.324	−1.046	
Span 3		−0.404	−1.397	−0.961	2.357	6.926	6.020	−12.946	8.107	
Max Mt		−13.810	19.034		−16.085	15.696		−18.270	8.413	
Min Mt		−11.923	12.517		−6.283	6.207		−11.389	7.061	
FR Sh		17.000	17.000		12.000	12.000		16.000	16.000	
El Sh		0.190	1.046		2.245	2.139		1.751	−0.465	
Sum Sh		17.190	18.046		14.245	14.139		17.751	15.535	

TABLE A3
Moment Distribution by Hardy Cross's Method (1930) for dead load case of Worked Example A1 (Exact Method)

	C1	TB1		C2	TB2		C3	TB3		C4
Section	800 × 400	750 × 400		400 × 400	600 × 400		400 × 400	750 x 400		400 × 400
Span (m)		6.8			4.4			6.4		
K	0.011378	0.008272	0.008272	0.005689	0.006545	0.006545	0.005689	0.008789	0.008789	0.005689
DF	0.5790	0.4210	0.4034	0.2774	0.3192	0.3113	0.2706	0.4181	0.6071	0.3929
DLFEM		−68.204	68.204		−27.476	27.476		−71.180	68.613	
Dist	39.490	28.714	−16.430	−11.298	−13.001	13.605	11.826	18.273	−41.655	−26.958
CO		−8.215	14.357		6.803	−6.500		−20.827	9.136	
Dist	4.756	3.458	−8.536	−5.870	−6.754	8.507	7.395	11.426	−5.547	−3.590
CO		−4.268	1.729		4.254	−3.377		−2.773	5.713	
Dist	2.471	1.797	−2.413	−1.660	−1.910	1.915	1.664	2.571	−3.468	−2.245
CO		−1.207	0.898		0.957	−0.955		−1.734	1.286	
Dist	0.699	0.508	−0.749	−0.515	−0.592	0.837	0.728	1.124	−0.781	−0.505
CO		−0.374	0.254		0.419	−0.296		−0.390	0.562	
Dist	0.217	0.158	−0.271	−0.187	−0.215	0.214	0.186	0.287	−0.341	−0.221
CO		−0.136	0.079		0.107	−0.107		−0.171	0.144	
Dist	0.079	0.057	−0.075	−0.051	−0.059	0.087	0.075	0.116	−0.087	−0.056
CO		−0.037	0.029		0.043	−0.030		−0.044	0.058	
Dist	0.022	0.016	−0.029	−0.020	−0.023	0.023	0.020	0.031	−0.035	−0.023
Sum	47.733	−47.733	57.047	−19.600	−37.447	41.397	21.894	−63.291	33.598	−33.598

equilibrium. For the live load shear, the sum of moments is calculated by adding the maximum moment on one end and the minimum moment at the other end for the maximum shear.

A comparison is then carried out by Hardy Cross's method (1930) on the dead load case as shown in Table A3.

Comparison shows that the errors are within 10%.

Once the support moments are analyzed, the span moments can be determined as if the beams are statically determinate structures.

To further minimize computation efforts, there was a practice of eliminating the rotational restraints offered by the interior columns. It was considered acceptable as the moments on both sides of the columns tend to cancel out. As such, the moment distribution table is further simplified as in Table A4.

WORKED EXAMPLE A2 TO DEMONSTRATE THE PLANE FRAME ANALYSIS TO LATERAL LOAD BY HAND CALCULATION (APPROXIMATE METHOD)

The following 3-bay 3-storeys frame in Figure A.A4 under lateral loads is to be analyzed under the assumption that the beams are infinitely stiff. In the frame, the interior column I values are two times that of the exterior columns and the E values are the same for all columns and beams. With S_s as the cumulated floor shear, the shear are $S_s \dfrac{I_c}{2I_c + 2\times 2I_c} = \dfrac{S_s}{6}$ and $S_s \dfrac{2I_c}{2I_c + 2\times 1.5I_c} = \dfrac{S_s}{3}$ on the exterior columns and interior columns respectively. The beams are of the same section so that their stiffnesses are inversely proportional to their lengths.

The procedures of the analysis are described as follows:

(i) Below each of the floors, the cumulated floor shear is distributed onto the columns in accordance with their shear stiffnesses in terms of I/H^3 where H is the height of the column.

(ii) The moment of each column is then calculated by $S\times H/2$ where S is the column shear. This is under the assumption that the column heads or ends are perfectly restrained from rotation, which implies the floor is infinitely stiff as compared with the columns. For example, at the second storey where the cumulated shear is 135 + 45 = 180 kN, the shear distributed onto the exterior columns are $180\times 1/(1+2+2+1) = 180\times 1/6 = 30\text{kN}$ and $180\times 2/6 = 60\text{kN}$ for the interior columns.

(iii) The sum of column moment at a joint will be distributed to the adjoining beam in accordance with the beam's stiffness (4*EI/L* where *L* is the beam span length) to achieve equilibrium. For example, at the third joint (counting from the left) at the storey, where the sum of the column moment (upper and lower columns) is $90 + 120 = 210\text{kNm}$, the moment distributed to the

TABLE A4
Table for the Simplified Moment Distribution of the Continuous Beam (Approximate Method) with Interior Columns as Pinned Supports

	C1	TB1		TB2		TB3		C4
Section	**800 x 400**	**750 x 400**		**600 x 400**		**750 x 400**		**400 x 400**
Span (m)		**6.8**		**4.4**		**6.4**		
K	0.011378	0.008272	0.008272	0.006545	0.006545	0.008789	0.008789	0.005689
DF	0.5790	0.4210	0.5583	0.4417	0.4268	0.5732	0.6071	0.3929
DLFEM		−68.204	68.204	−27.476	27.476	−71.180	68.613	
Dist		−19.039	14.357	−5.863	6.068	−20.827	20.400	
Sum		−87.243	82.561	−33.339	33.544	−92.007	89.013	
Span 1		−50.514	36.467	−36.467	−10.451	10.451	2.053	
Span 2		5.388	18.613	−18.613	19.227	−19.227	−3.777	
Span 3		−3.173	−10.962	10.962	39.269	−39.269	34.973	
Sum		−48.299	44.118	−44.118	48.044	−48.044	33.249	
FR Sh		60.180	60.180	37.776	37.776	66.014	62.454	
El Sh		0.615	−0.615	−0.892	0.892	2.312	−2.312	
Sum Sh		60.795	59.565	36.884	38.668	68.325	60.142	
LLFEM		−19.267	19.267	−10.982	10.982	−17.067	17.067	
Dist		−3.886	4.056	−1.709	1.753	−5.181	3.568	
Sum		−23.153	23.322	−12.691	12.735	−22.247	20.634	
Span 1		−13.405	10.301	−10.301	−2.952	2.952	0.580	
Span 2		2.051	7.085	−7.085	7.299	−7.299	−1.434	
Span 3		−0.767	−2.651	2.651	9.495	−9.495	8.107	
Max Mt		−14.173	17.387	−17.387	16.795	−16.795	8.687	
Min Mt		−11.354	7.651	−4.435	4.347	−6.543	6.673	
FR Sh		17.000	17.000	12.000	12.000	16.000	16.000	
El Sh		0.959	0.887	2.964	2.809	1.581	0.335	
Sum Sh		17.959	17.887	14.964	14.809	17.581	16.335	

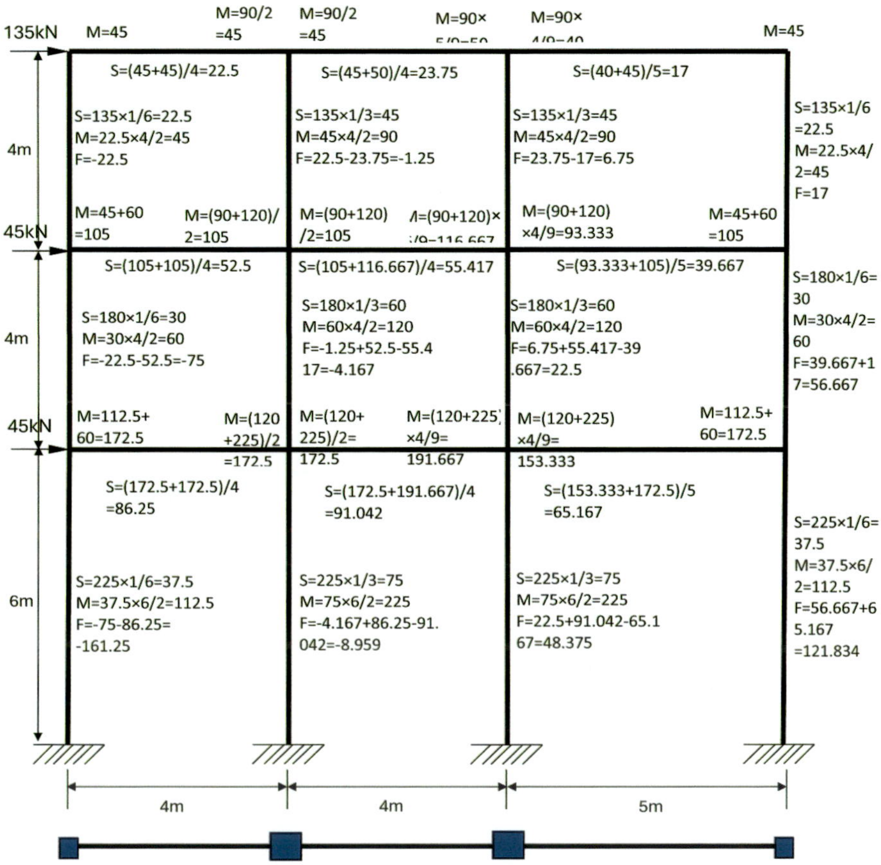

FIGURE A.A4 Plane frame analysis to lateral load (approximate method by hand calculation).

left beam is $210\times1/4\left(1/4+1/5\right)=210\times5/9=116.667\text{kNm}$ and to the right is $90\times1/5\left(1/4+1/5\right)=210\times4/9=93.333\text{kNm}$.

(iv) The beam shear is calculated by the sum of the end moments divided by its span. For example, for the mid beam at the second storey where the end moments are 105 and 116.67, the beam shear is $\left(105+116.67\right)/4=55.417$ kN.

(v) The column axial loads are simply worked out by the summation of the axial loads from above and the net beam shears.

For the Worked Example A2, equilibrium check is carried out at the base level:

Horizontal equilibrium:

Total external load $=135+45+45=225\text{kN}$;

Total column shear $=37.5+75+75+37.5=225\text{kN}$

Vertical equilibrium:

Total external load = 0

Total column vertical load $= -161.25 - 8.959 + 48.375 + 121.384 = 0$

Rotational equilibrium about the base level, taking moment about the base of the left column:

Total external moment $= 135 \times 14 + 45 \times 10 + 45 \times 6 = 2610\text{kNm}$

By column reactions

$= 112.5 + 225 + 225 + 112.5 - 8.959 \times 4 + 48.375 \times 8 + 121.834 \times 13 = 2610\text{kNm}.$

REFERENCES

Hardy Cross (1930) *Analysis of Continuous Frames by Distributing Fixed End Moments.* Transactions of the American Society of Civil Engineers, Papers and Discussions.

Reynolds, E.C. and Steedman, J.C. (1988) Reinforced Concrete Designer's Handbook, 10th Edition. E & FN Spon, Taylor & Francis Group.

Volokh, K.Y. (2002) On Foundations of the Hardy Cross Method. International Journal of Solids and Structures, 39(16), 4197–4200. doi:10.1016/S0020-7683(02)00345-1.

Appendix B Derivation of the Stiffness Matrix of a Strut under Axial Load

The shape function of a strut of length L is

$$v = a_0 + a_1 x + a_2 x^2 + a_3 x^3 \tag{B1}$$

$$\theta_z = \frac{dv}{dx} = a_1 + 2a_2 x + 3a_3 x^2 \tag{B2}$$

where v is the lateral displacement at x from end 1; a_0, a_1, a_2, a_3 are constants depending on the values of v at both ends.

When $x = 0$, $v = v_1$; $\dfrac{\partial v}{\partial x} = \theta_{z1}$; $x = L$, $v = v_2$, $\dfrac{\partial v}{\partial x} = \theta_{z2}$

Solving with $a_0 = v_1$; $a_1 = \theta_{z1}$ and substituting and solving:

$$a_2 = \frac{3v_2 - 3v_1 - 2\theta_{z1}L - \theta_{z2}L}{L^2};\; a_3 = \frac{2v_1 - 2v_2 + 2\theta_{z1}L + \theta_{z2}L}{L^3}$$

$$\therefore v = v_1 + \theta_{z1}x + \frac{3v_2 - 3v_1 - 2\theta_{z1}L - \theta_{z2}L}{L^2}x^2 + \frac{2v_1 - 2v_2 + 2\theta_{z1}L + \theta_{z2}L}{L^3}x^3 \tag{B3}$$

Re-arranging

$$v = \left(1 - 3\frac{x^2}{L^2} + 2\frac{x^3}{L^3}\right)v_1 + \left(x - 2\frac{x^2}{L} + \frac{x^3}{L^2}\right)\theta_{z1} + \left(3\frac{x^2}{L^2} - 2\frac{x^3}{L^3}\right)v_2 + \left(\frac{-x^2}{L} + \frac{x^3}{L^3}\right)\theta_{z2} \tag{B4}$$

If the strut is under an axial load P, the energy functional can be written as

$$\Pi = \frac{1}{2}\int_0^L EI_z\left(\frac{\partial^2 v}{\partial x^2}\right)^2 dx - S_{y1}v_1 - M_{z1}\theta_{z1} - S_{y2}v_2 - M_{z2}\theta_{z2} - \frac{1}{2}P\delta \tag{B5}$$

DOI: 10.1201/9781003636380-16

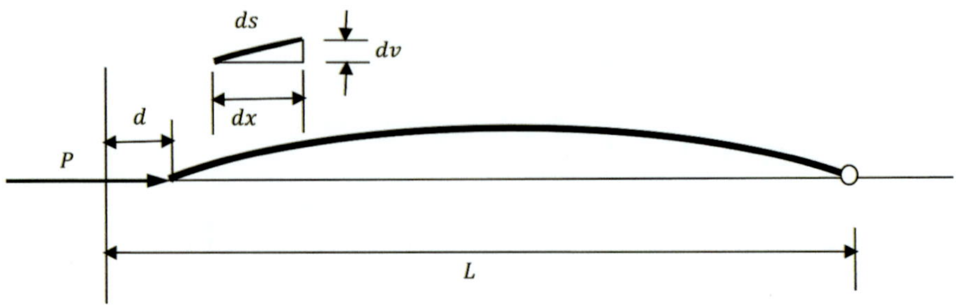

FIGURE A.B1 Shortening of strut under effect of axial load.

where δ is the shortening of the strut under the axial load P so that the work done is $\frac{1}{2}P\delta$.

δ is determined as follows:

$$ds = \sqrt{dx^2 + dv^2} = dx\sqrt{1+\left(\frac{dv}{dx}\right)^2}$$

$$L = \int_0^L ds = \int_\delta^L dx\sqrt{1+\left(\frac{dv}{dx}\right)^2} = \int_0^L dx\sqrt{1+\left(\frac{dv}{dx}\right)^2} - \int_0^\delta dx\sqrt{1+\left(\frac{dv}{dx}\right)^2}$$

$\approx \int_0^L \left(1+\frac{1}{2}\left(\frac{dv}{dx}\right)^2\right)dx - \int_\delta^L \left(1+\frac{1}{2}\left(\frac{dv}{dx}\right)^2\right)dx$ (binomial expansion ignoring orders higher than $\left(\frac{dv}{dx}\right)^2$ which are very small).

$$L = \int_0^L [1]dx + \int_0^L \frac{1}{2}\left(\frac{dv}{dx}\right)^2 dx - \int_0^\delta [1]dx - \int_0^\delta \frac{1}{2}\left(\frac{dv}{dx}\right)^2 dx$$

$$L = L + \int_0^L \frac{1}{2}\left(\frac{dv}{dx}\right)^2 dx - \delta - \int_0^\delta \frac{1}{2}\left(\frac{dv}{dx}\right)^2 dx$$

$$\delta = \int_0^L \frac{1}{2}\left(\frac{dv}{dx}\right)^2 dx - \int_0^\delta \frac{1}{2}\left(\frac{dv}{dx}\right)^2 dx \tag{B6}$$

δ is taken as $\int_0^L \frac{1}{2}\left(\frac{dv}{dx}\right)^2 dx$ as $\int_0^\delta \frac{1}{2}\left(\frac{dv}{dx}\right)^2 dx$ is small.

$$\delta = \int_0^L \frac{1}{2}\left(\frac{dv}{dx}\right)^2 dx \tag{B7}$$

Substituting into Equation B.5, the energy functional is

$$\Pi = \frac{1}{2}\int_0^L EI_z\left(\frac{\partial^2 v}{\partial x^2}\right)^2 dx - S_{y1}v_1 - M_{z1}\theta_{z1} - S_{y2}v_2 - M_{z2}\theta_{z2} - \frac{P}{2}\int_0^L\left(\frac{dv}{dx}\right)^2 dx \quad \text{(B8)}$$

$$\frac{1}{2}\int_0^L EI_z\left(\frac{\partial^2 v}{\partial x^2}\right)^2 dx = \frac{2EI_z}{L^3}\left[L^2\left(\theta_{z1}{}^2 + \theta_{z1}\theta_{z2} + \theta_{z2}{}^2\right) + 3L\left(v_1 - v_2\right)\left(\theta_{z1} + \theta_{z2}\right) + 3\left(v_1 - v_2\right)^2\right] \quad \text{(B9)}$$

and

$$\frac{P}{2}\int_0^L\left(\frac{dv}{dx}\right)^2 dx = \frac{P}{30L}\left[2L^2\left(\theta_{z1}{}^2 + \theta_{z2}{}^2\right) + 18\left(v_1{}^2 + v_2{}^2\right) + 3\left(v_1 - v_2\right)\left(\theta_{z1} + \theta_{z2}\right) - L^2\theta_{z1}\theta_{z2} - 36v_1v_2\right] \quad \text{(B10)}$$

Substituting Equations B.9 and B.10 into Equation B.8. For Π to be minimum with respect to $v_1, v_2, \theta_{z1}, \theta_{z2}$, we list

$$\frac{\partial \Pi}{\partial v_1} = 0$$

$$\Rightarrow \frac{12EI_z}{L^3}v_1 + \frac{6EI_z}{L^2}\theta_{z1} - \frac{12EI_z}{L^3}v_2 + \frac{6EI_z}{L^2}\theta_{z2} - P\left(\frac{6}{5L}v_1 + \frac{1}{10}\theta_{z1} - \frac{6}{5L}v_2 + \frac{1}{10}\theta_{z2}\right) = S_{y1} \quad \text{(B11)}$$

$$\frac{\partial \Pi}{\partial \theta_{z1}} = 0$$

$$\Rightarrow \frac{6EI_z}{L^2}v_1 + \frac{4EI_z}{L}\theta_{z1} - \frac{6EI_z}{L^2}v_1 + \frac{2EI_z}{L}\theta_{z2} - P\left(\frac{1}{10}v_1 + \frac{2L}{15}\theta_{z1} - \frac{1}{10}v_2 - \frac{L}{30}\theta_{z2}\right) = M_{x1} \quad \text{(B12)}$$

$$\frac{\partial \Pi}{\partial v_2} = 0$$

$$\Rightarrow \frac{-12EI_z}{L^3}v_1 - \frac{6EI_z}{L^2}\theta_{z1} + \frac{12EI_z}{L^3}v_2 - \frac{6EI_z}{L^2}\theta_{z2} - P\left(\frac{-6}{5L}v_1 - \frac{1}{10}\theta_{z1} + \frac{6}{5L}v_2 - \frac{1}{10}\theta_{z2}\right) = S_{y1} \quad \text{(B13)}$$

$$\frac{\partial \Pi}{\partial \theta_{z2}} = 0$$

$$\Rightarrow \frac{6EI_z}{L^2}v_1 + \frac{2EI_z}{L}\theta_{z1} - \frac{6EI_z}{L^2}v_1 + \frac{2EI_z}{L}\theta_{z2} - P\left(\frac{1}{10}v_1 - \frac{L}{30}\theta_{z1} - \frac{1}{10}v_2 + \frac{2L}{15}\theta_{z2}\right) = M_{x1} \quad \text{(B14)}$$

With inclusion of the axial load and axial deformation $\left(u_1 - u_2\right)\frac{EA}{L} = F_x$, we can write in matrix form as

$$\left(k_e + Pk_g\right)\Delta = F \tag{B15}$$

where

$$k_e = \begin{bmatrix} EA/L & 0 & 0 & -EA/L & 0 & 0 \\ 0 & 12EI_z/L^3 & 6EI_z/L^2 & 0 & -12EI_z/L^3 & 6EI_z/L^2 \\ 0 & 6EI_z/L^2 & 4EI_z/L & 0 & -6EI_z/L^2 & 2EI_z/L \\ -EA/L & 0 & 0 & EA/L & 0 & 0 \\ 0 & -12EI_z/L^3 & -6EI_z/L^2 & 0 & 12EI_z/L^3 & -6EI_z/L^2 \\ 0 & 6EI_z/L^2 & 2EI_z/L & 0 & -6EI_z/L^2 & 4EI_z/L \end{bmatrix} \tag{B16}$$

$$k_g = \frac{-1}{30L}\begin{bmatrix} 0 & 0 & 0 & 0 & 0 & 0 \\ 0 & 36 & 3L & 0 & -36 & 3L \\ 0 & 3L & 4L^2 & 0 & -3L & -L^2 \\ 0 & 0 & 0 & 0 & 0 & 0 \\ 0 & -36 & -3L & 0 & 36 & -3L \\ 0 & 3L & -L^2 & 0 & -3L & 4L^2 \end{bmatrix} \tag{B17}$$

$\boldsymbol{k}_e$ is the elastic stiffness matrix we are familiar with while k_g is the geometric matrix used to modify the stiffness of the strut under axial load. Basically the stiffness of the strut represented by $\left(\boldsymbol{k}_e + P\boldsymbol{k}_g\right)$ will be reduced when P is positive (in compression) and increased verse versa (Belytschko et al. 2000).

$\boldsymbol{k}_e$ and $\boldsymbol{k}_g$ can be similarly expanded for a strut in 3-dimensions with six degrees of freedom. The 3-dimensional versions of k_e and k_g are listed as follows:

$$k_e = \begin{bmatrix}
\frac{EA}{L} & 0 & 0 & 0 & 0 & 0 & \frac{-EA}{L} & 0 & 0 & 0 & 0 & 0 \\
0 & \frac{12EI_z}{L^3} & 0 & 0 & 0 & \frac{6EI_z}{L^2} & 0 & \frac{-12EI_z}{L^3} & 0 & 0 & 0 & \frac{6EI_z}{L^2} \\
0 & 0 & \frac{12EI_y}{L^3} & 0 & \frac{6EI_y}{L^2} & 0 & 0 & 0 & \frac{-12EI_y}{L^3} & 0 & \frac{-6EI_y}{L^2} & 0 \\
0 & 0 & 0 & \frac{GJ}{L} & 0 & 0 & 0 & 0 & 0 & \frac{-GJ}{L} & 0 & 0 \\
0 & 0 & \frac{-6EI_y}{L^2} & 0 & \frac{4EI_y}{L} & 0 & 0 & 0 & \frac{6EI_y}{L^2} & 0 & \frac{2EI_y}{L} & 0 \\
0 & \frac{6EI_z}{L^2} & 0 & 0 & 0 & \frac{4EI_z}{L} & 0 & \frac{-6EI_z}{L^2} & 0 & 0 & 0 & \frac{2EI_z}{L} \\
\frac{-EA}{L} & 0 & 0 & 0 & 0 & 0 & \frac{EA}{L} & 0 & 0 & 0 & 0 & 0 \\
0 & \frac{-12EI_z}{L^3} & 0 & 0 & 0 & \frac{-6EI_z}{L^2} & 0 & \frac{12EI_z}{L^3} & 0 & 0 & 0 & \frac{-6EI_z}{L^2} \\
0 & 0 & \frac{-12EI_y}{L^3} & 0 & \frac{-6EI_y}{L^2} & 0 & 0 & 0 & \frac{12EI_y}{L^3} & 0 & \frac{6EI_y}{L^2} & 0 \\
0 & 0 & 0 & \frac{-GJ}{L} & 0 & 0 & 0 & 0 & 0 & \frac{GJ}{L} & 0 & 0 \\
0 & 0 & \frac{-6EI_y}{L^2} & 0 & \frac{2EI_y}{L} & 0 & 0 & 0 & \frac{6EI_y}{L^2} & 0 & \frac{4EI_y}{L} & 0 \\
0 & \frac{-6EI_z}{L^2} & 0 & 0 & 0 & \frac{2EI_z}{L} & 0 & \frac{-6EI_z}{L^2} & 0 & 0 & 0 & \frac{4EI_z}{L}
\end{bmatrix} \tag{B18}$$

$$\boldsymbol{k_g} = \frac{-1}{30L}\begin{bmatrix}
0 & 0 & 0 & 0 & 0 & 0 & 0 & 0 & 0 & 0 & 0 & 0 \\
0 & 36 & 0 & 0 & 0 & 3L & 0 & -36 & 0 & 0 & 0 & 3L \\
0 & 0 & 36 & 0 & -3L & 0 & 0 & 0 & -36 & 0 & -3L & 0 \\
0 & 0 & 0 & 0 & 0 & 0 & 0 & 0 & 0 & 0 & 0 & 0 \\
0 & 0 & -3L & 0 & 4L^2 & 0 & 0 & 0 & 3L & 0 & -L^2 & 0 \\
0 & 3L & 0 & 0 & 0 & 4L^2 & 0 & -3L & 0 & 0 & 0 & -L^2 \\
0 & 0 & 0 & 0 & 0 & 0 & 0 & 0 & 0 & 0 & 0 & 0 \\
0 & -36 & 0 & 0 & 0 & -3L & 0 & 36 & 0 & 0 & 0 & -3L \\
0 & 0 & -36 & 0 & 3L & 0 & 0 & 0 & 36 & 0 & 3L & 0 \\
0 & 0 & 0 & 0 & 0 & 0 & 0 & 0 & 0 & 0 & 0 & 0 \\
0 & 0 & -3L & 0 & -L^2 & 0 & 0 & 0 & 3L & 0 & 4L^2 & 0 \\
0 & 3L & 0 & 0 & 0 & -L^2 & 0 & -3L & 0 & 0 & 0 & 4L^2
\end{bmatrix} \tag{B19}$$

$\boldsymbol{k}_e$ and $\boldsymbol{k}_g$ can also be similarly transformed when the strut is at certain orientations to the X, Y, and Z axes which are to be used with connection of the displacement and force matrices Δ and $\boldsymbol{F}$ as shown:

$$\Delta = \begin{bmatrix} u_1 & v_1 & w_1 & \theta_{x1} & \theta_{y1} & \theta_{z1} & u_2 & v_2 & w_2 & \theta_{x2} & \theta_{y2} & \theta_{z2} \end{bmatrix}^T$$

$$\boldsymbol{F} = \begin{bmatrix} P_1 & S_{x1} & S_{y1} & M_{x1} & M_{y1} & M_{z1} & P_2 & S_{x2} & S_{y2} & M_{x2} & M_{y2} & M_{z2} \end{bmatrix}^T$$

REFERENCE

Belytschko, T., Liu, W.K., Moran, B. and Elkhodary, K. (2000) *Nonlinear Finite Elements for Continua and Structures*. Hoboken: John Wiley & Sons.

Appendix C Structural Analysis for a Beam Member under the Formation of Plastic Hinge

With the development of a plastic hinge during the analysis, the analysis for a beam member needs to be modified, which will be discussed in this section. The underlying principle and approach is

(1) The analysis is carried out in a stepwise manner. First-order elastic analysis is first carried out to identify sections where the working moments would exceed the respective pre-determined elastic limits termed plastic moments M_p.
(2) At a section where the plastic moment has been exceeded, the stiffness at the section is turned to "zero". Consider a 2-dimensional strut member whose initial elastic stiffness is as follows:

$$\boldsymbol{k}_e = \begin{bmatrix} \frac{AE}{L} & 0 & 0 & \frac{-AE}{L} & 0 & 0 \\ 0 & \frac{12EI}{L^3} & \frac{6EI}{L^2} & 0 & \frac{-12EI}{L^3} & \frac{6EI}{L^2} \\ 0 & \frac{6EI}{L^2} & \frac{4EI}{L} & 0 & \frac{-6EI}{L^2} & \frac{2EI}{L} \\ \frac{-AE}{L} & 0 & 0 & \frac{AE}{L} & 0 & 0 \\ 0 & \frac{-12EI}{L^3} & \frac{-6EI}{L^2} & 0 & \frac{12EI}{L^3} & \frac{-6EI}{L^2} \\ 0 & \frac{6EI}{L^2} & \frac{2EI}{L} & 0 & \frac{-6EI}{L^2} & \frac{4EI}{L} \end{bmatrix} \tag{C1}$$

where E is the elastic modulus, A is the cross-section area, I is the second moment of area of the section, and L is the length.

If a plastic hinge is formed at either end or both of a member, by Aktaş E. and Korkmaz A. (2007), its stiffness will be modified as follows:

DOI: 10.1201/9781003636380-17

$$\boldsymbol{k}_{p1} = \begin{bmatrix} \frac{AE}{L} & 0 & 0 & \frac{-AE}{L} & 0 & 0 \\ 0 & \frac{3EI}{L^3} & 0 & 0 & \frac{-3EI}{L^3} & \frac{3EI}{L^2} \\ 0 & 0 & 0 & 0 & 0 & 0 \\ \frac{-AE}{L} & 0 & 0 & \frac{AE}{L} & 0 & 0 \\ 0 & \frac{-3EI}{L^3} & 0 & 0 & \frac{3EI}{L^3} & \frac{-3EI}{L^2} \\ 0 & \frac{3EI}{L^2} & 0 & 0 & \frac{-3EI}{L^2} & \frac{3EI}{L} \end{bmatrix} \tag{C2}$$

$$\boldsymbol{k}_{p2} = \begin{bmatrix} \frac{AE}{L} & 0 & 0 & \frac{-AE}{L} & 0 & 0 \\ 0 & \frac{3EI}{L^3} & \frac{3EI}{L^2} & 0 & \frac{-3EI}{L^3} & 0 \\ 0 & \frac{3EI}{L^2} & \frac{3EI}{L} & 0 & \frac{-3EI}{L^2} & 0 \\ \frac{-AE}{L} & 0 & 0 & \frac{AE}{L} & 0 & 0 \\ 0 & \frac{-3EI}{L^3} & \frac{-3EI}{L^2} & 0 & \frac{3EI}{L^3} & 0 \\ 0 & 0 & 0 & 0 & 0 & 0 \end{bmatrix} \tag{C3}$$

If both ends are pinned, the element being an axial load member. That is, it possesses no bending and shear stiffness. The stiffness matrix will contain only the AE/L entries with all others equal to zero.

$$\boldsymbol{k}_{p0} = \begin{bmatrix} \frac{AE}{L} & 0 & 0 & \frac{-AE}{L} & 0 & 0 \\ 0 & 0 & 0 & 0 & 0 & 0 \\ 0 & 0 & 0 & 0 & 0 & 0 \\ \frac{-AE}{L} & 0 & 0 & \frac{AE}{L} & 0 & 0 \\ 0 & 0 & 0 & 0 & 0 & 0 \\ 0 & 0 & 0 & 0 & 0 & 0 \end{bmatrix} \tag{C4}$$

(3) A factor $R_L = M_p / M_e$ (where M_e is the moment obtained from elastic analysis at the section exceeding M_p) is multiplied to the applied load $\boldsymbol{P}$ such that the section moment becomes M_p.
(4) The remaining loads are then applied to the structure modified by $\boldsymbol{k}_{p1}$ and/or $\boldsymbol{k}_{p2}$ as "plastic analysis".

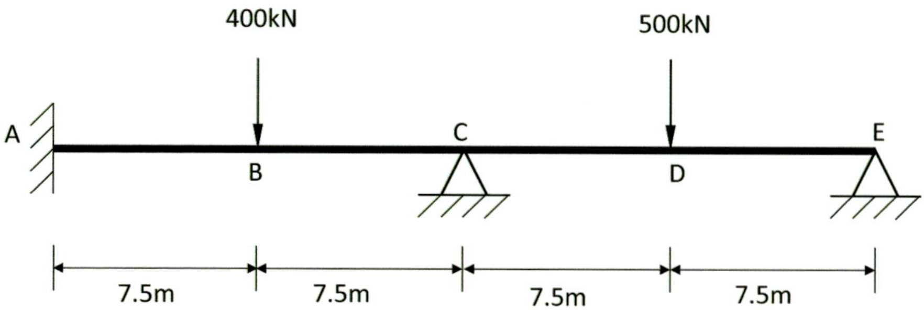

FIGURE A.C1 Continuous beam in Worked Example C1.

(5) If there are still moments exceeding M_p after the plastic analysis, repeat step 1 again until no further exceedance over M_p. There may be instability if too many sections moments exceeding M_p. The final results of the summation of that of the two models.

<u>Worked Example C1</u>

Consider a continuous reinforced concrete beam ABCDE under two point loads as shown in Figure A.C1. The own weights of the beams are ignored. The beams are of section 500(w) × 600(d). The elastic modulus of concrete is $E = 26{,}400{,}000\,\text{kN/m}^2$, $I = 0.5 \times 0.6^3 / 12 = 0.009\text{m}^4$. $M_p = 1260\text{kNm}$

The elastic stiffness matrix of the continuous beam of four spans (equal span of $L = 7.5\text{m}$) is constructed and assembled as follows. The axial deformation has been ignored as the beam does not resist axial load.

$$\boldsymbol{k}_e = \begin{bmatrix} \frac{12EI}{L^3} & \frac{6EI}{L^2} & \frac{-12EI}{L^3} & \frac{6EI}{L^2} & 0 & 0 & 0 & 0 & 0 & 0 \\ \frac{6EI}{L^2} & \frac{4EI}{L} & \frac{6EI}{L^2} & \frac{2EI}{L} & 0 & 0 & 0 & 0 & 0 & 0 \\ \frac{-12EI}{L^3} & \frac{-6EI}{L^2} & \frac{24EI}{L^3} & 0 & \frac{-12EI}{L^3} & \frac{6EI}{L^2} & 0 & 0 & 0 & 0 \\ \frac{6EI}{L^2} & \frac{2EI}{L} & 0 & \frac{8EI}{L} & \frac{-6EI}{L^2} & \frac{2EI}{L} & 0 & 0 & 0 & 0 \\ 0 & 0 & \frac{-12EI}{L^3} & \frac{-6EI}{L^2} & \frac{24EI}{L^3} & 0 & \frac{-12EI}{L^3} & \frac{6EI}{L^2} & 0 & 0 \\ 0 & 0 & \frac{6EI}{L^2} & \frac{2EI}{L} & 0 & \frac{8EI}{L} & \frac{-6EI}{L^2} & \frac{2EI}{L} & 0 & 0 \\ 0 & 0 & 0 & 0 & \frac{-12EI}{L^3} & \frac{-6EI}{L^2} & \frac{24EI}{L^3} & 0 & \frac{-12EI}{L^3} & \frac{6EI}{L^2} \\ 0 & 0 & 0 & 0 & \frac{6EI}{L^2} & \frac{2EI}{L} & 0 & \frac{8EI}{L} & \frac{-6EI}{L^2} & \frac{2EI}{L} \\ 0 & 0 & 0 & 0 & 0 & 0 & \frac{-12EI}{L^3} & \frac{-6EI}{L^2} & \frac{12EI}{L^3} & \frac{-6EI}{L^2} \\ 0 & 0 & 0 & 0 & 0 & 0 & \frac{6EI}{L^2} & \frac{2EI}{L} & \frac{-6EI}{L^2} & \frac{8EI}{L} \end{bmatrix} \tag{C5}$$

The stiffness matrix is modified to the support conditions as follows:

$$
\boldsymbol{S_e} = \begin{bmatrix}
1 & 0 & 0 & 0 & 0 & 0 & 0 & 0 & 0 & 0 \\
0 & 1 & 0 & 0 & 0 & 0 & 0 & 0 & 0 & 0 \\
0 & 0 & \frac{24EI}{L^3} & 0 & 0 & \frac{6EI}{L^2} & 0 & 0 & 0 & 0 \\
0 & 0 & 0 & \frac{8EI}{L} & 0 & \frac{2EI}{L} & 0 & 0 & 0 & 0 \\
0 & 0 & 0 & 0 & 1 & 0 & 0 & 0 & 0 & 0 \\
0 & 0 & \frac{6EI}{L^2} & \frac{2EI}{L} & 0 & \frac{8EI}{L} & \frac{-6EI}{L^2} & \frac{2EI}{L} & 0 & 0 \\
0 & 0 & 0 & 0 & 0 & \frac{-6EI}{L^2} & \frac{24EI}{L^3} & 0 & 0 & \frac{6EI}{L^2} \\
0 & 0 & 0 & 0 & 0 & \frac{2EI}{L^2} & 0 & \frac{8EI}{L} & 0 & \frac{2EI}{L} \\
0 & 0 & 0 & 0 & 0 & 0 & 0 & 0 & 1 & 0 \\
0 & 0 & 0 & 0 & 0 & 0 & \frac{6EI}{L^2} & \frac{2EI}{L} & 0 & \frac{8EI}{L}
\end{bmatrix} \quad \text{(C6)}
$$

$$
= \begin{bmatrix}
1 & 0 & 0 & 0 & 0 & 0 & 0 & 0 & 0 & 0 \\
0 & 1 & 0 & 0 & 0 & 0 & 0 & 0 & 0 & 0 \\
0 & 0 & 13516.8 & 0 & 0 & 25344 & 0 & 0 & 0 & 0 \\
0 & 0 & 0 & 253440 & 0 & 63360 & 0 & 0 & 0 & 0 \\
0 & 0 & 0 & 0 & 1 & 0 & 0 & 0 & 0 & 0 \\
0 & 0 & 25344 & 63360 & 0 & 253440 & -25344 & 63360 & 0 & 0 \\
0 & 0 & 0 & 0 & 0 & -25344 & 13516.8 & 0 & 0 & 25344 \\
0 & 0 & 0 & 0 & 0 & 63360 & 0 & 253440 & 0 & 63360 \\
0 & 0 & 0 & 0 & 0 & 0 & 0 & 0 & 1 & 0 \\
0 & 0 & 0 & 0 & 0 & 0 & 25344 & 63360 & 0 & 126720
\end{bmatrix}
$$

By $\boldsymbol{S_e}.\Delta = \boldsymbol{P}$

where $\boldsymbol{P} = \begin{bmatrix} 0 & 0 & 400 & 0 & 0 & 0 & 500 & 0 & 0 & 0 \end{bmatrix}^T$

Solving

$$\Delta = \begin{bmatrix} 0 & 0 & -0.0184995 & -0.0014796 & 0 & 0.0059186 & 0.0813802 & 0.0029593 & 0 & -0.0177557 \end{bmatrix}^T$$

The individual stiffness matrices of the four spans are identical at

$$\boldsymbol{k}_e = \begin{bmatrix} \frac{12EI}{L^3} & \frac{6EI}{L^2} & \frac{-12EI}{L^3} & \frac{6EI}{L^2} \\ \frac{6EI}{L^2} & \frac{4EI}{L} & \frac{-6EI}{L^2} & \frac{2EI}{L} \\ \frac{-12EI}{L^3} & \frac{-6EI}{L^2} & \frac{12EI}{L^3} & \frac{-6EI}{L^2} \\ \frac{6EI}{L^2} & \frac{2EI}{L} & \frac{-6EI}{L^2} & \frac{4EI}{L} \end{bmatrix} = \begin{bmatrix} 6758.4 & 25344 & -6758.4 & 25344 \\ 25344 & 126720 & -25344 & 63360 \\ -6758.4 & -25344 & 6758.4 & -25344 \\ 25344 & 63360 & -25344 & 126720 \end{bmatrix} \quad \text{(C7)}$$

Multiplying $\boldsymbol{k}_e$ to its nodal displacement gives the internal forces of the four spans

$$\text{Span AB: } \begin{bmatrix} 6758.4 & 25344 & -6758.4 & 25344 \\ 25344 & 126720 & -25344 & 63360 \\ -6758.4 & -25344 & 6758.4 & -25344 \\ 25344 & 63360 & -25344 & 126720 \end{bmatrix} \begin{bmatrix} 0 \\ 0 \\ -0.0184995 \\ -0.0014796 \end{bmatrix} = \begin{bmatrix} -162.5 \\ -562.5 \\ 162.5 \\ -656.25 \end{bmatrix}$$

$$\text{Span BC: } \begin{bmatrix} 6758.4 & 25344 & -6758.4 & 25344 \\ 25344 & 126720 & -25344 & 63360 \\ -6758.4 & -25344 & 6758.4 & -25344 \\ 25344 & 63360 & -25344 & 126720 \end{bmatrix} \begin{bmatrix} -0.0184995 \\ -0.0014796 \\ 0 \\ 0.0059186 \end{bmatrix} = \begin{bmatrix} 237.5 \\ 656.25 \\ -237.5 \\ 1125 \end{bmatrix}$$

$$\text{Span CD: } \begin{bmatrix} 6758.4 & 25344 & -6758.4 & 25344 \\ 25344 & 126720 & -25344 & 63360 \\ -6758.4 & -25344 & 6758.4 & -25344 \\ 25344 & 63360 & -25344 & 126720 \end{bmatrix} \begin{bmatrix} 0 \\ 0.0059186 \\ 0.0813802 \\ 0.0029593 \end{bmatrix} = \begin{bmatrix} -325 \\ -1125 \\ 325 \\ -1312.5 \end{bmatrix}$$

$$\text{Span DE: } \begin{bmatrix} 6758.4 & 25344 & -6758.4 & 25344 \\ 25344 & 126720 & 25344 & 63360 \\ -6758.4 & -25344 & 6758.4 & -25344 \\ 25344 & 63360 & -25344 & 126720 \end{bmatrix} \begin{bmatrix} 0.0813802 \\ 0.0029593 \\ 0 \\ -0.0177557 \end{bmatrix} = \begin{bmatrix} 175 \\ 1312.5 \\ -17.5 \\ 0 \end{bmatrix}$$

Plot of the bending moment diagram is as follows in Figure A.C2:

So, only the moment at D of $M_e = 1312.5\text{kNm}$ has exceeded $M_p = 1260\text{kNm}$ where a plastic hinge will be formed.

The factor $R_L = M_p / M_e = 1260/1312.5 = 0.96$ is applied to reduce the applied load so that M_p will not be exceeded. By elastic analysis, the member forces (and also the deflections) all become $R_L = 0.96$ of their original values as

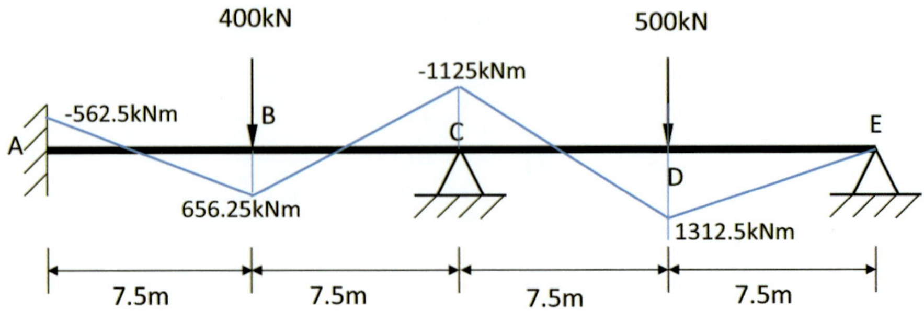

FIGURE A.C2 Bending moment diagram at full load and elastic stage.

$$\text{Span AB: } F_{RAB} = 0.96\begin{bmatrix} -162.5 \\ -562.5 \\ 162.5 \\ -565.25 \end{bmatrix} = \begin{bmatrix} -156 \\ -540 \\ 156 \\ -630 \end{bmatrix};$$

$$\text{Span BC: } F_{RBC} = 0.96\begin{bmatrix} 237.5 \\ 656.25 \\ -237.5 \\ 1125 \end{bmatrix} = \begin{bmatrix} 228 \\ 630 \\ -228 \\ 1080 \end{bmatrix}$$

$$\text{Span CD: } F_{RCD} = 0.96\begin{bmatrix} -325 \\ -1125 \\ 325 \\ -1312.5 \end{bmatrix} = \begin{bmatrix} -312 \\ -1080 \\ 312 \\ -1260 \end{bmatrix}$$

$$\text{Span DE: } F_{RDE} = 0.96\begin{bmatrix} 175 \\ 1312.5 \\ -175 \\ 0 \end{bmatrix} = \begin{bmatrix} 168 \\ 1260 \\ -168 \\ 0 \end{bmatrix}$$

The bending moments are shown in Figure A.C3 under 0.96 of the applied load.

Node D has reached the plastic state and cannot take up any further moment. Therefore, for addition of the remaining loads, it has to be taken as a pinned connection with the model shown in Figure A.C4.

So the stiffness of Spans CD and DE has to be modified by Equations C3 and C2 respectively as

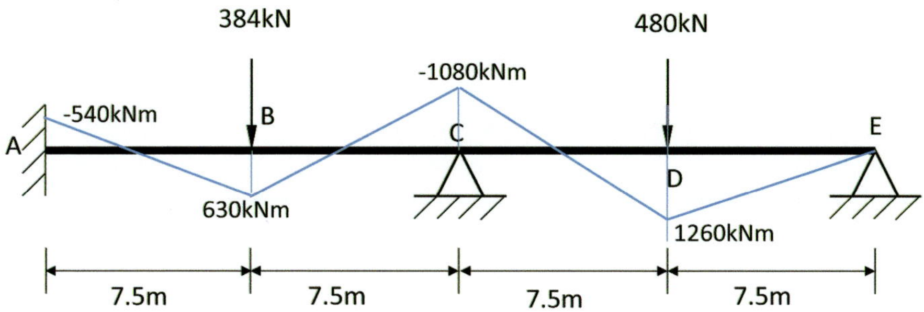

FIGURE A.C3 Introducing a plastic hinge at D and the applied load revised to 0.96 of their original values.

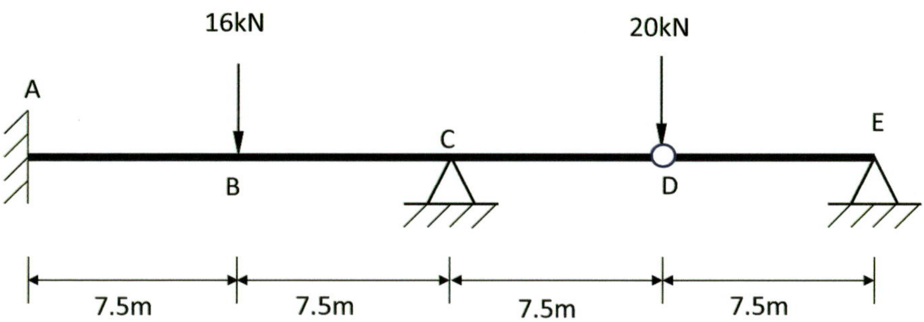

FIGURE A.C4 Plastic model modified from the elastic model.

$$
\boldsymbol{k}_{pCD} = \begin{bmatrix} \frac{3EI}{L^3} & \frac{3EI}{L^2} & \frac{-3EI}{L^3} & 0 \\ \frac{3EI}{L^2} & \frac{3EI}{L} & \frac{-3EI}{L^2} & 0 \\ \frac{-3EI}{L^3} & \frac{-3EI}{L^2} & \frac{3EI}{L^3} & 0 \\ 0 & 0 & 0 & 1 \end{bmatrix} = \begin{bmatrix} 1689.6 & 12672 & -1689.6 & 0 \\ 12672 & 95040 & -12672 & 0 \\ -1689.6 & -12672 & 1689.6 & 0 \\ 0 & 0 & 0 & 1 \end{bmatrix}
$$

$$
\boldsymbol{k}_{pDE} = \begin{bmatrix} \frac{3EI}{L^3} & 0 & \frac{-3EI}{L^3} & \frac{3EI}{L^2} \\ 0 & 1 & 0 & 0 \\ \frac{-3EI}{L^3} & 0 & \frac{3EI}{L^3} & \frac{-3EI}{L^2} \\ \frac{3EI}{L^2} & 0 & \frac{-3EI}{L^2} & \frac{3EI}{L} \end{bmatrix} = \begin{bmatrix} 1689.6 & 0 & -1689.6 & 12672 \\ 0 & 1 & 0 & 0 \\ -1689.6 & 0 & 1689.6 & -12672 \\ 12672 & 0 & -12672 & 95040 \end{bmatrix}
$$

The entries of 1 in cell (6, 6) of $\boldsymbol{k}_{pCD}$ and Cell (3, 3) **of** $\boldsymbol{k}_{pDE}$ are added to prevent the assembled matrix from being singular. Alternatively, the rows and columns of them can be eliminated from matrix assemblage.

The stiffness of the spans AB and BC remain unchanged as they do not possess plastic hinges. The matrices of the 4 spans are assembled in the usual manner. After modification to support and the hinge connection, the plastic matrix becomes:

$$\boldsymbol{S}_p = \begin{bmatrix} 1 & 0 & 0 & 0 & 0 & 0 & 0 & 0 & 0 & 0 \\ 0 & 1 & 0 & 0 & 0 & 0 & 0 & 0 & 0 & 0 \\ 0 & 0 & \frac{24EI}{L^3} & 0 & 0 & \frac{6EI}{L^2} & 0 & 0 & 0 & 0 \\ 0 & 0 & 0 & \frac{8EI}{L} & 0 & \frac{2EI}{L} & 0 & 0 & 0 & 0 \\ 0 & 0 & 0 & 0 & 1 & 0 & 0 & 0 & 0 & 0 \\ 0 & 0 & \frac{6EI}{L^2} & \frac{2EI}{L} & 0 & \frac{7EI}{L} & \frac{-3EI}{L^2} & 0 & 0 & 0 \\ 0 & 0 & 0 & 0 & 0 & \frac{-3EI}{L^2} & \frac{6EI}{L^2} & 0 & 0 & \frac{3EI}{L^2} \\ 0 & 0 & 0 & 0 & 0 & 0 & 0 & 253440 & 1 & 0 \\ 0 & 0 & 0 & 0 & 0 & 0 & 0 & 0 & 0 & 0 \\ 0 & 0 & 0 & 0 & 0 & 0 & \frac{3EI}{L^2} & 0 & 0 & \frac{3EI}{L} \end{bmatrix}$$

$$= \begin{bmatrix} 1 & 0 & 0 & 0 & 0 & 0 & 0 & 0 & 0 & 0 \\ 0 & 1 & 0 & 0 & 0 & 0 & 0 & 0 & 0 & 0 \\ 0 & 0 & 13516.8 & 0 & 0 & 25344 & 0 & 0 & 0 & 0 \\ 0 & 0 & 0 & 253440 & 0 & 63360 & 0 & 0 & 0 & 0 \\ 0 & 0 & 0 & 0 & 1 & 0 & 0 & 0 & 0 & 0 \\ 0 & 0 & 25344 & 633360 & 0 & 221760 & -12672 & 0 & 0 & 0 \\ 0 & 0 & 0 & 0 & 0 & -12670 & 3379.2 & 0 & 0 & 12672 \\ 0 & 0 & 0 & 0 & 0 & 0 & 0 & 1 & 0 & 0 \\ 0 & 0 & 0 & 0 & 0 & 0 & 0 & 0 & 1 & 0 \\ 0 & 0 & 0 & 0 & 0 & 0 & 12672 & 0 & 0 & 95040 \end{bmatrix}$$

Solving against the "remaining load" of

where $\boldsymbol{P}_r = \begin{bmatrix} 0 & 0 & 16 & 0 & 0 & 0 & 20 & 0 & 0 & 0 \end{bmatrix}^T$, the displacement matrix is

$$\Delta_p = \begin{bmatrix} 0 & 0 & -0.0023674 & -0.0004735 & 0 & 0.0018939 & 0.0260417 & 0 & 0 & -0.0034722 \end{bmatrix}^T$$

Multiplying the displacement to the individual matrices

$$\text{Span AB:}\begin{bmatrix}6758.4 & 25344 & -6758.4 & 25344\\ 25344 & 126720 & -25344 & 63360\\ -6758.4 & -25344 & 6758.4 & -25344\\ 25344 & 63360 & -25344 & 126720\end{bmatrix}\begin{bmatrix}0\\0\\-0.0023674\\-0.0004735\end{bmatrix}=\begin{bmatrix}4\\30\\-4\\0\end{bmatrix}$$

$$\text{Span BC:}\begin{bmatrix}6758.4 & 25344 & -6758.4 & 25344\\ 25344 & 126720 & -25344 & 63360\\ -6758.4 & -25344 & 6758.4 & -25344\\ 25344 & 63360 & -25344 & 126720\end{bmatrix}\begin{bmatrix}-0.0023674\\-0.0004735\\0\\0.0018939\end{bmatrix}=\begin{bmatrix}20\\0\\-20\\150\end{bmatrix}$$

$$\text{Span CD:}\begin{bmatrix}1689.6 & 12672 & -1689.6 & 0\\ 12672 & 95040 & -12672 & 0\\ -1689.6 & -12672 & 1689.6 & 0\\ 0 & 0 & 0 & 1\end{bmatrix}\begin{bmatrix}0\\0.0018939\\0.0260417\\0\end{bmatrix}=\begin{bmatrix}-20\\-150\\20\\0\end{bmatrix}$$

$$\text{Span DE:}\begin{bmatrix}1689.6 & 0 & -1689.6 & 12672\\ 0 & 1 & 0 & 0\\ -1689.6 & 0 & 1689.6 & -12672\\ 12672 & 0 & -12672 & 95040\end{bmatrix}\begin{bmatrix}0.0260417\\0\\0\\-0.0034722\end{bmatrix}=\begin{bmatrix}0\\0\\0\\0\end{bmatrix}$$

For the remaining load, the bending moment diagram is sketched as in Figure A.C5.

The final result is the summation of both models as sketched in Figure A.C6. There is no exceedance over $M_p = 1260\text{kNm}$.

If however, the plastic hinge is decreased to $M_p = 1207.5\text{kNm}$, so $M_p / M_e = 1207.5/1312.5 = 0.92$. Similarly by elastic analysis, the member forces (and also the deflections) all become $R_L = 0.92$ of their original values as

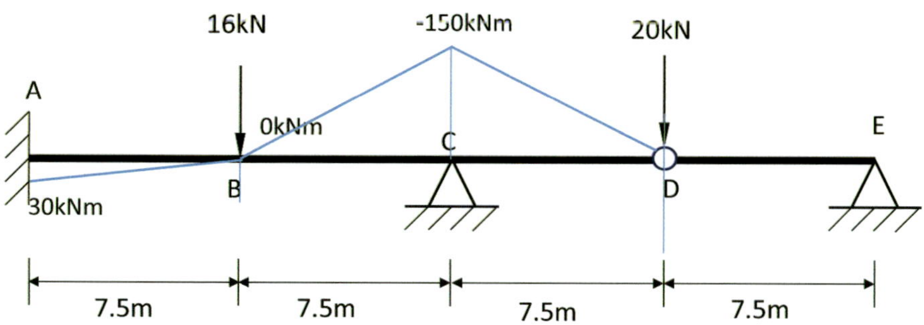

FIGURE A.C5 Bending moment diagram due to "remaining load".

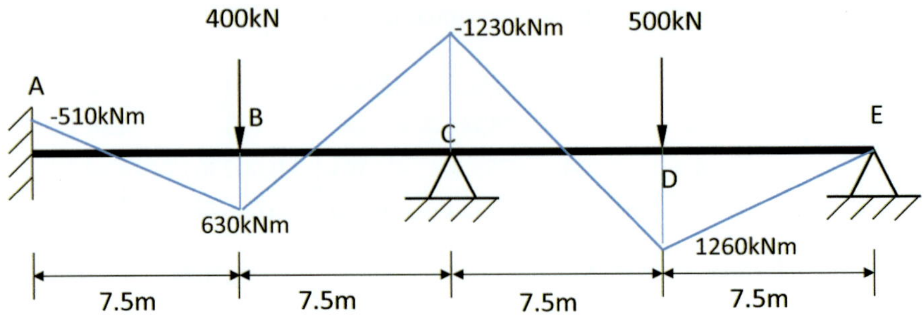

FIGURE A.C6 Final bending boment diagram.

$$\text{Span AB: } F_{RAB} = 0.92\begin{bmatrix} -162.5 \\ -562.5 \\ 162.5 \\ -656.25 \end{bmatrix} = \begin{bmatrix} -149.5 \\ -517.5 \\ 149.5 \\ -603.75 \end{bmatrix};$$

$$\text{Span BC: } F_{RBC} = 0.92\begin{bmatrix} 237.5 \\ 656.25 \\ -237.5 \\ 1125 \end{bmatrix} = \begin{bmatrix} 218.5 \\ 603.75 \\ -218.5 \\ 1035 \end{bmatrix};$$

$$\text{Span CD: } F_{RCD} = 0.92\begin{bmatrix} -325 \\ -1125 \\ 325 \\ -1312.5 \end{bmatrix} = \begin{bmatrix} -299 \\ -1035 \\ 299 \\ -1207.5 \end{bmatrix};$$

$$\text{Span DE: } F_{RDE} = 0.92\begin{bmatrix} 175 \\ 1312.5 \\ -175 \\ 0 \end{bmatrix} = \begin{bmatrix} 161 \\ 1207.5 \\ -161 \\ 0 \end{bmatrix}$$

The bending moments are shown in Figure A.C7 under 0.92 of the applied load.

For the remaining load, Node D has reached the plastic state and cannot take up any further moment. Therefore, for addition of the remaining loads, it has to be taken as a pinned connection with the model shown in Figure A.C8.

In Figure C8, the additional support moment at C is $-40 \times 7.5 = -300\,\text{kNm}$ which increases the support moment at C to $-1035 - 300 = -1335\,\text{kNm} > M_p = 1207.5$ kNm. As the moment of resistance cannot resist the final moment at C and there can be no moment redistribution, the beam fails.

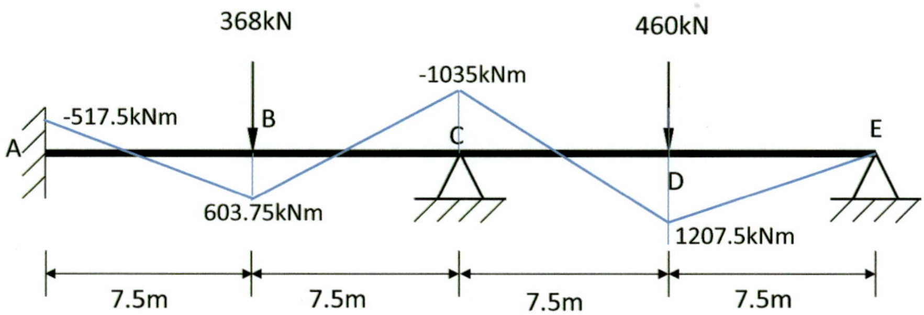

FIGURE A.C7 Introducing a plastic hinge at D and the applied load revised to 0.96 of their original values.

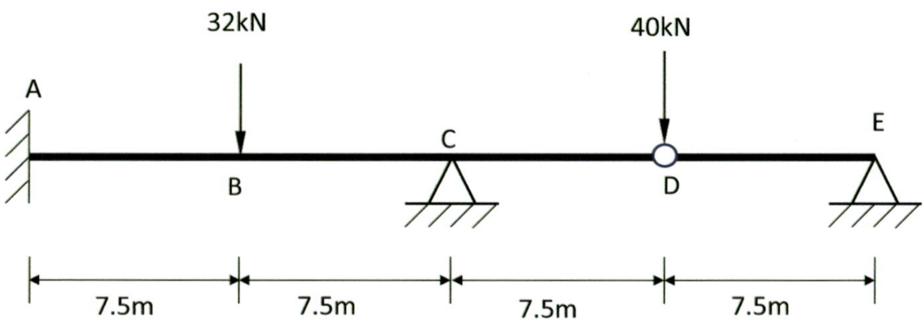

FIGURE A.C8 Plastic model modified from the elastic model.

REFERENCE

Aktaş, E. and Korkmaz, A. (2007) Plastic Hinge Analysis Using Reduced Basis Approximations. *Proceedings of the International Symposium on Advances in Earthquake and Structural Engineering*, Süleyman Demirel University, Isparta–Antalya, Turkey, 24–26 October.

Appendix D
Brief Description of the Finite Element Method, Its Application and Illustration by the Formulation of Some Elements

D.1 INTRODUCTION

This appendix intends to present a brief outline of the finite element method and its applications, under the illustration by the formulation of a few common structural elements. The aim is to allow the readers who are less familiar with the finite element method to have a quick grasp of its applications and limitations. Nevertheless, readers may refer to other references for more rigorous proofs, derivations and in-depth discussions which are not well covered in this book.

D.2 BASIC PRINCIPLES AND FORMULATIONS

Briefly speaking, by the finite element method in structural analysis, a structure as a continuum such as a plate structure is "discretized" into a number of elements of finite sizes interconnected by "nodes" at the elements' corners and/or edges. The stiffness of each discretized element is formulated and the whole structure is "assembled" in the same manner as a frame structure for analysis. The derivation of the stiffness of each element is based on an assumed "shape function" (or displacement function as alternatively called) which governs the deflection mode of the element, i.e. whether the mode is linear, quadratic or cubic etc. as governed by the degree of the polynomial assumed in the shape function. Unlike a line element where the conventional formulation of its stiffness actually obeys a cubic parabola which is exact to the elastic theory, the accuracy of other elements such as plate or shell depends very much on the shape functions with respect to the pattern of load applied. Nevertheless, the accuracy will generally increase if the structure is more finely divided (meshed) where greater freedom of movement is allowed by the provision of greater number of nodes or using a higher order polynomial. This is the famous "h-p criterion" with h standing for "fine meshing" and p standing for increasing the order of polynomial

DOI: 10.1201/9781003636380-18

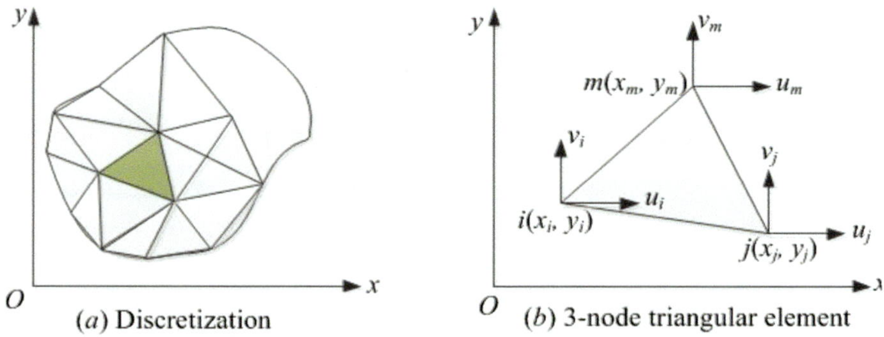

FIGURE A.D1 Discretization of an in-plane element in finite element analysis.

approximation in the element. In general, the use of higher polynomial function is more effective than the use of finer mesh at the same computer resources. The use of higher order element is, however, subject to the limitations of mid-sides problems in some cases (but not all) and mesh generation. For the modern computer system with 64-bit CPU and 16 or 32 GB RAM, the use of finer but low order mesh with easier mesh generation seems to be favoured by many commercial programs. The limitations of the low order elements in commercial program has been clearly illustrated by Cheng and Law (2008). More details about finite element formulation are given by Zienkiewicz et al. (2005) and Cheng et al. (2020).

The discretization or meshing of the finite element method is illustrated in Figure A.D1. In general, the use of a finer mesh is more useful if the solution domain is irregular. Another important engineering consideration is the choice of super-elements defined by the engineers. Some commercial programs allow the users to define super-elements without nodal compatibility. If regular element generation is chosen by the users, there will be nodal incompatibility between some of the generated elements. While the use of regular instead of distorted element can provide higher numerical accuracy in element sense, there will be discontinuity in the results due to nodal incompatibility, and the global accuracy may be even worse than that using distorted element. For the PLATE program developed by Cheng for thin/thick plate analysis, Cheng provides an algorithm to subdivide the super-elements automatically so that there will be no nodal incompatibility. The problem of some commercial programs in mesh generation for highly irregular problems is totally eliminated by the addition of such a pre-processing which is automatically performed.

The derivation of the stiffness matrix of a single element in finite element analysis is outlined as follows, under the illustration of a 4-node rectangular element with corner nodes 1, 2, 3, and 4 undergoing in-plane loads with in-plane deflections in the x and y directions.

The derivation starts with the assumption of a "shape function" (or displacement function) $\boldsymbol{N}(\boldsymbol{x}, \boldsymbol{y})$, which defines deflection $\boldsymbol{u}$ within the element in accordance with the movements of its corner or side nodes $\boldsymbol{a}^e$ that can be connected with other elements. So it follows that $\boldsymbol{u} = \boldsymbol{N}(\boldsymbol{x}, \boldsymbol{y})$ should be so formulated that it will give

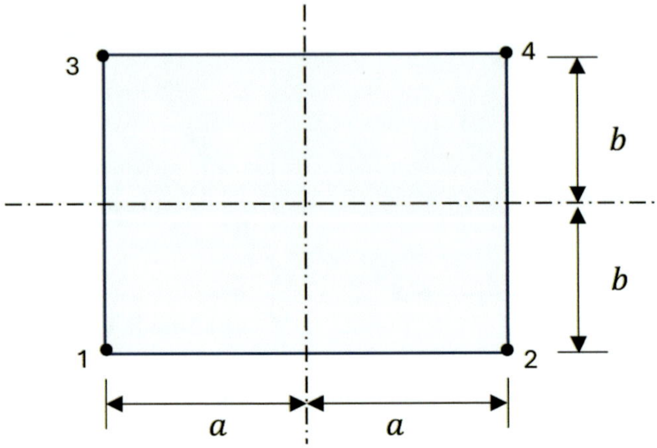

FIGURE A.D 2 A 4-node plane stress element.

$$u = N_1(x,y)a^e{}_1 + N_2(x,y)a^e{}_2 + \ldots + N_n(x,y)a^e{}_n \tag{D1}$$

where n is the number of nodes.

As $u = a^e{}_1$ at (x_1, y_1), $u = a^e{}_2$ at (x_2, y_2) etc., for the 4-node rectangle in Figure A.D2, we can formulate for node 1 where its coordinates are$(-a,-b)$

$$N_1(x,y) = \frac{(x-a)(y-b)}{(-a-a)(-b-b)} = \frac{(x-a)(y-b)}{4ab} = \frac{1}{4}\left(1-\frac{x}{a}\right)\left(1-\frac{y}{b}\right) \tag{D2}$$

The readers can easily verify that $N_1(x,y) = 1$ at $(-a,-b)$ and 0 at other nodes which fulfil the requirements of giving $u = a^e{}_1$. The formulation is based on

$$N_i(x,y) = N_{ix}(x)N_{iy}(y) \tag{D3}$$

where with n nodes on one side with the same x-coordinates

$$N_{ix}(x) = \frac{(x-x_1)(x-x_2)\ldots(x-x_{i-1})(x-x_{i+1})\ldots(x-x_n)}{(x_i-x_1)(x_i-x_2)\ldots(x_i-x_{i-1})(x_i-x_{i+1})\ldots(x_i-x_n)} \tag{D4}$$

which is simply the continuous multiplication of $(x-x_k)$ for $k=1$ to n with the omission of $(x-x_i)$ in the numerator and (x_i-x_k) in the denominator.

Similarly for $N_{iy}(y)$

$$N_{iy}(y) = \frac{(y-y_1)(y-y_2)\ldots(y-y_{i-1})(y-y_{i+1})\ldots(y-y_n)}{(y_i-y)(y_i-y_2)\ldots(y_i-y_{i-1})(y_i-y_{i+1})\ldots(y_i-y_n)} \tag{D5}$$

The so formed equation is a polynomial known as the Lagrange polynomial.

The N functions of other nodes are formulated similarly. Summing up all the 4 nodes for the rectangular element in Figure A.D2,

$$\begin{cases} N_1(x,y) = \dfrac{1}{4}\left(1-\dfrac{x}{a}\right)\left(1-\dfrac{y}{b}\right) & N_3(x,y) = \dfrac{1}{4}\left(1-\dfrac{x}{a}\right)\left(1+\dfrac{y}{b}\right) \\ N_2(x,y) = \dfrac{1}{4}\left(1+\dfrac{x}{a}\right)\left(1-\dfrac{y}{b}\right) & N_4(x,y) = \dfrac{1}{4}\left(1+\dfrac{x}{a}\right)\left(1+\dfrac{y}{b}\right) \end{cases} \tag{D6}$$

So we can write the deflection u_x and u_y in x and y directions as

$$u_x = N_1(x,y)u_{x1} + N_2(x,y)u_{x2} + N_3(x,y)u_{x3} + N_4(x,y)u_{x4} \tag{D7}$$

$$u_y = N_1(x,y)u_{y1} + N_2(x,y)u_{y2} + N_3(x,y)u_{y3} + N_4(x,y)u_{y4} \tag{D8}$$

where u_{xi} and u_{yi} are the nodal displacements of node i $(i = 1,2,3,4)$.

With the displacement functions formed, we can form the strain displacement matrix $\boldsymbol{B}$ by differentiating the displacement with respect to x and y as

$$\boldsymbol{B} = \partial \boldsymbol{N} \tag{D9}$$

$$\text{So that } \varepsilon = \boldsymbol{B}\boldsymbol{a}^e \tag{D10}$$

For the plane stress problem, the direct and shear strains are

$$\varepsilon_x = \frac{\partial u_x}{\partial x} = \frac{\partial N}{\partial x} a^e{}_x \tag{D11}$$

$$\varepsilon_y = \frac{\partial u_y}{\partial y} = \frac{\partial N}{\partial y} a^e{}_y \tag{D12}$$

$$\gamma_{xy} = \frac{\partial u_x}{\partial y} + \frac{\partial u_y}{\partial x} = \frac{\partial N}{\partial y} a^e{}_x + \frac{\partial N}{\partial x} a^e{}_y \tag{D13}$$

When expressed in matrix form for the 4 node element, it is

$$\boldsymbol{\mu} = \begin{bmatrix} \dfrac{\partial N_1}{\partial x} & 0 & \dfrac{\partial N_2}{\partial x} & 0 & \dfrac{\partial N_3}{\partial x} & 0 & \dfrac{\partial N_4}{\partial x} & 0 \\ 0 & \dfrac{\partial N_1}{\partial y} & 0 & \dfrac{\partial N_2}{\partial y} & 0 & \dfrac{\partial N_3}{\partial y} & 0 & \dfrac{\partial N_4}{\partial y} \\ \dfrac{\partial N_1}{\partial y} & \dfrac{\partial N_1}{\partial x} & \dfrac{\partial N_2}{\partial y} & \dfrac{\partial N_2}{\partial x} & \dfrac{\partial N_3}{\partial y} & \dfrac{\partial N_3}{\partial x} & \dfrac{\partial N_4}{\partial y} & \dfrac{\partial N_4}{\partial x} \end{bmatrix} \begin{bmatrix} u_{x1} \\ u_{y1} \\ u_{x2} \\ u_{y2} \\ u_{x3} \\ u_{y3} \\ u_{x4} \\ u_{y4} \end{bmatrix} \tag{D14}$$

That is,
$$\boldsymbol{B} = \begin{bmatrix} \frac{\partial N_1}{\partial x} & 0 & \frac{\partial N_2}{\partial x} & 0 & \frac{\partial N_3}{\partial x} & 0 & \frac{\partial N_4}{\partial x} & 0 \\ 0 & \frac{\partial N_1}{\partial y} & 0 & \frac{\partial N_2}{\partial y} & 0 & \frac{\partial N_3}{\partial y} & 0 & \frac{\partial N_4}{\partial y} \\ \frac{\partial N_1}{\partial y} & \frac{\partial N_1}{\partial x} & \frac{\partial N_2}{\partial y} & \frac{\partial N_2}{\partial x} & \frac{\partial N_3}{\partial y} & \frac{\partial N_3}{\partial x} & \frac{\partial N_4}{\partial y} & \frac{\partial N_4}{\partial x} \end{bmatrix} \quad \text{(D15)}$$

for the 4 node element.

We may now relate stress σ to strain ε in matrix form by an elastic stress-strain matrix $\boldsymbol{D}$ as

$$\sigma = \boldsymbol{D}\varepsilon \quad \text{(D16)}$$

In the plane stress problem, with μ as the Poisson's ratio, it is expressed as

$$\sigma = \frac{E}{1-\mu^2}\begin{bmatrix} 1 & \mu & 0 \\ \mu & 1 & 0 \\ 0 & 0 & (1-\mu)/2 \end{bmatrix}\begin{bmatrix} \varepsilon_x \\ \varepsilon_y \\ \gamma_{xy} \end{bmatrix} \quad \text{(D17)}$$

where E is the Young's modulus and μ is the Poisson's ratio.

That is,
$$\boldsymbol{D} = \frac{E}{1-\mu^2}\begin{bmatrix} 1 & \mu & 0 \\ \mu & 1 & 0 \\ 0 & 0 & (1-\mu)/2 \end{bmatrix} \quad \text{(D18)}$$

The total strain energy in the element is

$$\int \boldsymbol{\varepsilon}^T \boldsymbol{\sigma} d\boldsymbol{V} = \int \boldsymbol{\varepsilon}^T \boldsymbol{D}\boldsymbol{\varepsilon} d\boldsymbol{V} = \int \boldsymbol{a}^{eT} \boldsymbol{B}^T \boldsymbol{D}\boldsymbol{B}\boldsymbol{a}^e d\boldsymbol{V} = \boldsymbol{a}^{eT}\left(\int \boldsymbol{B}^T \boldsymbol{D}\boldsymbol{B} d\boldsymbol{V}\right)\boldsymbol{a}^e \quad \text{(D19)}$$

The total strain energy is equal to the external work done by the forces at the nodes which is

$$P_1 a^e{}_1 + P_2 a^e{}_3 + \ldots + P_n a^e{}_n \quad \text{(D20)}$$

If expressed in matrix form, it is

$$\boldsymbol{a}^{eT}\boldsymbol{P} = P_1 a^e{}_1 + P_2 a^e{}_3 + \ldots + P_n a^e{}_n \quad \text{(D21)}$$

Equating Equations D19 and D21,

$$\boldsymbol{a}^{eT}\left(\int \boldsymbol{B}^T \boldsymbol{D}\boldsymbol{B} d\boldsymbol{V}\right)\boldsymbol{a}^e = \boldsymbol{a}^{eT}\boldsymbol{P}$$

$$\Rightarrow \left(\int \boldsymbol{B}^T \boldsymbol{D}\boldsymbol{B} d\boldsymbol{V}\right)\boldsymbol{a}^e = \boldsymbol{P} \quad \text{(D22)}$$

By analogy with the basic equation in stiffness analysis as $\boldsymbol{K}.\Delta = \boldsymbol{P}$, $\int \boldsymbol{B}^T \boldsymbol{DB}dV$ in Equation D22 can be taken as the "stiffness matrix" $\boldsymbol{K}$ of the element as $\boldsymbol{a}^e$ is the nodal displacement (Δ) matrix and $\boldsymbol{P}$ is the external force matrix. So, it can be written as

$$\boldsymbol{K}.\Delta = \boldsymbol{P} \Rightarrow \int \boldsymbol{B}^T \boldsymbol{DB}dV.\Delta = \boldsymbol{P}\left(\boldsymbol{K} = \int \boldsymbol{B}^T \boldsymbol{DB}dV\right) \tag{D23}$$

Returning to the 4 node element where $\boldsymbol{B}$ is given in Equation D15 and $\boldsymbol{D}$ is given in Equation D18 and $\boldsymbol{B}$ can be worked out with the $\boldsymbol{N}$ functions listed in Equation D6, the matrix $\boldsymbol{B}^T\boldsymbol{DB}$ can be calculated as a 8 × 8 matrix. Each of the entry in the resultant matrix can be integrated from $-a$ to a and from $-b$ to b to form the final stiffness matrix. Take an example of the array in (1, 1) in $\boldsymbol{B}^T\boldsymbol{DB}$ which is $\frac{(b-y)^2}{16ab} + \frac{(a-x)^2}{32ab}(1-\mu)$, the integration of which is

$$\int_0^t \int_{-b}^b \int_{-a}^a \frac{(b-y)^2}{16ab} + \frac{(a-x)^2}{32ab}(1-\mu)\,dxdydz = \frac{a^2(1-\mu)+2b^2}{6ab}t$$

where t is the thickness of the membrane.

So the stiffness matrix of the rectangular element is

$$K = \frac{Et}{1-\mu^2}\begin{bmatrix} K_{11} & K_{12} \\ K_{21} & K_{22} \end{bmatrix} \tag{D24}$$

where

$$K_{11} = \begin{bmatrix} \frac{a^2(1-\mu)+2b^2}{6ab} & \frac{1+\mu}{8} & \frac{a^2(1-\mu)-4b^2}{12ab} & \frac{-1+3\mu}{8} \\ \frac{1+\mu}{8} & \frac{2a^2+b^2(1-\mu)}{6ab} & \frac{1-3\mu}{8} & \frac{a^2-b^2(1-\mu)}{6ab} \\ \frac{a^2(1-\mu)-4b^2}{12ab} & \frac{1-3\mu}{8} & \frac{a^2(1-\mu)+2b^2}{6ab} & \frac{-1-\mu}{8} \\ \frac{-1+3\mu}{8} & \frac{a^2-b^2(1-\mu)}{6ab} & \frac{-1-\mu}{8} & \frac{2a^2+b^2(1-\mu)}{6ab} \end{bmatrix}$$

$$K_{12} = \begin{bmatrix} \frac{-a^2(1-\mu)+b^2}{6ab} & \frac{1-3\mu}{8} & \frac{-a^2(1-\mu)-2b^2}{12ab} & \frac{-1-\mu}{8} \\ \frac{-1+3\mu}{8} & \frac{-4a^2+b^2(1-\mu)}{12ab} & \frac{-1-\mu}{8} & \frac{-2a^2-b^2(1-\mu)}{12ab} \\ \frac{-a^2(1-\mu)-2b^2}{12ab} & \frac{1+\mu}{8} & \frac{-a^2(1-\mu)+b^2}{6ab} & \frac{-1+3\mu}{8} \\ \frac{1+\mu}{8} & \frac{-2a^2-b^2(1-\mu)}{12ab} & \frac{1-3\mu}{8} & \frac{-4a^2+b^2(1-\mu)}{12ab} \end{bmatrix}$$

$$K_{21} = K_{12}^{\,T}$$

$$K_{22} = \begin{bmatrix} \dfrac{a^2(1-\mu)+2b^2}{6ab} & \dfrac{-1-\mu}{8} & \dfrac{a^2(1-\mu)-4b^2}{12ab} & \dfrac{1-3\mu}{8} \\ \dfrac{-1-\mu}{8} & \dfrac{2a^2+b^2(1-\mu)}{6ab} & \dfrac{-1+3\mu}{8} & \dfrac{a^2-b^2(1-\mu)}{6ab} \\ \dfrac{a^2(1-\mu)-4b^2}{12ab} & \dfrac{-1+3\mu}{8} & \dfrac{a^2(1-\mu)+2b^2}{6ab} & \dfrac{1+\mu}{8} \\ \dfrac{1-3\mu}{8} & \dfrac{a^2-b^2(1-\mu)}{6ab} & \dfrac{1+\mu}{8} & \dfrac{2a^2+b^2(1-\mu)}{6ab} \end{bmatrix}$$

The stiffness matrix of the rectangular element is similar to that of a strut member. In simulating a structure, the elements can be joined together as shown in Figure A.D1. The stiffness matrix of the whole structure can be formed by assembly of the various elements in a manner similar to a frame structure made up of line members.

Following Equation D23, Δ can be solved which gives the deformation of the structure. Applying Equation D10, the strain ε matrix can be obtained by $\varepsilon = \boldsymbol{B}.\Delta$ and then using Equation D16, the stresses at any point within the structure can be solved by $\sigma = \boldsymbol{D}\varepsilon$.

To demonstrate the above procedure, the 4 node element is used to simulate a concrete wall of length 3 m, thickness 0.4 m and height 12 m under two lateral point loads each of 1000 kN at its top corner as shown in Figure A.D3. The wall is fixed at its bottom level and it comes a cantilever wall under the two lateral point loads. The Young's modulus of the concrete is taken as 25000000 kN/m² and the Poisson's ratio is taken as 0.2.

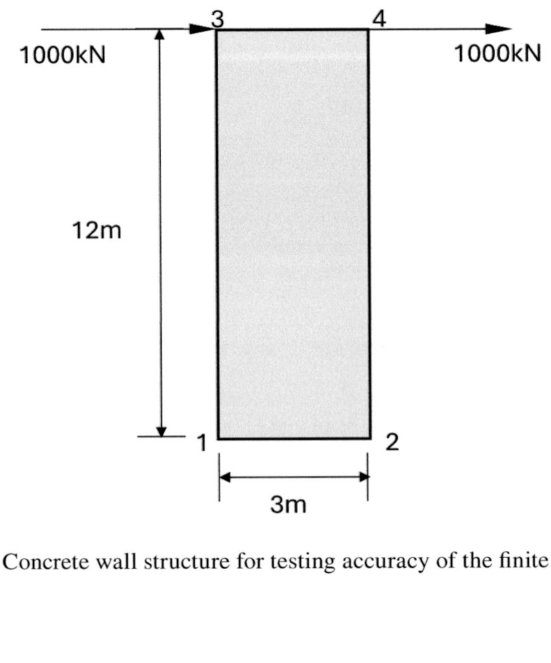

FIGURE A.D3 Concrete wall structure for testing accuracy of the finite element.

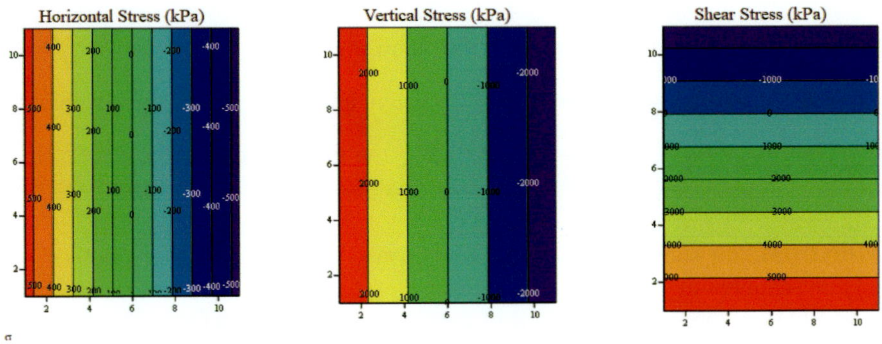

FIGURE A.D4 Stress contours of the bottom element of wall formed by 4 node single rectangle elements.

The stiffness matrix $\boldsymbol{K}$ in Equation D24 can be formed by substituting the values of $a = 1.5$, $b = 6$, $t = 0.4$, $E = 25000000$ and $\mu = 0.2$. $\boldsymbol{K}$ is then modified to the support conditions (nodes 1 and 2 are restrained) by retaining only K_{22} so that the stiffness matrix is $\boldsymbol{S} = \boldsymbol{K}_{22}$ modified to support conditions. The applied load matrix is

$$\boldsymbol{P} = \begin{bmatrix} 1000 & 0 & 1000 & 0 \end{bmatrix}^T \tag{D25}$$

Then the deflection can be solved as :

$$\Delta = \boldsymbol{S}^{-1}.\boldsymbol{P} \tag{D26}$$

Numerically, $\Delta = \begin{bmatrix} 0.006902 & 0.001245 & 0.006902 & -0.001245 \end{bmatrix}^T$ which are the lateral and vertical deflections of nodes 3 and 4 so that the lateral deflections of them are 0.006902 m and node 3 upward deflects by 0.001245 m and node 4 deflects downward by the same magnitude.

The strain displacement matrix in Equation D15 is now taken and worked with the nodal displacement obtained as $\Delta\left(= a^e\right)$. Using Equations D16 to D18, the stresses at any point of the element can be calculated as $\sigma = \boldsymbol{D}.\boldsymbol{B}.\Delta$ and plotted in Figure A.D4.

It will be seen in D.3 of this appendix that the results are rough and erratic under the use of a single element to simulate this structure. However, based on the development of this approach and using a greater number of elements (higher degrees of meshing of the structure) and/or more sophisticated elements developed, much better performance can be achieved. The use of an incompatible element with the addition of a bending term can greatly improve the accuracy of the solution. However, this is not commonly adopted in commercial programs.

D.3 PERFORMANCE OF THE FINITE ELEMENT UNDER DIFFERENT SCENARIOS

From the worked example simulated by a single 4 node plane stress element as discussed in Section D.2, the deflection is first examined. For comparison with the maximum lateral deflection by the cantilever beam theory as $\Delta_{beam} = \dfrac{PL^3}{3EI}$ where L is the length of the cantilever beam and EI is its flexural rigidity as $\Delta_{beam} = \dfrac{2000 \times 12^3}{3 \times 25000000 \times \left(\dfrac{0.4 \times 3^3}{12}\right)} = 0.0512\text{m}$, the error is very large (0.006912 m versus 0.0512 m), showing that the element formed by the finite element method with only one element of 4 nodes is "over-stiff".

In addition, the linear variations of the stresses as shown in Figure A.D4 are due to the limitations imposed by the shape functions which must be linear in both directions. The magnitudes are also very different from that determined by the beam theory. Considering the vertical stress, the extreme values in the finite element are only ±2703 kN/m² along the vertical edges. However, the extreme vertical stress determined by the beam theory is $M / Z = 24000 / 0.6 = 40000\text{kN/m}^2$ where M is the moment = $2000 \times 12 = 24000\text{kNm}$ and Z is the elastic modulus $= 0.4 \times 3^2 / 6 = 0.6$ m³. So the error is also very large.

The wall is then meshed into two elements by stacking two elements with $b = 6\text{m}$ on each other. The matrix is "assembled" in the same manner as line elements in a frame structure. In this case, nodes 3 and 4 of the lower element are joined to nodes 1 and 2 of the upper element. That is, $\boldsymbol{K}_{22}$ or the lower element is added to $\boldsymbol{K}_{11}$ of the upper element in the assembled stiff matrix. The top corners lateral deflections become 0.0196 m which are closer to the beam theory at 0.0512 m as compared with 0.0069 m in the single element analysis.

The wall is further meshed to 5, 10, and 20 elements similarly. The top corner lateral deflections are respectively 0.0407 m, 0.048 m, and 0.0502 m which are approaching to that of the beam theory deflection at 0.0512 m.

The stresses at the bottom element in the model meshed to 20 number of equal elements are similarly worked out and plotted as in Figure A.D5.

The vertical stress at the extreme fibre is ±38386 kN/m² which is close to the beam model at 40000 kN/m². Yet the convergence is not very ideal as it requires meshing with more than 10 elements to achieve reasonable accuracy. So, a more

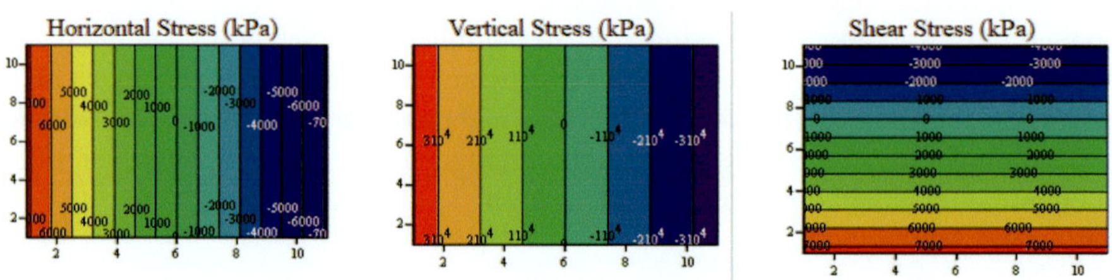

FIGURE A.D5 Stress contours of the bottom element of the wall formed by 4 node rectangle elements meshed to 20.

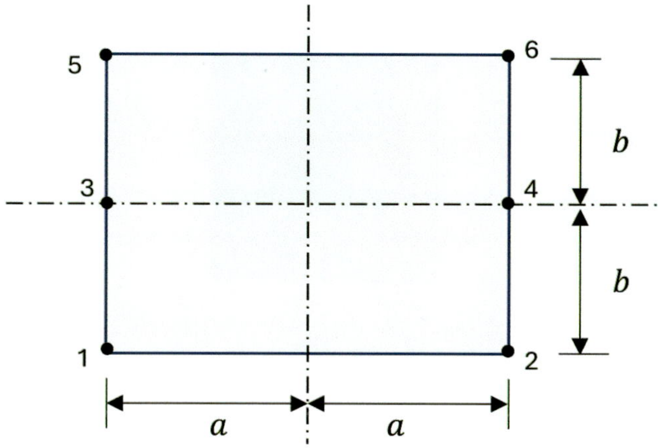

FIGURE A.D6 6-node plane stress element – 4 corner nodes and 2 mid-side nodes.

sophisticated element is developed with 6 nodes – 2 additional mid-side nodes as in Figure A.D6.

The shape functions are first formulated by Equations D4 and D5, for simplicity in the formulation, set $\xi = \frac{x}{a}, \eta = \frac{y}{b}$ such as $d\xi = \frac{1}{a}dx,\ d\eta = \frac{1}{b}dy$

$$\begin{cases} N_1 = \frac{-1}{4}(1-\xi)\eta(1-\eta) & N_4 = \frac{1}{2}(1+\xi)(1-\eta^2) \\ N_2 = \frac{-1}{4}(1+\xi)\eta(1-\eta) & N_5 = \frac{1}{4}(1-\xi)\eta(1+\eta) \\ N_3 = \frac{1}{2}(1-\xi)(1-\eta^2) & N_6 = \frac{1}{4}(1+\xi)\eta(1+\eta) \end{cases} \quad \text{(D27)}$$

From these shape functions, it can be seen that the displacement follows quadratic variations in the vertical direction and linear in the horizontal direction.

The B-matrix will be of size 3 × 12 is

$$B = \begin{bmatrix} \frac{1}{a}\frac{dN_1}{d\xi} & 0 & \frac{1}{a}\frac{dN_2}{d\xi} & 0 & \frac{1}{a}\frac{dN_3}{d\xi} & 0 & \frac{1}{a}\frac{dN_4}{d\xi} & 0 & \frac{1}{a}\frac{dN_5}{d\xi} & 0 & \frac{1}{a}\frac{dN_6}{d\xi} & 0 \\ 0 & \frac{1}{b}\frac{dN_1}{d\eta} & 0 & \frac{1}{b}\frac{dN_2}{d\eta} & 0 & \frac{1}{b}\frac{dN_3}{d\eta} & 0 & \frac{1}{b}\frac{dN_4}{d\eta} & 0 & \frac{1}{b}\frac{dN_5}{d\eta} & 0 & \frac{1}{b}\frac{dN_6}{d\eta} \\ \frac{1}{b}\frac{dN_1}{d\eta} & \frac{1}{a}\frac{dN_1}{d\xi} & \frac{1}{b}\frac{dN_2}{d\eta} & \frac{1}{a}\frac{dN_2}{d\xi} & \frac{1}{b}\frac{dN_3}{d\eta} & \frac{1}{a}\frac{dN_3}{d\xi} & \frac{1}{b}\frac{dN_4}{d\eta} & \frac{1}{a}\frac{dN_4}{d\xi} & \frac{1}{b}\frac{dN_5}{d\eta} & \frac{1}{a}\frac{dN_5}{d\xi} & \frac{1}{b}\frac{dN_6}{d\eta} & \frac{1}{a}\frac{dN_6}{d\xi} \end{bmatrix}$$

(D28)

The stiffness matrix, which is of array size 12, × 12 is similarly formed numerically by matrix multiplication in accordance with Equation D23. Similarly, meshing at a

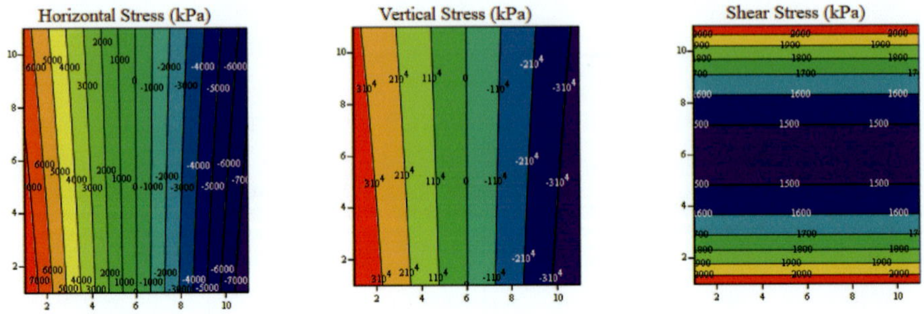

FIGURE A.D7 Stress contours of the bottom element of the wall formed by 6 node rectangle elements meshed to 10.

single element, 2, 3, 5, and 10 elements are tried, and the top deflections are determined respectively at 0.044 m, 0.0503 m, 0.0509 m, 0.051 m, 0.0511 m (compared with the beam deflection at 0.0512 m), and the vertical extreme fibre stresses are respectively at ±28772 kPa, ±37576 kPa, ±39170 kPa, ±39805 kPa, ±39974 kPa (compared with the beam theory at ±40000 kPa). So, the convergence is much better than the 4-node element as it practically converges at the meshing of 3. The stress contours for the lowest element at the meshing at 10 are also plotted at Figure A.D7. It should be noted that the contours for the direct stresses are curved (at quadratic) as dictated by the shape functions.

The next element to be examined is an 8-node element with 2 side nodes at the middle third at each side as shown in Figure A.D8.

In accordance with Equations D4 and D5, the shape functions are :

$$\begin{cases} N_1 = \dfrac{1}{32}(1-\xi)(1-\eta)(9\eta^2-1); & N_5 = \dfrac{9}{32}(1-\xi)(1-\eta^2)(1-3\eta) \\ N_2 = \dfrac{1}{32}(1+\xi)(1-\eta)(9\eta^2-1); & N_6 = \dfrac{9}{32}(1+\xi)(1-\eta^2)(1-3\eta) \\ N_3 = \dfrac{1}{32}(1-\xi)(1+\eta)(9\eta^2-1); & N_7 = \dfrac{9}{32}(1-\xi)(1-\eta^2)(1+3\eta) \\ N_4 = \dfrac{1}{32}(1+\xi)(1+\eta)(9\eta^2-1); & N_8 = \dfrac{9}{32}(1+\xi)(1-\eta^2)(1+3\eta) \end{cases} \quad \text{(D29)}$$

From Equation D29, it can be seen that the displacement follows a cubic parabolic relation in the vertical direction and linear in the horizontal direction as dictated by the shape functions. The analytical results show that, with a single element, the lateral deflection at the top level is 0.0511 m which is almost exactly equal to that of the beam deflection. The vertical extreme fibre stresses are 40000 kPa which is exactly identical to that of the beam analysis. By the vertical stress results, it can be concluded that this element converges at the single element without meshing. The horizontal and vertical stress contours are plotted in Figure A.D9. The shear stress

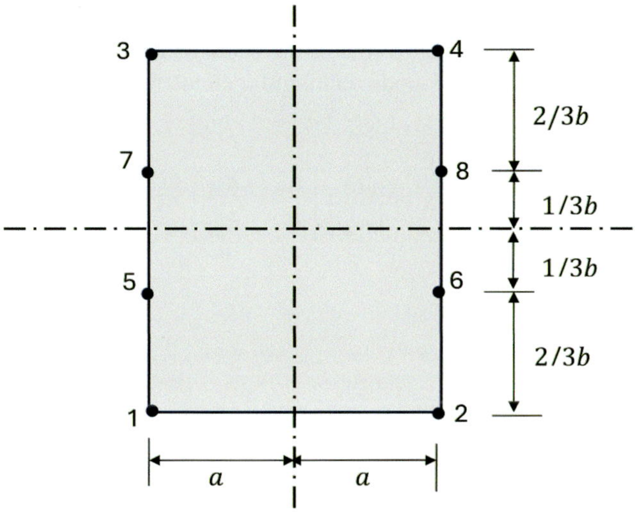

FIGURE A.D8 8-node plane stress element – 4 corner nodes and 4 mid-side nodes.

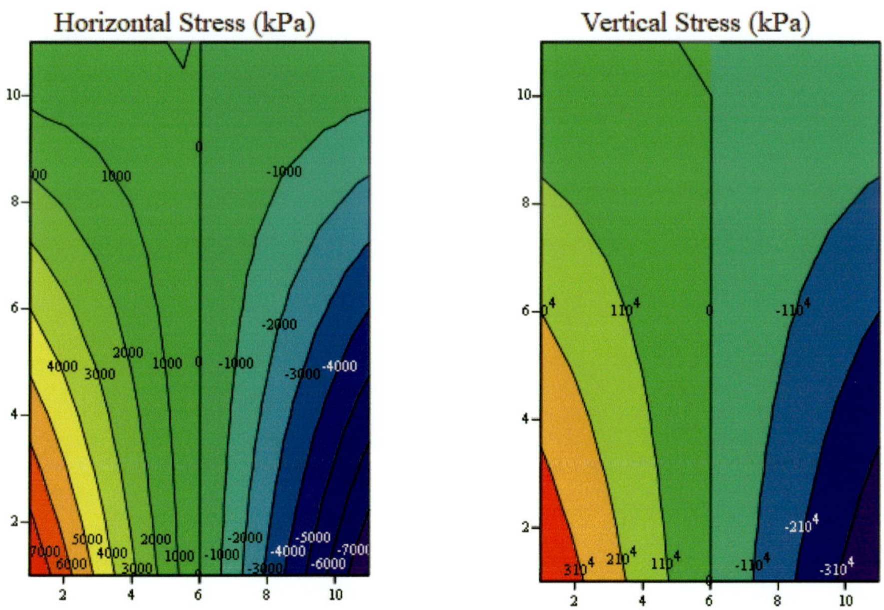

FIGURE A.D9 Stress contours of the wall formed by 8-node rectangle element without meshing.

is not required to be plotted as it is at the constant value of 1667 kPa throughout the element which is also the average stress of the beam as $2000/(0.4\times 3)=1667\,\text{kPa}$.

The stiffness matrix of the 8-node rectangular element as formulated by a similar procedure is listed as follows:

$$K=\frac{Et}{1-\mu^2}\begin{bmatrix} K_{11} & K_{12} \\ K_{21} & K_{22} \end{bmatrix} \tag{D30}$$

where

$$K_{11}:=\begin{bmatrix}
\frac{259\cdot a^2\cdot(1-\mu)+32\cdot b^2}{420\cdot a\cdot b} & \frac{1+\mu}{8} & \frac{259\cdot a^2\cdot(1-\mu)-64\cdot b^2}{840\cdot a\cdot b} & \frac{-1+3\cdot\mu}{8} & \frac{-91\cdot a^2\cdot(1-\mu)+19\cdot b^2}{1680\cdot a\cdot b} & \frac{7-21\cdot\mu}{320} & \frac{-91\cdot a^2\cdot(1-\mu)-38\cdot b^2}{3360\cdot a\cdot b} & \frac{-7\cdot(1+\mu)}{320} \\
\frac{1+\mu}{8} & \frac{259\cdot a^2+8\cdot b^2\cdot(1-\mu)}{210\cdot a\cdot b} & \frac{1-3\cdot\mu}{8} & \frac{259\cdot a^2-16\cdot b^2\cdot(1-\mu)}{420\cdot a\cdot b} & \frac{-7+21\cdot\mu}{320} & \frac{-364\cdot a^2+19\cdot b^2\cdot(1-\mu)}{3360\cdot a\cdot b} & \frac{-7\cdot(1+\mu)}{320} & \frac{-182\cdot a^2-19\cdot b^2\cdot(1-\mu)}{3360\cdot a\cdot b} \\
\frac{259\cdot a^2\cdot(1-\mu)-64\cdot b^2}{840\cdot a\cdot b} & \frac{1-3\cdot\mu}{8} & \frac{259\cdot a^2\cdot(1-\mu)+32\cdot b^2}{420\cdot a\cdot b} & \frac{-1-\mu}{8} & \frac{-91\cdot a^2\cdot(1-\mu)-38\cdot b^2}{3360\cdot a\cdot b} & \frac{7\cdot(1+\mu)}{320} & \frac{-91\cdot a^2\cdot(1-\mu)+19\cdot b^2}{1680\cdot a\cdot b} & \frac{-7\cdot(1-3\cdot\mu)}{320} \\
\frac{-1+3\cdot\mu}{8} & \frac{259\cdot a^2-16\cdot b^2\cdot(1-\mu)}{420\cdot a\cdot b} & \frac{-1-\mu}{8} & \frac{259\cdot a^2+8\cdot b^2\cdot(1-\mu)}{210\cdot a\cdot b} & \frac{7+7\cdot\mu}{320} & \frac{-182\cdot a^2-19\cdot b^2\cdot(1-\mu)}{3360\cdot a\cdot b} & \frac{7-21\cdot\mu}{320} & \frac{-364\cdot a^2+19\cdot b^2\cdot(1-\mu)}{3360\cdot a\cdot b} \\
\frac{-91\cdot a^2\cdot(1-\mu)+19\cdot b^2}{1680\cdot a\cdot b} & \frac{-7+21\cdot\mu}{320} & \frac{-91\cdot a^2\cdot(1-\mu)-38\cdot b^2}{3360\cdot a\cdot b} & \frac{7\cdot(1+\mu)}{320} & \frac{259\cdot a^2\cdot(1-\mu)+32\cdot b^2}{420\cdot a\cdot b} & \frac{-1-\mu}{8} & \frac{259\cdot a^2\cdot(1-\mu)-64\cdot b^2}{840\cdot a\cdot b} & \frac{1-3\cdot\mu}{8} \\
\frac{7-21\cdot\mu}{320} & \frac{-364\cdot a^2+19\cdot b^2\cdot(1-\mu)}{3360\cdot a\cdot b} & \frac{7\cdot(1+\mu)}{320} & \frac{-182\cdot a^2-19\cdot b^2\cdot(1-\mu)}{3360\cdot a\cdot b} & \frac{-1-\mu}{8} & \frac{259\cdot a^2+8\cdot b^2\cdot(1-\mu)}{210\cdot a\cdot b} & \frac{-1+3\cdot\mu}{8} & \frac{259\cdot a^2-16\cdot b^2\cdot(1-\mu)}{420\cdot a\cdot b} \\
\frac{-91\cdot a^2\cdot(1-\mu)-38\cdot b^2}{3360\cdot a\cdot b} & \frac{-7\cdot(1+\mu)}{320} & \frac{-91\cdot a^2\cdot(1-\mu)+19\cdot b^2}{1680\cdot a\cdot b} & \frac{7\cdot(1-3\cdot\mu)}{320} & \frac{259\cdot a^2\cdot(1-\mu)-64\cdot b^2}{840\cdot a\cdot b} & \frac{-1+3\cdot\mu}{8} & \frac{259\cdot a^2\cdot(1-\mu)+32\cdot b^2}{420\cdot a\cdot b} & \frac{1+\mu}{8} \\
\frac{-7\cdot(1+\mu)}{320} & \frac{-182\cdot a^2-19\cdot b^2\cdot(1-\mu)}{3360\cdot a\cdot b} & \frac{-7+21\cdot\mu}{320} & \frac{-364\cdot a^2+19\cdot b^2\cdot(1-\mu)}{3360\cdot a\cdot b} & \frac{1-3\cdot\mu}{8} & \frac{259\cdot a^2-16\cdot b^2\cdot(1-\mu)}{420\cdot a\cdot b} & \frac{1+\mu}{8} & \frac{259\cdot a^2+8\cdot b^2\cdot(1-\mu)}{210\cdot a\cdot b}
\end{bmatrix}$$

$$K_{12}:=\begin{bmatrix}
\frac{-441\cdot a^2\cdot(1-\mu)+33\cdot b^2}{560\cdot a\cdot b} & \frac{57\cdot(1-3\mu)}{320} & \frac{-441\cdot a^2\cdot(1-\mu)-66\cdot b^2}{1120\cdot a\cdot b} & \frac{-57\cdot(1+\mu)}{320} & \frac{63\cdot a^2\cdot(1-\mu)-6\cdot b^2}{280\cdot a\cdot b} & \frac{-3\cdot(1-3\cdot\mu)}{40} & \frac{63\cdot a^2\cdot(1-\mu)+12\cdot b^2}{560\cdot a\cdot b} & \frac{3\cdot(1+\mu)}{40} \\
\frac{-57\cdot(1-3\mu)}{320} & \frac{-1764\cdot a^2+33\cdot b^2\cdot(1-\mu)}{1120\cdot a\cdot b} & \frac{-57\cdot(1+\mu)}{320} & \frac{(-882)\cdot a^2-33\cdot b^2\cdot(1-\mu)}{1120\cdot a\cdot b} & \frac{3\cdot(1-3\cdot\mu)}{40} & \frac{126\cdot a^2-3\cdot b^2\cdot(1-\mu)}{280\cdot a\cdot b} & \frac{3\cdot(1+\mu)}{40} & \frac{63\cdot a^2+3\cdot b^2\cdot(1-\mu)}{280\cdot a\cdot b} \\
\frac{-441\cdot a^2\cdot(1-\mu)-66\cdot b^2}{1120\cdot a\cdot b} & \frac{57\cdot(1+\mu)}{320} & \frac{-441\cdot a^2\cdot(1-\mu)+33\cdot b^2}{560\cdot a\cdot b} & \frac{-57\cdot(1-3\mu)}{320} & \frac{63\cdot a^2\cdot(1-\mu)+12\cdot b^2}{560\cdot a\cdot b} & \frac{-3\cdot(1+\mu)}{40} & \frac{63\cdot a^2\cdot(1-\mu)-6\cdot b^2}{280\cdot a\cdot b} & \frac{3\cdot(1-3\cdot\mu)}{40} \\
\frac{57\cdot(1+\mu)}{320} & \frac{-882\cdot a^2-33\cdot b^2\cdot(1-\mu)}{1120\cdot a\cdot b} & \frac{57\cdot(1-3\mu)}{320} & \frac{-1764\cdot a^2+33\cdot b^2\cdot(1-\mu)}{1120\cdot a\cdot b} & \frac{-3\cdot(1+\mu)}{40} & \frac{63\cdot a^2+3\cdot b^2\cdot(1-\mu)}{280\cdot a\cdot b} & \frac{-3\cdot(1-3\cdot\mu)}{40} & \frac{126\cdot a^2-3\cdot b^2\cdot(1-\mu)}{280\cdot a\cdot b} \\
\frac{63\cdot a^2\cdot(1-\mu)-6\cdot b^2}{280\cdot a\cdot b} & \frac{3-9\cdot\mu}{40} & \frac{63\cdot a^2\cdot(1-\mu)+12\cdot b^2}{560\cdot a\cdot b} & \frac{-3\cdot(1+\mu)}{40} & \frac{-441\cdot a^2\cdot(1-\mu)+33\cdot b^2}{560\cdot a\cdot b} & \frac{-57\cdot(1-3\mu)}{320} & \frac{-441\cdot a^2\cdot(1-\mu)-66\cdot b^2}{1120\cdot a\cdot b} & \frac{57\cdot(1+\mu)}{320} \\
\frac{-3\cdot(1-3\cdot\mu)}{40} & \frac{126\cdot a^2-3\cdot b^2\cdot(1-\mu)}{280\cdot a\cdot b} & \frac{-3\cdot(1+\mu)}{40} & \frac{63\cdot a^2+3\cdot b^2\cdot(1-\mu)}{280\cdot a\cdot b} & \frac{57\cdot(1-3\mu)}{320} & \frac{-1764\cdot a^2+33\cdot b^2\cdot(1-\mu)}{1120\cdot a\cdot b} & \frac{57\cdot(1+\mu)}{320} & \frac{-882\cdot a^2-33\cdot b^2\cdot(1-\mu)}{1120\cdot a\cdot b} \\
\frac{63\cdot a^2\cdot(1-\mu)+12\cdot b^2}{560\cdot a\cdot b} & \frac{3\cdot(1+\mu)}{40} & \frac{63\cdot a^2\cdot(1-\mu)-6\cdot b^2}{280\cdot a\cdot b} & \frac{-3\cdot(1-3\cdot\mu)}{40} & \frac{-441\cdot a^2\cdot(1-\mu)-66\cdot b^2}{1120\cdot a\cdot b} & \frac{-57\cdot(1+\mu)}{320} & \frac{-441\cdot a^2\cdot(1-\mu)+33\cdot b^2}{560\cdot a\cdot b} & \frac{57\cdot(1-3\mu)}{320} \\
\frac{3\cdot(1+\mu)}{40} & \frac{63\cdot a^2+3\cdot b^2\cdot(1-\mu)}{280\cdot ab} & \frac{3\cdot(1-3\cdot\mu)}{40} & \frac{126\cdot a^2-3\cdot b^2\cdot(1-\mu)}{280\cdot a\cdot b} & \frac{-57\cdot(1+\mu)}{320} & \frac{-882\cdot a^2-33\cdot b^2\cdot(1-\mu)}{1120\cdot a\cdot b} & \frac{-57\cdot(1-3\mu)}{320} & \frac{-1764\cdot a^2+33\cdot b^2\cdot(1-\mu)}{1120\cdot a\cdot b}
\end{bmatrix}$$

$$K_{21}=K_{12}{}^T$$

$$K_{22}:=\begin{bmatrix}
\frac{126\cdot a^2\cdot(1-\mu)+27\cdot b^2}{70\cdot a\cdot b} & 0 & \frac{63\cdot a^2\cdot(1-\mu)-27\cdot b^2}{70\cdot a\cdot b} & 0 & \frac{-693\cdot a^2\cdot(1-\mu)-27\cdot b^2}{560\cdot a\cdot b} & \frac{81\cdot(1-3\cdot\mu)}{320} & \frac{-693\cdot a^2\cdot(1-\mu)+54\cdot b^2}{1120\cdot a\cdot b} & \frac{-81\cdot(1+\mu)}{320} \\
0 & \frac{504\cdot a^2+27\cdot b^2\cdot(1-\mu)}{140\cdot a\cdot b} & 0 & \frac{252\cdot a^2-27\cdot b^2\cdot(1-\mu)}{140\cdot a\cdot b} & \frac{-81\cdot(1-3\mu)}{320} & \frac{-2772\cdot a^2-27\cdot b^2\cdot(1-\mu)}{1120\cdot a\cdot b} & \frac{-81\cdot(1+\mu)}{320} & \frac{-1386\cdot a^2+27\cdot b^2\cdot(1-\mu)}{1120\cdot a\cdot b} \\
\frac{63\cdot a^2\cdot(1-\mu)-27\cdot b^2}{70\cdot a\cdot b} & 0 & \frac{126\cdot a^2\cdot(1-\mu)+27\cdot b^2}{70\cdot a\cdot b} & 0 & \frac{-693\cdot a^2\cdot(1-\mu)+54\cdot b^2}{1120\cdot a\cdot b} & \frac{81\cdot(1+\mu)}{320} & \frac{-693\cdot a^2\cdot(1-\mu)-27\cdot b^2}{560\cdot a\cdot b} & \frac{-81\cdot(1-3\mu)}{320} \\
0 & \frac{252\cdot a^2-27\cdot b^2\cdot(1-\mu)}{140\cdot a\cdot b} & 0 & \frac{504\cdot a^2+27\cdot b^2\cdot(1-\mu)}{140\cdot a\cdot b} & \frac{81\cdot(1+\mu)}{320} & \frac{-1386\cdot a^2+27\cdot b^2\cdot(1-\mu)}{1120\cdot a\cdot b} & \frac{81\cdot(1-3\mu)}{320} & \frac{-2772\cdot a^2-27\cdot b^2\cdot(1-\mu)}{1120\cdot a\cdot b} \\
\frac{-693\cdot a^2\cdot(1-\mu)-27\cdot b^2}{560\cdot a\cdot b} & \frac{-81\cdot(1-3\mu)}{320} & \frac{-693\cdot a^2\cdot(1-\mu)+54\cdot b^2}{1120\cdot a\cdot b} & \frac{81\cdot(1+\mu)}{320} & \frac{126\cdot a^2\cdot(1-\mu)+27\cdot b^2}{70\cdot a\cdot b} & 0 & \frac{63\cdot a^2\cdot(1-\mu)-27\cdot b^2}{70\cdot a\cdot b} & 0 \\
\frac{81\cdot(1-3\cdot\mu)}{320} & \frac{-2772\cdot a^2-27\cdot b^2\cdot(1-\mu)}{1120\cdot a\cdot b} & \frac{81\cdot(1+\mu)}{320} & \frac{-1386\cdot a^2+27\cdot b^2\cdot(1-\mu)}{1120\cdot a\cdot b} & 0 & \frac{504\cdot a^2+27\cdot b^2\cdot(1-\mu)}{140\cdot a\cdot b} & 0 & \frac{252\cdot a^2-27\cdot b^2\cdot(1-\mu)}{140\cdot a\cdot b} \\
\frac{-693\cdot a^2\cdot(1-\mu)+54\cdot b^2}{1120\cdot a\cdot b} & \frac{-81\cdot(1+\mu)}{320} & \frac{-693\cdot a^2\cdot(1-\mu)-27\cdot b^2}{560\cdot a\cdot b} & \frac{81\cdot(1-3\mu)}{320} & \frac{63\cdot a^2\cdot(1-\mu)-27\cdot b^2}{70\cdot a\cdot b} & 0 & \frac{126\cdot a^2\cdot(1-\mu)+27\cdot b^2}{70\cdot a\cdot b} & 0 \\
\frac{-81\cdot(1+\mu)}{320} & \frac{-1386\cdot a^2+27\cdot b^2\cdot(1-\mu)}{1120\cdot a\cdot b} & \frac{-81\cdot(1-3\mu)}{320} & \frac{-2772\cdot a^2-27\cdot b^2\cdot(1-\mu)}{1120\cdot a\cdot b} & 0 & \frac{252\cdot a^2-27\cdot b^2\cdot(1-\mu)}{140\cdot a\cdot b} & 0 & \frac{504\cdot a^2+27\cdot b^2\cdot(1-\mu)}{140\cdot a\cdot b}
\end{bmatrix}$$

These equations are manipulated in the software Mathcad for processing. For the above numerical examples, the converged displacement at the top level of the cantilever wall is 0.0511 m which is very close to that by the beam theory at 0.0512 m. However, it should be noted that the beam theory has not taken the Poisson effect into account as the Poisson's ratio is not a design parameter. The finite element analysis is more correct in this respect. Nevertheless, as the displacement function of all three elements are linear, vertical stress distributions in the horizontal direction are also linear, by which the extreme fibre stress follows that of the beam theory at 40000 kPa in the worked examples. So convergence can also be indicated by this stress.

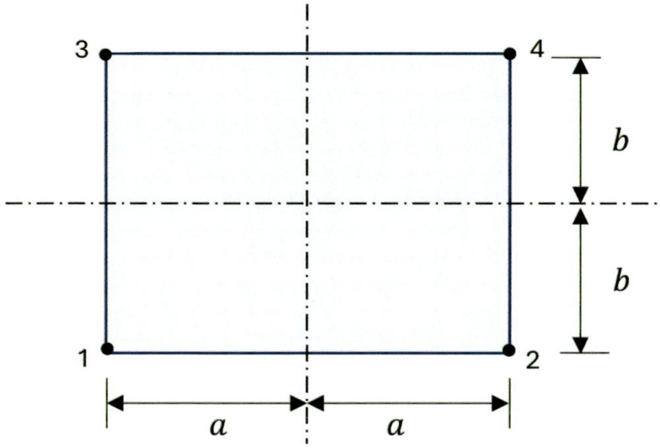

FIGURE A.D10 Rectangular plate bending element.

D.4 THIN PLATE BENDING ELEMENT – NO SHEAR DEFORMATION

While the "plane stress" element discussed in the foregoing deals with in-plane loads – loads acting in the plane of the plate structure, another finite element worth discussing is the "plate bending" element which caters for out-of-plane loads. The element is used to simulate slab elements generally. The stiffness matrix of a rectangular element shown in Figure A.D10 is derived as follows:

A displacement function for a rectangular element can be defined for the out-of-plane displacement w as:

$$w = \alpha_1 + \alpha_2 x + \alpha_3 y + \alpha_4 x^2 + \alpha_5 xy + \alpha_6 y^2 + \alpha_7 x^3 + \alpha_8 x^2 y + \alpha_9 xy^2 + \alpha_{10} y^3 + \alpha_{11} x^3 y + \alpha_{12} xy^3 \tag{D31}$$

The rotations about X and Y axes are as follows:

$$\theta_x = -\frac{\partial w}{\partial y} = -\alpha_3 - \alpha_5 x - 2\alpha_6 y - \alpha_8 x^2 - 2\alpha_9 xy - 3\alpha_{10} y^2 - \alpha_{11} x^3 - 3\alpha_{12} xy^2 \tag{D32}$$

$$\theta_y = \frac{\partial w}{\partial x} = \alpha_2 + 2\alpha_4 x + \alpha_5 y + 3\alpha_7 x^2 + 2\alpha_8 xy + \alpha_9 y^2 + 3\alpha_{11} x^2 y + \alpha_{12} y^3 \tag{D33}$$

So the displacement function can be listed generally as

$$\Delta = \begin{bmatrix} w \\ \theta_x \\ \theta_y \end{bmatrix} = \begin{bmatrix} w \\ -\dfrac{\partial w}{\partial y} \\ \dfrac{\partial w}{\partial x} \end{bmatrix}$$

$$=\begin{bmatrix} 1 & x & y & x^2 & xy & y^2 & x^3 & x^2y & xy^2 & y^3 & x^3y & xy^3 \\ 0 & 0 & -1 & 0 & -x & -2y & 0 & -x^2 & -2xy & -3y^2 & -x^3 & -3xy^2 \\ 0 & 1 & 0 & 2x & y & 0 & 3x^2 & 2xy & y^2 & 0 & 3x^2y & y^3 \end{bmatrix} \begin{bmatrix} \alpha_1 \\ \alpha_2 \\ \alpha_3 \\ \alpha_4 \\ \alpha_5 \\ \alpha_6 \\ \alpha_7 \\ \alpha_8 \\ \alpha_9 \\ \alpha_{10} \\ \alpha_{11} \\ \alpha_{12} \end{bmatrix}$$

$$\text{Or } \Delta = \boldsymbol{B\alpha} \tag{D34}$$

where

$$\boldsymbol{B} = \begin{bmatrix} 1 & x & y & x^2 & xy & y^2 & x^3 & x^2y & xy^2 & y^3 & x^3y & xy^3 \\ 0 & 0 & -1 & 0 & -x & -2y & 0 & -x^2 & -2xy & -3y^2 & -x^3 & -3xy^2 \\ 0 & 1 & 0 & 2x & y & 0 & 3x^2 & 2xy & y^2 & 0 & 3x^2y & y^3 \end{bmatrix}$$

and $\boldsymbol{\alpha} = \lfloor \alpha_1 \quad \alpha_2 \quad \alpha_3 \quad \alpha_4 \quad \alpha_5 \quad \alpha_6 \quad \alpha_7 \quad \alpha_8 \quad \alpha_9 \quad \alpha_{10} \quad \alpha_{11} \quad \alpha_{12} \rfloor^T$

For the corner nodes where $x = \pm a$ and $y = \pm b$

$$\Delta = \begin{bmatrix} 1 & -a & -b & (-a)^2 & (-a)(-b) & (-b)^2 & (-a)^3 & (-a)^2(-b) & (-a)(-b)^2 & (-b)^3 & (-a)^3(-b) & (-a)(-b)^3 \\ 0 & 0 & -1 & 0 & -(-a) & -2(-b) & 0 & -(-a)^2 & -2(-a)(-b) & -3(-b)^2 & -(-a)^3 & -3(-a)(-b)^2 \\ 0 & 1 & 0 & 2(-a) & -b & 0 & 3(-a)^2 & 2(-a)(-b) & (-b)^2 & 0 & 3(-a)^2(-b) & (-b)^3 \\ 1 & a & -b & a^2 & a(-b) & (-b)^2 & a^3 & a^2(-b) & a(-b)^2 & (-b)^3 & a^3(-b) & a(-b)^3 \\ 0 & 0 & -1 & 0 & -a & -2(-b) & 0 & -a^2 & -2a(-b) & -3(-b)^2 & -a^3 & -3a(-b)^2 \\ 0 & 1 & 0 & 2a & -b & 0 & 3a^2 & 2a(-b) & (-b)^2 & 0 & 3a^2(-b) & (-b)^3 \\ 1 & -a & b & (-a)^2 & (-a)b & b^2 & (-a)^3 & (-a)^2 b & (-a)b^2 & b^3 & (-a)^3 b & (-a)b^3 \\ 0 & 0 & -1 & 0 & -(-a) & -2b & 0 & -(-a)^2 & -2(-a)b & -3b^2 & -(-a)^3 & -3(-a)b^2 \\ 0 & 1 & 0 & 2(-a) & b & 0 & 3(-a)^2 & 2(-a)b & b^2 & 0 & 3a^2(-b) & b^3 \\ 1 & a & b & a^2 & ab & b^2 & a^3 & a^2b & ab^2 & b^3 & a^3b & ab^3 \\ 0 & 0 & -1 & 0 & -a & -2b & 0 & -a^2 & -2ab & -3b^2 & -a^3 & -3ab^2 \\ 0 & 1 & 0 & 2a & b & 0 & 3a^2 & 2ab & b^2 & 0 & 3a^2b & b^3 \end{bmatrix} \begin{bmatrix} \alpha_1 \\ \alpha_2 \\ \alpha_3 \\ \alpha_4 \\ \alpha_5 \\ \alpha_6 \\ \alpha_7 \\ \alpha_8 \\ \alpha_9 \\ \alpha_{10} \\ \alpha_{11} \\ \alpha_{12} \end{bmatrix}$$

$$\Delta = \boldsymbol{A}.\boldsymbol{\alpha} \tag{D35}$$

where

$$A=\begin{bmatrix}1 & -a & -b & a^2 & ab & b^2 & -a^3 & -a^2b & -ab^2 & -b^3 & a^3b & ab^3\\ 0 & 0 & -1 & 0 & a & 2b & 0 & -a^2 & -2ab & -3b^2 & a^3 & 3ab^2\\ 0 & 1 & 0 & -2a & -b & 0 & 3a^2 & 2ab & b^2 & 0 & -3a^2b & -b^3\\ 1 & a & -b & a^2 & -ab & b^2 & a^3 & -a^2b & ab^2 & -b^3 & -a^3b & -ab^3\\ 0 & 0 & -1 & 0 & -a & 2b & 0 & -a^2 & 2ab & -3b^2 & -a^3 & -3ab^2\\ 0 & 1 & 0 & 2a & -b & 0 & 3a^2 & -2ab & b^2 & 0 & -3a^2b & -b^3\\ 1 & -a & b & a^2 & -ab & b^2 & -a^3 & a^2b & -ab^2 & b^3 & -a^3b & -ab^3\\ 0 & 0 & -1 & 0 & a & -2b & 0 & -a^2 & 2ab & -3b^2 & a^3 & 3ab^2\\ 0 & 1 & 0 & -2a & b & 0 & 3a^2 & -2ab & b^2 & 0 & -3a^2b & b^3\\ 1 & a & b & a^2 & ab & b^2 & a^3 & a^2b & ab^2 & b^3 & a^3b & ab^3\\ 0 & 0 & -1 & 0 & -a & -2b & 0 & -a^2 & -2ab & -3b^2 & -a^3 & -3ab^2\\ 0 & 1 & 0 & 2a & b & 0 & 3a^2 & 2ab & b^2 & 0 & 3a^2b & b^3\end{bmatrix}$$

$$A^{-1}=\begin{bmatrix}\frac{1}{4} & \frac{b}{8} & -\frac{a}{8} & \frac{1}{4} & \frac{b}{8} & \frac{a}{8} & \frac{1}{4} & \frac{-b}{8} & \frac{-a}{8} & \frac{1}{4} & \frac{-b}{8} & \frac{a}{8}\\ \frac{-3}{8a} & \frac{-b}{8a} & \frac{1}{8} & \frac{3}{8a} & \frac{b}{8a} & \frac{1}{8} & \frac{-3}{8a} & \frac{b}{8a} & \frac{1}{8} & \frac{3}{8a} & \frac{-b}{8a} & \frac{1}{8}\\ \frac{-3}{8b} & \frac{-1}{8} & \frac{a}{8b} & \frac{-3}{8b} & \frac{-1}{8} & \frac{-a}{8b} & \frac{3}{8b} & \frac{-1}{8} & \frac{-a}{8b} & \frac{3}{8b} & \frac{-1}{8} & \frac{a}{8b}\\ 0 & 0 & \frac{1}{8a} & 0 & 0 & \frac{-1}{8a} & 0 & 0 & \frac{1}{8a} & 0 & 0 & \frac{-1}{8a}\\ \frac{1}{2ab} & \frac{1}{8a} & \frac{-1}{8b} & \frac{-1}{2ab} & \frac{-1}{8a} & \frac{-1}{8b} & \frac{-1}{2ab} & \frac{1}{8a} & \frac{1}{8b} & \frac{1}{2ab} & \frac{-1}{8a} & \frac{1}{8b}\\ 0 & \frac{-1}{8b} & 0 & 0 & \frac{-1}{8b} & 0 & 0 & \frac{1}{8b} & 0 & 0 & \frac{1}{8b} & 0\\ \frac{1}{8a^3} & 0 & \frac{-1}{8a^2} & \frac{-1}{8a^3} & 0 & \frac{-1}{8a^2} & \frac{1}{8a^3} & 0 & \frac{-1}{8a^2} & \frac{-1}{8a^3} & 0 & \frac{-1}{8a^2}\\ 0 & 0 & \frac{-1}{8ab} & 0 & 0 & \frac{1}{8ab} & 0 & 0 & \frac{1}{8ab} & 0 & 0 & \frac{-1}{8ab}\\ 0 & \frac{1}{8ab} & 0 & 0 & \frac{-1}{8ab} & 0 & 0 & \frac{-1}{8ab} & 0 & 0 & \frac{1}{8ab} & 0\\ \frac{1}{8b^3} & \frac{1}{8b^2} & 0 & \frac{1}{8b^3} & \frac{1}{8b^2} & 0 & \frac{-1}{8b^3} & \frac{1}{8b^2} & 0 & \frac{-1}{8b^3} & \frac{1}{8b^2} & 0\\ \frac{-1}{8a^3b} & 0 & \frac{1}{8a^2b} & \frac{1}{8a^3b} & 0 & \frac{1}{8a^2b} & \frac{1}{8a^3b} & 0 & \frac{-1}{8a^2b} & \frac{-1}{8a^3b} & 0 & \frac{-1}{8a^2b}\\ \frac{-1}{8ab^3} & \frac{-1}{8ab^2} & 0 & \frac{1}{8ab^3} & \frac{1}{8ab^2} & 0 & \frac{1}{8ab^3} & \frac{-1}{8ab^2} & 0 & \frac{-1}{8ab^3} & \frac{1}{8ab^2} & 0\end{bmatrix}$$

$$\boldsymbol{\alpha}=\boldsymbol{A}^{-1}.\Delta \qquad \text{(D36)}$$

The "stresses" are taken as the bending and twisting moments per unit width which are

$$M_x = -\int_{-0.5t}^{0.5t} \sigma_y z dz = \frac{Et^3}{12\left(1-\mu^2\right)}\left(\frac{\partial^2 w}{\partial y^2} + \mu\frac{\partial^2 w}{\partial x^2}\right)$$

$$M_y = +\int_{-0.5t}^{0.5t} \sigma_x z dz = \frac{-Et^3}{12\left(1-\mu^2\right)}\left(\mu\frac{\partial^2 w}{\partial y^2} + \frac{\partial^2 w}{\partial x^2}\right)$$

$$M_{xy} = -\int_{-0.5t}^{0.5t} \tau_{xy} z dz = \frac{Et^3}{12\left(1-\mu^2\right)}\left(1-\mu\right)\left(\frac{\partial^2 w}{\partial x \partial y}\right)$$

$$M_{yx} = \int_{-0.5t}^{0.5t} \tau_{xy} z dz = \frac{-Et^3}{12\left(1-\mu^2\right)}\left(1-\mu\right)\left(\frac{\partial^2 w}{\partial x \partial y}\right)$$

$$M_{xy} = -M_{yx}$$

$$\begin{bmatrix} M_x \\ M_y \\ M_{xy} \end{bmatrix} = \frac{Et^3}{12\left(1-\mu^2\right)} \begin{bmatrix} 1 & -\mu & 0 \\ -\mu & 1 & 0 \\ 0 & 0 & \frac{1-\mu}{2} \end{bmatrix} \begin{bmatrix} \frac{\partial^2 w}{\partial y^2} \\ -\frac{\partial^2 w}{\partial x^2} \\ 2\frac{\partial^2 w}{\partial x \partial y} \end{bmatrix} \tag{D37}$$

where $\frac{\partial^2 w}{\partial y^2} = \frac{\partial \theta_x}{\partial y}$ with θ_x as rotation about the X-axis

$-\frac{\partial^2 w}{\partial x^2} = \frac{\partial \theta_x}{\partial y}$ with θ_y as rotation about the Y-axis

$2\frac{\partial^2 w}{\partial x \partial y}$ is the absolute magnitude of $\frac{\partial \theta_x}{\partial y}$ and $\frac{\partial \theta_y}{\partial x}$.

$\frac{\partial^2 w}{\partial y^2}$ and $-\frac{\partial^2 w}{\partial x^2}$ are respectively the bending curvatures about x and y axes and $2\frac{\partial^2 w}{\partial x \partial y}$ are the sum of absolute twisting curvatures about x and y axes. These equations are based on the plane remains plane assumption and this formulation is termed the "Kirchhoff Plate" theory.

The strain energy in an element of dimension *dxdy* is therefore

$$E = \frac{1}{2}\left[M_x dx \left(\frac{\partial^2 w}{\partial y^2} dy \right) + M_y dy \left(-\frac{\partial^2 w}{\partial x^2} dx \right) + M_{xy} dy \frac{\partial^2 w}{\partial x \partial y} dx + \left(-M_{yx} dx \frac{\partial^2 w}{\partial x \partial y} dy \right) \right]$$

$$= \frac{1}{2}\left[M_x \frac{\partial^2 w}{\partial y^2} + M_y \frac{\partial^2 w}{\partial x^2} + 2M_{xy} \frac{\partial^2 w}{\partial x \partial y} \right] dxdy$$

(D38)

So $\boldsymbol{M} = \begin{bmatrix} M_x \\ M_y \\ M_{xy} \end{bmatrix}$ as the bending and twisting moment per unit width is the stress function and $\boldsymbol{\kappa} = \begin{bmatrix} \dfrac{\partial^2 w}{\partial y^2} \\ -\dfrac{\partial^2 w}{\partial x^2} \\ 2\dfrac{\partial^2 w}{\partial x \partial y} \end{bmatrix}$ is the curvature is the strain function.

Equation D38 can be written as

$$\boldsymbol{E} = \frac{1}{2} \boldsymbol{M}^T \boldsymbol{\kappa} \tag{D39}$$

Referring to the displacement function listed in Equation D31, the strain function can be formed:

Furthermore, $\dfrac{Et^3}{12\left(1-\mu^2\right)} \begin{bmatrix} 1 & -\mu & 0 \\ -\mu & 1 & 0 \\ 0 & 0 & \dfrac{1-\mu}{2} \end{bmatrix}$ in Equation D37 is the $\boldsymbol{D}$ matrix relating stress and strain.

$$\boldsymbol{\kappa} = \boldsymbol{B\alpha} = \begin{bmatrix} 0 & 0 & 0 & 0 & 0 & 2 & 0 & 0 & 2x & 6y & 0 & 6xy \\ 0 & 0 & 0 & -2 & 0 & 0 & -6x & -2y & 0 & 0 & -6xy & 0 \\ 0 & 0 & 0 & 0 & 2 & 0 & 0 & 4x & 4y & 0 & 6x^2 & 6y^3 \end{bmatrix} \begin{bmatrix} \alpha_1 \\ \alpha_2 \\ \alpha_3 \\ \alpha_4 \\ \alpha_5 \\ \alpha_6 \\ \alpha_7 \\ \alpha_8 \\ \alpha_9 \\ \alpha_{10} \\ \alpha_{11} \\ \alpha_{12} \end{bmatrix} \tag{D40}$$

By Equation D36, $\boldsymbol{\kappa} = \boldsymbol{B}.\boldsymbol{A}^{-1}.\Delta$

$$\boldsymbol{M} = \boldsymbol{D}.\boldsymbol{\kappa} = \boldsymbol{D}.\boldsymbol{B}.\boldsymbol{A}^{-1}.\Delta \tag{D41}$$

The strain energy is then

$$\int_{-b}^{b}\int_{-a}^{a} \kappa^T.Mdxdy = \int_{-b}^{b}\int_{-a}^{a} \Delta^T A^{-1} B^T.D.B.A^{-1}.\Delta dxdy$$

$$= \Delta^T \left(A^{-1}\right)^T \left(\int_{-b}^{b}\int_{-a}^{a} B^T.D.Bdxdy \right) A^{-1}\Delta \tag{D42}$$

Taking the external load as $P = \begin{pmatrix} S & M_x & M_y \end{pmatrix}^T$ where S is the vertical shear, M_x and M_y are the moments about the X-axis and Y-axis respectively. For the 4 nodes,

$P = \begin{pmatrix} S_1 & M_{x1} & M_{y1} & S_2 & M_{x2} & M_{y2} & S_3 & M_{x3} & M_{y3} & S_4 & M_{x4} & M_{y4} \end{pmatrix}^T$ and Δ defined in Equation D34, $\Delta^T P$ is the external work done. Equating, we have

$$\Delta^T P = \Delta^T \left(A^{-1}\right)^T \left(\int_{-b}^{b}\int_{-a}^{a} B^T.D.Bdxdy \right) A^{-1}\Delta$$

$$\Rightarrow P = \left[\left(A^{-1}\right)^T \left(\int_{-b}^{b}\int_{-a}^{a} B^T.D.Bdxdy \right) A^{-1} \right] \Delta$$

So $\left(A^{-1}\right)^{T}\left(\int_{-b}^{b}\int_{-a}^{a} B^{T}.D.Bdxdy\right)A^{-1}$ is the stiffness matrix denoted by

$$K=\left(A^{-1}\right)^{T}\left(\int_{-b}^{b}\int_{-a}^{a} B^{T}.D.Bdxdy\right)A^{-1} \tag{D43}$$

With the above derivation, K can be analytically formed as

$$K=\frac{Et}{1-\mu^{2}}\begin{bmatrix} K_{11} & K_{12} \\ K_{21} & K_{22} \end{bmatrix} \tag{D44}$$

where

$$K_{11}:=\begin{bmatrix}
\frac{7-2\cdot\mu}{10\cdot a\cdot b}+\frac{a}{b^3}+\frac{b}{a^3} & \frac{1+4\cdot\mu}{10\cdot a}+\frac{a}{b^2} & \frac{-1-4\cdot\mu}{10\cdot b}-\frac{b}{a^2} & \frac{-7+2\cdot\mu}{10\cdot a\cdot b}+\frac{a}{2b^3}-\frac{b}{a^3} & \frac{-1-4\cdot\mu}{10\cdot a}+\frac{a}{2b^2} & \frac{-1+\mu}{10\cdot b}-\frac{b}{a^2} \\
\frac{1+4\cdot\mu}{10a}+\frac{a}{b^2} & \frac{4a}{3\cdot b}+\frac{4\cdot b(1-\mu)}{15\cdot a} & -\mu & \frac{-1-4\cdot\mu}{10\cdot a}+\frac{a}{2b^2} & \frac{-4\cdot b(1-\mu)}{15a}+\frac{2\cdot a}{3\cdot b} & 0 \\
\frac{-1-4\cdot\mu}{10\cdot b}-\frac{b}{a^2} & -\mu & \frac{4\cdot a\cdot(1-\mu)}{15\cdot b}+\frac{4\cdot b}{3\cdot a} & \frac{1-\mu}{10\cdot b}+\frac{b}{a^2} & 0 & \frac{-a\cdot(1-\mu)}{15\cdot b}+\frac{2\cdot b}{3\cdot a} \\
\frac{-7+2\cdot\mu}{10\cdot a\cdot b}+\frac{a}{2b^3}-\frac{b}{a^3} & \frac{-1-4\cdot\mu}{10a}+\frac{a}{2b^2} & \frac{1-\mu}{10\cdot b}+\frac{b}{a^2} & \frac{7-2\cdot\mu}{10\cdot a\cdot b}+\frac{a}{b^3}+\frac{b}{a^3} & \frac{1+4\cdot\mu}{10\cdot a}+\frac{a}{b^2} & \frac{1+4\cdot\mu}{10\cdot b}+\frac{b}{a^2} \\
\frac{-1-4\cdot\mu}{10a}+\frac{a}{2b^2} & \frac{-4\cdot b\cdot(1-\mu)}{15\cdot a}+\frac{2a}{3\cdot b} & 0 & \frac{1+4\cdot\mu}{10\cdot a}+\frac{a}{b^2} & \frac{4\cdot a}{3\cdot b}+\frac{4\cdot b(1-\mu)}{15\cdot a} & \mu \\
\frac{-1+\mu}{10\cdot b}-\frac{b}{a^2} & 0 & \frac{-a\cdot(1-\mu)}{15\cdot b}+\frac{2\cdot b}{3\cdot a} & \frac{1+4\cdot\mu}{10\cdot b}+\frac{b}{a^2} & \mu & \frac{4\cdot a\cdot(1-\mu)}{15\cdot b}+\frac{4\cdot b}{3\cdot a}
\end{bmatrix}$$

$$K_{12}:=\begin{bmatrix}
\frac{-7+2\cdot\mu}{10\cdot a\cdot b}+\frac{b}{2a^3}-\frac{a}{b^3} & \frac{-1+\mu}{10\cdot a}-\frac{a}{b^2} & \frac{1+4\cdot\mu}{10\cdot b}-\frac{b}{2a^2} & \frac{7-2\cdot\mu}{10\cdot a\cdot b}-\frac{b}{2a^3}-\frac{a}{2b^3} & \frac{1-\mu}{10\cdot a}-\frac{a}{2b^2} & \frac{1-\mu}{10\cdot b}-\frac{b}{2a^2} \\
\frac{1-\mu}{10\cdot a}+\frac{a}{b^2} & \frac{-b\cdot(1-\mu)}{15\cdot a}+\frac{2\cdot a}{3\cdot b} & 0 & \frac{-1+\mu}{10\cdot a}+\frac{a}{2b^2} & \frac{b\cdot(1-\mu)}{15\cdot a}+\frac{a}{3\cdot b} & 0 \\
\frac{1+4\cdot\mu}{10\cdot b}-\frac{b}{2a^2} & 0 & \frac{-4\cdot a\cdot(1-\mu)}{15\cdot b}+\frac{2\cdot b}{3\cdot a} & \frac{-1+\mu}{10\cdot b}+\frac{b}{2a^2} & 0 & \frac{a\cdot(1-\mu)}{15\cdot b}+\frac{1}{3}\cdot\frac{b}{a} \\
\frac{7-2\cdot\mu}{10\cdot a\cdot b}-\frac{b}{2a^3}-\frac{a}{2b^3} & \frac{1-\mu}{10\cdot a}-\frac{a}{2b^2} & \frac{-1+\mu}{10\cdot b}+\frac{b}{2a^2} & \frac{-7+2\cdot\mu}{10\cdot a\cdot b}+\frac{b}{2a^3}-\frac{a}{b^3} & \frac{-1+\mu}{10\cdot a}-\frac{a}{b^2} & \frac{-1-4\cdot\mu}{10\cdot b}+\frac{b}{2a^2} \\
\frac{-1+\mu}{10\cdot a}+\frac{a}{2b^2} & \frac{b\cdot(1-\mu)}{15\cdot a}+\frac{a}{3\cdot b} & 0 & \frac{1-\mu}{10\cdot a}+\frac{a}{b^2} & \frac{-b\cdot(1-\mu)}{15\cdot a}+\frac{2}{3}\cdot\frac{a}{b} & 0 \\
\frac{1-\mu}{10\cdot b}-\frac{b}{2a^2} & 0 & \frac{a\cdot(1-\mu)}{15\cdot b}+\frac{b}{3\cdot a} & \frac{-1-4\cdot\mu}{10\cdot b}+\frac{b}{2a^2} & 0 & \frac{-4\cdot a\cdot(1-\mu)}{15\cdot b}+\frac{2\cdot b}{3\cdot a}
\end{bmatrix}$$

$$K_{21}=K_{12}{}^{T}$$

$$K_{22}:=\begin{bmatrix}
\frac{7-2\cdot\mu}{10\cdot a\cdot b}+\frac{a}{b^3}+\frac{b}{a^3} & \frac{-1-4\cdot\mu}{10\cdot a}-\frac{a}{b^2} & \frac{-1-4\cdot\mu}{10\cdot b}-\frac{b}{a^2} & \frac{-7+2\cdot\mu}{10\cdot a\cdot b}+\frac{a}{2\cdot b^3}-\frac{b}{a^3} & \frac{-a}{2\cdot b^2}+\frac{1+4\cdot\mu}{10\cdot a} & \frac{-1+\mu}{10\cdot b}-\frac{b}{a^2} \\
\frac{-1-4\cdot\mu}{10\cdot a}-\frac{a}{b^2} & \frac{4\cdot a}{3\cdot b}+\frac{4\cdot b\cdot(1-\mu)}{15\cdot a} & \mu & \frac{-a}{2\cdot b^2}+\frac{1+4\cdot\mu}{10\cdot a} & \frac{-4\cdot b\cdot(1-\mu)}{15\cdot a}+\frac{2\cdot a}{3\cdot b} & 0 \\
\frac{-1-4\cdot\mu}{10\cdot b}-\frac{b}{a^2} & \mu & \frac{4\cdot a\cdot(1-\mu)}{15\cdot b}+\frac{4\cdot b}{3\cdot a} & \frac{1-\mu}{10\cdot b}+\frac{b}{a^2} & 0 & \frac{-a\cdot(1-\mu)}{15\cdot b}+\frac{2\cdot b}{3\cdot a} \\
\frac{-7+2\cdot\mu}{10\cdot a\cdot b}+\frac{a}{2\cdot b^3}-\frac{b}{a^3} & \frac{-a}{2\cdot b^2}+\frac{1+4\cdot\mu}{10\cdot a} & \frac{1-\mu}{10\cdot b}+\frac{b}{a^2} & \frac{7-2\cdot\mu}{10\cdot a\cdot b}+\frac{a}{b^3}+\frac{b}{a^3} & \frac{-1-4\cdot\mu}{10\cdot a}-\frac{a}{b^2} & \frac{1+4\cdot\mu}{10\cdot b}+\frac{b}{a^2} \\
\frac{-a}{2\cdot b^2}+\frac{1+4\cdot\mu}{10\cdot a} & \frac{-4\cdot b\cdot(1-\mu)}{15\cdot a}+\frac{2\cdot a}{3\cdot b} & 0 & \frac{-1-4\cdot\mu}{10\cdot a}-\frac{a}{b^2} & \frac{4\cdot a}{3\cdot b}+\frac{4\cdot b\cdot(1-\mu)}{15\cdot a} & -\mu \\
\frac{-1+\mu}{10\cdot b}-\frac{b}{a^2} & 0 & \frac{-a\cdot(1-\mu)}{15\cdot b}+\frac{2\cdot b}{3\cdot a} & \frac{1+4\cdot\mu}{10\cdot b}+\frac{b}{a^2} & -\mu & \frac{4\cdot a\cdot(1-\mu)}{15\cdot b}+\frac{4\cdot b}{3\cdot a}
\end{bmatrix}$$

As different from the plane stress elements where the basic shape functions are linear for the 4 node element, that of the plate bending element is up to cubic as revealed in Equation D31. The performance in convergence should therefore be better.

A numerical trial is performed on a plate element of $a = 1.8\text{m}$; $b = 1.2\text{m}$ $t = 0.8$ m, i.e., plan dimension of 1.8 m by 1.2 m and thickness 0.8 m. The Young's modulus is taken as 25000000 kN/m² and the Poisson's ratio $\mu = 0.2$. The stiffness matrix is first built by Equation D44. Nodes 1 and 2, as shown in Figure A.D10, are restrained from vertical translation and rotation, i.e., only K_{22} is retained in modifying the stiffness matrix to support conditions. So the plate becomes a cantilever plate.

A vertical load of 1000 kN is applied to each of nodes 3 and 4 (Figure A.D10). The deflections at nodes 3 and 4 after analysis are 0.00232 m. Comparing with a cantilever beam of section $2a$ x t and length $2b = 2.4\text{m}$ under 2000 kN at its free end with deflection $\dfrac{PL^3}{3EI} = \dfrac{2000 \times 2.4^3}{3 \times 25000000 \times \left(3.6 \times 0.8^3/12\right)} = 0.0024\text{m}$, the analysis by the finite element based on this 4 node element is quite close.

The bending moments M_x and M_y are also calculated by Equation D41 with the deflection determined; the contours of them are shown in Figure A.D11. The maximum bending moment along the edge of nodes 1 and 2 is constant at 1333.33 kNm/m along the edge of nodes 1 and 2, giving a total bending moment of $1333.33 \times 3.6 = 4800\text{kNm}$.

Comparing with the cantilever beam, the support moment is $2000 \times 2.4 = 4800$ kNm; the results by the finite element method is exact. So the element converges at a single element.

The convergence can be explained by the basic shape function indicated by Equation D31 which is a cubic parabolic relation. The cubic parabolic relation is exact for a beam element under point load. In fact, the difference in deflection as revealed by the foregoing calculation is due to the Poisson effect as exhibited by the plate element through the use of Poisson's ratio. Identical results in deflection will be arrived at if Poisson's ratio is set to 0.

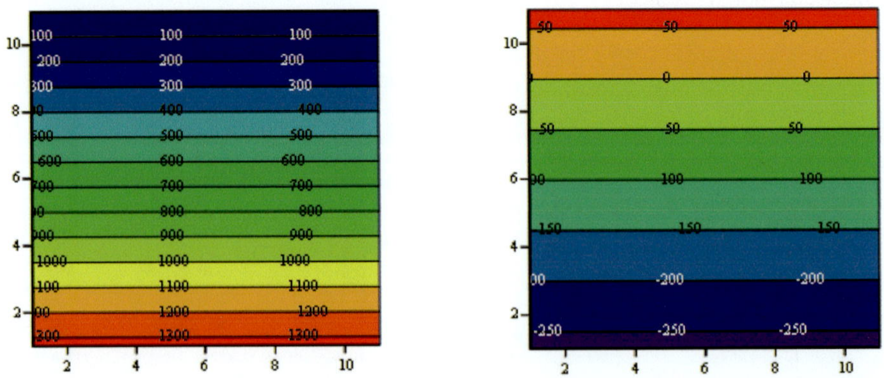

FIGURE A.D11 Mx and my contours of the plate bending element.

However, the aforementioned element is purely flexural with no account for the shear deformation, which may be substantial for a thick structure such as a pile cap and transfer plate. More advanced elements have been developed. Examples are formulations by Law and Cheng (2015), Cheng and Law (2008) and Law et al. (2011). In each of these elements, the shear stiffness matrix is added to the flexural stiffness matrix for analysis.

D.5 RECTANGULAR SHEAR DEFORMABLE BEAM PLATE BASED ON SHEAR DEFORMABLE BEAM

The Kirchhoff plate developed in Section D.4 does not allow out-of-plane shear deformation, which may not be adequately accurate for a thick plate structure with significant out-of-plane shear deformation that would affect out-of-plane deformation and subsequently load distribution. Cheng and Law (2006) has developed a shear deformable plate bending element which will be described in detail in this sub-section.

The formulation of a shear deformable beam, as shown in Figure A.D12 is first given, to which the shear deformable plate would be referred to. In the figure, the total angular rotation of a section as $\frac{d\omega}{ds}$ is the sum of bending and shear deformation as $\psi + \gamma$. The distance along the beam is also normalized as $r = s / L$.

ω as the out of plane deformation of the beam is simulated by a third degree polynomial.

$$\omega = \alpha_0 + \alpha_1 r + \alpha_2 r^2 + \alpha_3 r^3 \tag{D45}$$

$$\psi = \beta_0 + \beta_1 r + \beta_2 r^2 \tag{D46}$$

$$\frac{d\omega}{ds} = \frac{1}{L}\frac{d\omega}{dr} = \psi + \gamma \tag{D47}$$

Under constant shear force within the beam, the shear strain is constant at

$$\gamma = \gamma_0 \tag{D48}$$

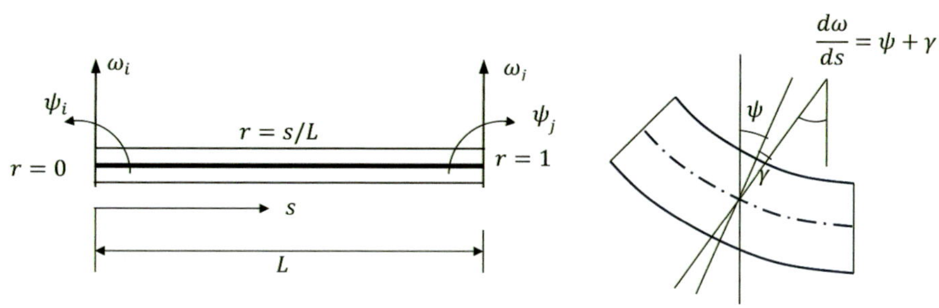

FIGURE A.D12 Shear deformable beam.

Under the boundary conditions of $r = 0$, $\omega = \omega_i$, $\psi = \psi_i$; $r = 1$, $\omega = \omega_j$, $\psi = \psi_j$, combining Equations D45 to D48 for solving the α and β coefficients, we may list

$$\omega = (1-r)^2(1+2r)\omega_i + r^2(3-2r)\omega_j + Lr(1-r)^2\psi_i + Lr^2(-1+r)\psi_j + Lr(1-r)(1-2r)\gamma_0 \tag{D49}$$

$$\psi = \frac{6r}{L}(-1+r)\omega_i + \frac{6r}{L}(1-r)\omega_j + (1-4r+3r^2)\psi_i + r(-2+3r)\psi_j + 6r(-1+r)\gamma_0 \tag{D50}$$

Bending energy of the shear deformable beam per unit width is

$$U_B = \int_0^L \frac{M^2}{2EI}dx = \frac{EIL}{2}\int_0^L\left(\frac{M}{EI}\right)^2 dr = \frac{EIL}{2}\int_0^L \kappa^2 dr = \frac{Et^3}{12}\frac{L}{2}\int_0^L\left(-\frac{d\psi}{Ldr}\right)^2 dr \tag{D51}$$

where κ is the curvature given by $\kappa = -\frac{d\psi}{ds} = -\frac{1}{L}\frac{d\psi}{ds}$ and t is the beam depth.

By differentiating Equation D49

$$\frac{1}{L}\frac{d\psi}{ds} = \frac{-6+12r}{L^2}\omega_i + \frac{6-12r}{L^2}\omega_j + \frac{-4+6r}{L}\psi_i + \frac{-2+6r}{L}\psi_j + \frac{-6+12r}{L}\gamma_0 \tag{D52}$$

Substituting into Equation D51, integrating and re-arranging, U_B can be expressed as

$$U_B = \frac{Et^3}{12}\frac{L}{2}\left[\left(\frac{-\psi_i}{L} + \frac{\psi_j}{L}\right)^2 + \frac{1}{12}\left(\frac{12}{L^2}\omega_i + \frac{-12}{L^2}\omega_j + \frac{6}{L}\psi_i + \frac{6}{L}\psi_j + \frac{12}{L}\gamma_0\right)^2\right]$$

(D53)

Shear energy U_S can be expressed as

$$U_S = \int_0^L \frac{S^2}{2GA_r}dx = \frac{GA_rL}{2}\int_0^L\left(\frac{S}{GA_r}\right)^2 dr = \frac{5EtL}{24(1+\mu)}\gamma_0^2 \tag{D54}$$

Total energy is

$$U = U_B + U_S \tag{D55}$$

where μ is the Poisson's ratio and A_r as the area for shear energy determination is taken as 5/6 of the actual area because of the parabolic stress distribution of the shear stress across the section.

By applying the minimum energy principle, the equilibrium point is obtained by minimizing the total energy with respect to all the dependent parameters. So, one can perform $\frac{\partial U}{\partial \gamma_0} = \frac{\partial U_B}{\partial \gamma_0} + \frac{\partial U_S}{\partial \gamma_0} = 0$ on Equation D55. After substituting from Equations D52 and D53, differentiating, and re-arranging,

$$\gamma_0 = \frac{-2\delta}{L}\omega_i + \frac{2\delta}{L}\omega_j - \delta\psi_i - \delta\psi_j \tag{D56}$$

$$\delta = \frac{6t^2(1+\mu)}{5L^2 + 12(1+\mu)t^2} \tag{D57}$$

Let the displacement vector be $\begin{bmatrix} \omega_i & \psi_i & \omega_j & \psi_j \end{bmatrix}$, by Equation D56, the strain displacement B matrix for shear is

$$B_S = \begin{bmatrix} \frac{-2\delta}{L} & -\delta & \frac{2\delta}{L} & -\delta \end{bmatrix} \tag{D58}$$

By substituting Equation D55 into Equation D51 and re-arranging, the strain displacement B matrix for bending is

$$B_b = \begin{bmatrix} \frac{6(1-2r)(1-2\delta)}{L^2} & \frac{1+3(1-2r)(1-2\delta)}{L} & \frac{-6(1-2r)(1-2\delta)}{L^2} & \frac{-1+3(1-2r)(1-2\delta)}{L} \end{bmatrix} \tag{D59}$$

So, the stiffness matrix for shear is $K = K_s + K_b$ with

$$K_s = \int_0^1 B_S{}^T G A_r B_s L dr = GA_r \begin{bmatrix} 4\delta^2/L & 2\delta^2 & -4\delta^2/L & 2\delta^2 \\ 2\delta^2 & \delta^2 L & -2\delta^2 & \delta^2 L \\ -4\delta^2/L & -2\delta^2 & 4\delta^2/L & -2\delta^2 \\ 2\delta^2 & \delta^2 L & -2\delta^2 & \delta^2 L \end{bmatrix} \tag{D60}$$

$$K_b = EI \begin{bmatrix} \frac{12}{L^3}(1-2\delta)^2 & \frac{6}{L^2}(1-2\delta)^2 & \frac{-12}{L^3}(1-2\delta)^2 & \frac{6}{L^2}(1-2\delta)^2 \\ \frac{6}{L^2}(1-2\delta)^2 & \frac{4}{L}(1-3\delta+3\delta^2) & \frac{-6}{L^2}(1-2\delta)^2 & \frac{2}{L}(1-6\delta+6\delta^2) \\ \frac{-12}{L^3}(1-2\delta)^2 & \frac{-6}{L^2}(1-2\delta)^2 & \frac{12}{L^3}(1-2\delta)^2 & \frac{-6}{L^2}(1-2\delta)^2 \\ \frac{6}{L^2}(1-2\delta)^2 & \frac{2}{L}(1-6\delta+6\delta^2) & \frac{-6}{L^2}(1-2\delta)^2 & \frac{4}{L}(1-3\delta+3\delta^2) \end{bmatrix}$$

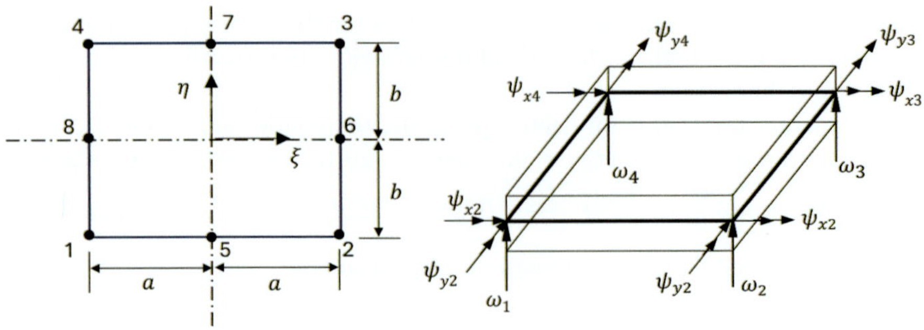

FIGURE A.D13 Pseudo 8-node thick plate element and degrees of freedom.

If shear deformation is ignored with $\delta \rightarrow 0$, the stiffness converges to that of the classical beam theory.

At $r = 1/2$, substituting into Equations D59 and D56 for γ_0

$$\psi_{r=\frac{1}{2}} = \frac{-3}{L}(1-2\delta)\omega_i + \frac{3}{L}(1-2\delta)\omega_j + \left(\frac{-1}{4}+\frac{3}{2}\delta\right)\psi_i + \left(\frac{-1}{4}+\frac{3}{2}\delta\right)\psi_j \quad \text{(D61)}$$

Consider a rectangular plate of corner nodes 1, 2, 3, 4, side lengths 2a and 2b as shown in Figure A.D13. The plate also comprises fictitious nodes 5, 6, 7, and 8 which will be eliminated so that only the corner nodes will be retained in the stiffness matrix finally formed.

The nodal displacement matrix is

$$a^e = \begin{bmatrix} \omega_1 & \psi_{x1} & \psi_{y1} & \omega_2 & \psi_{x2} & \psi_{y2} & \omega_3 & \psi_{x3} & \psi_{y3} & \omega_4 & \psi_{x4} & \psi_{y4} \end{bmatrix}^T$$

By Equation D56, the shear strain along the edges $\overline{12}$, $\overline{23}$, $\overline{34}, \overline{41}$ are

$$\begin{cases} \gamma_{s12} = -\dfrac{\delta_a}{a}\omega_1 + \dfrac{\delta_a}{a}\omega_2 + \delta_a\psi_{y1} + \delta_a\psi_{y2} \\ \gamma_{s23} = -\dfrac{\delta_b}{b}\omega_2 + \dfrac{\delta_b}{b}\omega_3 - \delta_b\psi_{x2} - \delta_b\psi_{x3} \\ \gamma_{s34} = -\dfrac{\delta_a}{a}\omega_4 + \dfrac{\delta_a}{a}\omega_3 + \delta_a\psi_{y4} + \delta_a\psi_{y3} \\ \gamma_{s41} = -\dfrac{\delta_b}{b}\omega_1 + \dfrac{\delta_b}{b}\omega_4 - \delta_b\psi_{y1} - \delta_b\psi_{y4} \end{cases} \quad \text{(D62)}$$

$$\delta_a = \frac{(t/2a)^2(1+\mu)}{5/6+2(t/2a)^2(1+\mu)} = \frac{3t^2(1+\mu)}{10a^2+6t^2(1+\mu)}; \quad \delta_b = \frac{3t^2(1+\mu)}{10b^2+6t^2(1+\mu)} \quad \text{(D63)}$$

and t is the thickness of the plate, μ is the Poisson's ratio.

The shear strain γ within the element can be expressed as

$$\begin{cases} \gamma_x = \gamma_{x1} N_1^{\,0} + \gamma_{x2} N_2^{\,0} + \gamma_{x3} N_3^{\,0} + \gamma_{x4} N_4^{\,0} \\ \gamma_y = \gamma_{y1} N_1^{\,0} + \gamma_{y2} N_2^{\,0} + \gamma_{y3} N_3^{\,0} + \gamma_{y4} N_4^{\,0} \end{cases} \tag{D64}$$

where $N_1^{\,0} = \dfrac{(1-\xi)(1-\eta)}{4}$; $\quad N_2^{\,0} = \dfrac{(1+\xi)(1-\eta)}{4}$; $\quad N_3^{\,0} = \dfrac{(1+\xi)(1+\eta)}{4}$;

$N_4^{\,0} = \dfrac{(1-\xi)(1+\eta)}{4}$

with $\xi = x/a$ and $\eta = y/b$.

Substituting Equations D62 and D63 into Equation D64 and simplifying, we may express γ_x and γ_y in terms of the nodal displacement and form the B matrix for shear given by $\gamma = B_s a^e$.

$$B_s = \begin{bmatrix} \frac{-(1-\eta)\delta_a}{2a} & 0 & \frac{(1-\eta)\delta_a}{2} & \frac{(1-\eta)\delta_a}{2a} & 0 & \frac{(1-\eta)\delta_a}{2} & \frac{(1+\eta)\delta_a}{2a} & 0 & \frac{(1+\eta)\delta_a}{2} & \frac{-(1+\eta)\delta_a}{2a} & 0 & \frac{(1+\eta)\delta_a}{2} \\ \frac{-(1-\xi)\delta_b}{2b} & \frac{-(1-\xi)\delta_b}{2} & 0 & \frac{-(1+\xi)\delta_b}{2b} & \frac{-(1+\xi)\delta_b}{2} & 0 & \frac{(1+\xi)\delta_b}{2b} & \frac{-(1+\xi)\delta_b}{2} & 0 & \frac{(1-\xi)\delta_b}{2b} & \frac{-(1-\xi)\delta_b}{2} & 0 \end{bmatrix} \tag{D65}$$

The D matrix for shear is

$$D_s = G\frac{5}{6}t\begin{bmatrix} 1 & 0 \\ 0 & 1 \end{bmatrix} = \frac{5Et}{12(1+\mu)}\begin{bmatrix} 1 & 0 \\ 0 & 1 \end{bmatrix} \tag{D66}$$

The stiffness matrix for shear is formed by expanding and integrating the following expression

$$K_s^{\,e} = \int_{-b}^{b}\int_{-a}^{a} B_s^{\,T} D_s B_s \,dx dy = ab\int_{-1}^{1}\int_{-1}^{1} B_s^{\,T} D_s B_s \,d\xi d\eta \text{'} \tag{D67}$$

In the formulation of the bending stiffness matrix, normal (in the vertical plane perpendicular to the edge) and tangential rotations (in the vertical plane containing the edge) of the mid-sides nodes 5, 6, 7, 8 at the 4 edges as shown in Figure A.D13 are first determined. Equation D60 will be used to formulate the tangential rotations whilst the normal rotations will be taken as the average of the end nodes. ψ_{x5} and ψ_{y5} are listed as illustrations while others can be similarly formed:

$$\psi_{x5} = \frac{1}{2}(\psi_{x1} + \psi_{x1}) \tag{D68a}$$

$$\psi_{y5} = \frac{3}{L}(1-2\delta)\omega_1 - \frac{3}{L}(1-2\delta)\omega_2 + \left(\frac{-1}{4} + \frac{3}{2}\delta\right)\psi_1 + \left(\frac{-1}{4} + \frac{3}{2}\delta\right)\psi_2 \tag{D68b}$$

The bending energy for the pseudo 8 node element is determined by the classical Q8 element shape functions as:

$$\psi_x = \sum_{i=1}^{8} N_{ix}\psi_{xi}; \quad \psi_y = \sum_{i=1}^{8} N_{iy}\psi_{yi} \tag{D69}$$

By Equations D4 and D5,

$$\begin{cases} N_1 = \frac{1}{4}\xi(1-\xi)\eta(1-\eta) & N_5 = \frac{-1}{2}(1-\xi^2)\eta(1-\eta) \\ N_2 = \frac{-1}{4}\xi(1+\xi)\eta(1-\eta) & N_6 = \frac{1}{2}\xi(1+\xi)(1-\eta^2) \\ N_3 = \frac{1}{4}\xi(1+\xi)\eta(1+\eta) & N_7 = \frac{1}{2}(1-\xi^2)\eta(1+\eta) \\ N_4 = \frac{-1}{4}\xi(1-\xi)\eta(1+\eta) & N_8 = \frac{-1}{2}\xi(1-\xi)(1-\eta^2) \end{cases} \tag{D70}$$

Applying static condensation to the pseudo 8 node element by eliminating ψ_{x5} to ψ_{x8} and ψ_{y5} to ψ_{y8} on Equation D69, we may list

$$\psi_x = \frac{-3(1-2\delta_b)}{4b}N_8\omega_1 + \left[N_1 + \frac{1}{2}N_5 + \left(\frac{-1}{4} + \frac{3}{2}\delta_b\right)N_8\right]\psi_{x1} - \frac{3(1-2\delta_b)}{4b}N_6\omega_2$$

$$\left[N_2 + \frac{1}{2}N_5 + \left(\frac{-1}{4} + \frac{3}{2}\delta_b\right)N_6\right]\psi_{x2} + \frac{3(1-2\delta_b)}{4b}N_6\omega_3 + \left[N_3 + \frac{1}{2}N_7 + \left(\frac{-1}{4} + \frac{3}{2}\delta_b\right)N_6\right]\psi_{x3}$$
$$+ \frac{3(1-2\delta_b)}{4b}N_8\omega_4 + \left[N_4 + \frac{1}{2}N_7 + \left(\frac{-1}{4} + \frac{3}{2}\delta_b\right)N_8\right]\psi_{x4}$$

(D71)

$$\psi_y = \frac{3(1-2\delta_a)}{4a}N_5\omega_1 + \left[N_1 + \frac{1}{2}N_8 + \left(\frac{-1}{4} + \frac{3}{2}\delta_a\right)N_5\right]\psi_{y1} - \frac{3(1-2\delta_a)}{4a}N_5\omega_2$$

$$\left[N_2 + \frac{1}{2}N_6 + \left(\frac{-1}{4} + \frac{3}{2}\delta_b\right)N_5\right]\psi_{y2} - \frac{3(1-2\delta_a)}{4a}N_7\omega_3$$
$$+\left[N_3 + \frac{1}{2}N_6 + \left(\frac{-1}{4} + \frac{3}{2}\delta_b\right)N_7\right]\psi_{y3} + \frac{3(1-2\delta_a)}{4a}N_7\omega_4 + \left[N_4 + \frac{1}{2}N_8 + \left(\frac{-1}{4} + \frac{3}{2}\delta_b\right)N_7\right]\psi_{y4}$$

(D72)

The curvature matrix of the element is given by

$$\kappa = \begin{bmatrix}\kappa_x & \kappa_y & \kappa_{xy}\end{bmatrix}^T = \begin{bmatrix}\frac{\partial \psi_x}{\partial y} & \frac{\partial \psi_y}{\partial x} & \frac{\partial \psi_y}{\partial y} - \frac{\partial \psi_x}{\partial x}\end{bmatrix}^T \tag{D73}$$

By appropriately differentiating Equations D71 and D72, the B matrix for bending as equated by $\kappa = B_B a^e$ would be $B_B = \begin{bmatrix} B_{B1} & B_{B2} & B_{B3} & B_{B4} \end{bmatrix}$ where B_{B1} is presented as illustration as follows while B_{B2} to B_{B4} can be similarly formed:

$$B_{B1} = \begin{bmatrix} \dfrac{3(1-2\delta_b)(1-\xi)\eta}{4b^2} & \dfrac{(1-\xi)(3\eta-1-6\eta\delta_b)}{4b} & 0 \\ \dfrac{-3(1-2\delta_a)(1-\eta)\xi}{4a^2} & 0 & \dfrac{(1-\eta)(3\xi-1-6\delta_{a\xi})}{4a} \\ \dfrac{3\left[(1-2\delta_a)(1-\xi^2)+(1-2\delta_b)(1-\eta^2)\right]}{-8ab} & \dfrac{(1-\eta)\left[-3\eta-1+6\delta_b(1+\eta)\right]}{8a} & \dfrac{(1-\xi)\left[3\xi+1-6\delta_a(1+\xi)\right]}{8b} \end{bmatrix} \tag{D74}$$

$$\text{As} \begin{bmatrix} M_x \\ M_y \\ M_{xy} \end{bmatrix} = D_\kappa, \quad D_b = \begin{bmatrix} 1 & -\mu & 0 \\ -\mu & 1 & 0 \\ 0 & 0 & \dfrac{1-\mu}{2} \end{bmatrix} \tag{D75}$$

The stiffness matrix for bending can be formed by expanding and integrating the following equation

$$K_b^{\ e} = \int_{-b}^{b}\int_{-a}^{a} B_B^{\ T} D_b B_B dxdy = ab\int_{-1}^{1}\int_{-1}^{1} B_B^{\ T} D_b B_B d\xi d\eta \tag{D76}$$

The total stiffness of the plate will be

$$K = K_s^{\ e} + K_b^{\ e} \tag{D77}$$

For general quadrilateral thick plate element, the formulation is more complicated but can provide better accuracy as compared with some commercial programs. Interested readers can refer to the works by Cheng and Law (2008).

D.6 THREE-DIMENSIONAL ELEMENTS (BRICK)

Three-dimensional elements or "brick" elements have also been developed for simulating structures in 3-dimensions. The strains in each brick element comprise linear strains $\varepsilon_{x,}\ \varepsilon_{y,}\ \varepsilon_{z,}$ in the x, y and z directions and shear strains $\gamma_{xy,}\ \gamma_{yz,}\ \gamma_{xz,}$ in the xy, yz and xz planes. The stresses comprise direct stresses $\sigma_{x,}\ \sigma_{y,}\ \sigma_{z,}$ and shear stress $\tau_{xy,}\ \tau_{yz,}\ \tau_{xz}$. Figure A.D14 shows an arch wall structure being simulated by an assembly of many bricks.

A popular "brick" element is an 8-node rectangular prism element is as shown in Figure A.D15.

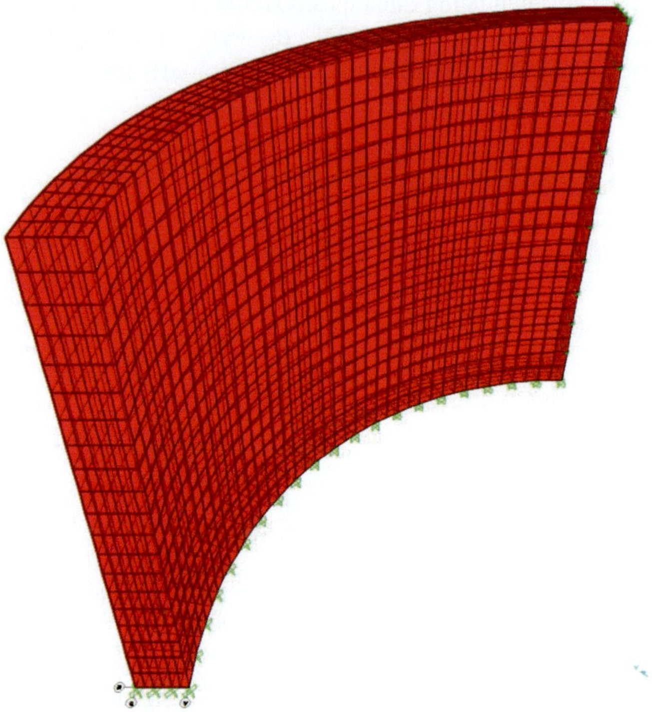

FIGURE A.D14 An arched wall structure being simulated by 3-dimensional "brick" elements.

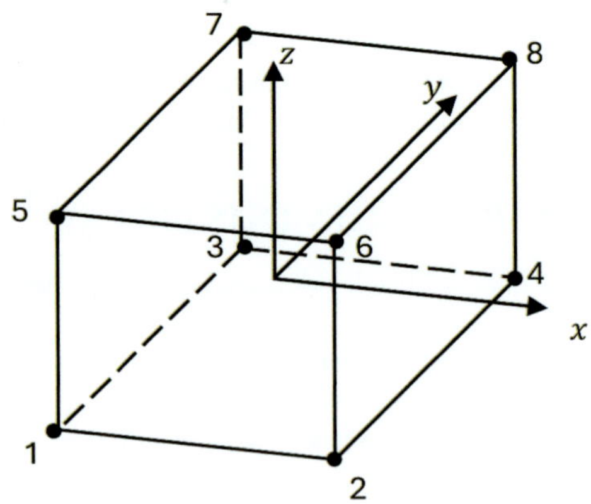

FIGURE A.D15 Rectangular Prism Element

In Figure A.D15, the shape functions are similarly established in the same principle as the 2-dimensional plate element. They are in the form of

$$N_i = \frac{1}{8}\left(1 \pm \frac{x}{a}\right)\left(1 \pm \frac{y}{b}\right)\left(1 \pm \frac{z}{c}\right) \tag{D78}$$

where $i = 1,2,3,4,5,6,7,8$ and the length, width and height of the brick in the x, y, z directions are respectively $2a$, $2b$ and $2c$. Similar to the 2-dimensional elements, the $\pm$ signs are so chosen that $N_i = 1$ at node i and 0 at others.

With u, v, w as the displacement in the x, y and z direction respectively and $u = \sum_i^8 N_i u_i$; $v = \sum_i^8 N_i v_i$ and $w = \sum_i^8 N_i w_i$ and u_i, v_i and w_i as the nodal displacements in the x, y, z directions at node i, the strain is

$$\varepsilon = \begin{bmatrix} \varepsilon_x \\ \varepsilon_y \\ \varepsilon_z \\ \gamma_{xy} \\ \gamma_{yz} \\ \gamma_{xz} \end{bmatrix} = \begin{bmatrix} \frac{\partial u}{\partial x} \\ \frac{\partial v}{\partial y} \\ \frac{\partial w}{\partial z} \\ \frac{\partial u}{\partial y} + \frac{\partial v}{\partial x} \\ \frac{\partial v}{\partial z} + \frac{\partial w}{\partial y} \\ \frac{\partial w}{\partial x} + \frac{\partial u}{\partial z} \end{bmatrix} \tag{D79}$$

The strain displacement matrix is a 6 × 24 matrix of the followings:

$$B = \begin{bmatrix} \frac{\partial N_1}{\partial x} & 0 & 0 & & & \frac{\partial N_8}{\partial x} & 0 & 0 \\ 0 & \frac{\partial N_1}{\partial y} & 0 & & & 0 & \frac{\partial N_8}{\partial y} & 0 \\ 0 & 0 & \frac{\partial N_1}{\partial z} & & & 0 & 0 & \frac{\partial N_8}{\partial z} \\ \frac{\partial N_1}{\partial y} & \frac{\partial N_1}{\partial x} & 0 & \cdots & \cdots & \frac{\partial N_8}{\partial y} & \frac{\partial N_8}{\partial x} & 0 \\ 0 & \frac{\partial N_1}{\partial z} & \frac{\partial N_1}{\partial y} & & & 0 & \frac{\partial N_8}{\partial z} & \frac{\partial N_8}{\partial y} \\ \frac{\partial N_1}{\partial z} & 0 & \frac{\partial N_1}{\partial x} & & & \frac{\partial N_8}{\partial z} & 0 & \frac{\partial N_8}{\partial x} \end{bmatrix} \tag{D80}$$

And the stress-strain matrix

$$D = \frac{E(1-\mu)}{(1+\mu)(1-2\mu)}\begin{bmatrix} 1 & \mu/(1-\mu) & \mu/(1-\mu) & 0 & 0 & 0 \\ \mu/(1-\mu) & 1 & \mu/(1-\mu) & 0 & 0 & 0 \\ \mu/(1-\mu) & \mu/(1-\mu) & 1 & 0 & 0 & 0 \\ 0 & 0 & 0 & \frac{1-2\mu}{2(1-\mu)} & 0 & 0 \\ 0 & 0 & 0 & 0 & \frac{1-2\mu}{2(1-\mu)} & 0 \\ 0 & 0 & 0 & 0 & 0 & \frac{1-2\mu}{2(1-\mu)} \end{bmatrix} \quad \text{(D81)}$$

The stiffness matrix of the brick element can be formed as

$$K = \int_{-c}^{c}\int_{-b}^{b}\int_{-a}^{a} B^T DBdxdydz \quad \text{(D82)}$$

Detail formulation is left to the readers who are interested. As there are 8 nodes and each with 3 degrees of freedom, the stiffness matrix will be of size 24 × 24.

As linear behaviour is dictated by the assumed shape functions listed in Equation D78, large degree of meshing is expected for high accuracy as similar to the 4 node plane stress element. So more sophisticated elements, e.g., a rectangular prism element with mid-side nodes may be used. As the matrix will be very large and the algebraic manipulation in the formation of the stiff matrix will also be extremely complicated, numerical methods instead of algebraic methods are often used in the formulation.

D.7 USE OF NON-RECTANGULAR ELEMENT

For general use, the rectangular element obviously has restrictions in fitting into structures of non-rectangular shape. To solve the problem, non-rectangular elements are used which are termed "isoparametric elements". Non-rectangular elements are useful when the structure is not regular, though their accuracy is slightly less than that for the regular element. Consider a 4-node quadrilateral element in Figure A.D16(a) which can be transformed to a rectangular one in D16(b) with normalized coordinates. The process has been discussed in Cheng and Law (2024).

Figure A.D16(a) is defined in the Cartesian coordinate system while Figure A.D16(b) is defined in the local coordinate system within a specific domain with ξ and η both within the range of –1 and 1.

The displacement at any point $P(x, y)$ is

$$x = \sum_{i=1}^{m} N'_i x_i \quad \text{and} \quad y = \sum_{i=1}^{m} N'_i y_i \quad \text{(D83)}$$

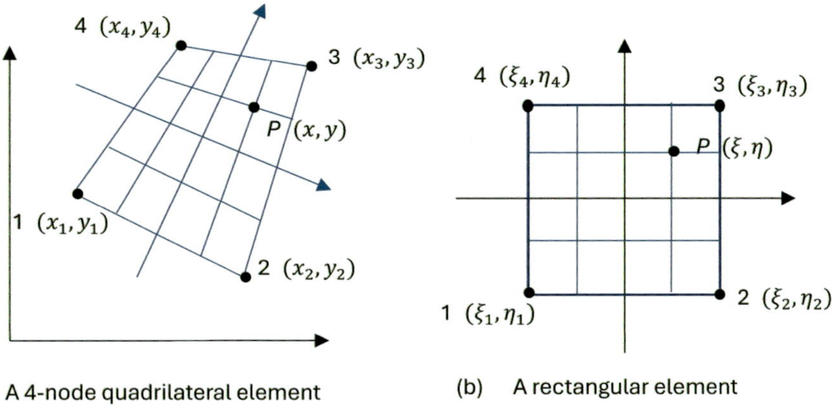

FIGURE A.D16 Isoparametric transformation

where (x, y) are the coordinates in the Cartesian i-th node in the local coordinate system, N'_i is the interpolation function of the i-th node in the local coordinate system, and m is the number of nodes chosen to transform the coordinates. For the 4 node element, Equation D83 can be written in matrix form as

$$\begin{bmatrix} x \\ y \end{bmatrix} = \begin{bmatrix} N_1 & 0 & N_2 & 0 & N_3 & 0 & N_4 & 0 \\ 0 & N_1 & 0 & N_2 & 0 & N_3 & 0 & N_4 \end{bmatrix} \begin{bmatrix} x_1 \\ y_1 \\ x_2 \\ y_2 \\ x_3 \\ y_3 \\ x_4 \\ y_4 \end{bmatrix} \tag{D84}$$

The element displacement model can be written as

$$\begin{bmatrix} u \\ v \end{bmatrix} = \begin{bmatrix} N_1 & 0 & N_2 & 0 & N_3 & 0 & N_4 & 0 \\ 0 & N_1 & 0 & N_2 & 0 & N_3 & 0 & N_4 \end{bmatrix} \begin{bmatrix} u_1 \\ v_1 \\ u_2 \\ v_2 \\ u_3 \\ v_3 \\ u_4 \\ v_4 \end{bmatrix} \tag{D85}$$

where $J = \begin{bmatrix} N_1 & 0 & N_2 & 0 & N_3 & 0 & N_4 & 0 \\ 0 & N_1 & 0 & N_2 & 0 & N_3 & 0 & N_4 \end{bmatrix}$ is called the Jacobi matrix. For the 4 node,

$$\begin{cases} N_1 = \frac{1}{4}(1-\xi)(1-\eta) & N_3 = \frac{1}{4}(1-\xi)(1+\eta) \\ N_2 = \frac{1}{4}(1+\xi)(1-\eta) & N_4 = \frac{1}{4}(1+\xi)(1+\eta) \end{cases} \tag{D86}$$

Since the shape functions are expressed in the natural coordinates, the derivative and integration transformation relationships are required when an iso-parametric element is used.

By the law of partial differential,

$$\begin{cases} \frac{\partial N_i}{\partial \xi} = \frac{\partial N_i}{\partial x}\frac{\partial x}{\partial \xi} + \frac{\partial N_i}{\partial y}\frac{\partial y}{\partial \xi} \\ \frac{\partial N_i}{\partial \eta} = \frac{\partial N_i}{\partial x}\frac{\partial x}{\partial \eta} + \frac{\partial N_i}{\partial y}\frac{\partial y}{\partial \eta} \end{cases} \tag{D87}$$

Or in matrix form

$$\begin{bmatrix} \frac{\partial N_i}{\partial \xi} \\ \frac{\partial N_i}{\partial \eta} \end{bmatrix} = \begin{bmatrix} \frac{\partial x}{\partial \xi} & \frac{\partial y}{\partial \xi} \\ \frac{\partial x}{\partial \eta} & \frac{\partial y}{\partial \eta} \end{bmatrix} \begin{bmatrix} \frac{\partial N_i}{\partial x} \\ \frac{\partial N_i}{\partial y} \end{bmatrix} = J \begin{bmatrix} \frac{\partial N_i}{\partial x} \\ \frac{\partial N_i}{\partial y} \end{bmatrix} \tag{D88}$$

Where the Jacobian $J = \begin{bmatrix} \frac{\partial x}{\partial \xi} & \frac{\partial y}{\partial \xi} \\ \frac{\partial x}{\partial \eta} & \frac{\partial y}{\partial \eta} \end{bmatrix} = \begin{bmatrix} \sum_{i=1}^{4} \frac{\partial N_i}{\partial \xi} x_i & \sum_{i=1}^{4} \frac{\partial N_i}{\partial \xi} y_i \\ \sum_{i=1}^{4} \frac{\partial N_i}{\partial \eta} x_i & \sum_{i=1}^{4} \frac{\partial N_i}{\partial \eta} y_i \end{bmatrix}$

$$= \begin{bmatrix} \frac{\partial N_1}{\partial \xi} & \frac{\partial N_2}{\partial \xi} & \frac{\partial N_3}{\partial \xi} & \frac{\partial N_4}{\partial \xi} \\ \frac{\partial N_1}{\partial \eta} & \frac{\partial N_2}{\partial \eta} & \frac{\partial N_3}{\partial \eta} & \frac{\partial N_4}{\partial \eta} \end{bmatrix} \begin{bmatrix} x_1 & y_1 \\ x_2 & y_2 \\ x_3 & y_3 \\ x_4 & y_4 \end{bmatrix}$$

The inverse of Equation D88 gives

$$\begin{bmatrix} \frac{\partial N_i}{\partial x} \\ \frac{\partial N_i}{\partial y} \end{bmatrix} = J^{-1} \begin{bmatrix} \frac{\partial N_i}{\partial \xi} \\ \frac{\partial N_i}{\partial \eta} \end{bmatrix} \tag{D89}$$

The areas in the Cartesian coordinate system and the natural coordinate system are related by

$$dA = dxdy = |J|d\xi d\eta \tag{D90}$$

where $|\boldsymbol{J}|$ is the determinant of the Jacobian matrix $\boldsymbol{J}$. Therefore, the element stiffness matrix and equivalent nodal load matrix can be transformed to the global coordinates using

$$K = \int_{-1}^{1}\int_{-1}^{1} B^T DBt|J|d\xi d\eta \tag{D91}$$

For the 4 node element

$$J = \begin{bmatrix} \frac{-1}{4}(1-\eta) & \frac{1}{4}(1-\eta) & \frac{-1}{4}(1-\eta) & \frac{1}{4}(1-\eta) \\ \frac{-1}{4}(1+\xi) & \frac{-1}{4}(1+\xi) & \frac{1}{4}(1+\xi) & \frac{1}{4}(1+\xi) \end{bmatrix} \begin{bmatrix} x_1 & y_1 \\ x_2 & y_2 \\ x_3 & y_3 \\ x_4 & y_4 \end{bmatrix}$$

$$= \frac{1}{4}\begin{bmatrix} (1-\eta)(-x_1+x_2-x_3+x_4) & (1-\eta)(-y_1+y_2-y_3+y_4) \\ (1-\xi)(-x_1-x_2+x_3+x_4) & (1-\xi)(-y_1-y_2+y_3+y_4) \end{bmatrix} \tag{D92}$$

$$|J| = \frac{1}{4}(1-\eta)(1-\xi)\big[(-x_1+x_2-x_3+x_4)(-y_1-y_2+y_3+y_4)$$
$$-(-x_1-x_2+x_3+x_4)(-y_1+y_2-y_3+y_4)\big] \tag{D93}$$

D.8 ELIMINATION OF INTERIOR NODES

For the 8 node element shown in Figure A.D8, if it is intended to use only nodes no. 1, 2, 3, and 4 as external nodes to join other elements or to form supports while keeping nodes no. 5, 6, 7, and 8 as internal nodes without joining to other elements, the nodes no. 5, 6, 7, and 8 may be eliminated in forming the global stiffness matrix. Such an elimination can reduce the number of nodes participating in the global analysis which can save time and storage in computation. However, the number of nodes to be eliminated and the number of nodes to remain have to be identical.

As suggested by Zienkiewicz et al. (2005), the elimination can be performed as follows. The original matrix is to be partitioned into 4 matrices as $K = \begin{bmatrix} K_{11} & K_{12} \\ K_{21} & K_{22} \end{bmatrix}$ where $\boldsymbol{K}_{11}$ contains the 4 nodes to be retained. Then the matrix with the internal nodes eliminated is formed by

$$\boldsymbol{K}_r = \boldsymbol{K}_{11} - \boldsymbol{K}_{12}\left(\boldsymbol{K}_{22}\right)^{-1}\boldsymbol{K}_{21} \tag{D94}$$

D.9 SAMPLE COMPUTER CODING

Cheng has developed many finite element programs for teaching and research, mainly in Fortran 90. To help readers to acquire actual knowledge in the computer implementation, some sample codes developed by Cheng are given below for reference. These codes are taken out from a finite element seepage program with a free surface.

ASSEMBLAGE OF GLOBAL STIFFNESS MATRIX

```
SUBROUTINE ASSEM(X,Y,MT,CON,LD,KMAT,FIX,BOUND,EMN,
& ITER,GAU,FIX1,INFIN)
INTEGER TNN,TNE,TNB,ENN,NLD,TNEB,TMT
INTEGER*1 FIX(TNN),GAU(TNE,4),FIX1(TNN),INFIN(TNE)
REAL X(TNN),Y(TNN),MT(TMT,2)
REAL*4 KMAT(NLD),BOUND(TNN)
INTEGER*2 CON(TNE,ENN),EMN(TNE)
INTEGER*4 LD(TNN)
COMMON /KMAI/ TNN,TNE,TNB,ENN,NLD,TMT,TNEB,IP,IAIR
COMMON /KELM/ ELM(8,8)
COMMON/DEF/SHAPE(8)
WRITE (*,*) 'ASSEMBLING STIFFNESS MATRICES'
WRITE (*,*) 'ITERATION ',ITER
CALL ZERO(KMAT,NLD)
II = 0
DO 10 I=1,TNE
!ESCAPE COMPUTATION WITH AIR ELEMENT
IF (EMN(I) .LT. 0) THEN
 II = II + 1
 GOTO 10
ENDIF
CALL ZERO2(ELM,8,8)
DO 20 K=1,4
IF (GAU(I,K) .EQ. 1) GO TO 20
IF (INFIN(I) .EQ. 0) CALL DSHAPE(K) ! call the shape function
IF (INFIN(I) .EQ. 1) CALL DSHAPE1(K)
CALL AELM(I,X,Y,MT,CON,K,EMN,INFIN(I)) ! form element
stiffness matrix
 20 CONTINUE
CALL GOBL(I,CON,LD,KMAT) ! form global stiffness matrix using
skyline storage method
10 CONTINUE
JJ = 0
DO I = 1,TNN
IF (KMAT(LD(I)) .LT. 1.0E-20) THEN
FIX(I) = 1
FIX1(I) = 1
BOUND(I)=0
JJ = JJ+1
ENDIF
END DO
RETURN
END
```

Forming the element stiffness matrix for a Gaussian point

```
SUBROUTINE AELM(W,X,Y,MT,CON,K,EMN,IJ)
INTEGER*1 IJ
INTEGER TNN,TNE,TNB,ENN,NLD,TMT,K,W
REAL X(TNN),Y(TNN),MT(TMT,2),BMAT(2,8),AJACOB(4),DET
INTEGER*2 CON(TNE,ENN),EMN(TNE)
COMMON /KMAI/ TNN,TNE,TNB,ENN,NLD,TMT,TNEB,IP,IAIR
COMMON /ABC/ DNR(8), DNS(8)
COMMON /KELM/ ELM(8,8)
COMMON/DEF/SHAPE(8)
C
IF (IP .EQ. 1) THEN
XI = 0.
IF (IJ .EQ. 0) CALL ASHAPE(K)
IF (IJ .EQ. 1) CALL ASHAPE1(K)
DO I = 1,ENN
XI = XI+SHAPE(I)*X(CON(W,I))*2.*3.1415927
END DO
ELSE
XI = 1.
ENDIF
CALL ZERO(AJACOB,4)
DO 10 I=1,ENN
AJACOB(1)=DNR(I)*X(CON(W,I))+AJACOB(1)
AJACOB(2)=DNR(I)*Y(CON(W,I))+AJACOB(2)
AJACOB(3)=DNS(I)*X(CON(W,I))+AJACOB(3)
AJACOB(4)=DNS(I)*Y(CON(W,I))+AJACOB(4)
10 CONTINUE
DET=AJACOB(1)*AJACOB(4)-AJACOB(2)*AJACOB(3) ! Jacobian
IF (DET .LT. 0.) THEN
WRITE (*,*) 'NEGATIVE JACOBIAN FOR',W,K
STOP
ENDIF
IF (IJ .EQ. 1) CALL DSHAPE(K)
DO 20 JJ=1,ENN! form B matrix
BMAT(1,JJ)= (AJACOB(4)*DNR(JJ)-AJACOB(2)*DNS(JJ))/DET
BMAT(2,JJ)=(-AJACOB(3)*DNR(JJ)+AJACOB(1)*DNS(JJ))/DET
20 CONTINUE
DO 30 I=1,ENN
DO 30 J=1,ENN ! Numerical integration to form the element
stiffness matrix
ELM(I,J)=DET*MT(EMN(W),2)*BMAT(2,I)*BMAT(2,J)*XI+
1 MT(EMN(W),1)*BMAT(1,I)*BMAT(1,J)*DET+ELM(I,J)*XI
30 CONTINUE
RETURN
END
```

D.10 CONCLUDING REMARKS

The principle and actual practice in the formulation of the Finite Element Method are outlined in the foregoing. Generally, accuracy depends on the degrees of the shape function commonly formulated by polynomials as adopted and the degree of meshing. If the shape function chosen is appropriate for the structural action to be investigated, even a single element can converge to the exact solution.

Nevertheless, a high degree of the shape function may, apart from the complexity involved in formulation, may sometimes lead to numerical instability. With the high power of hardware available today, the high degree of meshing allowed under reasonable time and capacity in computation may compensate for the simple shape functions so chosen.

There are many other tricks in program implementation which include mesh generation, adoption of band matrix storage, skyline profile storage and the corresponding equation solver. The solution of the system equation is typically the Gaussian elimination based on band matrix or skyline profile, while the use of iterative methods adopting the preconditioned conjugate gradient method or similar has also been in the past. Sample solvers for band matrix and skyline profile and preconditioned conjugate gradient iterative solvers can be obtained from Cheng at natuerymc@yahoo.com.hk for learning purposes. Iterative solver is still popular in DDA or similar analysis. Parallel processing for finite element solver is available in some commercial programs for modern CPU with multiple process. Cheng developed some teaching code using OpenMP and found that if the dimension of the matrix exceeds 1000 (case dependent), the use of parallel processing can save the overall computer time.

REFERENCES

Cheng, Y.M. and Law, C.W. (2008) Development of a New and Efficient Thick Plate Element. Structural Engineering and Mechanics, 29(3), 327–354.

Cheng, Y.M., Law, C.W. and Liu, L.L. (2024) Analysis, Design and Construction of Foundations, 2nd Edition. CRC Press, Taylor & Francis Group.

Cheng, Y.M., Wang, J.H., Li, L. and Fung, W.H. (2020) Frontier in Civil Engineering, Vol. 3: Numerical Methods and Implementation in Geotechnical Engineering. Benjamin Press.

Law, C.W. and Cheng, Y.M. (2015) Improved Thick Plate Analysis and Design. The Hong Kong Institution of Engineers Transactions, 13(2), 49–58.

Law C.W. and Cheng Y.M. (2015), Studies on the P-delta effects of piles embedded in cohesionless soil, *HKIE Transaction*, 22(3), 134–152.

Law, C.W., Cheng, Y.M. and Yang, Y. (2011) Problems with Some Common Plate Bending Elements and the Development of a Pseudo Higher Order Plate Bending Element. The Hong Kong Institution of Engineers Transactions, 19(1), 12–21.

Zienkiewicz, O.C., Taylor, R.L. and Zhu, J.Z. (2005) The Finite Element Method: Its Basis and Fundamentals, 6th Edition. Elsevier.

Appendix E
A Worked Example on Determination of Lateral Seismic Load in accordance with the Response Spectrum Method

Worked Example E1

Consider the following three-storey one-bay plane frame.

The following procedures are adopted:

(i) Find the structure's fundamental natural periods:

The same plane frame structure has been analyzed in Chapter 7 as Worked Example 7.3. The dynamic properties are copied from Chapter 7 as follows:

The first three highest eigenvalues with eigenvectors showing movements in the lateral directions are:

(1) $\lambda_1 = 0.003071014$; period $T = 2\pi\sqrt{\lambda} = 0.3827$ sec

(2) $\lambda_2 = 0.000178739$; period $T = 2\pi\sqrt{\lambda} = 0.0840$ sec

(3) $\lambda_3 = 0.000027629$; period $T = 2\pi\sqrt{\lambda} = 0.03303$ sec

The corresponding eigenvectors are (same as Worked Example 7.3):

$$\phi_1 = \begin{bmatrix} -0.1520382 \\ -0.3679630 \\ -0.5779608 \end{bmatrix}; \phi_2 = \begin{bmatrix} 0.4472364 \\ -0.3203594 \\ 0.3804339 \end{bmatrix}; \phi_3 = \begin{bmatrix} -0.3476058 \\ 0.5072724 \\ -0.1857971 \end{bmatrix}$$

The mode shapes are plotted as follows:

The sum of the modal mass participation of the three modes is equal to the sum of the masses $29400 \times 2 + 44100 = 102900$ kg. So, the three modes are used.

DOI: 10.1201/9781003636380-19

(ii) Selection of a response spectrum:
The response spectrum from ASCE7-16 is chosen with the followings parameters:
$S_{DS} = 2.35\text{m/sec}^2$; $S_{D1} = 0.56\text{m/sec}^2$;

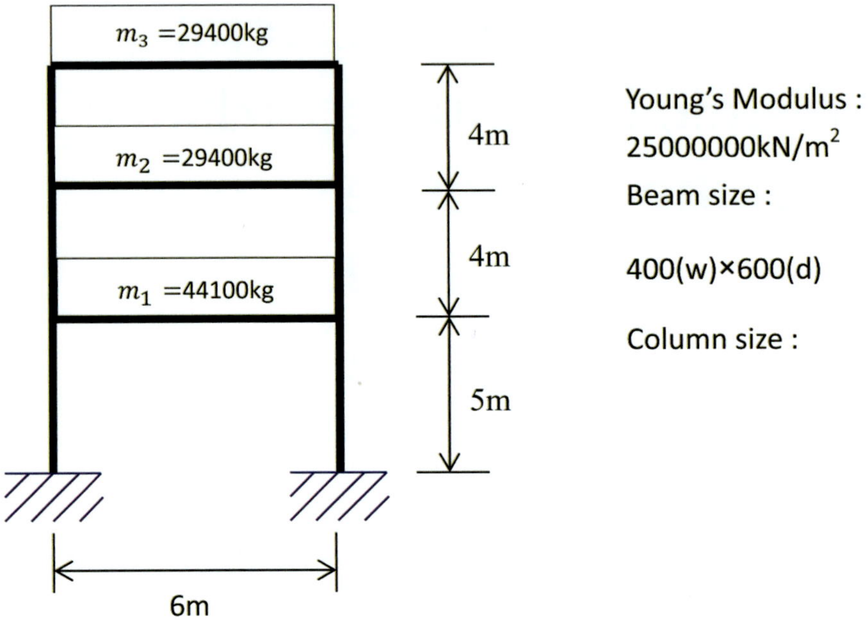

FIGURE A.E1 Portal frame for determination of seismic load in accordance with the response spectrum method.

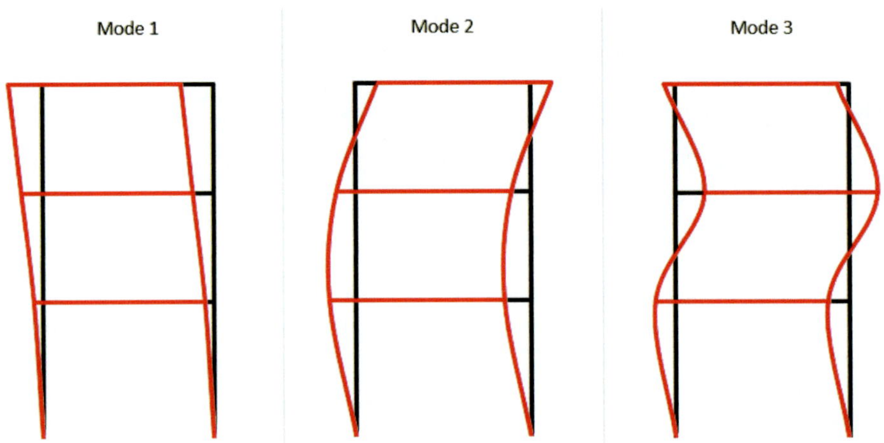

FIGURE A.E2 Mode shapes of the frame.

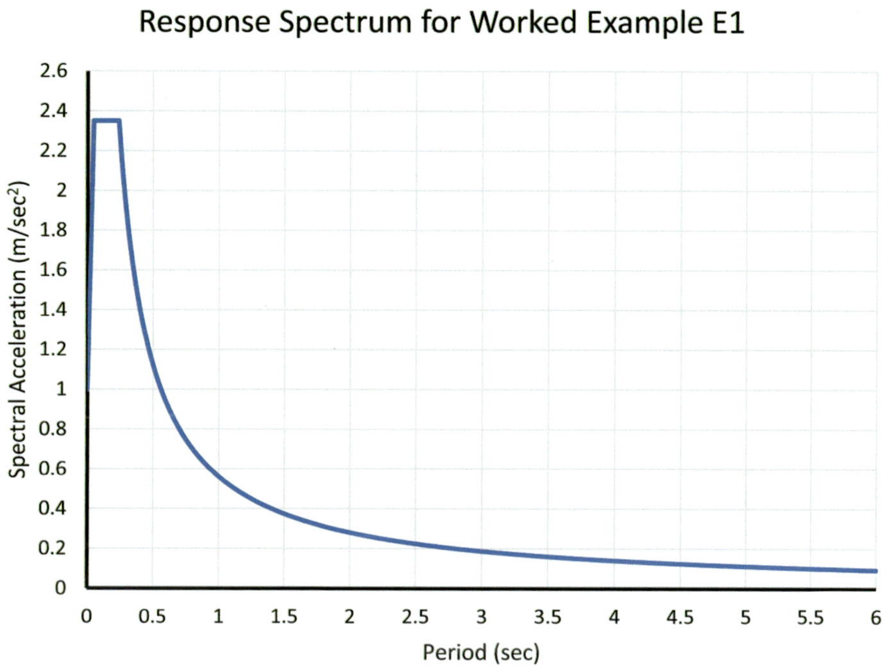

FIGURE A.E3 Response spectrum used in Worked Example E1.

$$T_0 = 0.2\left(S_{D1} / S_{DS}\right) = 0.0477\,\text{sec}$$

$$T_S = S_{D1} / S_{DS} = 0.238\,\text{sec}$$

PGA = 1sec; $T_L = 6\,\text{sec}$

The curve beyond T_S is taken as S_{D1} / T
So, the response spectrum is constructed as follows:
The spectral acceleration of the three modes based on their periods as read from Figure A.E3 in m/sec^2 are (1) $S_{a1} = 1.463$; (2) $S_{a2} = 2.35$; (3) $S_{a3} = 1.9355$.

(iii) The modal acceleration factors of the three modes are calculated as follows:
Mode 1

Storey	M	ϕ	$M\phi$	$M\phi^2$	$M\Gamma_1\phi S_{a1}$
1	44.1	−0.1520382	−6.7048829	1.0193981	22.8476
2	29.4	−0.367963	−10.818111	3.9806642	36.8639
3	29.4	−0.5779608	−16.992046	9.8207362	57.9022
		Sum	−34.51504	14.820798	

The modal participation factor is $\Gamma_1 = -34.51504 / 14.820798 = -2.3288$

Mode 2

Storey	M	ϕ	$M\phi$	$M\phi^2$	$M\Gamma_2\phi S_{a2}$
1	44.1	−0.4472364	−19.723124	8.8208982	51.7167
2	29.4	−0.3203594	−9.4185652	3.0173255	24.6968
3	29.4	0.3804339	11.184757	4.2550607	−29.3280
		Sum	−17.956932	16.093284	

The modal participation factor is $\Gamma_2 = -17.956932 / 16.09328 = -1.1158$

Mode 3

Storey	M	ϕ	$M\phi$	$M\phi^2$	$M\Gamma_3\phi S_{a3}$
1	44.1	−0.3476058	−15.329417	5.328595	12.5390
2	29.4	0.5072724	14.913808	7.5653626	−12.1990
3	29.4	−0.1857971	−5.4624338	1.0149042	4.4681
		Sum	−5.8780435	13.908862	

The modal participation factor is $\Gamma_3 = -5.8780435 / 13.908862 = -0.4226$

It should be noted that the Γ_s calculated will be used to calculate the seismic shear force of the modes at the storeys in the last column for the three modes in the above tables.

(iv) The final shears on the storeys are summed by the square roof of sum of squares (SRSS) method for the three modes as

(1) Storey one $\sqrt{22.8476^2 + 51.7167^2 + 12.5390^2} = 57.9125\text{kN}$

(2) Storey two $\sqrt{36.8639^2 + 24.6968^2 + (-12.1990)^2} = 46.0184\text{kN}$

(3) Storey three $\sqrt{57.9022^2 + (-29.3280)^2 + 4.4681^2} = 65.0597\text{ kN}$

More discussion about the response spectrum method is given by Chopra (2001), Saatcioglu and Humar (2003).

REFERENCES

Chopra, A. (2001) Dynamics of Structures, 2nd Edition. Prentice Hall.

Saatcioglu, M. and Humar, J. (2003) Dynamic Analysis of Buildings for Earthquake Resistant Design. Canadian Journal of Civil Engineering, 30, 338–359.

Appendix F Equations for Determination of Stress-Strain Relation of Unconfined and Confined Concrete to Attard and Setunge (1996)

F.1 INTRODUCTION

The equations have been widely used by researchers to investigate behaviour of concrete under unconfined and confined conditions, as shown in Figure F.1 extracted from figure 8 by Attard and Setunge (1996) in which some basic symbols are defined.

F.2 DEFINITIONS OF SYMBOLS

$f_{c'}$ is the concrete cylinder strength in MPa
f_{cu} is the concrete cube strength in MPa
f_0 is the peak concrete strength in the stress-strain relation under confined or unconfined condition for design in MPa
f_t is the concrete tensile strength in MPa
f_r is the lateral confinement stress in MPa
f_{sp} is the cylinder split strength in MPa
E_c is the Young's modulus of concrete
ρ is the density of concrete in kg/m^3

F.3 EQUATIONS

From Attard and Setunge's (1996) giving the main equations and Pam et al. (2001) which supplements an equation relating concrete strain at concrete's unconfined cylinder strength listed in the Reference, the following equations are listed:

DOI: 10.1201/9781003636380-20

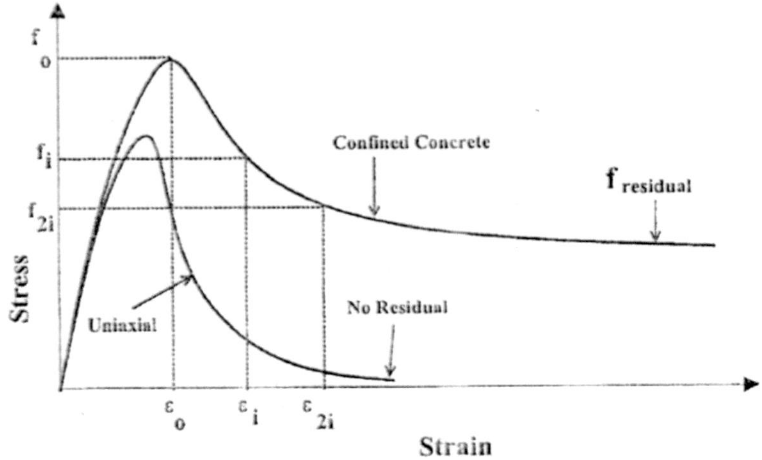

Fig. 8—Typical stress-strain relationship for confined and unconfined concrete

FIGURE F.1 Typical stress-strain relationship for confined and unconfined concrete.

$f_c = 0.72 f_{cu}$ or $0.9 f_{c'}$ (Peak Concrete Strength under unconfined condition. It can otherwise be defined by user by other relations)

$$k = 1.25\left(1 + 0.062\frac{f_r}{f_c}\right)(f_c)^{-0.21} \quad \text{(18) of Attard \& Setunge}$$

$f_t = 0.9 f_{sp}$ (f_t is the concrete tensile strength)

(f_{sp} is the split cylinder result)

$f_{sp} = 0.32(f_c)^{0.67}$ MPa (No silica fume) (19) of Attard & Setunge

$f_{sp} = 0.62\sqrt{f_c}$ MPa (with silica fume) (19) of Attard & Setunge

f_t as the tensile strength of concrete may be independently assessed by user

$f_0 = f_c\left(1 + \dfrac{f_r}{f_t}\right)^k$ ($f_0 = f_c$ under no confinement, $f_r = 0$) (17) of Attard & Setunge

$E_c = \left(3320\sqrt{f_c} + 6900\right)\left(\dfrac{\rho}{2320}\right)^{1.5}$ (or other relation defined by user) (8) of Attard & Setunge

$$\varepsilon_c = 4.11 f_{co}^{\ 0.75} / E_c \quad \text{(4b) of Pam et al.}$$

$$\varepsilon_0 = \varepsilon_c\left[1 + (17 - 0.06 f_c)\left(\frac{f_r}{f_c}\right)\right] \quad \text{(20) of Attard \& Setunge}$$

Combining

$$\varepsilon_0 = 4.11 f_{co}^{\ 0.75} / E_c \left(1 + 17 - 0.06 f_c\right)\left(\frac{f_r}{f_c}\right) \quad \text{(20 of Attard \& Setunge)}$$

$$\varepsilon_i = \varepsilon_0 \left[2 + \frac{\dfrac{\varepsilon_{ic}}{\varepsilon_c} - 2}{1 + 1.12\left(\dfrac{f_r}{f_c}\right)^{0.26}}\right] \quad \text{(23 of Attard \& Setunge)}$$

$$\frac{\varepsilon_{ic}}{\varepsilon_c} = 2.5 - 0.3\ln\left(f_c\right) \quad \text{(11 of Attard \& Setunge)}$$

Combining (23) of Attard & Setunge and (11) of Attard & Setunge

$$\varepsilon_i = \varepsilon_0 \left[2 + \frac{2.5 - 0.3 ln\left(f_c\right) - 2}{1 + 1.12\left(\dfrac{f_r}{f_c}\right)^{0.26}}\right] \quad \text{(11) and (23) of Attard \& Setunge}$$

$$\varepsilon_{2i} = 2\varepsilon_i - \varepsilon_0 \quad \text{(page 440 of Attard \& Setunge)}$$

$$\frac{f_{ic}}{f_c} = 1.41 - 0.17\ln\left(f_c\right) \quad \text{(12) of Attard \& Setunge}$$

$$f_i = \left(\frac{\dfrac{f_{ic}}{f_c} - 1}{5.06\left(\dfrac{f_r}{f_c}\right)^{0.57} + 1} + 1\right) f_o \quad \text{(22) of Attard \& Setunge}$$

Combining (12) and (22) of Attard & Setunge

$$f_{2i} = \left(\frac{\dfrac{f_{2ic}}{f_c} - 1}{6.35\left(\dfrac{f_r}{f_c}\right)^{0.62} + 1} + 1\right) f_o \quad \text{(24) of Attard \& Setunge}$$

$$\frac{f_{2ic}}{f_c} = 1.45 - 0.25\ln\left(f_c\right) \quad \text{(25) of Attard \& Setunge}$$

Combining (24) and (25) of Attard & Setunge

$$f_{2i} = \left(\frac{1.45 - 0.25\ln(f_c) - 1}{6.35\left(\frac{f_r}{f_c}\right)^{0.62} + 1} + 1 \right) f_o$$

$$A_1 = E_c \varepsilon_0 / f_0 \qquad \text{(2) of Attard \& Setunge}$$

$$B_1 = \frac{(A_1 - 1)^2}{0.55} - 1 \qquad \text{(2) of Attard \& Setunge}$$

(by setting $f_{pl} = 0.45 f_o$ and $\alpha = 1$ as suggested)
$C = A - 2$; $D = B + 1$ (3) of Attard & Setunge)

$$A_2 = \left[\frac{\varepsilon_{2i} - \varepsilon_i}{\varepsilon_o} \right] \left[\frac{\varepsilon_{2i} E_i}{f_o - f_i} - \frac{4\varepsilon_i E_{2i}}{f_o - f_{2i}} \right] \text{(13 of Attard \& Setunge)}$$

where $E_i = \dfrac{f_i}{\varepsilon_i}$ and $E_{2i} = \dfrac{f_{2i}}{\varepsilon_{2i}}$

$$B_2 = (\varepsilon_i - \varepsilon_{2i}) \left[\frac{E_i}{f_o - f_i} - \frac{4\varepsilon_i E_{2i}}{f_o - f_{2i}} \right] \text{(14 of Attard \& Setunge)}$$

$C_2 = A_2 - 2$; $D_2 = B_2 + 1$ (3) of Attard & Setunge)

$$Y = \frac{AX + BX^2}{1 + CX + DX^2} \qquad \text{(1) of Attard \& Setunge)}$$

where $Y = \dfrac{f}{f_o}$ $X = \dfrac{\varepsilon}{\varepsilon_o}$

For $\varepsilon \le \varepsilon_0$, $A = A_1$, $B = B_1$, $C = C_1$, $D = D_1$, this is the ascending (pre-peak) portion.

For $\varepsilon > \varepsilon_0$, $A = A_2$, $B = B_2$, $C = C_2$, $D = D_2$, this is the descending (post-peak) portion.

REFERENCES

Attard, M.M. and Setunge, S. (1996) The Stress Strain Relationship of Confined and Unconfined Concrete. ACI Materials Journal, 93(5), 432–442.

Pam, H.J., Kwan, A.K.H. and Ho, J.C.M. (2001) Post-Peak Behavior and Flexural Ductility of Doubly Reinforced Normal- and High-Strength Concrete Beams. Structural Engineering and Mechanics, 12(5), 459–474.

Appendix G Formulae and Worked Examples on Structural Design against Fire

G.1 FIRE CURVES

Equations for the fire curves adopted by EN1991-1-2:2023 (2023)

Standard time-temperature curve: $\theta_g = 20 + 345\log_{10}(8t+1)$;
External fire curve: $\theta_g = 660\left(1 - 0.687e^{-0.32t} - 0.313e^{-3.8t}\right) + 20$;
Hydrocarbon curve: $\theta_g = 1080\left(1 - 0.325e^{-0.167t} - 0.675e^{-2.5t}\right) + 20$

The HCM (modified hydrocarbon curve) Tunnel Curve is $\theta_g = 1280\left(1 - 0.325e^{-0.167t} - 0.675e^{-2.5t}\right) + 20$

The RWS (Rijkswaterstaat) Tunnel Curve is given by

Time (min)	0	3	5	10	30	60	90	120	180	240
Temperature (°C)	20	890	1140	1200	1300	1350	1300	1200	1200	1200

The RABT-ZTV (Train)

Time (min)	0	5	60	170
Temperature (°C)	15	1200	1200	15

The RABT-ZTV (Car)

Time (min)	0	5	30	140
Temperature (°C)	15	1200	1200	15

DOI: 10.1201/9781003636380-21

ASTM-E117

Time (min)	0	5	10	30	60	120	240
Temperature (°C)	0	538	704	840	927	1010	1093

G.2 CONSTRUCTING TEMPERATURE CONTOURS IN ACCORDANCE WITH THE SIMPLIFIED DESIGN METHOD IN SECTION 7 OF BSEN1992-1-2:2023 (2023)

G2.1 For Structure with One Face Exposed to Standard Fire, the Temperature Increase with Time Is from Equation 7.2 of BSEN1992-1-2:2023 (2023)

$$\theta_1(x,t) = 345\log_{10}\left(\frac{7(t-\Delta t)}{60}+1\right)\exp\left(-x\sqrt{\frac{k}{t}}\right) \tag{G1}$$

where t is the duration of the standard fire (in seconds), $t \geq 1800\,\text{sec}$

x is the distance from the exposed face (in m)

$\Delta t = 720\,\text{sec}$, representing a delay between the temperature in the fire compartment and the concrete surface temperature as an approximation for the effects of convection and radiation.

$k = 3\times10^6\,\text{sec/m}^2$.

The resulting temperature field is

$$\theta(x,t) = \theta_1(x,t) + 20^o C \tag{G2}$$

G2.2 For Structure with Two Opposite Faces Exposed to Standard Fire, the Temperature Increase with Time Is from Equation 7.4 of BSEN1992-1-2:2023 (2023)

$$\theta_2(y,t) = \theta_1(y,t) + \theta_1(b-y,t)(\text{Case A}) \tag{G3}$$

$$\theta_2(z,t) = \theta_1(z,t) + \theta_1(h-z,t)(\text{Case B}) \tag{G4}$$

where the thickness is b, h , y , z are defined in Figure A.G1 extracted from Figure 7.2 of BSEN1992-1-2:2023 (2023). The coordinate system defined are applicable to the scenarios for structures with three and four faces exposed to fire to be discussed.

The resulting temperature field for Case A and Case B are respectively

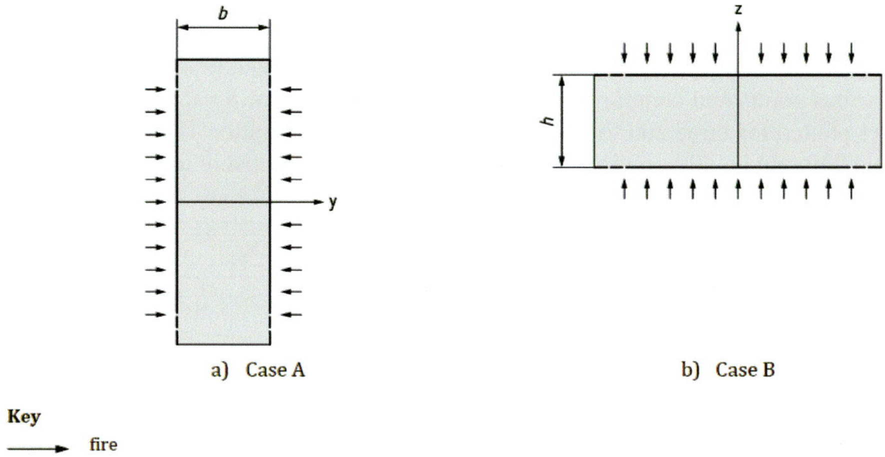

FIGURE A.G1 Definition of symbols for Equations G2 and G3 extracted from figure 7.2 of BSEN1992-1-2:2023 (2023).

$$\theta(y,t) = \theta_2(y,t) + 20^o C \quad \text{(G5)}$$

$$\theta(z,t) = \theta_2(z,t) + 20^o C \quad \text{(G6)}$$

G2.3 For Structure with Three Faces Exposed to Standard Fire, Temperature Increase Is to Be Determined by Equations G7 and G8.

$$\theta(y,z,t) = \theta_2(y,t) + \theta_1(z,t) - \frac{\theta_2(y,t).\theta_1(z,t)}{\theta_1(0,t)} + \Delta\theta(y',z',t) + 20 \quad \text{(G7)}$$

$$\text{where } \Delta\theta(y',z',t) = \left(345\log_{10}\left(\frac{8t}{60}+1\right) - \theta_1(0,t)\right)\frac{(a_c - y')(a_c - z')}{a_c^{\ 2}} \quad \text{(G8)}$$

with $a_c = 0.04\,\text{m}$ for fire duration up to 60 min and $a_c = 0.1\text{m}$ for fire duration longer than 60 min. y' and z' are coordinates at the corners exposed to fire to account for the reduction due to local effects of convection and radiation at the beginning of the fire. In this case, $\Delta\theta(y',z',t)$ applies to two corners exposed to fire.

G2.4 For Structure with Four Faces Exposed to Standard Fire, Temperature Increase Is to Be Determined by Equation G8.

$$\theta(y,z,t) = \theta_2(y,t) + \theta_2(z,t) - \frac{\theta_2(y,t).\theta_2(z,t)}{\theta_1(0,t)} + \Delta\theta(y',z',t) + 20 \quad \text{(G9)}$$

$\Delta\theta(y',z',t)$ is similarly defined by Equation G8 which applies to the 4 corners.

G2.5 Temperature Contours

By the use of equations in Sections G2.3 and G2.4, temperatures are calculated at various points and contour plots for a concrete structure of $b = 0.5\text{m}$ and $h = 0.6\text{m}$ are plotted for three and four faces exposed with time duration 120 min (7200 sec) and 240 min (14400 sec) in Figures A.G2 and A.G3 respectively for illustration.

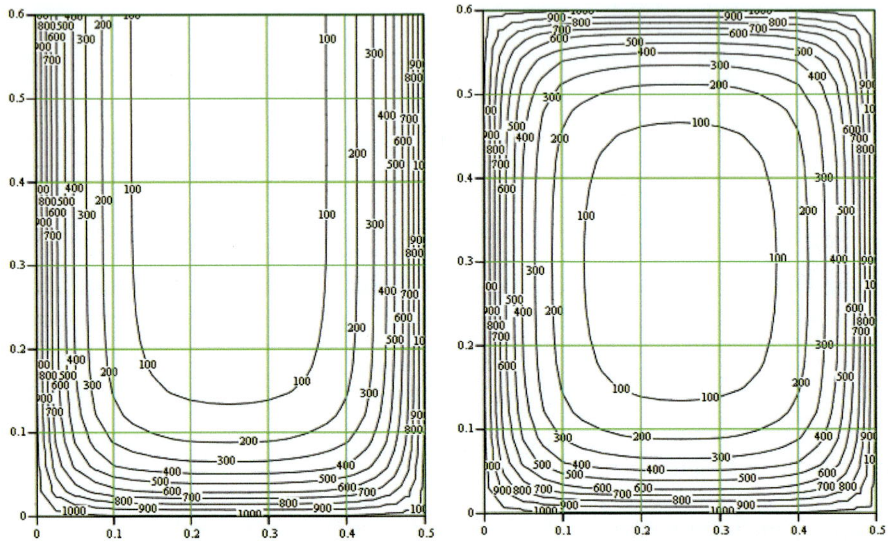

FIGURE A.G2 Temperature contours for three-sides (left) and four sides (right) exposed structural elements with duration of 120 min.

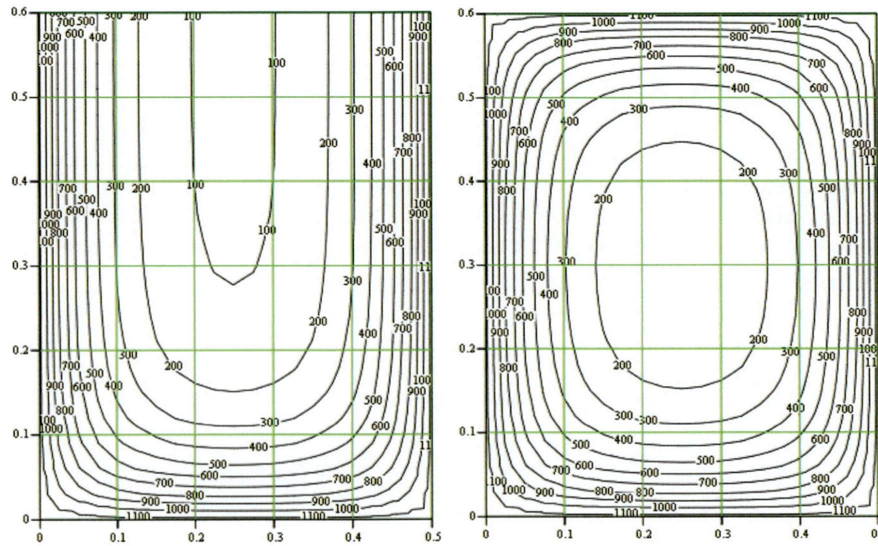

FIGURE A.G3 Temperature contours for three-sides (left) and four sides (right) exposed structural elements with duration of 240 min.

G.3 WORKED EXAMPLE G.1 FOR "GENERAL DESIGN RULES" (SECTION 6.2 OF BSEN1992-1-2:2003 (2023)) – MODIFICATION OF TABULATED DATA

Consider a beam with reinforcement of 4H32 so that $A_{s,prov} = 3217\text{mm}^2$ while the required reinforcement is $A_{s,req} = 2805\text{mm}^2$. $f_{yk} = 500\text{N/mm}^2$, $\eta_{fi} = 0.7$, $\gamma_s = 1.15$ (NOTE 1 of Section 6.2(3) of BSEN 1992-1-1:2023 (2023)).

By Equation 6.1 of BSEN 1992-1-2:2023 (2023),

$$\sigma_{s,fi} = \eta_{fi}\frac{f_{yk}}{\gamma_s}\times\frac{A_{s,req}}{A_{s,prov}} = 0.7\times\frac{500}{1.15}\times\frac{2805}{3217} = 265.37\text{N/mm}^2.$$

Taking $\sigma_{s,fi}$ as reaching the strength limit as $f_{se,\theta}$, the factor of $f_{se,\theta} / f_{yk} = 0.53$, the critical temperature for hot rolled steel or cold worked steel is $\theta_{cr} = 528°\text{C}$ according to table 5.3 of BSEN 1992-1-2:2023 (2023) (obtained by interpolation).

By Equation 6.3 of BSEN 1992-1-2:2023 (2023), the adjustment in the axis distance (concrete cover to reinforcement) in mm is

$$\Delta a = 0.1(500 - \theta_{cr}) = 0.1(500 - 528) = -2.8\text{mm}.$$

This adjustment of axis distance shall be employed to modify the axis distance listed the relevant tables (table 6.6 or 6.7) in section 6 of BSEN 1992-1-2:2023 (2023) in accordance with the R criterion (R 30 or R 60 etc.) required. For example, if the beam is simply supported with beam width 400 mm requiring R 180, the required axis distance is 65 mm according to table 6.6 of BSEN 1992-1-2:2023 (2023), the axis distance can be reduced to $65 - 2.8 = 62.2\text{mm}$.

G.4 WORKED EXAMPLE G.2 – FOR CALCULATION OF THE BENDING RESISTANCE OF A BEAM WHOSE TENSION SIDE OF THE CROSS-SECTION IS EXPOSED TO FIRE WITH $A_{s,prov}/A_{s,req}$ < 1.3

In accordance with section 7.3.3.2 of BSEN1992-1-2:2023 (2023), for slabs and beams resisting bending whose tension side is exposed to fire, the resistance may be calculated by disregarding the strength reduction of concrete by considering the full cross-section, unless the depth of concrete in compression exceeds $x = 0.25d$. Consider a simply supported reinforced concrete beam of size 600(h) × 500(b), singly reinforced. The concrete mix is C30/37 with reinforcement 4H32 ($A_{s,prov} = 3217$ mm²). The design fire rating is R240. The beam is exposed to fire on 3 sides.

Design data: $b = 500\text{mm}$; $h = 600\text{mm}$ $a = 75\text{mm}$; $f_{ck} = 30$ N/mm²;

$$f_{yk} = 500\text{N/mm}^2;\ A_{s,prov} = 3217\text{mm}^2.$$

$$d = h - a = 600 - 75 = 525\text{mm}$$

$$M_{Ed} = 500\text{kNm}$$

$$\eta = 1,\ \alpha_{cc} = 0.85,\ \gamma_s = 1.5$$

$A_{s,req}$ is first to be calculated to resist M_{Ed} in accordance with BSEN1991-1-1:2023 (2023). Reference is made to Moseley et al. (2012):

$$K = \frac{M_{Ed}}{f_{ck}bd^2} = \frac{500 \times 10^6}{30 \times 500 \times 525^2} = 0.121$$

$$z = d\left[0.5 + \sqrt{0.25 - \frac{0.5K}{\eta\alpha_{cc}/\gamma_s}}\right] = 525\left[0.5 + \sqrt{0.25 - \frac{K}{1.134}}\right] = 461.28\text{mm}$$

As $z/d = 0.879 \Rightarrow x/d = (1-0.879) \times 2 = 0.242 < 0.25$, the check method is applicable.

$$A_{s,req} = \frac{M_{Ed}}{0.87 f_{yk} z} = 2492\text{mm}^2$$

Applying Equation 7.18 of BSEN1992-1-2:2023 (2023) reproduced as Equation G10 in this Appendix for the calculation of the moment of resistance after fire $M_{Rd,fi}$,

$$M_{Rd,fi} = \frac{\gamma_s}{\gamma_{s,fi}} \frac{\sum_1^{n_{st}} f_{sy,\theta,i}}{n_{st} f_{yk}} M_{Ed} \frac{A_{s,prov}}{A_{s,req}} \tag{G10}$$

The material factors are $\gamma_s = 1.15$ under non-fire condition and $\gamma_{s,fi} = 1$ (corresponds to $\gamma_{M,fi}$ under fire condition as per section 4.5 of BSEN1992-1-2:2023 (2023))

$f_{sy,\theta,i}$ which is the strength of each reinforcing bar at elevated temperature of the 4 reinforcing bars depends on temperature. It can be estimated from the temperature contour prepared in Figure A.G4 or with its known coordinates be calculated by Equations G4 and G5 directly.

So, the temperatures of the four bars are

$$\theta_3(y,z,t) = \theta_3(0.075, 0.075, 14400) = 645.21°\text{C}$$

$$\theta_3(y,z,t) = \theta_3(0.075, 0.192, 14400) = 448.84°\text{C}$$

$$\theta_3(y,z,t) = \theta_3(0.075, 0.308, 14400) = 448.84°\text{C}$$

$$\theta_3(y,z,t) = \theta_3(0.075, 0.425, 14400) = 645.21°\text{C}$$

$f_{sy,\theta,i}$ for each bar can be estimated from Table 8.2 in Chapter 8 which is Table 5.3 of BSEN1992-1-2:2023 (2023) where $f_{sy,\theta,i}/f_{yk} = 0.237, 0.728, 0.728, 0.237$.

So, substituting into Equation G7

$$M_{Rd,fi} = \frac{\gamma_s}{\gamma_{s,fi}} \frac{\sum_1^{n_{st}} f_{sy,\theta,i}}{n_{st} f_{yk}} M_{Ed} \frac{A_{s,prov}}{A_{s,req}} = \frac{1.15}{1} \frac{2(0.237 + 0.728) \times 500}{4 \times 500} 500 \frac{3217}{2492} = 358.15\text{kNm}$$

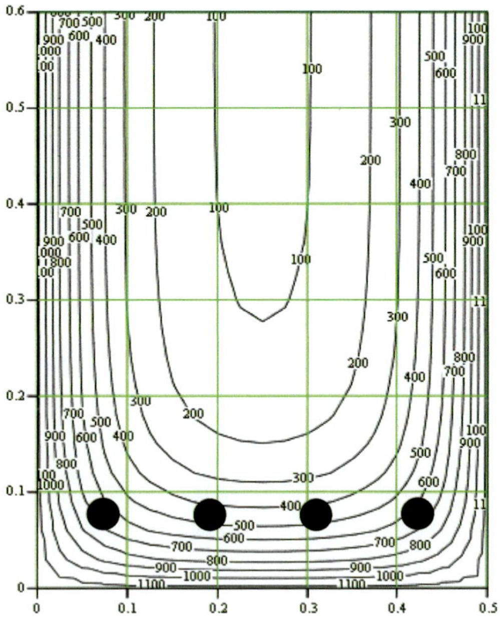

FIGURE A.G4 Temperature contour and reinforcing bar for Worked Example.

As $358.15 > \eta_{fi} M_{Ed} = 0.7 \times 500 = 350$ kNm, the design complies with the fire resistance required in this Worked Example G.2.

G5 WORKED EXAMPLE G.3 – FOR CALCULATION OF THE BENDING RESISTANCE OF A BEAM WITH REDUCTION OF CROSS-SECTION

In accordance with the design approach in section 7.3.2 of BSEN1992-1-2:2023 (2023), a reduction of the cross-section due to fire should be assumed with a rim zone at the faces exposed to fire. The rim is of width a_z given by Equation 7.15 of BSEN1992-1-2:2023 (2023) reproduced as Equation G11 as follows:

$$a_z = \begin{cases} 0.011\sqrt{1+\dfrac{t-27}{27}.\sqrt{\dfrac{w}{0.0125}}} & \text{for } 0.075 \le w < 0.2 \\ 0.011\sqrt{1+4\dfrac{t-27}{27}} & \text{for } w \ge 0.2 \end{cases} \quad \text{(G11)}$$

w is practically dimension of the section under fire as shown in figure 7.5 of BSEN1992-1-2:2023 (2023).

Using Worked Example G2, $w = b = 0.5 > 0.2$, so

$$a_z = 0.011\sqrt{1+4\frac{t-27}{27}} = 0.011\sqrt{1+4\frac{240-27}{27}} = 0.06276.$$

So, the beam width should effectively be reduced by $2a_z = 2\times 0.06276 = 0.1253$ m. For beam depth, the reduction is also by $a_z = 0.06276$.

For beam depth, concrete is not considered in providing strength as the bottom side of the beam is under tension. So the assumption of section reduction does not affect the design.

The beam width is, however, reduced by 0.1253 m so that the beam width under fire becomes $0.5 - 0.1253 = 0.3745\,\text{m}$.

The temperature at the centre at θ_M is calculated by Equation G7 to be 96.34°C. So, the strength factor in accordance with Figure 8.20 in Chapter 8 is 1 which implies no deduction of concrete strength. Re-working the $A_{s,req}$,

$$K = \frac{M_{Ed}}{f_{ck}bd^2} = \frac{500\times 10^6}{30\times 374.5\times 525^2} = 0.1615$$

$$z = d\left[0.5 + \sqrt{0.25 - \frac{0.5K}{\eta\alpha_{cc}/\gamma_s}}\right] = 525\left[0.5 + \sqrt{0.25 - \frac{K}{1.134}}\right] = 434.72\text{mm}$$

$$A_{s,req} = \frac{M_{Ed}}{0.87 f_{yk} z} = 2644\text{mm}^2$$

Using Equation G10, $M_{Rd,fi}$ can be calculated to be 428.51 kNm.

REFERENCES

British Standard Institution (2023) *Eurocode BSEN 1992-1-1:2023 Eurocode 2 – Design of Concrete Structures Part 1-1: General Rules and Rules for Buildings, Bridges and Civil Engineering Structures*. British Standard Institution

British Standard Institution (2023) *Eurocode BSEN 1992-1-2:2023 Eurocode 2 – Design of Concrete Structures Part 1-2: Structural Fire Design*. British Standard Institution

Mosley, W.H., Bungey, J. and Hulse, R. (2012) Reinforced Concrete Design to Eurocode 2. Palgrave Macmillan.

Appendix H Reinforcement Design of Rectangular Column to Eurocode BSEN1992-1-1:2023

H.1 STRESS/STRAIN RELATIONSHIP OF CONCRETE AND STEEL

By Equation (8.4) of the Eurocode, the concrete and reinforcement stress-strain relations are mathematically expressed in Sections H.1 and H.2 and graphically by Figure A.H1.

$$\sigma_{cd} = \begin{cases} f_{cd}\left[1-\left(1-\dfrac{\varepsilon_{cd}}{0.002}\right)^2\right] & \text{where } 0 \le \varepsilon_{cd} \le 0.0035 \\ f_{cd} & \text{where } \varepsilon_{cd} \ge 0.002 \end{cases} \tag{H1}$$

$$\sigma_s = \begin{cases} f_{yk} & \text{where} \quad \varepsilon_s > \gamma_s f_{yk} / E_s \\ E_s \varepsilon_s & \text{where} \quad \gamma_s f_{yk} / E_s \ge \varepsilon_s \ge -f_{yk} / E_s \; E_s = 200000\text{N/mm}^2 \\ -f_{yk} & \text{where} \quad \varepsilon_s < -\gamma_s f_{yk} / E_s \end{cases} \tag{H2}$$

H.2 DESIGN PRINCIPLES AND ASSUMPTIONS

In accordance with the Eurocode, the design check for the rectangular column section with pre-determined section and reinforcement contents is based on check of uniaxial bending in the two directions separately by which the moments of resistance in the two directions at $M_{Rdy,N}$ and $M_{Rdz,N}$ under the design axial of N_{Ed} are obtained. The equation (8.2) of the Eurocode (extracted as equation 9.5 in Chapter 9) is then checked as follows:

$$\left(\frac{|M_{Edz}|}{M_{Rdz,N}}\right)^{\alpha_N} + \left(\frac{|M_{Edy}|}{M_{Rdy,N}}\right)^{\alpha_N} \le 1.0 \tag{H3}$$

DOI: 10.1201/9781003636380-22

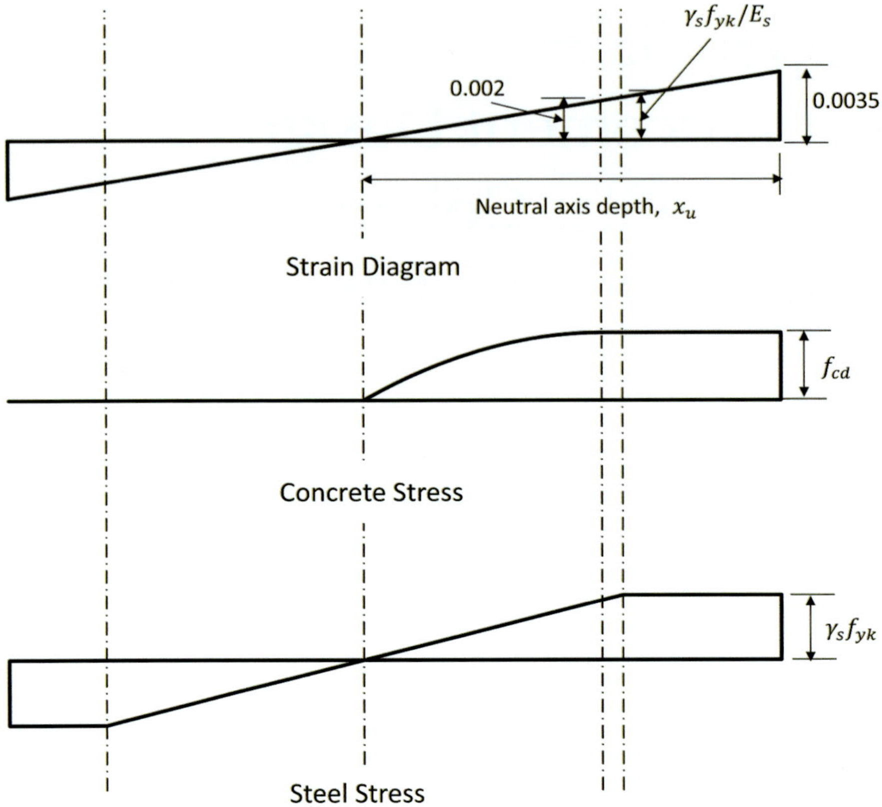

FIGURE A.H1 Stress-strain diagram for RC design of column.

where α_N is to be read and interpolated from the following table with $N_{Rd,0}$ as the axial load capacity of the section without moment.

$\lvert N_{Ed}\rvert / N_{Rd,0}$	0.1	0.7	1.0
α_N	1.0	1.5	2.0

H.3 CONTRIBUTION FROM CONCRETE

Considering $\varepsilon_{cd} = \dfrac{0.0035}{x_u} u = 1.75 \dfrac{u}{x_u}$, the concrete stress can be expressed by the horizontal coordinate u with $u = 0$ at the neutral axis

$$\sigma_{cd} = \begin{cases} f_{cd}\left[1-\left(1-1.75\dfrac{u}{x_u}\right)^2\right] & \text{where} \quad 0 \le u \le \dfrac{4}{7}x_u \\ f_{cd} \quad \text{where} \quad u \ge \dfrac{4}{7}x_u \end{cases} \tag{H4}$$

Let b be the width and h be the depth of the column section.

Case 1 – When $x_u \le h$, Entire Compression Zone within Column Section

Total concrete force= $F_c = f_{cd}b\dfrac{4x_u}{7}\times\dfrac{2}{3}+f_{cd}b\dfrac{3x_u}{7}=\dfrac{17}{21}f_{cd}bx_u$ (H5)

Total concrete Moment about neutral axis =

$$M_{cn} = f_{cd}b\frac{8x_u}{21}\times\left(\frac{5}{8}\times\frac{4x_u}{7}\right)+f_{cd}b\frac{3x_u}{7}\times\left(\frac{4x_u}{7}+\frac{3x_u}{7}\times\frac{1}{2}\right)$$

$$= f_{cd}b\frac{20x_u^2}{147}+f_{cd}b\frac{33x_u^2}{98}=\frac{139}{294}f_{cd}bx_u^2 \tag{H6}$$

Centre of load is $\dfrac{M_{cn}}{F_c}=\dfrac{\dfrac{139}{294}f_{cd}bx_u^2}{\dfrac{17}{21}f_{cd}bx_u}=\dfrac{139}{238}x_u$ from neutral axis or $\dfrac{99}{238}x_u$ from extreme end.

Converting to Centre of Section

$$M_c = F_c\times\left(\frac{h}{2}-\frac{99}{238}x_u\right)=\frac{17}{21}f_{cd}bx_u\left(\frac{h}{2}-\frac{99}{238}x_u\right) \tag{H7}$$

Case 2 – $\boldsymbol{x_u \ge h \ge \frac{3}{7}x_u}$ Parabolic Stress Profile Partially within Column Section

$$F_c = \int_{x_u-h}^{h-\frac{3}{7}x_u} f_{cd}b\left[1-\left(1-\frac{0.0035}{0.002x_u}u\right)^2\right]du+f_{cd}b\frac{3x_u}{7}=$$

$$= \int_{x_u-h}^{h-\frac{3}{7}x_u} f_{cd}b\left[1-\left(1-\frac{1.75u}{x_u}\right)^2\right]du+f_{cd}b\frac{3x_u}{7} \tag{H8}$$

After integration and simplification,

$$F_c = f_{cd}b\left(\frac{27}{336}x_u+\frac{7}{16}h+\frac{21}{16x_u}h^2-\frac{49}{48x_u^2}h^3\right) \tag{H9}$$

Moment about Centre of Section

$$M_c = \int_{x_u-h}^{h-\frac{3}{7}x_u} f_{cd}b\left[1-\left(1-\frac{0.0035}{0.002x_u}u\right)^2\right]\left(u-\left(x_u-\frac{h}{2}\right)\right)du + f_{cd}b\frac{3x_u}{7}\left(\frac{h}{2}-\frac{3x_u}{14}\right) \tag{H10}$$

After integration and simplification

$$\begin{aligned} M_c &= f_{cd}b\left[\frac{-1911}{10976}hx_u+\frac{5481}{65856}x_u^{\,2}-\frac{1029}{4704}\frac{h^3}{x_u}+\frac{2401}{9408}\frac{h^4}{x_u^{\,2}}\right]+f_{cd}b\frac{3x_uh}{14}-f_{cd}b\frac{9x_u^{\,2}}{98} \\ &= f_{cd}b\left[\frac{441}{10976}hx_u-\frac{567}{65856}x_u^{\,2}-\frac{1029}{4704}\frac{h^3}{x_u}+\frac{2401}{9408}\frac{h^4}{x_u^{\,2}}\right] \end{aligned} \tag{H11}$$

Case 3 – $\boldsymbol{x_u \geq \frac{3}{7}x_u}$ Only the Rectangular Stress Profile within Column Section

$$F_c = f_{cd}bx_u \tag{H12}$$

$$M_c = 0 \tag{H13}$$

H.4 CONTRIBUTION FROM THE STEEL REINFORCEMENT

The Steel reinforcements are considered as individual bars. With their known locations, each of their strains ε_{si} can be found from Figure AH.1 under a pre-determined neutral axis depth of x_u and subsequently stress σ_{si} and forces $F_{si} = \sigma_{si}A_{si}$. The total force and moment contributed by the steel reinforcements are given by Equations H13 and H14 where x_{si} is the location of the bar from the neutral axis.

$$F_s = \Sigma\sigma_{si}A_{si} \tag{H14}$$

$$M_s = \Sigma\sigma_{si}A_{si}x_{si} \tag{H15}$$

H.5 DESIGN PROCEDURE

The design procedures can be as follows which should be performed by computer program or spreadsheets:

(1) Pre-determine the reinforcements contents and arrangements in the column section.

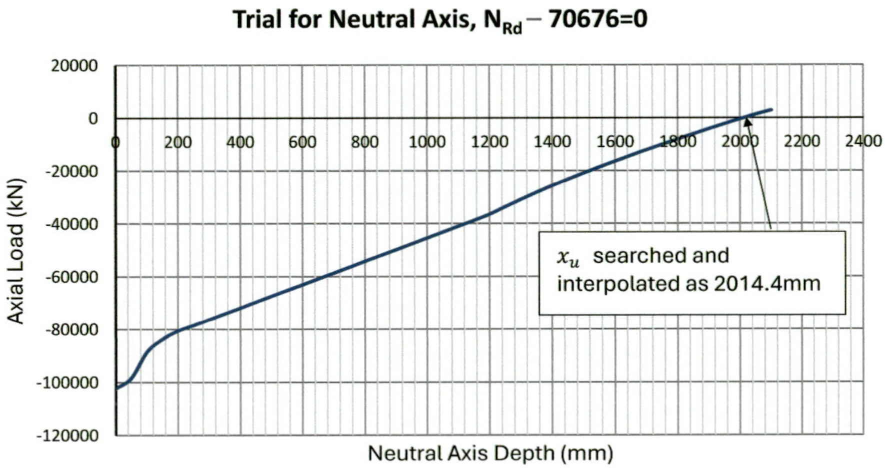

FIGURE A.H2 Demonstration of search for x_u

(2) Initially assume a neutral depth in the y-direction which can be very small at the beginning by which the total axial load capacity $N_{Rd} = N_{Rd,c} + N_{Rd,s}$ can be obtained as per equations in H.3 and H.4. Increase the neutral axis depth at small increments until the N_{Rd} just exceeds the design load capacity N_{Ed} or $N_{Rd} - N_{Ed} > 0$. Using linear interpolation with the previous value, obtain

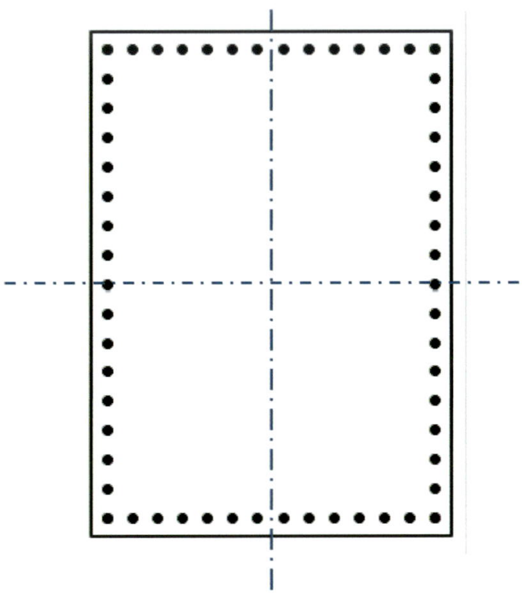

FIGURE A.H3 Column section for Worked Example H1.

TABLE H1
The Design Load and Checking Procedures

	P (kN)	M_x (kNm)	M_y (kNm)	x_{uy} (mm)	$M_{Rdy,N}$ (kNm)	x_{uz} (mm)	$M_{Rdz,N}$ (kNm)	N_{Ed} / N_{Ed0}	α_N	(8.2) of Eurocode
Load Comb 1	73582	−467.2	−18.118	2102.6	15215	1585.3	11568	0.7835	1.6392	0.0033
Load Comb 2	58013	170.88	−12.084	1688.6	24534	1272.9	18734	0.6178	1.4315	0.0008
Load Comb 3	66721	−959.06	−24.492	1908.1	19707	1438.1	15000	0.7105	1.5175	0.0102
Load Comb 4	59234	−12034	3221.7	1717.9	23920	1295	18256	0.6308	1.4423	0.4532
Load Comb 5	65500	11246	−3258.3	1876.2	20442	1414.1	15567	0.6975	1.4979	0.5046
Load Comb 6	55536	−2494.5	3298.5	1630.3	25734	1229	19666	0.5914	1.4095	0.118
Load Comb 7	69198	1706.3	−3335.1	1974.2	18147	1487.8	13799	0.7369	1.5614	0.1338
Load Comb 8	72218	5477.4	−4242.6	2059.5	16138	1552.4	12270	0.769	1.615	0.3545
Load Comb 9	52516	−6265.5	4206	1561.5	27120	1177.1	20745	0.5592	1.3827	0.242
Load Comb 10	61941	251.31	−36.624	1785.1	22477	1345.5	17138	0.6596	1.4663	0.0015
Load Comb 11	72099	−1067	−51.1	2055.9	16219	1549.7	12331	0.7678	1.6129	0.0126
Load Comb 12	63365	−13988	3736.1	1821.1	21683	1372.6	16524	0.6747	1.479	0.6339
Load Comb 13	70676	13172	−3823.9	2014.4	17174	1518.2	13054	0.7526	1.5877	0.7986
Load Comb 14	59051	−2858.3	3825.7	1713.5	24014	1291.6	18329	0.6288	1.4407	0.1512
Load Comb 15	74989	2042.6	−3913.4	2150.8	14257	1622.1	10836	0.7985	1.6642	0.223
Load Comb 16	78513	6442.2	−4972.2	2291.9	11837	1730.3	8982.7	0.8361	1.7268	0.7099
Load Comb 17	55527	−7257.9	4884.4	1630.1	25739	1228.8	19669	0.5913	1.4094	0.3083
Load Comb 18	42792	367.83	−24.092	1358	31112	1023.4	23890	0.4557	1.2964	0.0033
Load Comb 19	52951	−950.43	−38.568	1571.2	26926	1184.5	20594	0.5639	1.3865	0.0099
Load Comb 20	44216	−13871	3748.7	1385.9	30570	1044.7	23454	0.4708	1.309	0.4461
Load Comb 21	51527	13289	−3811.3	1539.5	27558	1160.6	21087	0.5487	1.3739	0.4625
Load Comb 22	39902	−2741.8	3838.3	1303	32186	981.92	24740	0.4249	1.2708	0.1374
Load Comb 23	55841	2159.2	−3900.9	1637.4	25590	1234.3	19554	0.5946	1.4122	0.1331
Load Comb 24	59364	6558.7	−4959.6	1721.1	23853	1297.4	18203	0.6321	1.4435	0.3082
Load Comb 25	36379	−7141.3	4897	1239.3	33456	933.85	25744	0.3874	1.2395	0.2753

an accurate neutral axis depth x_u by which $N_{Rd} = N_{Ed}$ as demonstrated in Figure A.H2 where the neutral axis depth is searched for $N_{Ed} = 70676\,\text{kN}$.

(3) Using x_u obtained in step 2, calculate the moment of resistance $M_{Rdy,N}$ as the sum of moments of resistance contributed by concrete and steel as per equations in Sections H.3 and H.4.

(4) Repeat the procedures in steps 2 and 3 for the z-direction.

(5) Check the Equation H3 listed under Section H.2. The section is adequate if it satisfies the equation. Otherwise revise section and repeat steps from 1 to 5.

H.6 WORKED EXAMPLE

Worked Example H1 whose section is identical to Worked Example 9.1 in Chapter 9 is performed to illustrate the foregoing procedures. The column is 1500 mm by 2000 mm with 58H40 bars as shown in Figure A.H3. The concrete grade is C35/45. Concrete cover to main bars is 50 mm. The section is to be checked for 25 load cases as listed in Table H1.

As the maximum value checked for Equation 8.2 of the Eurocode is 0.7986 less than unity, the column section is adequate for the load cases.

The column N_{Rd}–M_{Rd} Curves are shown in Figure 9.9 in Chapter 9.

REFERENCE

British Standard Institution (2023) Eurocode BSEN 1992-1-2:2023 *Eurocode 2 - Design of Concrete Structures Part 1-2: Structural Fire Design*. British Standard Institution.

Index

R

S

T

V

U

W

Z

For Product Safety Concerns and Information please contact our EU representative GPSR@taylorandfrancis.com Taylor & Francis Verlag GmbH, Kaufingerstraße 24, 80331 München, Germany

Batch number: 10615481

Printed by Printforce, the Netherlands